Advances in Experimental Medicine and Biology

Volume 789

For further volumes:
http://www.springer.com/series/5584

Sabine Van Huffel • Gunnar Naulaers
Alexander Caicedo • Duane F. Bruley
David K. Harrison
Editors

Oxygen Transport to Tissue XXXV

Springer

Editors
Sabine Van Huffel
Department of Electrical Engineering
and iMinds Future Health Department
KU Leuven
Leuven, Belgium

Alexander Caicedo
Department of Electrical Engineering
and iMinds Future Health Department
KU Leuven
Leuven, Belgium

David K. Harrison
Microvascular Measurements
Lorenzen, Italy

Gunnar Naulaers
Neonatal Intensive Care Unit
KU Leuven
University Hospitals
Leuven, Belgium

Duane F. Bruley
Synthesizer, Inc.
Ellicott City, MD, USA

ISSN 0065-2598
ISBN 978-1-4614-7256-8 ISBN 978-1-4614-7411-1 (eBook)
DOI 10.1007/978-1-4614-7411-1
Springer New York Heidelberg Dordrecht London

Library of Congress Control Number: 2013941477

Printed on acid-free paper

Springer is part of Springer Science+Business Media (www.springer.com)

The 40th ISOTT Conference President, Sabine Van Huffel, would like to dedicate this volume in memory of two outstanding members:

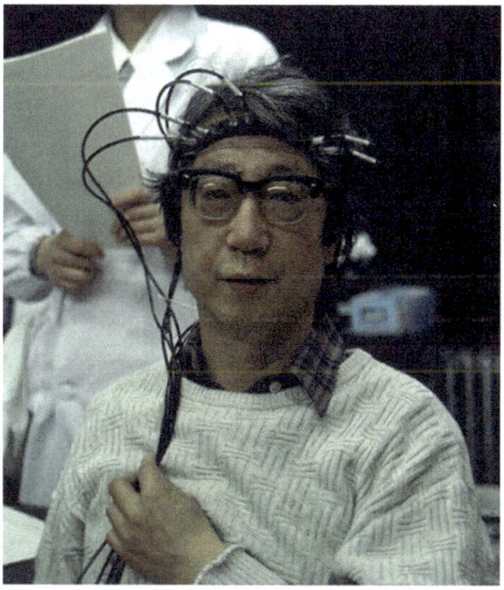

Dr. Mamoru Tamura *was born in Sapporo, Japan, in 1943 and died in Beijing, China, on August 7, 2011. He was not only one of the great pioneers in biomedical optics and functional near-infrared spectroscopy but also one of the key persons in ISOTT. In particular, he was the honorary president of the ISOTT 2008 meeting in his hometown Sapporo, Japan. During the ISOTT 2012 meeting, Professor Eiji Takahashi presented a memorial lecture to honor Dr. Mamoru Tamura. A memorial note is included in this volume.*

Ludwig Schleinkofer *(born on April 27, 1952) worked for more than 30 years as Director of R&D at Hamamatsu Deutschland GmbH. He was one of the leading and globally recognized experts in medical applications of near-infrared spectroscopy. Since 2004, he was member of the ISOTT society and served in the executive committee from 2008 to 2011. He suddenly passed away on July 18, 2012. With Ludwig, we have lost an important member of ISOTT: he was not only a long-lasting major sponsor of many former meetings but also a person who was actively supporting the ideas and activities of ISOTT. Many of us shared thoughts with him while drinking several glasses of good wine. With Ludwig, we lose an important supporter and a good friend.*

Preface

The 40th Annual Meeting of the International Society on Oxygen Transport to Tissue (ISOTT) was held on August 19–24, 2012, at the Novotel hotel in Bruges, Belgium. The historic city of Bruges, called "The Venice of the North," provided an ideal venue to combine an outstanding scientific program with ample opportunities to visit this medieval environment and appreciate the beauty of its architecture, the charm of its channels and boot tours while tasting one of the many famous Belgian beers.

Sabine Van Huffel, President of ISOTT 2012, organized the meeting cochaired by Gunnar Naulaers. One hundred participants, of which 28 were PhD students and 63 principal investigators or postdoctoral researchers, attended the meeting.

The meeting included 11 invited lectures, consisting of nine keynotes, one memorial lecture for Mamoru Tamura presented by Eiji Takahashi, and one Kovach lecture delivered by Peter Vaupel. The oral presentations were organized around seven topics related with oxygen transport to tissue. These topics were as follows: clinical applications, muscle oxygenation, cancer, measurement technologies, oxygen transport modeling and near-infrared spectroscopy (NIRS), cell metabolism, and brain oxygenation. Each topic was presented by one or two invited speakers and a series of contributed talks. In total, there were 59 contributed oral presentations and 29 poster presentations.

This year's annual meeting comprised a special Ph.D. student program consisting of three specific Ph.D. treats:

1. A poster flash presentation by promising local Ph.D. talents (six posters) on Monday
2. A Meet-the-Experts lunch on Tuesday, during which PhD students met ISOTT experts from the scientific committee
3. Finally, on Thursday, ISOTT's Presidents' Panel Discussion, held by past, present, and future ISOTT presidents, addressing questions related with oxygen transport to tissue

Social events included a welcome walking dinner reception on Sunday night. On Monday, there was a guided tour at the City Hall of Bruges followed by a Belgian beer discovery and, finally, dinner at the romantic Maximiliaan Van Oostenrijk restaurant. The highlight of ISOTT 2012 was certainly the gala dinner held at Bruges Concert Hall on Thursday, the day of the award ceremony, along with a dance party.

The annual meetings bring together scientists from various fields (physiology, mathematics, biomedicine, biology, chemistry, physics, engineering, etc.) in a unique international forum. Traditionally, ISOTT conferences are a place where an atmosphere of interaction is created, where many questions are asked after each presentation, and where lively discussions occur at a high scientific level. This vivid interaction is the main motivation for members to participate and gain new ideas and knowledge in the broad field of oxygen transport to tissue. The social events offer ample opportunities to network and form friendships, making each meeting a real success.

The society was deeply saddened to learn recently of the death of David Maguire, a longtime member and former president of ISOTT. A tribute to his life and scientific contributions will be paid during the 2013 ISOTT meeting and will be included in the next volume of *Oxygen Transport to Tissue*.

Leuven, Belgium

Sabine Van Huffel, Ph.D.
President ISOTT 2012

Acknowledgements

As President of the 2012 Meeting of the International Society on Oxygen Transport to Tissue, held on August 19–24, 2012, in Bruges, Belgium, I would like to gratefully acknowledge the support of our sponsors:

Industrial Sponsors

HAMAMATSU
PHOTON IS OUR BUSINESS

Hamamatsu Photonics, www.hamamatsu.com

COVIDIEN
positive results for life

Covidien, www.covidien.com

OXFORD
OPTRONIX

Oxford Optronix, www.oxford-optronix.com

Quality in Products and Services, http://www.qps-nv.be

AD Instruments, www.adinstruments.com

Moor Instruments, gb.moor.co.uk

Academic Sponsors

Fonds Wetenschappelijk Onderzoek, www.fwo.be

Arenberg Doctoral School of Science, Engineering and Technology. set.kuleuven.be/phd

Institute of Electrical and Electronic Engineers, www.ieee.org

Fonds de la Recherche Scientifique, www2.frs-fnrs.be

IEEE-EMBS Benelux Chapter, www.embs-chapter.be

International Society on Oxygen Transport to Tissue, www.isott.org

In addition, ISOTT 2012 was endorsed by the ESMRMB:

European Society for Magnetic Resonance in Medicine and Biology, www.esmrmb.org

Organization ISOTT 2012

Participants at ISOTT 2012

Local Organizing Committee

Sabine Van Huffel	KU Leuven, Belgium
Gunnar Naulaers	University Hospitals Leuven, Belgium
Peter Hespel	KU Leuven, Belgium
Bart Meyns	University Hospitals Leuven, Belgium
Patrick Wouters	University Hospital Gent, Belgium
Bernard Gallez	Université Catholique de Louvain, Belgium
Carine Michiels	Facultés Universitaires Notre-Dame de la Paix, Namur, Belgium

Scientific Committee

Sabine Van Huffel	KU Leuven, Belgium
Gunnar Naulaers	University Hospitals Leuven, Belgium
Peter Hespel	KU Leuven, Belgium
Bart Meyns	University Hospitals Leuven, Belgium
Patrick Wouters	University Hospital Gent, Belgium
Bernard Gallez	Université Catholique de Louvain, Belgium
Carine Michiels	FUND Namur, Belgium
William Welch	Georgetown University, USA
Clare Elwell	University College London, UK
Eiji Takahashi	Saga University, Japan
Harold Swartz	Dartmouth Medical School, Hanover, USA
Martin Wolf	University Hospital Zurich, Switzerland

ISOTT Officers and Executive Committee

Sabine Van Huffel (President, Leuven, Belgium)
Oliver Thews (Secretary, Halle, Germany)
Peter E. Keipert (Treasurer, San Diego, USA)
Duane F. Bruley (Knisely Award Committee, Ellicott City, USA)
William Welch (Past President, Washington, DC, USA)
Harold Swartz (President-Elect, Hanover, USA)
Donard Buerk (Philadelphia, USA)
Chris E. Cooper (Colchester Essex, UK)
Lin Li (Philadelphia, USA)
Michelle A. Puchowicz (Cleveland, USA)
Rammohan Maikala (Hopkinton, USA)
Kasuto Masamoto (Tokyo, Japan)
Masaomi Nangaku (Tokyo, Japan)
Clare Elwell (London, UK)
Ursula Wolf (Bern, Switzerland)

Newly Elected Members of the Executive Committee

Jerry D. Glickson (Philadelphia, USA)
Howard J. Halpern (Chicago, USA)
Terence Leung (London, UK)
Frederik Palm (Uppsala, Sweden)

Awards

The Melvin H. Knisely Award

The Melvin H. Knisely Award was established in 1983 to honor Dr. Knisely's accomplishments in the field of the transport of oxygen and other metabolites and anabolites in the human body. Over the years, he has inspired many young investigators, and this award is to honor his enthusiasm for assisting and encouraging young scientists and engineers in various disciplines. The award is to acknowledge outstanding young investigators. This award was first presented during the banquet of the 1983 annual conference of ISOTT in Ruston, Louisiana. The award includes a Melvin H. Knisely plaque and a cash prize.

Melvin H. Knisely Award Recipients

1983 Antal G. Hudetz (Hungary)
1984 Andras Eke (Hungary)
1985 Nathan A. Bush (USA)
1986 Karlfried Groebe (Germany)
1987 Isumi Shibuya (Japan)
1988 Kyung A. Kang (Korea/USA)
1989 Sanja Batra (Canada)
1990 Stephen J. Cringle (Australia)
1991 Paul Okunieff (USA)
1992 Hans Degens (The Netherlands)
1993 David A. Benaron (USA)
1994 Koen van Rossem (Belgium)
1995 Clare E. Elwell (UK)
1996 Sergei A. Vinogradov (USA)
1997 Chris E. Cooper (UK)

1998 Martin Wolf (Switzerland)
1999 Huiping Wu (USA)
2000 Valentina Quaresima (Italy)
2001 Fahmeed Hyder (Bangladesh)
2002 Geofrey De Visscher (Belgium)
2003 Mohammad Nadeem Khan (USA)
2004 Fredrick Palm (Sweden)
2005 Nicholas Lintell (Australia)
2007 Ilias Tachtsidis (UK)
2008 Kazuto Masamoto (Japan)
2009 Rossana Occhipinti (USA)
2010 Sebastiano Cicco (Italy)
2011 Mei Zhang (USA)
2012 Takahiro Igarashi (Japan)

The Dietrich W. Lübbers Award

The Dietrich W. Lübbers Award was established in honor of Professor Lübbers's long-standing commitment, interest, and contributions to the problems of oxygen transport to tissue and to the society. This award was first presented in 1994 during the annual conference of ISOTT in Istanbul, Turkey.

Dietrich W. Lübbers Award Recipients

1994 Michael Dubina (Russia)
1995 Philip E. James (UK/USA)
1996 Resit Demit (Germany)
1997 Juan Carlos Chavez (Peru)
1998 Nathan A. Davis (UK)
1999 Paola Pichiule (USA)
2000 Ian Balcer (USA)
2001 Theresa M. Busch (USA)
2002 Link K. Korah (USA)
2003 James J. Lee (USA)
2004 Richard Olson (Sweden)
2005 Charlotte Ives (UK)
2006 Bin Hong (China/USA)
2007 Helga Blockx (Belgium)
2008 Joke Vanderhaegen (Belgium)
2009 Matthew Bell (UK)

2010 Alexander Caicedo Dorado (Belgium)
2011 Malou Friederich (Sweden)
2012 Maria Papademetriou (UK)

The Britton Chance Award

The Britton Chance Award was established in honor of Professor Chance's longstanding commitment, interest, and contributions to the science and engineering aspects of oxygen transport to tissue and to the society. This award was first presented in 2004 during the annual conference of ISOTT in Bari, Italy.

Britton Chance Award Recipients

2004 Derek Brown (Switzerland)
2005 James Lee (USA)
2006 Hanzhu Jin (China/USA)
2007 Eric Mellon (USA)
2008 Jianting Wang (USA)
2009 Jessica Spires (USA)
2010 Ivo Trajkovic (Switzerland)
2011 Alexander Caicedo Dorado (Belgium)
2012 Felix Scholkmann (Switzerland)

The Duane F. Bruley Travel Award

The Duane F. Bruley Travel Award was established in 2003 and first presented by ISOTT at its 2004 annual conference in Bari, Italy. This award was created to provide travel funds for student researchers in all aspects and areas of oxygen transport to tissue. The award signifies Dr. Bruley's interest in encouraging and supporting young researchers to maintain the image and quality of research associated with the society. As a cofounder of ISOTT in 1973, Dr. Bruley emphasizes cross-disciplinary research among basic scientists, engineers, medical scientists, and clinicians. His pioneering work of constructing mathematical models for oxygen and other anabolite/metabolite transport in the microcirculation, employing computer solutions, was the first to consider system nonlinearities, time dependence, as well as multidimensional diffusion, convection, and reaction kinetics. It is hoped that receiving the Duane F. Bruley Travel Award will inspire students to excel in their research and will assist in securing future leadership for ISOTT.

Duane F. Bruley Travel Award Recipients

2004 Helga Blocks (Belgium), Jennifer Caddick (UK), Antonio Franco (Italy), Charlotte Ives (UK), Nicholas Lintell (Australia), Leonardo Mottola (Italy), Samin Rezania (USA/Iran), Ilias Tachtsidis (UK), Liang Tang (USA/China), Iyichi Sonoro (Japan)

2005 Robert Bradley (UK), Kathy Hsieh (Australia), Harald Oey (Australia), Jan Shah (Australia)

2006 Ben Gooch (UK), Ulf Jensen (Germany), Smruta Koppaka (USA), Daya Singh (UK), Martin Tisdall (UK), Bin Wong (USA), Kui Xu (USA)

2007 Dominique De Smet (Belgium), Thomas Ingram (UK), Nicola Lai (USA), Andrew Pinder (UK), Joke Vanderhaegen (Belgium)

2008 Sebastiano Cicco (Italy)

2009 Lei Gao (UK), Obinna Ndubuizu (USA), Joke Vanderhaegen (Belgium), Jianting Wang (USA)

2010 Zareen Bashir (UK), Martin Biallas (Switzerland), Takashi Eriguchi (Japan), Jack Honeysett (UK), Catalina Meßmer (USA), Tracy Moroz (UK), Yoshihiro Murata (Japan), Mark Muthalib (Australia)

2011 Catherine M. Hesford (UK), Luke S. Holdsworth (UK), Andreas Metz (Switzerland), Maria D. Papademetriou (UK), Patrik Persson (Sweden), Felix Scholkmann (Switzerland), Kouichi Yoshihara (Japan)

2012 Allan Al-armaghany (UK), Yuta Sekiguchi (Japan), Tharindi Hapuarachchi (UK), Rebecca Re (Italy), Ebba Sivertsson (Sweden), Ben Jones (UK), Andre Steimers (Germany), Malou Friederich-Persson (Sweden)

Kovach Lecture

The Kovach Lecture is presented periodically to honor a career dedicated to oxygenation research. Arisztid Kovach was a world-renowned cardiovascular physiologist and one of the early leaders of ISOTT. This lecture is dedicated to his remarkable scientific and teaching career.

Kovach Lecture Recipients

2011 John Severinghaus

2012 Peter Vaupel

Contents

Contributors

I would also like to thank those who helped make this meeting a success – Elsy Vermoesen, Liesbeth Van Meerbeek, Alexander Caicedo, Vladimir Matic, Ida Tassens, and Mimi Deprez – who excellently took care of the administrative and organizational sides of the conference, and I am grateful also to the superb staff of Novotel Bruges.

Reviewers

The editors would like to thank the following experts, who scientifically reviewed the papers:

Scientific reviewers	Affiliation
Alexander Caicedo	KU Leuven, Belgium
Pierre Danhier	Université Catholique de Louvain, Belgium
Geraldine De Preter	Université Catholique de Louvain, Belgium
Clare Elwell	University College London, UK
Arnab Ghosh	University College London, UK
Howard Halpern	University of Chicago, USA
David Harrison	Microvascular Measurements, St Lorenzen, Italy
Ingo Helmich	German Sports University Cologne, Germany
Catherine M. Hesford	University of Essex, UK
Louis Hoofd	BaLoH Software, Ooij, The Netherlands
Kyung Kang	University of Louisville, USA
Peter Keipert	Sangart Inc, San Diego, USA
Matthias Kohl-Bareis	University of Applied Sciences Koblenz, Germany
Christina Kolyva	University College London, UK
Joseph LaManna	Case Western Reserve University, Cleveland, USA

Terence Leung	University College London, UK
Lin Li	University of Pennsylvania, USA
Arnulf Mayer	University Medical Centre, Mainz, Germany
Avraham Mayevski	Bar Ilan University, Israel
Andreas Metz	University of Zurich, Switzerland
Gunnar Naulaers	KU Leuven, Belgium
Maria Papademetriou	University College London, UK
Michelle Puchowicz	Case Western Reserve University, Cleveland, USA
Kaoru Sakatani	Nihon University School of Medicine, Japan
Felix Scholkmann	University of Zurich, Switzerland
Ebba Sivertsson	Uppsula University, Sweden
Jessica Spires	Case Western Reserve University, Cleveland, USA
Harold Swartz	Dartmouth College, Hanover, USA
IliasTachtsidis	University College London, UK
Eiji Takahashi	Saga University, Japan
Oliver Thews	University of Halle, Germany
Sabine Van Huffel	KU Leuven, Belgium
Peter Vaupel	University Medical Centre, Mainz, Germany
Martin Wolf	University of Zurich, Switzerland
Ursula Wolf	University of Bern, Switzerland
Christopher Wolff	St Bartholomew's Hospital London, UK
Lurong Zhang	University of Florida, USA
Mei Zhang	University of Florida, USA

Technical Reviewers

Laraine Visser-Isles (Rotterdam, The Netherlands)
Eileen Harrison (St Lorenzen, Italy)

Chapter 1
Remembering Professor Mamoru Tamura

Eiji Takahashi

Abstract Dr. Mamoru Tamura (1943–2011) was the honorary president of the ISOTT 2008 meeting in Sapporo, Japan, and has made numerous contributions to biomedical optics and functional near-infrared spectrometry. This chapter briefly describes Dr. Tamura's scientific achievements and contributions to the society based on the "Memorial lecture about Mamoru Tamura's contributions to biomedical optics" in ISOTT 2012 in Bruges, Belgium.

1.1 Dr. Mamoru Tamura's Scientific Achievements and Contributions to the Society

Dr. Mamoru Tamura (Fig. 1.1) passed away on the 7th of August in 2011. He was not only one of the great pioneers in Biomedical Optics but also one of the key persons in ISOTT. Here, I briefly look back on his achievements in science and contributions to ISOTT.

Dr. Tamura was born in Sapporo, Japan, in 1943 and died in Beijing, China, in 2011. In his 68 years of life, he pursued the possibility of light as the tool for in vivo investigation of human functions, particularly in transport and utilization of oxygen in cells, tissues, organs, and the whole body.

After receiving a PhD from Hokkaido University in 1971, a big chance came to him: he had an opportunity to work in Dr. Britton Chance's lab in the University of Pennsylvania as a postdoc. In fact, this was a great encounter. From then on, we can find the influence of this scientific giant, Dr. Chance, on Dr. Tamura's research works in every aspect.

E. Takahashi (✉)
Advanced Technology Fusion, Graduate School of Science and Engineering, Saga University, Saga 840-8502, Japan
e-mail: eiji@cc.saga-u.ac.jp

S. Van Huffel et al. (eds.), *Oxygen Transport to Tissue XXXV*, Advances in Experimental Medicine and Biology 789, DOI 10.1007/978-1-4614-7411-1_1,

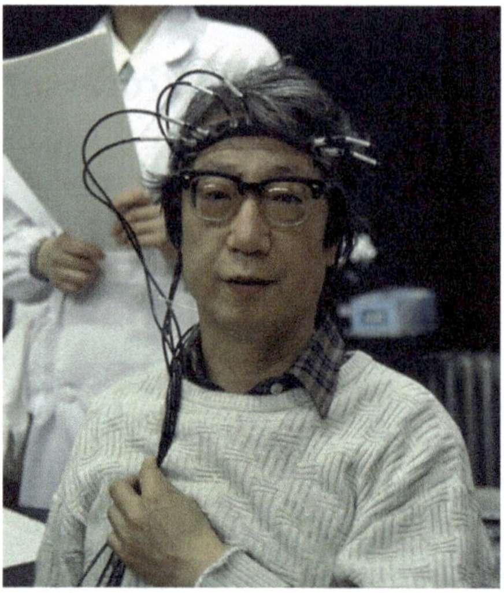

Fig. 1.1 Dr. Mamoru Tamura (1943–2011) (Photo by courtesy of Dr. Yoko Hoshi)

As soon as he came back from the USA, he published a paper [1] on oxygen transport to mitochondria in buffer-perfused rat heart, based on the work conducted in Dr. Chance's lab. Remarkably, in this paper we can already find two important keywords throughout his research, "oxygen" and "optical measurement." Here, he performed optical measurements of intracellular oxygen at Mb in cytosol and at cytochrome *c* oxidase in mitochondria in buffer-perfused rat hearts. Specifically, using a spectrophotometric technique in visible regions, he demonstrated coherence between cytochrome *c* oxidase oxidation and myoglobin oxygen saturation. The result clearly demonstrated significant gradients of oxygen between these two distinct intracellular compartments. Based on this experiment, he used to talk to me about the reason why hemoglobin and myoglobin have quite different P_{50} values for oxygen compared with that of cytochrome *c* oxidase.

After a while, he started to extend his in vitro oxygen measurement in buffer-perfused organs toward in vivo oxygen measurements in experimental animals and in humans. For this purpose, quite naturally, he utilized light in near-infrared regions. Figure 1.2 is the kind gift of Professor Marco Ferrari. This was taken when he visited Dr. Tamura's lab in 1989. In this photograph, Professor Ferrari noted, "in 1985 they started NIRS research in collaboration with Shimadzu."

In 1993, he published an important paper [2] in NIRS research where he demonstrated the ability of NIRS in functional neuroimaging in humans. Dr. Tamura and his lifelong colleague, Dr. Hoshi, demonstrated in vivo that blood oxygenation in the human brain and surrounding tissue can certainly be real time monitored by the NIRS technique. Importantly, these changes in oxygenation were related to functional activation of the human brain triggered by mental tasks.

In 1997, they published another important paper [3] in NIRS research. Using a blood-perfused rat brain model, they demonstrated that NIRS signals arising from mitochondria and hemoglobin can be distinguished and evaluated independently.

Fig. 1.2 Photo by courtesy of Professor Marco Ferrari

In this paper, they demonstrated coherence in the relationship between cytochrome *c* oxidase redox state in the brain tissue and hemoglobin oxygenation in blood (see Fig. 3 in ref. [3]). To me, it is quite interesting that in this 1997 paper, they presented a figure that is quite similar to one that can be found in the early work published about 20 years ago (see Fig. 6 in ref. [1]). It took almost 20 years of refining his optical technique so that the mitochondrial redox state could be measured in blood-perfused brain.

In Japan, important movements toward application of NIRS in clinical medicine were already underway in these years. In fact, a national project aiming at realization of optical CT for human use had started in Japan in 1992. The project team consisted of companies including Shimadzu and Hamamatsu, and the universities, and was supported by the Japanese government. Of course, Dr. Tamura was the leader of the team. Figure 1.3 is the first fruit of this project: a 64-channel time-resolved optical CT [4]. Reference [5] is the subsequent paper that reported applications of this optical CT in human subjects. This project proved that NIRS optical tomography is a new modality for neuropsychological studies in humans.

Figure 1.4 shows probably the first wearable NIRS used in a dog; the name of the dog is Tanne. This was Dr. Tamura's favorite photo.

He had been an active member of the ISOTT since the third official conference in Churchill College in Cambridge, UK, in 1977. Since then, he attended the meeting almost every year and enjoyed the familiar and free atmosphere of the conference. He was a member of the executive committee from 1999 to 2000 and probably from 1992 to 1993. He was the Honorary President of ISOTT 2008 held in his home town, Sapporo, Japan.

Fig. 1.3 A 64-channel time-resolved optical CT made in Japan (Photo by courtesy of Dr. Yoko Hoshi)

Fig. 1.4 Probably the first wearable NIRS system used in a dog (named Tanne) (Photo by courtesy of Dr. Yoko Hoshi)

Fig. 1.5 Dr. Chance and Dr. Tamura at the third annual meeting of Oxygen Dynamics Society Japan in Tokyo in 1998

(Please go to *http://isott.org/tl_files/isott/ISOTT%202008.jpg* and find Dr. Tamura).

He loved his hometown Sapporo, Hokkaido, very much and worked in Hokkaido University for 33 years. After officially retiring from Hokkaido University, he was invited as senior visiting professor in Tsinghua University in Beijing, China. He left Japan at the age of 65. He was very active also in China in building and organizing a new lab for biomedical optics with his Chinese and Japanese colleagues until his sudden death. He seeded this rapidly growing country with biomedical optics. Dr. Tamura contributed to developing a friendship between China and Japan through science on oxygen.

Since his stay in Dr. Chance's lab as a postdoc between 1971 and 1974, Dr. Tamura and Dr. Chance were bound each other by a close friendship. In 1996, we founded a Japanese organization like ISOTT, called Oxygen Dynamics Society Japan. Figure 1.5 was taken at the third annual meeting of the Oxygen Dynamics Society in Tokyo in 1998 where Dr. Chance was invited as a speaker. The title of his talk was "Measurements of energetics and oxygenation by noninvasive methods in medicine." I was deeply impressed that these two people shared the same thoughts on the science.

We lost the great scientist, Dr. Britton Chance, on the 16th of November in the year 2010. We lost another great scientist, Dr. Mamoru Tamura, just 1 year later.

Who was this man? I think that Dr. Tamura was a man who can dream and convey his dreams to people around him. He sowed the seeds and quietly watched the seeds grow.

Acknowledgments I thank many colleagues in ISOTT and Oxygen Dynamics Society Japan who talked with me about Dr. Tamura's life and science. Particularly, I thank Dr. Marco Ferrari, Dr. Yoko Hoshi, Dr. Kaoru Sakatani, and Dr. Takafumi Hamaoka for kindly supplying the valuable photographs. Also, many thanks to the President of ISOTT 2012, Prof. Sabine van Huffel, for giving me such a precious moment for remembering Dr. Tamura.

References

1. Tamura M, Oshino N, Chance B, Silver IA (1978) Optical measurements of intracellular oxygen concentration of rat heart in vitro. Arch Biochem Biophys 191(1):8–22
2. Hoshi Y, Tamura M (1993) Detection of dynamic changes in cerebral oxygenation coupled to neuronal function during mental work in man. Neurosci Lett 150:5–8
3. Hoshi Y, Hazeki O, Kakihana Y, Tamura M (1997) Redox behavior of cytochrome oxidase in the rat brain measured by near-infrared spectroscopy. J Appl Physiol 83:1842–1848
4. Eda H, Oda I, Ito Y et al (1999) Multichannel time-resolved optical tomographic imaging system. Rev Sci Instrum 70(3595):3595–3602
5. Hoshi Y, Oda I, Wada Y et al (2000) Visuospatial imagery is a fruitful strategy for the digit span backward task: a study with near-infrared optical tomography. Brain Res Cogn Brain Res 9(3):339–342

Part I
Hypoxia

Chapter 2
Increased Kidney Metabolism as a Pathway to Kidney Tissue Hypoxia and Damage: Effects of Triiodothyronine and Dinitrophenol in Normoglycemic Rats

Malou Friederich-Persson, Patrik Persson, Angelica Fasching, Peter Hansell, Lina Nordquist, and Fredrik Palm

Abstract Intrarenal tissue hypoxia is an acknowledged common pathway to end-stage renal disease in clinically common conditions associated with development of chronic kidney disease, such as diabetes and hypertension. In diabetic kidneys, increased oxygen metabolism mediated by mitochondrial uncoupling results in decreased kidney oxygen tension (PO_2) and contributes to the development of diabetic nephropathy. The present study investigated whether increased intrarenal oxygen metabolism *per se* can cause intrarenal tissue hypoxia and kidney damage, independently of confounding factors such as hyperglycemia and oxidative stress. Male Sprague–Dawley rats were untreated or treated with either triiodothyronine (T_3, 10 g/kg bw/day, subcutaneously for 10 days) or the mitochondria uncoupler dinitrophenol (DNP, 30 mg/kg bw/day, oral gavage for 14 days), after which in vivo kidney function was evaluated in terms of glomerular filtration rate (GFR, inulin clearance), renal blood flow (RBF, Transonic, PAH clearance), cortical PO_2 (Clark-type electrodes), kidney oxygen consumption (QO_2), and proteinuria. Administration of both T_3 and DNP increased kidney QO_2 and decreased PO_2 which resulted in proteinuria. However, GFR and RBF were unaltered by either treatment. The present study demonstrates that increased kidney metabolism per se can cause intrarenal tissue hypoxia which results in proteinuria. Increased kidney QO_2 and concomitantly reduced PO_2 may therefore be a mechanism for the development of chronic kidney disease and progression to end-stage renal disease.

M. Friederich-Persson (✉) • P. Persson • A. Fasching • P. Hansell • L. Nordquist
Department Medical Cell Biology, Div. Integrative Physiology, Uppsala University, Biomedical Center, D1:3, Husargatan 3, Box 571, SE-751 23 Uppsala, Sweden
e-mail: Malou.Friederich@mcb.uu.se

F. Palm
Department Medical Cell Biology, Div. Integrative Physiology, Uppsala University, Biomedical Center, D1:3, Husargatan 3, Box 571, SE-751 23, Uppsala, Sweden

Department of Medical and Health Sciences, Linköping University, Linköping, Sweden

S. Van Huffel et al. (eds.), *Oxygen Transport to Tissue XXXV*, Advances in Experimental Medicine and Biology 789, DOI 10.1007/978-1-4614-7411-1_2,

2.1 Introduction

Increased kidney oxygen consumption (QO_2) may result in decreased kidney tissue oxygen tension (PO_2), ultimately resulting in kidney damage. Importantly, intrarenal kidney tissue hypoxia is a proposed pathway to chronic kidney disease and end-stage renal disease in conditions such as diabetes and hypertension [1, 2]. To increase kidney QO_2, healthy rats were treated with either dinitrophenol (DNP) or triiodothyronine (T_3). DNP is a chemical uncoupler of mitochondria [3]. By decreasing the mitochondria membrane potential, the process of mitochondria uncoupling reduces formation of oxidative stress but increases mitochondria QO_2 [4]. Through the nuclear receptor superfamily, T_3 controls gene transcription and results in increased QO_2, mainly via mitochondrial pathways [5]. By utilizing DNP and T_3, we create two independent models of increased kidney QO_2 to investigate whether increased QO_2 per se can cause intrarenal tissue hypoxia and kidney damage.

2.2 Materials and Methods

All animal procedures were carried out according to the National Institutes of Health guidelines and approved by the local animal ethics committee. Normoglycemic male Sprague–Dawley rats (Charles River, Sulzfeld, Germany) received either no treatment, DNP administration (30 mg/kg bw dissolved in 1.5 % methyl cellulose) by oral gavage for14 days or T_3 (10 μg/kg bw/day) administered for 10 days by surgically implanted osmotic minipumps (Alzet Company, Cupertino, CA, USA). The animals receiving T_3 were simultaneously treated with candesartan (1 mg/kg bw in drinking water) in order to minimize the influence of thyroid hormone-induced renin release.

Animals were sedated with sodium thiobutabarbital (Inactin, 120 mg/kg bw i.p.) and placed on a servo-rectally controlled heating pad to maintain body temperature at 37 °C. Tracheotomy was performed to facilitate breathing, and polyethylene catheters were placed in the carotid artery and femoral vein to allow monitoring of mean arterial pressure (MAP, Statham P23dB, Statham Laboratories, Los Angeles, CA, USA), blood sampling, and infusion of saline (5 ml/kg bw/h). The left kidney was exposed by a subcostal flank incision and immobilized in a plastic cup. The left ureter and bladder were catheterized to allow for timed urine sampling and urinary drainage, respectively. After surgery, the animal was allowed to recover for 30 min followed by a 40-min experimental period at the end of which a blood sample was carefully collected from the renal vein in order to calculate arteriovenous differences.

Kidney cortex PO_2 was measured using Clark-type oxygen electrodes (Unisense, Aarhus, Denmark). Glomerular filtration rate (GFR) was measured by clearance of ^{3}H-inulin (185 kBq bolus followed by 185 kBq/kg bw/h; American Radiolabeled Chemicals, St Louis, MO, USA). Renal blood flow (RBF) was measured by Transonic flow probe (Transonic Systems Inc., Ithaca, NY, USA) or clearance by

^{14}C-para-aminohippuric acid (PAH, 185 kBq bolus followed by 185 kBq/kg bw/h; American Radiolabeled Chemicals). GFR was calculated as inulin clearance = ([inulin]$_{urine}$*urine flow)/[inulin]$_{plasma}$ and RBF with PAH clearance corrected for PAH extraction and the hematocrit. Total kidney QO_2 (μmol/min) was estimated from the arteriovenous difference in O_2 content (O_2ct = ([hemoglobin]*oxygen saturation*1.34 + PO_2*0.003)) *total RBF. Tubular Na^+ transport (T_{Na+}, μmol/min) was calculated as follows: $T_{Na+} = [P_{Na+}]$*GFR, where $[P_{Na+}]$ is plasma Na^+ concentration. T_{Na+} per Q_{O2} was calculated as T_{Na+}/Q_{O2}.

Statistical comparisons were performed using one-way analysis of variance with Dunnett's post hoc test. $p < 0.05$ was considered significant and all values are presented as mean ± SEM.

2.3 Results

A decreased body weight and increased kidney weight was observed in T_3-treated animals compared to controls. GFR, MAP, RBF, Na^+–excretion, and urinary flow rates were unaffected in both treated groups (Table 2.1). DNP and T_3 administration resulted in increased kidney QO_2 (Fig. 2.1) and decreased intrarenal kidney PO_2 (Fig. 2.2) compared to no treatment. Importantly, the reduced PO_2 resulted in increased proteinuria in both DNP and T_3-treated animals (Fig. 2.3).

Table 2.1 Systemic and kidney-specific parameters in control animals receiving either no treatment, dinitrophenol for 14 days, or triiodothyronine for 10 days

	Body weight (g)	Kidney weight (g)	Mean arterial pressure (mmHg)	Urine flow (μl/min)	Renal blood flow (ml/min)	Glomerular filtration rate (ml/min)	Sodium excretion (μmol/min)
No treatment	399 ± 10	1.4 ± 0.1	107 ± 2	3.2 ± 0.8	11.1 ± 0.7	1.6 ± 0.2	0.1 ± 0.1
Dinitrophenol	399 ± 7	1.4 ± 0.0	108 ± 7	4.1 ± 0.8	11.4 ± 1.1	2.1 ± 0.3	0.2 ± 0.1
Triiodothyronine	350 ± 7*	2.7 ± 0.1*	97 ± 4	4.4 ± 0.6	8.1 ± 1.3	1.8 ± 0.3	0.2 ± 0.2

*Denotes $p < 0.05$ versus control animals

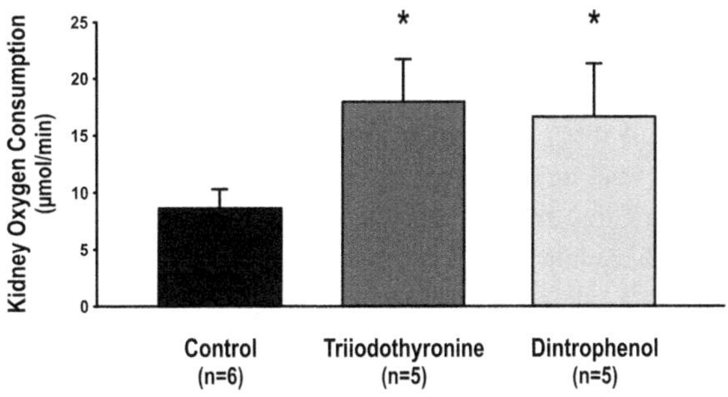

Fig. 2.1 Kidney oxygen consumption in control animals receiving either no treatment, dinitrophenol for 14 days, or triiodothyronine for 10 days. * denotes $p < 0.05$ versus control

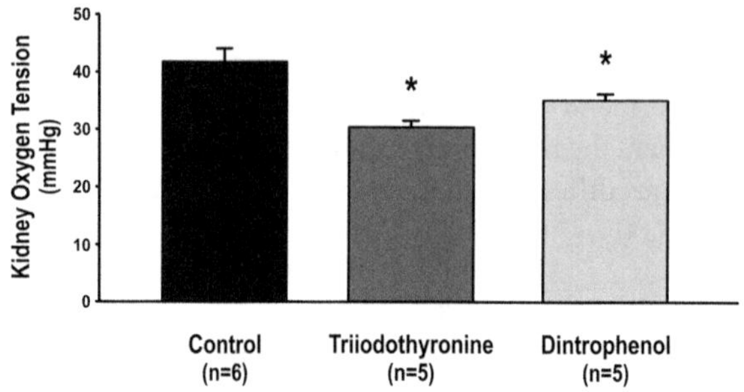

Fig. 2.2 Kidney oxygen tension in control animals receiving either no treatment, dinitrophenol for 14 days, or triiodothyronine for 10 days. * denotes $p<0.05$ versus control

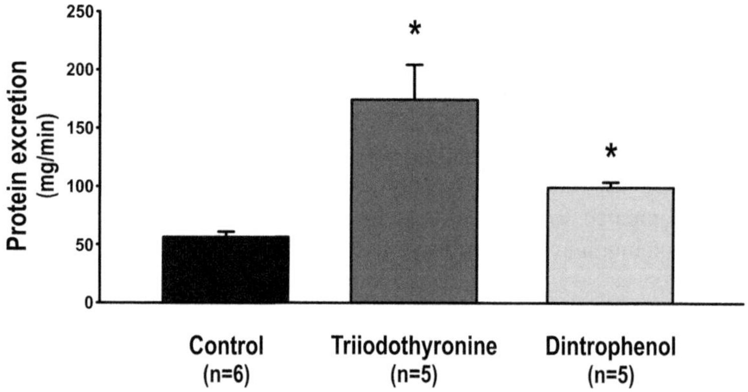

Fig. 2.3 Protein excretion in control animals receiving either no treatment, dinitrophenol for 14 days, or triiodothyronine for 10 days. * denotes $p<0.05$ versus control

2.4 Discussion

In the present study, administration of DNP and T_3 to normoglycemic rats increased kidney QO_2, decreased kidney PO_2, and increased proteinuria, demonstrating that increased kidney QO_2 is a pathway to kidney damage. Interestingly, treatment with DNP in patients has resulted in proteinuria [6, 7] and patients with untreated Grave's disease also develop proteinuria [8]. It may be argued that DNP is directly nephrotoxic as it is a known environmental toxin. However, this argument cannot apply to T_3 as it is an endogenous hormone. The use of T_3 to establish intrarenal hypoxia verifies that increased kidney QO_2 indeed ultimately results in kidney damage. As T_3 is known to affect the renal angiotensin system and cause hypertension [9], these animals were simultaneously treated with candesartan. It is also unlikely that the observed kidney damage is a result of ATP shortage rather than limited availability of oxygen as the Na^+ excretion is unchanged in both DNP and T_3-treated animals, a finding supported by others [10]. We propose that the mechanism to kidney damage is decreased kidney PO_2 resulting in intrarenal tissue hypoxia, leading to development of proteinuria.

In 1998, it was originally proposed by Fine et al., that the limitation of intrarenal oxygen availability is the key mechanism initiating kidney damage [11]. Since then, this theory has received considerable support, and intrarenal hypoxia is now an acknowledged pathway to chronic kidney disease and end-stage renal disease [1, 2, 12–14]. The oxygen level in the kidney is important because an increased RBF will increase the workload and, therefore, the kidney QO_2. This makes increased RBF a very inefficient way to increase renal PO_2, and the kidney is therefore vulnerable to alterations in either QO_2 or oxygen availability. This is further supported by studies reporting that Navajo Indians living at high altitude have increased incidence of end-stage renal disease compared to their corresponding control group at sea level [15]. Also, type 2 diabetic patients at high altitude have increased incidence of diabetic nephropathy despite similar blood pressure and metabolic parameters [16].

In the present study, the utilization of two independent models to create intrarenal tissue hypoxia both resulted in kidney damage evident as increased urinary protein leakage. This clearly demonstrated the importance of intrarenal PO_2 in maintaining normal kidney function.

References

1. Nangaku M (2006) Chronic hypoxia and tubulointerstitial injury: a final common pathway to end-stage renal failure. J Am Soc Nephrol 17(1):17–25
2. Mimura I, Nangaku M (2010) The suffocating kidney: tubulointerstitial hypoxia in end-stage renal disease. Nat Rev Nephrol 6(11):667–678
3. Parascandola J (1974) Dinitrophenol and bioenergetics: an historical perspective. Mol Cell Biochem 5(1–2):69–77
4. Echtay KS, Roussel D, St-Pierre J, Jekabsons MB, Cadenas S, Stuart JA et al (2002) Superoxide activates mitochondrial uncoupling proteins. Nature 415(6867):96–99
5. Wrutniak-Cabello C, Casas F, Cabello G (2001) Thyroid hormone action in mitochondria. J Mol Endocrinol 26(1):67–77
6. Beinhauer L (1934) Urticaria following the use of dinitrophenol. WV Med J 7:466–477
7. Imerman S, Imerman C (1936) Dinitrophenol poisoning with thrombocytopenia, granulopenia, anemia and purpura complicated by lung abscess. JAMA 106:1085–1088
8. Weetman AP, Tomlinson K, Amos N, Lazarus JH, Hall R, McGregor AM (1985) Proteinuria in autoimmune thyroid disease. Acta Endocrinol (Copenh) 109(3):341–347
9. Perez-Abud R, Rodriguez-Gomez I, Villarejo AB, Moreno JM, Wangensteen R, Tassi M, et al. Salt sensitivity in experimental thyroid disorders in rats. Am J Physiol Endocrinol Metab 301(2):E281–E287
10. Evans RG, Goddard D, Eppel GA, O'Connor PM. Stability of tissue PO_2 in the face of altered perfusion: a phenomenon specific to the renal cortex and independent of resting renal oxygen consumption. Clin Exp Pharmacol Physiol 38(4):247–254
11. Fine LG, Orphanides C, Norman JT (1998) Progressive renal disease: the chronic hypoxia hypothesis. Kidney Int Suppl 65:S74–S78
12. Nangaku M (2004) Hypoxia and tubulointerstitial injury: a final common pathway to end-stage renal failure. Nephron Exp Nephrol 98(1):e8–e12

13. Singh DK, Winocour P, Farrington K (2008) Mechanisms of disease: the hypoxic tubular hypothesis of diabetic nephropathy. Nat Clin Pract Nephrol 4(4):216–226
14. Palm F, Nordquist L. Renal tubulointerstitial hypoxia: cause and consequence of kidney dysfunction. Clin Exp Pharmacol Physiol 38(7):424–430
15. Hochman ME, Watt JP, Reid R, O'Brien KL (2007) The prevalence and incidence of end-stage renal disease in Native American adults on the Navajo reservation. Kidney Int 71(9):931–937
16. Sayarlioglu H, Erkoc R, Dogan E, Topal C, Algun E, Erem C et al (2005) Nephropathy and retinopathy in type 2 diabetic patients living at moderately high altitude and sea level. Ren Fail 27(1):67–71

Chapter 3
Hypoxia-Induced Cerebral Angiogenesis in Mouse Cortex with Two-Photon Microscopy

Kazuto Masamoto, Hiroyuki Takuwa, Yutaka Tomita, Haruki Toriumi, Miyuki Unekawa, Junko Taniguchi, Hiroshi Kawaguchi, Yoshiaki Itoh, Norihiro Suzuki, Hiroshi Ito, and Iwao Kanno

Abstract To better understand cellular interactions of the cerebral angiogenesis induced by hypoxia, a spatiotemporal dynamics of cortical microvascular restructuring during an exposure to continuous hypoxia was characterized with *in vivo* two-photon microscopy in mouse cortex. The mice were prepared with a closed cranial window over the sensory-motor cortex and housed in 8–9 % oxygen room for 2–4 weeks. Before beginning the hypoxic exposure, two-photon imaging of cortical microvasculature was performed, and the follow-up imaging was conducted weekly in the identical locations. We observed that 1–2 weeks after the onset of hypoxic exposure, a sprouting of new vessels appeared from the existing capillaries. An average emergence rate of the new vessel was 15 vessels per unit volume (mm^3). The highest emergence rate was found in the cortical depths of 100–200 μm, indicating no spatial uniformity among the cortical layers. Further, a leakage of fluorescent dye (sulforhodamine 101) injected into the bloodstream was not detected, suggesting that the blood brain barrier (BBB) was maintained. Future studies are needed to elucidate the roles of perivascular cells (e.g., pericyte, microglia, and astroglia) in a process of this hypoxia-induced angiogenesis, such as sprouting, growth, and merger with the existing capillary networks, while maintaining the BBB.

K. Masamoto, Ph.D. (✉)
Center for Frontier Science and Engineering, University of Electro-Communications, 1-5-1 Chofugaoka, Chofu, 182-8585 Tokyo, Japan

Molecular Imaging Center, National Institute of Radiological Sciences, Chiba, Japan
e-mail: masamoto@mce.uec.ac.jp

H. Takuwa • J. Taniguchi • H. Kawaguchi • H. Ito • I. Kanno
Molecular Imaging Center, National Institute of Radiological Sciences, Chiba, Japan

Y. Tomita • H. Toriumi • M. Unekawa • Y. Itoh • N. Suzuki
Department of Neurology, Keio University School of Medicine, Tokyo, Japan

S. Van Huffel et al. (eds.), *Oxygen Transport to Tissue XXXV*, Advances in Experimental Medicine and Biology 789, DOI 10.1007/978-1-4614-7411-1_3,

3.1 Introduction

Vascular structure in the brain is unique and specific to the brain. The inside of the cerebral microvessel is tightly sealed with endothelial cells, whereas the outside is wrapped with astroglial processes [1, 2]. These morphological features form the basis of the blood–brain barrier (BBB) which plays a major role in regulating the entry of substances into brain tissue in response to tissue demand.

Recently, we established an imaging system for repeatedly measuring cerebral microvasculature in mouse cortex and investigated the morphological changes induced by chronically exposed mild (8–9 % oxygen) and moderate (10–11 % oxygen) continuous hypoxia [3]. The cortical microvessels were imaged at identical locations repeatedly, and hypoxia-provoked changes of their length and diameter were characterized using two-photon microscopy. In this study, a drastic increase of the capillary diameter was found in the parenchymal cortex but less in the cortical surface [3]. A higher degree of the vascular restructuring was provoked under the 8–9 % oxygen condition relative to the 10–11 % one. These findings are in good agreement with previous reports that demonstrate the increase of microvascular density under chronic hypoxia [4–6].

In the present study, to further understand the cellular mechanisms of cerebrovascular restructuring induced by hypoxia, we determined spatiotemporal changes of microvascular networks, specifically focusing on cerebral angiogenesis. Two-photon microscopic imaging of cortical microvasculature was conducted at the identical locations of anesthetized mouse cortex repeatedly from the beginning of exposure to hypoxia (8–9 % oxygen) over 2–4 weeks. A change of microvascular networks was identified, and a number of angiogenesis provoked were quantified.

3.2 Materials and Methods

3.2.1 Animal Preparation

Animal use and experimental protocols were approved by the Institutional Animal Care and Use Committee in National Institute of Radiological Sciences and University of Electro-Communications in Tokyo. Four male C57BL/6 J mice (8–9 weeks; Japan SLC, Inc., Shizuoka) were used for the experiments. The animal was anesthetized with isoflurane (3–4 % for induction and 2 % for surgery), and craniotomy was performed. A diameter of 3.5-mm glass was attached to create a closed cranial window over the exposed cortex [7]. After recovery from surgery, the animal was caged in a normal room with a free access to water and food. For a chronic hypoxia experiment, the animal was housed in an 8–9 % oxygen room with nitrogen-balanced gas under normal pressure of the atmosphere (750–760 mmHg) and room temperature (25–26 °C). During this experiment, the animals were kept under hypoxic conditions except for periods of weekly cleaning of the oxygen-restricted room (2–3 h) and imaging experiments (about 3 h). On each imaging date, the animal was moved on a microscopic stage and exposed to atmosphere.

3.2.2 *Imaging Experiment*

The animal was anesthetized with isoflurane (3–4 % for induction and 1.0–1.2 % for imaging sessions in room air), and the animal's head was fixed on a stage of the microscope with a homemade attachment device. To visualize cortical microvasculature, sulforhodamine 101 (MP Biomedicals, Irvine, CA) dissolved in saline (10 mM) was injected intraperitoneally (8 μL/g body weight) to the animal, and the imaging with two-photon microscopy (TCS-SP5 MP, Leica Microsystems GmbH, Wetzlar, Germany) was performed on the cortex through a cranial window [8]. An excitation wavelength was 900 nm, while an emission signal was measured through a band-pass filter (610/75 nm). A single image was 1,024 by 1,024 pixels, and in-plane pixel size was 0.46 μm and acquired up to a depth of 0.3–0.8 mm from the cortical surface with a z-step size of 5 μm. Four to five locations were selected, avoiding the area with relatively large pial vessels (>100 μm in diameter). On the day beginning the exposure to hypoxia (day 0), pre-hypoxia control image was obtained, and follow-up imaging at identical locations was performed weekly.

All images were analyzed offline with LAS AF software (Leica Microsystems GmbH, Wetzlar, Germany), and a number of newly developed vessels were counted. A diameter of the new vessel was manually measured at the maximum cross section. To define an emergence rate of newly developed vessels per unit volume, the ratio of a total number of new vessels divided by the scanned volume over each 100-μm depth was measured.

3.3 Results and Discussion

Two-photon imaging of cortical microvasculature was typically achieved over depths of up to 800 μm from the cortical surface in mouse cortex [8]. In the present experiment, we observed that a depth detectability of the microvasculature varied depending on a location and time of the scan. To normalize this variation, a total volume scanned in each experiment was calculated in each animal by multiplying the area scanned by the maximum depth where the capillary structure was visually recognizable in the image. Figure 3.1 represents the depth-dependent changes of the scanned area at days 7 (Fig. 3.1a) and 14 (Fig. 3.1b) after exposure to hypoxia. A whole scanned area averaged from four animals was $0.88 \pm 0.10\ mm^2$ over depths of 0–150 μm at both days 7 and 14. In the deep cortical layers, however, a total of scanned areas (0.2–0.4 mm^2) decreased due to a lower signal-to-noise ratio.

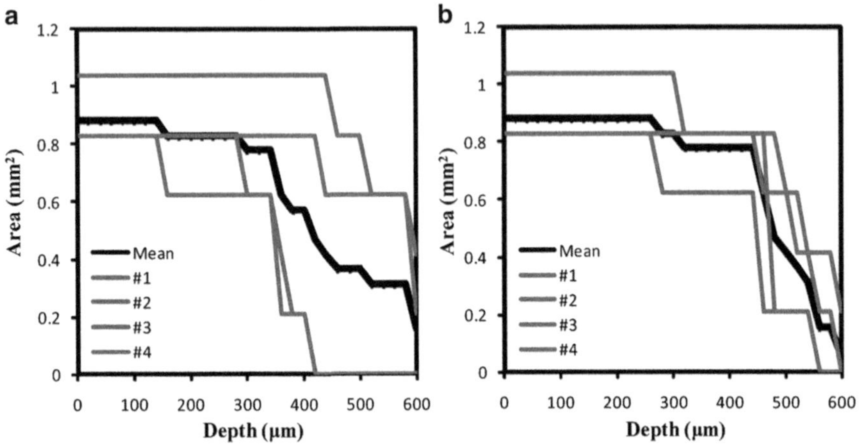

Fig. 3.1 A depth-dependent variation of total scanned areas with two-photon microscopy in four animals (#1–#4) measured at days 7 (**a**) and 14 (**b**) after exposure to hypoxia

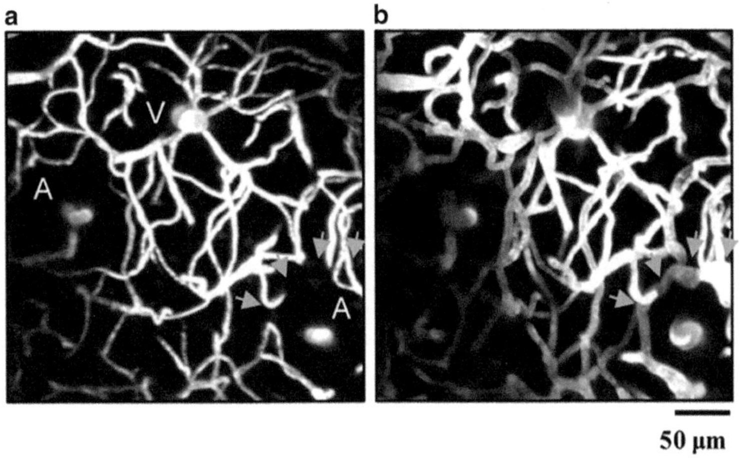

Fig. 3.2 A representative image of a new vessel development. Images of cortical microvasculature showed maximum intensity projection over depths of 20–260 μm from the cortical surface, measured at the identical location but different dates: pre-hypoxia (**a**) versus 14 days after exposure to hypoxia (**b**). Two major penetrating arterioles (*A*, *left middle* and *right bottom* in the image) and one vein (*V*, *center top* in the image) were identified, and dilation of capillaries was detected at day 14. Among them, a new vessel appeared around one of the arterioles (*arrows*)

One week after the exposure to hypoxia, parenchymal capillaries showed a remarkable increase of the diameter, and some locations had a new vessel formation. A new vessel that sprouted from existing capillaries was found on day 7 and also on day 14. Representative images that compared the same microvasculature at pre-hypoxia and day 14 after exposure to hypoxia are shown in Fig. 3.2. Because the image was sequentially taken over depths of 0–800 μm with a step size of 5 μm, misalignment that causes a virtually new vessel appearing from a non-scanned slice

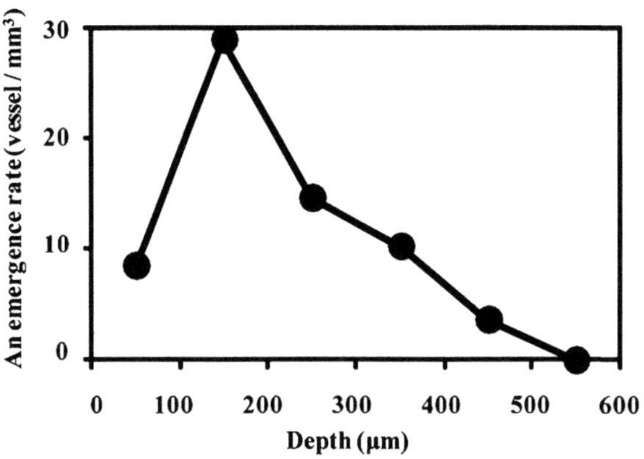

Fig. 3.3 An emergence rate of new vessels per unit volume was calculated every 100-μm depth over 600-μm depths from the cortical surface. The highest emergence rate was found at depths of 100–200 μm, and the deeper layers had fewer new vessels

can be ruled out. A mean diameter of the new vessel was 17 ± 7 μm ($n = 22$ vessels). This diameter was larger than we expected. In addition, we observed no detectable leakage of injected sulforhodamine from the new vessels, indicating that the BBB is preserved. In future studies, a role of vascular and perivascular cells, such as endothelium, pericyte, microglia, and astroglia, on maintaining the BBB in a process of hypoxia-induced angiogenesis should be determined.

In addition, a new vessel frequently appeared in the superficial cortical layers. The highest number of the emergence rate was found at the cortical depths of 100–200 μm, and the deeper cortex had a lower emergence rate, 8.5, 28.8, 14.6, 10.2, 3.6, and 0 vessels per unit volume (mm^3) over depths of 0–100, 100–200, 200–300, 300–400, 400–500, and 500–600 μm, respectively (Fig. 3.3). These layer dependences of the new vessel development were not agreed with previously known layer specificity of the hypoxia-vulnerable layers which were shown as cortical layers III and V/VI in the gerbil cortex [9]. The findings indicate variable cell populations that are sensitive to hypoxia and that produce the factors promoting angiogenesis over cortical layers.

3.4 Conclusions

Hypoxia-induced cerebral angiogenesis appeared from existing capillaries at days 7–14 after the onset of exposure to continuous hypoxia (8–9 % oxygen) in mouse cortex. The new vessel had an average diameter of 17 μm and the highest emergence rate was found in the cortical depths of 100–200 μm. Further studies are needed to elucidate the layer dependences of the hypoxia-induced angiogenesis and the cellular mechanisms, such as a sprouting, growth, and merger of new vessels, while maintaining the BBB.

Acknowledgments The authors thank Mr. Kouichi Yoshihara, Ryota Sakamoto, and Ryutaro Asaga for their help with the experiments. A part of this work was supported by Special Coordination Funds for Promoting Science and Technology, Japan (K.M.).

References

1. Abbott NJ, Rönnbäck L, Hansson E (2006) Astrocyte-endothelial interactions at the blood–brain barrier. Nat Rev Neurosci 7(1):41–53
2. Mathiisen TM, Lehre KP, Danbolt NC, Ottersen OP (2010) The perivascular astroglial sheath provides a complete covering of the brain microvessels: an electron microscopic 3D reconstruction. Glia 58(9):1094–1103
3. Yoshihara K, Takuwa H, Kanno I, Okawa S, Yamada Y, Masamoto K (2013) 3D Analysis of intracortical microvasculature during chronic hypoxia in mouse brains. Adv Exp Med Biol 765:357–363
4. LaManna JC, Vendel LM, Farrell RM (1992) Brain adaptation to chronic hypobaric hypoxia in rats. J Appl Physiol 72(6):2238–2243
5. Xu K, Lamanna JC (2006) Chronic hypoxia and the cerebral circulation. J Appl Physiol 100(2):725–730
6. Benderro GF, Lamanna JC (2011) Hypoxia-induced angiogenesis is delayed in aging mouse brain. Brain Res 1389:50–60
7. Tomita Y, Kubis N, Calando Y et al (2005) Long-term in vivo investigation of mouse cerebral microcirculation by fluorescence confocal microscopy in the area of focal ischemia. J Cereb Blood Flow Metab 25(7):858–867
8. Masamoto K, Tomita Y, Toriumi H et al (2012) Repeated longitudinal in vivo imaging of neuro-glio-vascular unit at the peripheral boundary of ischemia in mouse cerebral cortex. Neuroscience 212:190–200
9. Lin CS, Polsky K, Nadler JV, Crain BJ (1990) Selective neocortical and thalamic cell death in the gerbil after transient ischemia. Neuroscience 35(2):289–299

Chapter 4
Reduction of Cytochrome *c* Oxidase During Vasovagal Hypoxia-Ischemia in Human Adult Brain: A Case Study

Arnab Ghosh, Christina Kolyva, Ilias Tachtsidis, David Highton, Clare E. Elwell, and Martin Smith

Abstract Near-infrared spectroscopy (NIRS)-derived measurement of oxidized cytochrome *c* oxidase concentration ([oxCCO]) has been used as an assessment of the adequacy of cerebral oxygen delivery. We report a case in which a reduction in conscious level was associated with a reduction in [oxCCO]. Hypoxaemia was induced in a 31-year-old, healthy male subject as part of an ongoing clinical study. Midway through the hypoxaemic challenge, the subject experienced an unexpected vasovagal event with bradycardia, hypotension and reduced cerebral blood flow (middle cerebral artery blood flow velocity decrease from 70 to 30 cm s^{-1}) that induced a brief reduction in conscious level. An associated decrease in [oxCCO] was observed at 35 mm (−1.6 μM) but only minimal change (−0.1 μM) at 20-mm source-detector separation. A change in optical scattering was observed, but path length remained unchanged. This unexpected physiological event provides an unusual example of a severe reduction in cerebral oxygen delivery and is the first report correlating change in clinical status with changes in [oxCCO].

A. Ghosh (✉) • D. Highton
Neurocritical Care Unit, University College Hospitals, WC1N 3BG London, UK
e-mail: arnab.ghosh@ucl.ac.uk

C. Kolyva • I. Tachtsidis • C.E. Elwell
Medical Physics & Bioengineering, University College London WC1E 6BT London, UK

M. Smith
Neurocritical Care Unit, University College Hospitals, WC1N 3BG London, UK

Medical Physics & Bioengineering, University College London, WC1E 6BT London, UK

S. Van Huffel et al. (eds.), *Oxygen Transport to Tissue XXXV*, Advances in Experimental Medicine and Biology 789, DOI 10.1007/978-1-4614-7411-1_4,

4.1 Introduction

Cytochrome *c* oxidase is the final electron acceptor in the mitochondrial electron transport chain, and its oxidation state, measured by NIRS as [oxCCO], has thus been proposed as a marker of the adequacy of cerebral oxygen delivery (DO_2) [1]. However, prior studies of moderate hypoxia achieved only modest reductions in DO_2, leaving the relationship between DO_2 and Δ[oxCCO] unclear [2]. Furthermore, interpretation of [oxCCO] measurements is difficult as there are no adult human data regarding its total concentration or resting oxidation state.

We aimed to address some of these concerns with studies using a hybrid NIR spectrometer, the pHOS, to measure [oxCCO] and other optical parameters whilst modulating DO_2 in a cohort of healthy volunteers [3]. We report the case of a vasovagal event that occurred during one such study and explore the insights into cerebral physiology gleaned from this unexpected event.

4.2 Methods

A single case was selected from a larger group study. This study was approved by the Institutional Research Ethics Committee, and written informed consent was provided by the subject, a 31-year-old male who had been screened for pre-existing medical conditions, was selected as he suffered from vasovagal pre-syncope during a challenge that comprised induction of isocapnic hypoxia with a target arterial oxygen saturation (SpO_2) of 80 % [3].

The pHOS, described in more detail elsewhere [4], combines frequency domain (FD) and broadband (BB) components. Changes in chromophore concentration were estimated by using the changes in light attenuation as measured by the BB spectrometer, using the UCLn algorithm to resolve for three chromophores – oxyhaemoglobin (HbO_2), deoxyhaemoglobin (HHb) and oxCCO – between 780 and 900 nm. A fixed differential path length factor (DPF) of 6.26 [5] was used to enable comparison with previous studies. The FD component measured the absolute absorption and scattering coefficients (μa and μs, respectively) at discrete wavelengths 690, 750, 790 and 850 nm, allowing the estimation of DPF at each of these wavelengths using the diffusion approximation [6]. The pHOS optode was placed over the FP1 point on the right side of the forehead.

Other monitoring included beat-to-beat SpO_2, continuous non-invasive arterial blood pressure, and transcranial Doppler (TCD) ultrasonography was used to measure middle cerebral artery flow velocity (Vmca), insonating through the right temporal window, ipsilateral to pHOS monitoring. Estimated relative cerebral oxygen delivery ($ecDO_2$) was calculated as the product of changes in SpO_2 and Vmca (relative to their initial values) [2]. Synchronization between the pHOS and other monitors was performed by means of a signal voltage output by the frequency domain spectrometer for the length of recording; all data were resampled to a sample period of 3.2 s across the length of recording. Twenty-second data windows were selected for the reporting of summary data. Changes in hemoglobin species concentration are expressed as total hemoglobin ($\Delta[HbT]=\Delta[HbO_2]+\Delta[HHb]$) and hemoglobin difference ($\Delta[HbDiff]=\Delta[HbO_2]-\Delta[HHb]$). All analysis was carried out in Matlab® 2011b.

4.3 Results

Approximately 600 s after the commencement of recording, during the nadir of hypoxia, the subject suffered from sudden-onset bradycardia and hypotension (a typical vasovagal episode), resulting in a reduction in $ecDO_2$ to 41.5 % of baseline values (Table 4.1 and Fig. 4.1). During this time, the subject became briefly unresponsive. When this occurred, he was laid flat and the breathing circuit replaced with a Mapleson C circuit delivering high-flow oxygen.

Hemoglobin species showed a consistent pattern of change with a decrease in both [HbT] and [HbDiff] seen at all source-detector separations (Fig. 4.2). By contrast, [oxCCO] showed a source-detector separation-dependent decrease (Δ[oxCCO] −1.6 μM at 3.5 cm c.f. −0.1 μM at 2.0-cm source-detector separation), with larger decreases seen at further source-detector separations (Fig. 4.3). Comparison of Δ[oxCCO] (observed at the 3.5-cm detector) with $\Delta ecDO_2$ suggests a linear relationship (Fig. 4.4) between the two variables ($r^2=0.81$, $p<0.001$). There appeared to be a roughly 10 % decrease in the μs observed at all four discrete wavelengths, but this did not translate into an apparent change in DPF (Fig. 4.5).

Table 4.1 Mean values of systemic and cerebral physiological parameters observed during four observed phases

		Baseline	Hypoxia	Hypoxia-ischemia	Recovery
ABP (mmHg)		85.4	88.9	32.1	83.5
HR		78.0	117	41.3	101
Vmca (cm s^{-1})		70.4	73.9	34.1	57.8
SpO_2 (%)		100	72.6	86.3	99.4
$ecDO_2$ (%)		99.8	75.8	41.5	81.2
Δ[oxCCO] (μM)	3.5 cm	0	−0.74	−1.56	−0.13
	3.0 cm	0	−0.38	−0.62	0.008
	2.5 cm	0	−0.30	−0.55	0.045
	2.0 cm	0	−0.14	−0.16	0.23

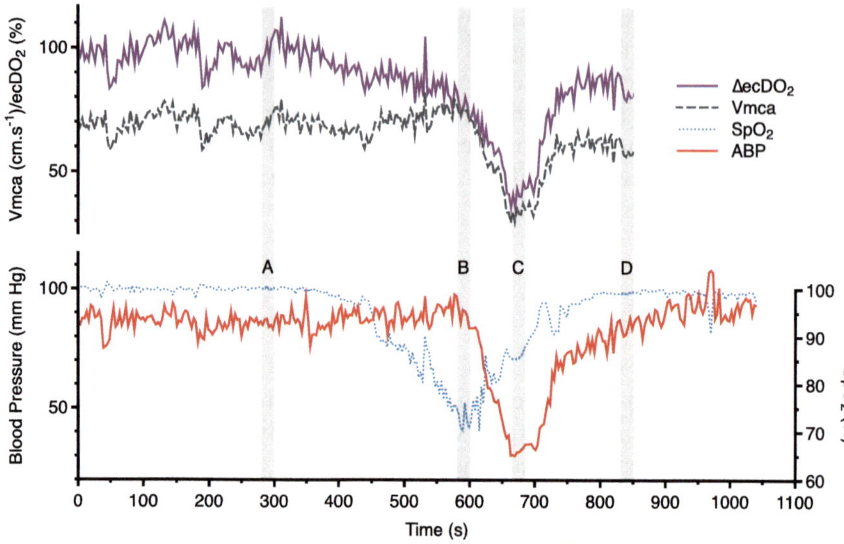

Fig. 4.1 Changes in systemic and physiological parameters. Data averaging periods *A* baseline, *B* hypoxia, *C* hypoxia-ischemia, *D* recovery

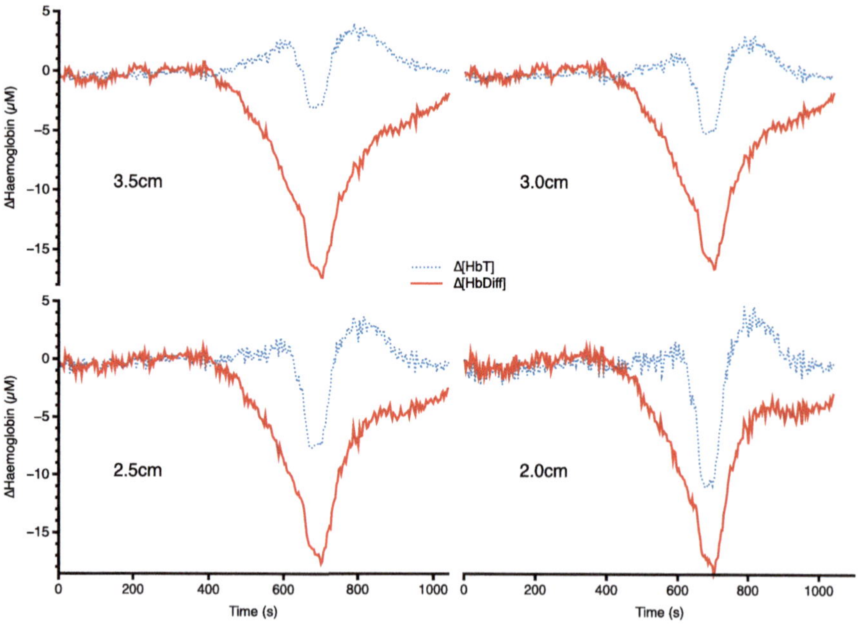

Fig. 4.2 Changes in hemoglobin concentrations at four source-detector separations

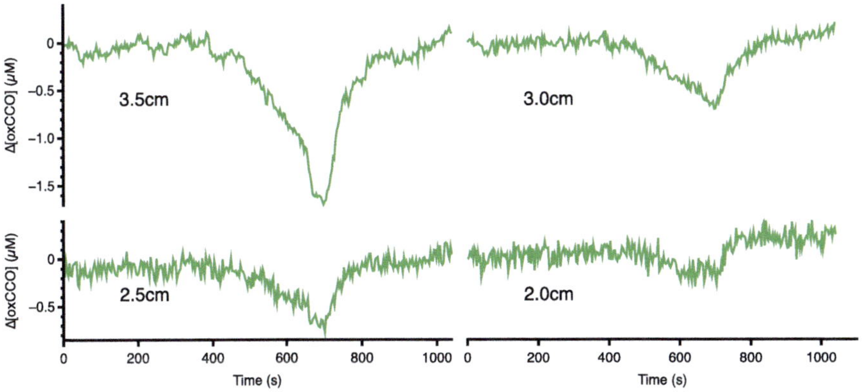

Fig. 4.3 Changes in [oxCCO] at four source-detector separations

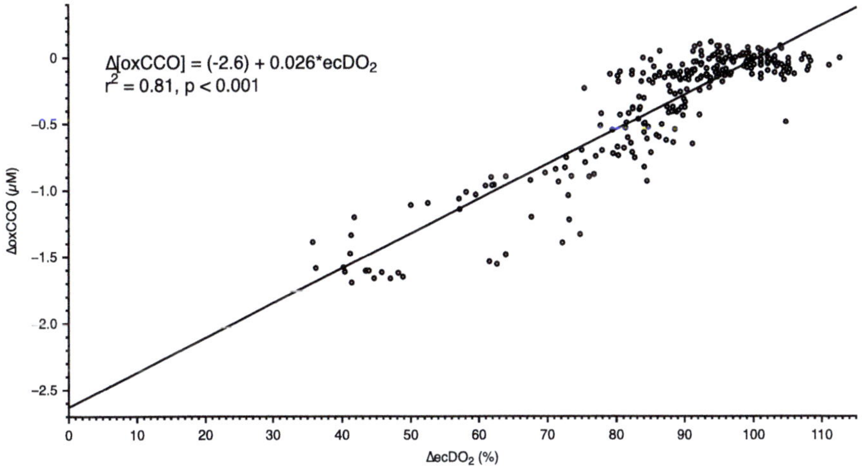

Fig. 4.4 Relationship between $ecDO_2$ and Δ[oxCCO] at 3.5 cm source-detector separation

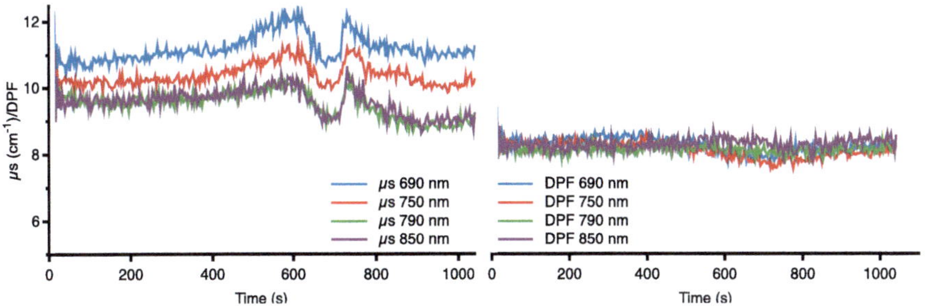

Fig. 4.5 Changes in μs and DPF

4.4 Discussion

We report a reduction in cranial [oxCCO] measured with the pHOS during a vasovagal pre-syncope in a healthy adult volunteer. This reduction was larger in further (3.5 cm) than closer (2.0 cm) source-detector separations. The pattern of [oxCCO] reduction has been related to neurological outcome following cardiopulmonary bypass [7], and asymptomatic reductions in [oxCCO] have been reported during moderate hypoxia [2], but this is the first change in [oxCCO] to be correlated to changes in conscious level and the largest change to be reported in humans.

Whilst hypoxia is a recognized cause of syncope, the occurrence of bradycardia and hypotension indicates a vasovagal etiology [8], in this case, provoked by – as the subject reported – the shock of seeing his reduced SpO_2 reading. This vasovagal response resulted in a profound reduction in cerebral blood flow (observed by both NIRS and TCD), and this led to a significant reduction in DO_2.

Prior experiments have achieved only modest reductions in DO_2, leaving questions unanswered about the relationship between DO_2 and Δ[oxCCO] [2]. In particular, it was unclear whether the relationship between DO_2 and Δ[oxCCO] was linear or whether there was a DO_2 threshold below which rapid CCO reduction occurs. However, our data (Fig. 4.4) suggest a linear relationship between Δ[oxCCO] and $\Delta ecDO_2$ ($r^2=0.81, p<0.001$). From this, the extrapolated value of Δ[oxCCO] of −2.6 μM with zero cerebral oxygen delivery is suggestive of a resting oxidized CCO concentration of around 2.6 μM. Although no reduction of this size has been reported in humans – understandably as a reduction in DO_2 to nothing is impractical in humans – these values are consistent with animal anoxia models [9, 10].

Given the distance dependence of the [oxCCO] but not hemoglobin changes, we consider spectral crosstalk to be unlikely. Similarly, the modest μs changes that were observed appeared consistent in their spectral expression across 690–850 nm and thus are unlikely to account for the [oxCCO] changes seen. DPF showed no significant qualitative change during the hypoxic or ischaemic phases of the study, although the absolute values were larger than those that were previously reported [5]; using the measured values of DPF rather than the fixed value of 6.26 would thus change the magnitude, but not the qualitative profile of our observed chromophore changes.

Whilst we are aware of the limitations of extrapolating from a single case report, these data, especially when considered in conjunction with the results of the larger cohort of (uncomplicated!) hypoxia and hypercapnia studies [3], further underline confidence in the ability to use NIRS to measure [oxCCO] as a marker of cellular energy status.

Acknowledgments This work was undertaken at University College London Hospitals and partially funded by the Department of Health's National Institute for Health Research Centers funding scheme, the Medical Research Council and Wellcome Trust.

References

1. Smith M, Elwell C (2009) Near-infrared spectroscopy: shedding light on the injured brain. Anesth Analg 108(4):1055–1057
2. Tisdall MM, Tachtsidis I, Leung TS, Elwell CE, Smith M (2007) Near-infrared spectroscopic quantification of changes in the concentration of oxidized cytochrome c oxidase in the healthy human brain during hypoxemia. J Biomed Opt 12(2):024002
3. Kolyva C, Ghosh A, Tachtsidis I, et al (2013) Dependence on NIRS source-detector spacing of cytochrome C oxidase response to hypoxia and hypercapnia in the adult brain. Adv Exp Med Biol.
4. Kolyva C, Tachtsidis I, Ghosh A et al (2012) Systematic investigation of changes in oxidized cerebral cytochrome c oxidase concentration during frontal lobe activation in healthy adults. Biomed Opt Express 3(10):2550–2566
5. Duncan A, Meek JH, Clemence M et al (1995) Optical pathlength measurements on adult head, calf and forearm and the head of the newborn infant using phase resolved optical spectroscopy. Phys Med Biol 40(2):295–304
6. Arridge SR, Cope M, Delpy DT (1992) The theoretical basis for the determination of optical pathlengths in tissue: temporal and frequency analysis. Phys Med Biol 37(7):1531–1560
7. Kakihana Y, Matsunaga A, Yasuda T, Imabayashi T, Kanmura Y, Tamura M (2008) Brain oxymetry in the operating room: current status and future directions with particular regard to cytochrome oxidase. J Biomed Opt 13(3):033001
8. Chen-Scarabelli C, Scarabelli TM (2004) Neurocardiogenic syncope. Br Med J 329(7461): 336–341
9. Cooper CE, Cope M, Springett R et al (1999) Use of mitochondrial inhibitors to demonstrate that cytochrome oxidase near-infrared spectroscopy can measure mitochondrial dysfunction noninvasively in the brain. J Cereb Blood Flow Metab 19(1):27–38
10. Cooper CE, Springett R (1997) Measurement of cytochrome oxidase and mitochondrial energetics by near-infrared spectroscopy. Philos Trans R Soc Lond B Biol Sci 352(1354): 669–676

Chapter 5
Increased HIF-1α and HIF-2α Accumulation, but Decreased Microvascular Density, in Chronic Hyperoxia and Hypercapnia in the Mouse Cerebral Cortex

Girriso F. Benderro, Constantinos P. Tsipis, Xiaoyan Sun, Youzhi Kuang, and Joseph C. LaManna

Abstract The partial pressure of oxygen in the brain parenchyma is tightly controlled, and normal brain function is delicately sensitive to continuous and controlled oxygen delivery. The objective of this study was to determine brain angiogenic adaptive changes during chronic normobaric *hyperoxia and hypercapnia* in mice. Four-month-old C56BL/6 J mice were kept in a normobaric chamber at 50 % O_2 and 2.5 % CO_2 for up to 3 weeks. Normoxic littermates were kept adjacent to the chamber and maintained on the same schedule. Physiological variables were measured at time points throughout the 3 weeks or when the mice were sacrificed. Freshly collected or fixed brain specimens were analyzed by Western blot analysis and immunohistochemistry (IHC). We found significant accumulation of hypoxia-inducible factors 1α and 2α (HIF-1α and HIF-2α) and increased expression of erythropoietin (EPO), cyclooxygenase-2 (COX-2), and angiopoietin-2 (Ang-2) in hyperoxia and hypercapnia. Conversely, vascular endothelial growth factor (VEGF), and VEGF receptor-2 (KDR/Flk-1), peroxisome proliferator-activated receptor gamma coactivator 1-α (PGC-1α), and prolyl hydroxylase-2 (PHD-2) expressions were decreased in hyperoxia and hypercapnia. Capillary density was significantly diminished by the end of the 3rd week of hyperoxia and hypercapnia as compared to control. We conclude that HIF-independent mechanisms contribute to brain capillary density modulation that is continuously adjusted in accordance with tissue oxygen tension.

G.F. Benderro
Department of Anatomy, Case Western Reserve University, Cleveland, OH, USA

C.P. Tsipis • X. Sun • Y. Kuang • J.C. LaManna (✉)
Department of Physiology and Biophysics, School of Medicine,
Case Western Reserve University, 10900 Euclid Ave, Cleveland, OH 44106, USA
e-mail: jcl4@case.edu

S. Van Huffel et al. (eds.), *Oxygen Transport to Tissue XXXV*, Advances in Experimental Medicine and Biology 789, DOI 10.1007/978-1-4614-7411-1_5,

5.1 Introduction

The main long-term mechanism by which the mammalian brain maintains its optimal continuous supply of oxygen and nutrients is by vascular remodeling [1–3]. Vascular remodeling (angiogenesis or regression) is a complex process that requires coordinated interaction of multiple vascular regulating factors among which HIF, VEGF, COX-2, and Ang-2 are crucial [2, 4, 5]. We previously showed that chronic hypoxia led to angiogenesis [3, 5, 6] and chronic hyperoxia led to capillary regression [7], suggesting a modulatory role for oxygen in determining capillary density. In this study we analyzed whether the vasodilation and respiratory acidosis effect of inspired CO_2 in hypercapnia coupled with hyperoxia have effects on expression of the main angiogenic growth factors and their activities on brain vascular remodeling.

5.2 Experimental Procedures

5.2.1 Exposure to Chronic Hyperoxia with Hypercapnia and Measurement of Blood Gas Content

Male 4-month-old C57BL/6 mice used for this study were purchased from Jackson Laboratories (Bar Harbor, ME, USA). The experimental protocol was approved by the IACUC. Experimental mice were placed in a normobaric chamber (Oxycycler™; BioSpherix Ltd., Lacona, NY), and a computer-regulated gas flow of 50 % O_2 and 2.5 % CO_2 (bal. N_2) was maintained for up to 3 weeks of exposure. Normoxic littermate control mice were housed in the same room adjacent to the chamber to match ambient conditions. Arterial blood pH, O_2, and CO_2 contents were measured by an ABL5 blood gas analyzer (Radiometer Medical, Copenhagen, Denmark) as described before [7].

5.2.2 Preparation of Whole Cell Lysates and Western Blot Analysis

Western blotting was done from freshly collected brain samples as described previously [3]. Briefly, samples of the cerebral cortex were homogenized in ice-cold RIPA lysis buffer. Homogenates were centrifuged, protein contents in the supernatants were determined, and proteins were separated by using SDS gel electrophoresis and transferred to nitrocellulose membranes (Bio-Rad, Hercules, CA). The specific primary antibodies used were HIF-1α and HIF-2α (1:500; R&D Systems, Minneapolis, MN); VEGF-A, FLK-1, and EPO (1:750; Santa Cruz Biotechnology, Santa Cruz, CA); Ang-2 (1:200; Millipore Co., Billerica, MA);

COX-2 (1:150, Cayman, Ann Arbor, MI); PGC-1α (1:750; Novus, Littleton, CO); β-actin (1:2,000; Santa Cruz, CA); and β-tubulin (1:3,000, Cell Signaling Tech., Beverly, MA). The membranes were washed with TBS-T, followed by incubation with the appropriate horseradish peroxidase-conjugated secondary antibodies (Millipore Co., Billerica, MA). Densitometry of scanned protein bands was measured and normalized to that of β-actin (optical density ratio) using ImageJ.

5.2.3 IHC and Determination of the Capillary Density

Tissue collection and IHC were done as described previously [3]. Briefly, mice were perfused transcardially with phosphate-buffered saline (PBS) followed by 4 % paraformaldehyde (PFA), for in vivo fixation of the tissues. The brain samples collected were embedded in paraffin, sectioned with a microtome, dried, and dehydrated. After antigen retrieval, quenching, and blocking, the sections were incubated with goat polyclonal primary antibody against GLUT-1 (1:200; Santa Cruz, CA), washed, and covered with biotinylated horse anti-goat secondary antibody (Vector Labs, Burlingame, CA). After cover slipping, a photo montage of the parietal cortex was created using an Aquos Q-image high-definition digital camera connected to a Nikon E600 Eclipse microscope with a 20X objective. Adobe Photoshop CS5 and ImageJ were used to count positively stained microvessels less than 20 μm in diameter to determine the capillary density (number per mm^2 of brain tissue) in the cortex. Statistical analysis was done as described previously [3].

5.3 Results

5.3.1 HIF-1α and HIF-2α Protein Accumulation and VEGF and Flk-1 Expression in the Brain

Since HIF-1α and HIF-2α are early responder transcription factors for induction of many genes involved in oxidative stress, adaptive responses, and angiogenic remodeling [8–10], we have examined the response of these transcription factors during hyperoxia and hypercapnia. Immunoblot assay demonstrated robust and significant ($P<0.05$) accumulation of both HIF-1α and HIF-2α (five- to sixfold) during the time of exposure compared to normoxic control (Fig. 5.1a). VEGF and its receptor (Flk-1) are critical factors during vascular remodeling [11, 12]. Western blot analysis showed diminished VEGF and Flk-1 protein expression throughout the periods of exposure. These decreases in VEGF and Flk-1 levels in chronic hyperoxia and hypercapnia were statistically significant ($P<0.05$) relative to their baseline expression during normoxia (Fig. 5.1b).

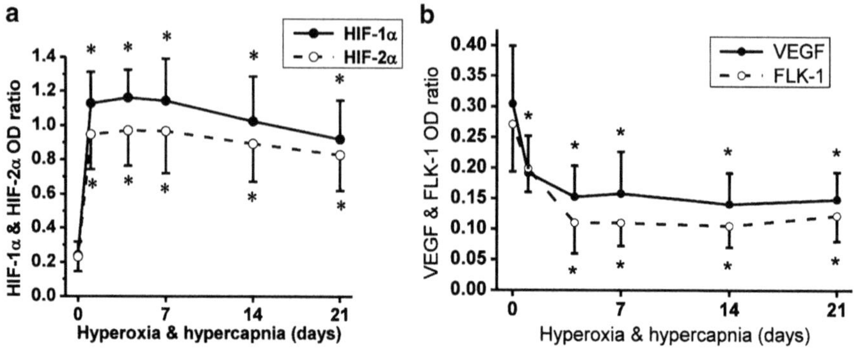

Fig. 5.1 HIF-1α and HIF-2α accumulation and VEGF and FLK-1 expression in cerebral cortex of mice in chronic hyperoxia and hypercapnia. (**a**) and (**b**): Western blot optical density (*OD*) ratio normalized to β-actin or β-tubulin in normoxia (0) and 1, 4, 7, 14, and 21 days of hyperoxia and hypercapnia $*p<0.05$ compared with normoxic control. Each value represents the mean ± SD from five mice per time point

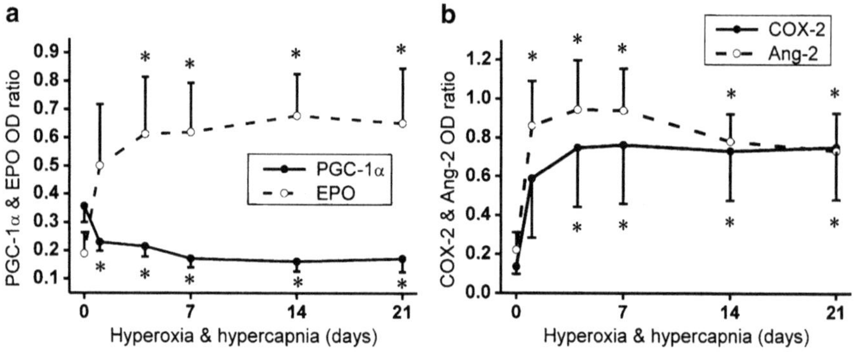

Fig. 5.2 Expression of PGC-1α, EPO, COX-2, and Ang-2 in mouse cerebral cortex during chronic hyperoxia and hypercapnia. (**a**) and (**b**) Western blot optical density ratio normalized to β-actin. $*P<0.05$ compared with normoxic values of each category. Values are mean ± SD. $n=5$ mice for EPO and PGC-1α and 4 mice for COX-2 and Ang-2 per time point

5.3.2 PGC-1α, EPO, COX-2, and Ang-2 Protein Expression

PGC-1α protein level was significantly decreased ($P<0.05$) in chronic hyperoxia and hypercapnia, compared to its relative expression in normoxia. PGC-1α, which is known to be an HIF-independent inducer of VEGF [13, 14], showed a similar trend of expression as compared with VEGF (Fig. 5.1b). Conversely, EPO protein level was significantly increased ($P<0.05$) in all time points of chronic hyperoxia and hypercapnia (Fig. 5.2a). It was demonstrated that EPO expression in the brain

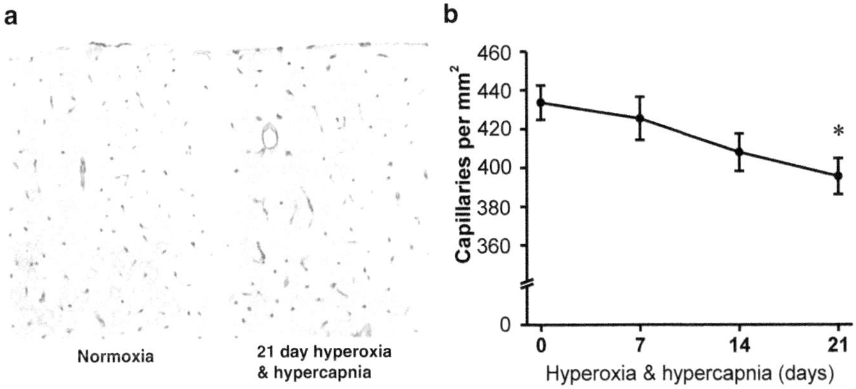

Fig. 5.3 Capillary density in mouse cerebral cortex in normoxic control and hyperoxia and hypercapnia. (**a**) GLUT-1-stained sections spanning part of the parietal cortex. (**b**) Capillary density (number per mm^2). Values are mean ± SE of mean, $n=4$ mice per time point

was mainly regulated by HIF-2α [15]. This may explain the similar trend in expression of HIF-2α (Fig. 5.1a) and EPO (Fig. 5.2a). HIF-independent induction of Ang-2 by COX-2 was well established [2, 5]. Western blot analysis showed a similar and strong expression of COX-2 and Ang-2 in hyperoxia and hypercapnia (Fig. 5.2b). Both COX-2 and Ang-2 protein expressions were significantly increased ($P<0.05$) throughout the duration of exposure.

5.3.3 *Capillary Density in the Cerebral Cortex and Arterial Blood Gas Content*

Cerebral capillaries were identified by GLUT-1 immunostaining, and their density was quantified by counting the number of positively stained capillaries per mm^2 (N/mm^2). Mice exposed to chronic hyperoxia and hypercapnia showed a moderate decrease in capillary density after the 1st and 2nd weeks, which became statistically significant ($P<0.05$) by the end of the 3rd week of exposure (Fig. 5.3a, b) as compared to normoxic baseline. The capillary density decreased from 434 ± 9/mm^2 in normoxia to 425 ± 11/mm^2 in the 1st week, 408 ± 10/mm^2 in the 2nd week, and 396 ± 9/mm^2 (about 9 % decrease) in the 3rd week of chronic hyperoxia and hypercapnia. We also recorded that the mice placed in hyperoxia and hypercapnia showed significant increase ($P<0.05$) in arterial blood O_2 and CO_2 contents (Table 5.1) and significant decrease ($P<0.05$) in arterial pH (respiratory acidosis) during exposure compared to normoxic littermates.

Table 5.1 Arterial blood gas parameters during normobaric hyperoxia (50 % O_2) and hypercapnia (2.5 % CO_2). Values are mean ± SD; *$P<0.05$ compared to normoxic value of each category; $n=8$ for normoxia and 6 for all hyperoxia and hypercapnia cases

		Hyperoxia and Hypercapnia (days)				
Variable	Normoxia	1	4	7	14	21
pH	7.35±0.05	7.17±0.08*	7.18±0.08*	7.19±0.09*	7.17±0.06*	7.23±0.04*
PO_2 (mmHg)	101±13	327±30*	322±25*	335±17*	327±19*	314±26*
PCO_2 (mmHg)	36±3	53±6*	46±6*	47±3*	47±3*	48±4*

5.4 Discussion

Vascular remodeling in the brain allows restoration of normal oxygen tension in the cerebral parenchyma during acclimatization to adverse environments [4, 11, 16]. It requires coordinated production and signaling interaction of multiple growth factors involved in angioplasticity [2]. During capillary regression, the balance between VEGF and Ang-2 signaling seems to play an important role [5]. In this study, we have observed progressive vascular regression in chronic hyperoxia and hypercapnia, which was consistent with decreased VEGF and increased Ang-2 protein expression. Different from our recent report on the effects of 50 % O_2 hyperoxia only [7], we report here a robust increase in HIF-1α and HIF-2α and arterial blood CO_2 content, alongside a significant decrease in arterial blood pH in hyperoxia with hypercapnia.

Elevated accumulation of HIF-1α and HIF-2α (five- to sixfold compared to normoxia) during hyperoxia and hypercapnia may be due to the combination of diminished PHD expression in hyperoxia [7] and diminished VHL activity in the hypercapnic acidic internal milieu [17]. However, accumulation of HIF-1α and HIF-2α failed to induce VEGF or its receptor (FLK-1). The reduction in VEGF expression trended similarly to PGC-1α expression, indicating HIF-independent transcriptional regulation of PGC-1α on VEGF expression [13, 14]. In relation to HIF-1α and HIF-2α accumulation, upregulation of EPO was observed. EPO has an HRE-binding domain and its induction is dependent on activation by HIF [8, 10]. EPO is known to be an endogenous mediator of neuroprotection in the CNS [9, 15]. It may be that the stabilization of HIF-1α and HIF-2α during hyperoxia and hypercapnia is essential for the neuroprotective response.

We observed a robust increase in the expression of both COX-2 and Ang-2 during hyperoxia and hypercapnia. Ang-2 is proangiogenic in the presence of VEGF and acts as an anti-angiogenic factor during vascular regression in the reduction or absence of VEGF [2, 5]. Hence, increased Ang-2 levels coupled with decreased VEGF may have compromised the integrity of some of the neurovascular unit, leading to microvascular regression and reduced capillary density during hyperoxia and hypercapnia. In conclusion, although more robust HIF-1α and HIF-2α expression in the brain, respiratory acidosis, and increased plasma CO_2 content was recorded during hyperoxia with hypercapnia, chronic hyperoxia with hypercapnia caused similar progressive capillary regression effect as hyperoxia alone [7] on the mouse cerebral cortex.

Acknowledgment This study was supported by NIH R01 NS38632.

References

1. Boero JA, Ascher J, Arregui A, Rovianen C, Woolsey TA (1999) Increased brain capillaries in chronic hypoxia. J Appl Physiol 86(4):1211–1219
2. Dore-Duffy P, LaManna JC (2007) Physiologic angiodynamics in the brain. Antioxid Redox Signal 9(9):1363–1371
3. Benderro GF, LaManna JC (2011) Hypoxia-induced angiogenesis is delayed in aging mouse brain. Brain Res 1389:50–60
4. Carmeliet P (2003) Angiogenesis in health and disease. Nat Med 9(6):653–660
5. Pichiule P, LaManna JC (2002) Angiopoietin-2 and rat brain capillary remodeling during adaptation and deadaptation to prolonged mild hypoxia. J Appl Physiol 93(3):1131–1139
6. Ndubuizu OI, Tsipis CP, Li A, LaManna JC (2010) Hypoxia-inducible factor-1 (HIF-1)-independent microvascular angiogenesis in the aged rat brain. Brain Res 1366:101–109
7. Benderro GF, Sun X, Kuang Y, Lamanna JC (2012) Decreased VEGF expression and microvascular density, but increased HIF-1 and 2alpha accumulation and EPO expression in chronic moderate hyperoxia in the mouse brain. Brain Res 1471:46–55
8. Fong GH (2008) Mechanisms of adaptive angiogenesis to tissue hypoxia. Angiogenesis 11(2):121–140
9. Mole DR, Ratcliffe PJ (2008) Cellular oxygen sensing in health and disease. Pediatr Nephrol 23(5):681–694
10. Semenza GL (2007) Life with oxygen. Science 318:62–64
11. Ferrara N, Gerber HP, LeCouter J (2003) The biology of VEGF and its receptors. Nat Med 9(6):669–676
12. Mizukami Y, Kohgo Y, Chung DC (2007) Hypoxia inducible factor-1 independent pathways in tumor angiogenesis. Clin Cancer Res 13(19):5670–5674
13. Arany Z, Foo SY, Ma Y et al (2008) HIF-independent regulation of VEGF and angiogenesis by the transcriptional coactivator PGC-1alpha. Nature 451(7181):1008–1012
14. Chinsomboon J, Ruas J, Gupta RK et al (2009) The transcriptional coactivator PGC-1alpha mediates exercise-induced angiogenesis in skeletal muscle. Proc Natl Acad Sci USA 106(50):21401–21406
15. Chavez JC, Baranova O, Lin J, Pichiule P (2009) The transcriptional activator hypoxia inducible factor 2 (HIF-2/EPAS-1) regulates the oxygen-dependent expression of erythropoietin in cortical astrocytes. J Neurosci 26(37):9471–9481
16. Dunn JF, Grinberg O, Roche M, Nwaigwe CI, Hou HG, Swartz HW (2000) Noninvasive assessment of cerebral oxygenation during acclimation to hypobaric hypoxia. J Cereb Blood Flow Metab 20(12):1632–1635
17. Mekhail K, Gunaratnam L, Bonicalzi ME, Lee S (2004) HIF activation by pH-dependent nucleolar sequestration of VHL. Nat Cell Biol 6(7):642–647

Chapter 6
Oxygen Delivery: The Principal Role of the Circulation

Christopher B. Wolff

Abstract Autoregulation of blood flow to most individual organs is well known. The balance of oxygen supply relative to the rate of oxygen consumption ensures normal function. There is less reserve as regards oxygen supply than for any other necessary metabolite or waste product so oxygen supply is flow dependent. Reduced rate of supply compromises tissue oxygenation long before any other substance. The present report reiterates evidence from earlier studies demonstrating that the rate of oxygen delivery (DO_2), for most individual tissues, is well sustained at a value bearing a ratio to oxygen consumption (VO_2) which is specific for the organ concerned. For the brain DO_2 is sustained at approximately three times the rate of oxygen consumption and for exercising skeletal muscle (below the anaerobic threshold), a ratio close to 1.5. The tissue-specific ratios are sustained in the face of alterations in local VO_2 and lowered arterial oxygen content (CaO_2). Tolerance varies between different organs. Hence, the role of the circulation is predominantly one of ensuring an adequate supply of oxygen. The precise values of the individual tissue DO_2:VO_2 ratios apply within physiological ranges which require further investigation.

6.1 Introduction

Blood flow regulation has been known for many years to be autoregulated for most, if not all, organ systems [1, 2]. Cardiac output is a result of the summation of blood flows through all individual tissues [2]. The purpose of this chapter is to illustrate the evidence that autoregulation sustains an individually appropriate rate of oxygen supply (oxygen delivery, DO_2) for most tissues.

C.B. Wolff (✉)
Clinical Pharmacology and Anaesthesia, William Harvey Research Institute, Barts and The London, London, UK
e-mail: chriswolff@doctors.org.uk

S. Van Huffel et al. (eds.), *Oxygen Transport to Tissue XXXV*, Advances in Experimental Medicine and Biology 789, DOI 10.1007/978-1-4614-7411-1_6,

6.2 Evidence for Constancy of the DO_2:VO_2 Ratio

Oxygen delivery (DO_2) for the majority of tissues bears a constant and tissue-specific ratio to the rate at which that tissue utilises oxygen (VO_2, its metabolic rate). The inverse, oxygen extraction ($E = VO_2/DO_2$) is, likewise, constant over a range of altered local metabolic rate. There is a range, varying between tissues, where constancy (of DO_2/VO_2) is sustained in the face of lowered arterial oxygen content (CaO_2) from either anaemia or hypoxia [3]. Constancy of the DO_2:VO_2 ratio, and hence E, is directly controlled by the tissue requirement for oxygen. Cardiac output is the total tissue blood flow largely determined by autoregulation of appropriate DO_2. "The heart puts out what it receives" [4]. A notable exception to DO_2 autoregulation is the renal flow autoregulation independent of SaO_2 [5].

6.3 Examples of Major Organ Constancy of the DO_2:VO_2 Ratio

6.3.1 Whole Body and Exercising Skeletal Muscle

For moderate exercise, cardiac output increases above the resting value in proportion to the increase in oxygen consumption. There is a proportional increase in oxygen delivery to the active muscle, with a constant DO_2-to-VO_2 ratio close to 1.5. This has been shown by analysis of the results from two exercise papers [6, 7]. Exercise of a small muscle group (quadriceps) gave a higher blood flow in both anaemia and hypoxia to sustain normal DO_2 for the moderate exercise involved (30 W to half maximal). The DO_2-to-VO_2 ratio was sustained with reduction in CaO_2 down to 60 % of normal [8].

6.3.2 Cerebral Autoregulation of Blood Flow and DO_2

Cerebral DO_2 is sustained at around three times VO_2 under hypoxic as well as under normal conditions at least down to 90 % saturation. This has been validated by the analysis of sea level and high altitude (3,810 m, 12,500 ft) values of Severinghaus et al. [9] by Wolff [10, 11]. Earlier studies show average global values of the DO_2-to-VO_2 ratio of 2.9 (SEM 1.1), and from PET scanning studies of individual cerebral tissue areas [12], the average value was 2.7 (0.1). Blood flows in local areas of the brain change in proportion to VO_2 thereby sustaining the DO_2-to-VO_2 ratio [13].

6.3.3 Coronary Arterial Oxygen Delivery to the Heart

For the heart Martinez et al. [14] reported blood flows, oxygen content and oxygen consumption to the right ventricle during air breathing and with three successively

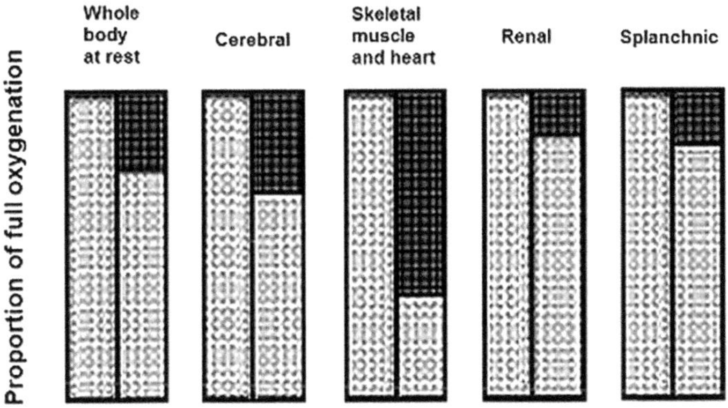

Fig. 6.1 Schematic illustrating the differing DO_2/VO_2 values for individual tissues during normal arterial oxygenation (L columns, proportion of full arterial saturation and content; R columns, venous). *Light sections* oxygenated, *dark sections* deoxygenated, true state fully mixed. Renal status involves autoregulation of plasma flow [5] (After Wolff [3])

lower oxygen levels in the inspirate. The work of the heart showed increased VO_2 from extra systemic circulation during hypoxia. Blood flow and oxygen delivery therefore increased. The DO_2-to-VO_2 ratio was sustained at 1.6 for all but the most severe hypoxia (below 60 % of normal CaO_2). The constancy of DO_2/VO_2 therefore applied here during a combination of metabolic rate change and progressive hypoxia [3].

6.3.4 General Summary

Specific oxygen delivery to consumption ratios in major organ systems are sustained very precisely in the face of changing metabolic rates even when oxygen content is lowered. Figure 6.1 illustrates the normal proportions of arterial and venous oxygen content for major organ systems. If we examine the DO_2-to-VO_2 ratio in detail, we can visualise the normal DO_2-to-VO_2 relationships from arterial and venous content values.

$$DO_2 / VO_2 = CO \times CaO_2 / [CO \times (CaO_2 - CvO_2)]$$

where CO is the cardiac output; CaO_2, arterial; and CvO_2, mixed venous oxygen content.

Since CO cancels out, we simply see that the DO_2-to-VO_2 ratio is the same as the CaO_2-to-(CaO_2–CvO_2) ratio.

The values for the splanchnic organs may differ from the 1/5 to 1/6 illustrated.

High oxygen extraction values for both skeletal muscle and heart are sustained at CaO_2 levels around as low as 60 % of normal. In contrast, oxygen delivery to the brain breaks down when arterial oxygen saturation (SaO_2) falls below around 90 % of normal [15]. The well-known property of most organs, to autoregulate blood flow

[1, 2, 16], requires an augmented description since, for most organs, it involves autoregulation of DO_2 rather than simply blood flow.

The gastrointestinal tract blood flow normally increases after food intake, tracking the sites of increased activity (secretion, absorption and smooth muscle contraction), thereby sustaining the normal DO_2:VO_2 ratio [17]. Normal function, as regards gastrointestinal blood flow regulation, is known to be highly vulnerable to various stresses [1]. For example, ileal mucosal blood flow in dogs subject to controlled haemorrhage falls off after trivial losses with no detectable systemic effects [18]. In clinical medicine, gastric tonometry reveals early impairment of mucosal blood flow in high-risk surgical patients where there is inadequate blood volume and is an important guide to the need to restore blood volume to normal [19, 20]. So long as autoregulation is within limits, resting global DO_2 remains constant.

Renal oxygen extraction is an exception to constancy of the DO_2:VO_2 ratio since low arterial oxygen content (CaO_2) does not alter renal blood flow. Oxygen extraction is increased when CaO_2 is low stimulating erythropoietin secretion [5].

6.4 Examples of Whole Body Constancy of the DO_2:VO_2 Ratio

6.4.1 *Exposure to 12 % Oxygen Inspirate*

Eight resting subjects were exposed to 12 % oxygen. The arterial oxygen saturation (SaO_2) fall varied between subjects. DO_2 (from CO, haemoglobin and SaO_2) was the same under hypoxic conditions as on air for all subjects [21].

6.4.2 *Longer-Term (Altitude) Hypoxia*

A high-altitude study involving eight young subjects started with an initial 5 days at 3,324 m (10,900 ft). They then made two ascents to around 5,000 m (16,400 ft) over a 28-day period. Heart rate and SaO_2 were recorded twice daily. Individual mean SaO_2 values varied between 81.9 % and 93.2 %. All but one subject (with the highest SaO_2) showed a highly significant negative correlation between heart rate and SaO_2 compatible with cardiac output compensation for low oxygen content [22].

6.4.3 *Periodic Breathing with Oscillating Cardiac Output*

In 1994 constant end-tidal PCO_2 ($PetCO_2$) was noted coupled with vigorous periodic breathing during mild exercise at 5,000 m (16,400 ft). Recent experimental observation showed periodic breathing with constant $PetCO_2$, in three subjects at similar altitude, despite the large swings in ventilation. Hence, the rate of arrival of

CO_2 at the lungs varied yet was cleared by the ventilation. The variation in the rate of arrival of CO_2 was a result of oscillatory cardiac output, confirmed in two of the subjects. The compensatory changes in blood flow appeared close to maintaining constant DO_2 [23].

6.5 Conclusions

The precision of the DO_2:VO_2 ratio (1/E) is important in that it has wide application. Important determinants of blood flow in cerebral, gastrointestinal, coronary and skeletal muscle tissues become constrained to asking how the matching of DO_2 to VO_2 was brought about. Evolution of multicellular organisms will have required a circulation particularly regulated for an adequate oxygen supply.

It is proposed that full awareness of DO_2/VO_2 constancy provides a good starting point for many studies involving the circulation.

References

1. Green HD, Rapela CE, Conrad MC (1963) Resistance (conductance) and capacitance phenomena in terminal vascular beds. In: Hamilton WF, Dow P (eds) Handbook of physiology, circulation. American Physiological Society, Washington, DC, pp 935–960
2. Guyton AC, Jones CE, Coleman TG (1973) Circulatory physiology: cardiac output and its regulation. WB Saunders Company, Philadelphia
3. Wolff CB (2007) Normal cardiac output, oxygen delivery and oxygen extraction. Adv Exp Med Biol 599:169–182
4. Starling EH (1965) The Linacre lecture on the law of the heart. In: Starling on the heart (facsimile reprints) with analysis and comment by Chapman CB, Mitchell JH. Dawsons, London, pp 121–147
5. Valtin H (1973) Ch 6 Renal hemodynamics and oxygen consumption. In: Renal function: mechanisms preserving fluid and solute balance in health and disease. Little, Brown and Company, Boston, pp 177–196
6. Koskolou MD, Roach RC, Calbet JA, Rådegran G, Saltin B (1997) Cardiovascular responses to dynamic exercise with acute anemia in humans. Am J Physiol 273(4 Pt 2):H1787–H1793
7. Roach RC, Koskolou MD, Calbet JAL, Saltin B (1999) Arterial O_2 content and tension in regulation of cardiac output and leg blood flow during exercise in humans. Am J Physiol 276 (2 Pt 2):H438–H445
8. Wolff CB (2003) Cardiac output, oxygen consumption and muscle oxygen delivery in submaximal exercise: normal and low O_2 states. Adv Exp Med Biol 510:279–284
9. Severinghaus JW, Chiodi H, Eger EI 2nd, Brandstater B, Hornbein TF (1966) Cerebral blood flow in man at high altitude. Circ Res 19(2):274–282
10. Wolff CB (2000) Cerebral blood flow and oxygen delivery at high altitude. High Alt Med Biol 1(1):33–38
11. Wolff CB, Barry P, Collier DJ (2002) Cardiovascular and respiratory adjustments at altitude sustain cerebral oxygen delivery – Severinghaus revisited. Comp Biochem Physiol A Mol Integr Physiol 132(1):221–229

12. Tyrrell P (1990) Tomographic measurement of cerebral blood flow and metabolism with positron emitting isotopes in man. In: Harper AM, Jennett S (eds) Cerebral blood flow and metabolism. Manchester University Press, Manchester, pp 90–107
13. Harper AM (1990) Physiological control of the cerebral circulation. In: Harper AM, Jennett S (eds) Cerebral blood flow and metabolism. Manchester University Press, Manchester, pp 4–26 (see Fig. 1.11, p 23)
14. Martinez RR, Setty S, Zong P, Tune JD, Downey HF (2005) Nitric oxide contributes to right coronary vasodilatation during systemic hypoxia. Am J Physiol 288(3):H1139–H1146
15. Wolff CB, Richardson N, Kemp O et al (2007) Near infra-red spectroscopy and arterial oxygen extraction at altitude. Adv Exp Med Biol 599:183–189
16. Carlson AJ, Johnson V (1948) The Machinery of the body, 3rd edn. The University of Chicago Press, Chicago
17. Folkow B, Neil E (1971) Circulation. Oxford University Press, London/Toronto
18. Guzman JA, Rosado AE, Kruse JA (2003) Dopamine-1 receptor stimulation impairs intestinal oxygen utilization during critical hypoperfusion. Am J Physiol 284(2):H668–H675
19. Mythen MG, Webb AR (1995) Perioperative plasma volume expansion reduces the incidence of gut mucosal hypoperfusion during cardiac surgery. Arch Surg 130(4):423–429
20. Lebuffe G, Robin E, Vallet B (2001) Gastric tonometry. Intensive Care Med 27(1):317–319
21. Bell M, Thake CD, Wolff CB (2011) Effect of inspiration of 12 % O_2 (balance N_2) on cardiac output, respiration, oxygen saturation and oxygen delivery. Adv Exp Med Biol 701:327–332
22. Brierley G, Parks T, Wolff CB (2012) The relationship of acute mountain sickness to arterial oxygen saturation at altitudes of 3324 to 5176 m. Adv Exp Med Biol 737:207–212
23. Wolff CB, Bell M, Thake CD, Collier DJ (2012) Oscillations in cardiac output in hypoxia with periodic breathing and constant end-tidal PCO_2 at high altitude (5000m). Adv Exp Med Biol 737:197–206

Chapter 7
Heart Rate Variability in Newborns with Hypoxic Brain Injury

Vladimir Matić, Perumpillichira J. Cherian, Devy Widjaja, Katrien Jansen, Gunnar Naulaers, Sabine Van Huffel, and Maarten De Vos

Abstract In neonatal intensive care units, there is a need for continuous monitoring of sick newborns with perinatal hypoxic ischemic brain injury (HIE). We assessed the utility of heart rate variability (HRV) in newborns with acute HIE undergoing simultaneous continuous EEG (cEEG) and ECG monitoring. HIE was classified using clinical criteria as well as visual grading of cEEG. Newborns were divided into two groups depending on the severity of the hypoxic injury and outcome. Various HRV parameters were compared between these groups, and significantly decreased HRV was found in neonates with severe HIE. As HRV is affected by many factors, it is difficult to attribute this difference solely to HIE. However, this study suggests that further investigation of HRV as a monitoring tool for acute neonatal hypoxic injury is warranted.

V. Matić (✉) • D. Widjaja • S. Van Huffel
Department of Electrical Engineering (ESAT-SCD), Katholieke Universiteit Leuven, Leuven, Belgium

iMinds Future Health Department, Leuven, Belgium
e-mail: Vladimir.Matic@esat.kuleuven.be

P.J. Cherian
Section of Clinical Neurophysiology, Department of Neurology, Erasmus MC, Rotterdam, the Netherlands

K. Jansen
Department of Pediatrics, University Hospital Gasthuisberg, Leuven, Belgium

G. Naulaers
Neonatal Intensive Care Unit, University Hospital Gasthuisberg, Leuven, Belgium

M. De Vos
Department of Electrical Engineering (ESAT-SCD), Katholieke Universiteit Leuven, Leuven, Belgium

iMinds Future Health Department, Leuven, Belgium

Department of Psychology, University of Oldenburg, Oldenburg, Germany

S. Van Huffel et al. (eds.), *Oxygen Transport to Tissue XXXV*, Advances in Experimental Medicine and Biology 789, DOI 10.1007/978-1-4614-7411-1_7,

7.1 Introduction

Perinatal hypoxic ischemic encephalopathy (HIE) represents one of the major causes of morbidity and mortality in newborns. The sequelae in survivors include cerebral palsy, sensory and cognitive problems, and epilepsy. Acute hypoxic brain injury is dynamic and evolves over time. Both recovery and worsening of the encephalopathy can occur. Monitoring of the brain function within 24–72 h postpartum is necessary to assess the evolution of the injury. This helps to guide treatment and predict neurodevelopmental outcome.

In clinical practice, several monitoring tools are used to assess the severity of hypoxic injury. A well-suited bedside tool for noninvasive assessment of brain function is the continuously recorded electroencephalogram (EEG). However, the expertise required to register and interpret EEG is not available in most neonatal intensive care units (NICUs). As an alternative, many NICUs use aEEG (amplitude integrated electroencephalogram) as a device to monitor brain function, due to its ease of application and interpretation [1, 2].

Apart from the evolution of the background EEG activity [1], it is also known that heart rate variability (HRV) may be affected in HIE. This can be due to direct effect of asphyxia on the heart or due to impaired central (cerebral and brain stem) regulation of heart rate, or both, and may help in prediction of outcome.

An earlier study investigated HRV in term asphyxiated newborns [3]. In that study HRV was examined 7 days postpartum. Therefore, although valuable for estimation of the outcome, that study does not show HRV properties at an early stage of the injury (first 3 days postpartum) when treatment interventions are expected to be maximally effective.

The aim of the present study is to examine the use of HRV obtained by continuous electrocardiogram (ECG) monitoring done simultaneously with continuous EEG registrations (cEEG) in patients with HIE as a diagnostic tool. We compare heart rate parameters between two groups of neonates, group I with mild HIE and group II with moderate to severe HIE, using long-term ECG recordings. Although many factors influence HRV parameters, our goal is to examine whether the differences in HRV parameters are sufficiently discriminative between these two groups of newborns. To test these differences, a linear discriminant analysis (LDA) classifier was used.

7.2 Dataset

All data were recorded at Sophia Children's Hospital, Erasmus MC (Rotterdam, the Netherlands). We studied 19 newborns with HIE and without recorded epileptic seizures. They were preselected from a larger set of 119 newborns who underwent continuous EEG video monitoring. Sampling frequency of the simultaneously recorded ECG signal was 256 Hz.

Continuous 2-h artifact-free ECG recordings were selected. In total, we analyzed 36 ECG segments recorded within 18–48 h after birth. The number of analyzed segments ranged from 1 to 4 per patient. The severity of encephalopathy was assessed using both clinical grading (Sarnat scores) and grades of background EEG activity [1]. The scoring of the background EEG was performed by an experienced clinical neurophysiologist (PJC). Newborns were divided into two groups: group I ($n=10$) with mild HIE and good outcome and group II ($n=9$) with moderate to severe HIE and poor outcome. Outcome was classified into good outcome [normal or mild disability – minimal abnormalities at neurological examination or mild cerebral palsy; the Gross Motor Functional Classification System (GMFCS 1–2)] and poor outcome [death or Mental Developmental Index (MDI) <70 or severe cerebral palsy (GMFCS 3–5)].

In these datasets we selected 2-h ECG recordings, as this was the longest artifact-free period we could find. This duration of ECG recordings was sufficient to accurately calculate and estimate long-term HRV parameters [4]. In total, we had 22 ECG segments from group I with a good outcome and 14 ECG segments from group II with a poor outcome.

7.3 Methods

To calculate HRV parameters the first step is to detect R peaks within the QRS complex of the ECG signal. We improved the accuracy of the HRV parameters by upsampling the ECG signal to 1,000 Hz and used it for further analysis. The Pan-Tompkins algorithm was applied for R peak detection. This way, we obtained a tachogram signal, also known as the RR interval. All 2-h ECG segments were visually inspected to ensure that there were no missed R peaks or false-positive R peak detections. In this way, we obtained an NN interval (normal-to-normal beat [4]).

Three groups of HRV parameters were calculated: time, geometry, and frequency domain parameters. All these parameters are calculated as proposed in [4]. As the first time domain parameter, mean NN interval is calculated. Variable SDNN is based on the standard deviation of NN time series and reflects the total power of the HRV spectra. By increasing the duration of the ECG recordings, the SDNN parameter increases as well. Therefore, it is important to compare this parameter using the ECG recordings of the same duration. The SDANN parameter reflects changes in NN time series using 5-min epochs. It represents the standard deviation of average NN intervals calculated over 5 min of recordings. Additionally, the SDNN index is calculated as a mean value of the 5-min standard deviations calculated over 2 h.

A commonly used time domain parameter is the RMSSD, the square root of the mean squared differences of successive NN intervals. pNN25 is the number of interval differences of successive NN intervals ≥ 25 ms divided by the total number of NN intervals and multiplied by 100. Finally, we calculated SDSD, the standard deviation of differences between adjacent NN intervals.

Additionally, we calculated standard geometric parameters as proposed in [4]. Triangular index, related to the histogram of the NN intervals and parameters SD1 and SD2 of the Poincaré plots, was calculated.

To calculate frequency parameters, we interpolated NN intervals using a 4-Hz sampling frequency. Thus, we obtained an equidistantly sampled signal. We then calculated power spectral density using Welch's periodogram.

Three frequency bands were defined (VLF, LF, and HF) of which the spectral power was expressed in absolute values (in ms^2) and in normalized units (n.u.) which represent the relative value of each power component in proportion to the total power minus the VLF component. The frequency bands were adapted to the newborn's heart rate: 0.01–0.04 Hz, 0.04–0.2 Hz, and 0.2–2 Hz, as proposed in [5].

The goal of this study is to examine whether HRV parameters from 2-h ECG are discriminative enough, i.e., to verify whether the outcome group can be predicted accurately using only parameters derived from the ECG signal. For this purpose, the leave-one-out method was applied in these experiments. We used ECG segments and corresponding HRV parameters from a single patient as an input to an LDA classifier. The classifier was trained using the HRV parameters obtained from the remaining patients. As the output, the classifier is expected to estimate the appropriate group of the patient, i.e., good or poor outcome.

Input features into the LDA classifier represented a subset of all calculated HRV parameters. As we did not know a priori which features would be the most discriminative, we combined features into different subsets to increase the discriminative power. The number of features was relatively low, and we were able to create subsets of various sizes and use an exhaustive combinatorial search to determine an optimal set of parameters.

7.4 Results

In total, 24 HRV parameters were calculated as proposed in [4], and they were used as input features into the LDA classifier. The best set of parameters achieved an accuracy of 80 % in discriminating the two groups of newborns. The most descriptive parameters were standard deviation of the heart rate signal (SDNN index), low-frequency power of the heart rate spectra (LF), and SD2 parameter of the Poincaré plot. Normalized values of these parameters are shown in Fig. 7.1. More importantly, there was a reduced HRV in the patient group with more severe HIE.

7.5 Discussion

In patients with HIE it is important to estimate the degree of hypoxic injury at an early stage. The most informative tool is the cEEG, and in this study we wanted to examine the use of continuously recorded ECG signal and HRV as a potential

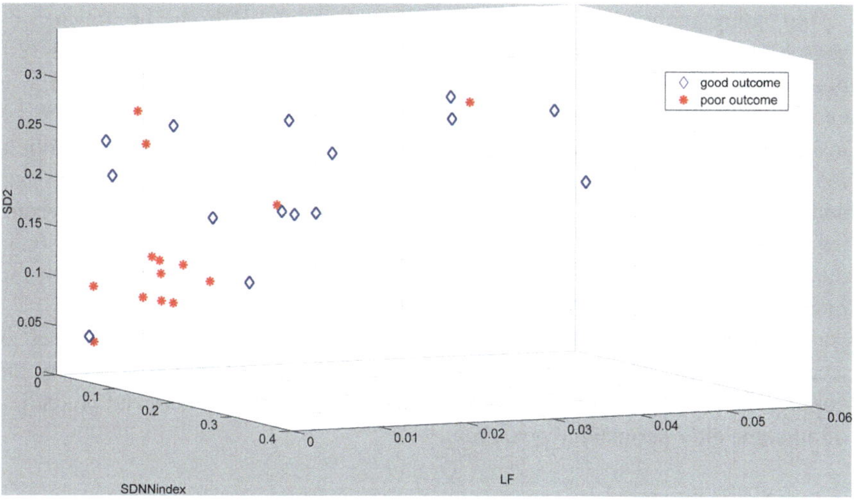

Fig. 7.1 Subset of HRV parameters that achieved the best discrimination between groups with good and poor outcome. All parameters are depicted using normalized values

additional tool. Advantage of the ECG signal is its well-defined morphology and straightforward and easy calculation of HRV parameters. A difficulty can arise in interpretation of the HRV parameters.

In this study, we tried to make our database as homogenous as possible by excluding patients who were treated with antiepileptic drugs. Many patients were excluded from the present study because the monitoring started more than 48 h postpartum or because the ECG recordings were corrupted by artifacts. Additionally, there is an inherent difficulty in making a new, better controlled study. Start of the ECG and EEG recordings is influenced by many medical and logistic factors that are often difficult to control.

In addition to hypoxic injury, HRV is influenced by many factors such as autonomic nervous system dysfunction, changes in blood pressure, temperature, respiration, and administered medications. Hence, in the present study it is difficult to draw definite conclusions about HIE as the major determinant of the HRV findings.

Despite these limitations, our results show that there is a significant relationship between the severity of HIE as graded clinically and by cEEG and HRV.

In [3], reduced HRV was also reported in term asphyxiated newborns 7 days after birth; in that study the Sarnat score was used to classify the patient groups. In our study, we used both the Sarnat score and the evolution of the background EEG to better classify encephalopathy. This is important because hypoxic injury usually evolves and brain function can recover or deteriorate over time. Additionally, we used the brain injury patterns seen on MRI scans done in the first week after birth and neurodevelopmental outcome in survivors at follow-up after 2 years. This made the classification of patients into two outcome groups more reliable.

For future study, more insight into HRV in perinatal HIE can be obtained by comparing HRV parameters at specific time points, such as 12, 24, 48, and 72 h postpartum. In that way maturational effects of the central nervous system as encountered in studies done after 1–2 weeks postpartum can be avoided. Such investigations would be valuable to confirm the possibility to use HRV as a tool to monitor the evolution of HIE recovery as well as to identify neonates requiring additional interventions or neuroprotection. In the case that correlations and dynamics do exist between HRV parameters and parameters derived from the cEEG (e.g., duration of inter-burst intervals), they can contribute to improve the assessment and monitoring of perinatal hypoxic brain injury.

Inclusion of a larger number of neonates with varying degrees of HIE, as well as studying HRV evolution over a longer period of time, would help to clarify the application of these parameters in the refinement of the assessment and prediction of outcome after perinatal hypoxic injury.

Acknowledgments This research is supported by Research Council KUL: GOA Ambiorics, GOA MaNet, CoE EF/05/006 Optimization in Engineering (OPTEC), IDO 05/010 EEG-fMRI, IDO 08/013 Autism, IOF-KP06/11 FunCopt; Flemish Government (FWO): PhD/postdoc grants, G.0302.07 (SVM), G.0341.07 (Data fusion), G.0427.10N (Integrated EEG-fMRI) research communities (ICCoS, ANMMM); IBBT; IWT: TBM070713-Accelero, TBM070706-IOTA3, TBM080658-MRI (EEG-fMRI), PhD Grants; Belgian Federal Science Policy Office: IUAP P6/04 (DYSCO, "Dynamical systems, control and optimization," 2007–2011); ESA PRODEX No 90348 (sleep homeostasis); and EU: FAST (FP6-MC-RTN-035801), Neuromath (COST-BM0601).

References

1. Cherian PJ (2010) Improvements in neonatal brain monitoring after perinatal asphyxia. Ph.D. thesis, Erasmus University, Rotterdam
2. De Vos M, Deburchgraeve W, Cherian PJ et al (2011) Automated artifact removal as preprocessing refines neonatal seizure detection. Clin Neurol 122(12):2345–2354
3. Aliefendioğlu D, Doğru T, Albayrak M, Dibekmısırlıoğlu E, Sanlı C (2012) Heart rate variability in neonates with hypoxic ischemic encephalopathy. Indian J Pediatr 79(11):1468–1472
4. Task Force of the European Society of Cardiology and the North American Society of Pacing and Electrophysiology (1996) Heart rate variability: standards of measurement, physiological interpretation and clinical use. Circulation 93(5):1043–1065
5. Doyle OM, Korotchikova I, Lightbody G, Marnane W, Kerins D, Boylan GB (2009) Heart rate variability during sleep in healthy term newborns in the early postnatal period. Physiol Meas 30(8):847–860

Part II
Brain Oxygenation

Chapter 8
Simultaneous Monitoring of Brain and Skin Oxygenation During Haemorrhagic Shock in Piglets

David F. Wilson and David K. Harrison

Abstract Phosphorescence quenching and visible lightguide spectrophotometry were used to measure brain cortex oxygen partial pressure and skin oxygen saturation, respectively, during stepwise haemorrhage and re-transfusion in four 4–7-day-old anaesthetised piglets. In three cases, the effect of administration of adrenalin (epinephrine) was investigated. Brain cortex partial pressure was measured using a conventional phosphorescence pO_2 probe ($bc_{loc}pO_2$) and using a self-contained phosphorescence microsensor ($bc_{micro}pO_2$). Peripheral tissue oxygen saturation was measured on the skin of the abdomen (abS_sO_2) and the distal right foreleg (flS_sO_2) using visible lightguide spectrophotometry. Haemorrhage of 65 ml reduced mean arterial blood pressure (MABP) from 75.5 ± 11.0 mmHg (mean ± standard deviation) to 42 ± 2.6 mmHg. Mean $bc_{loc}pO_2$ fell from 30.1 ± 3.1 to 13.1 ± 2.5 mmHg and mean $bc_{micro}pO_2$ fell from 33.8 ± 11.4 to 13.3 ± 9.5 mmHg. abS_sO_2 and flS_sO_2 values fell from 47.4 ± 8.1 % and 43.6 ± 10.9 %, respectively, to 21.9 ± 5.5 % and 23.8 ± 14.0 %. Infusion of adrenalin produced a mean transient increase in MABP to 137 ± 2.6 mmHg, falling to 75.7 ± 16.3 mmHg within 3 min. $bc_{loc}pO_2$ also increased to 24.1 ± 14.6 mmHg, but there were no significant changes in $bc_{micro}pO$, abS_sO_2 or flS_sO_2. Following reinfusion all parameters returned to values that were not statistically different from their pre-haemorrhage values. The dynamic recordings of all the oxygenation parameters indicated that they were sensitive indicators of the degree of haemorrhage during the experiments.

D.F. Wilson
Department of Biochemistry and Biophysics, University of Pennsylvania, Philadelphia, PA, USA

D.K. Harrison (✉)
Microvascular Measurements, Heilig-Kreuz-Strasse 19, 39030 St, Lorenzen, Italy
e-mail: Harrison.David.K@gmail.com

S. Van Huffel et al. (eds.), *Oxygen Transport to Tissue XXXV*, Advances in Experimental Medicine and Biology 789, DOI 10.1007/978-1-4614-7411-1_8,

8.1 Introduction

Numerous methods have been proposed for monitoring patients in haemorrhagic shock and for guiding resuscitation strategies. It is recognised that the monitoring of blood pressure does not necessarily detect the early onset of haemorrhage and that the measurement of peripheral and/or organ circulatory parameters may be more sensitive. Some recent techniques that have been proposed include near-infrared spectroscopy [1], hyperspectral imaging [2], electron paramagnetic resonance (EPR) [3], tonometry [4, 5] and transcutaneous pO_2 and pCO_2 [6] for monitoring muscle or skin oxygenation. Tonometry has also been proposed for the monitoring of pH in the gastrointestinal tract [7].

Clearly, maintaining the oxygen supply to the brain is a critical objective in any resuscitation strategy, and a number of studies have sought to monitor brain cortex oxygenation (pO_2 or SO_2) using optical or polarographic sensors [8–10].

The great majority of the above techniques are most suited for use in the hospital setting, and many are not adaptable for robustness or portability. However, portable, uncomplicated methods for assessing the status of trauma patients "in the field" enabling targeted treatment before they reach the emergency room could be of great value in improving survival rates in such circumstances.

In the pilot experiments described here, phosphorescence quenching [10, 11] and visible light spectroscopy (VLS) [12] probes were used to measure brain cortex oxygen partial pressure and skin tissue oxygen saturation simultaneously during stepwise haemorrhage and subsequent autologous transfusion. Since both of the techniques can be readily incorporated into robust, portable and easy-to-use instruments, the aim of these experiments was to assess whether these combined measurements could be used to detect early onset of shock, in advance of changes in blood pressure, during haemorrhage.

8.2 Methods

Brain cortex pO_2 was measured using oxygen-dependent phosphorescence quenching. For conventional measurements ($bc_{loc}pO_2$) [10] Oxyphor G4 was used, which has a 250 μs lifetime at zero oxygen and a quenching constant of 220 $mmHg^{-1}\ s^{-1}$. Approximately 15 μl of 100 μM solution in physiological saline was injected through the dura onto the brain surface and allowed to diffuse into the interstitial space. A self-contained microsensor [11] was also inserted into the striatum of the brain using a 21 gauge needle (which was then withdrawn), and parallel measurements were made ($bc_{micro}pO_2$). All $bcpO_2$ values were recorded continuously at 15-s intervals.

Skin SO_2 was measured using two lightguide spectrophotometers (Whitland Research RM200) [12, 13]. Two surface optodes were attached to the skin with double-sided tape, one to the abdomen (abS_sO_2) and the other to the back of the trotter of the right foreleg (flS_sO_2). abS_sO_2 values and, in three experiments only, flS_sO_2 values were recorded continuously at 2-s intervals.

Four piglets, aged 4–7 days and weighing 2.5–3.6 kg (mean 3.0 kg), were intubated and ventilated at an inspiratory oxygen fraction of 0.3. Isopropane and Fentanyl were used for anaesthesia. A 10-mm diameter burr hole was made in the skull to enable the insertion of a cortical pO_2 probe. Up to approximately 30 ml/kg blood was withdrawn (a haemorrhage equivalent to a loss of approximately 2 l in a human adult). In three piglets, on reaching maximum blood loss, adrenalin (0.1 mg/kg) was administered. The previously withdrawn blood was then reinfused, following which the piglets were euthanized. Arterial blood pressure was monitored continuously throughout the experiments and arterial blood gas and acid/base status measured at frequent intervals.

Results were tested for normal distribution using the Kolmogorov-Smirnov test and differences were subsequently compared using the Student *t*-test for paired values. A value of $p<0.05$ was considered to be statistically significant.

8.3 Results

Figure 8.1 shows the changes in MABP, $bc_{loc}pO_2$, $bc_{micro}pO_2$, abS_sO_2 and flS_sO_2 in a typical experiment. The mean results at various stages of haemorrhage are shown in Table 8.1. It can be seen that haemorrhage of 65 ml (achieved in all four piglets) produced a reduction in MABP from 75.5 ± 11.0 mmHg (mean ± standard deviation) to 42 ± 2.6 mmHg. The change in mean $bc_{loc}pO_2$ was from 30.1 ± 3.1 to 13.1 ± 2.5 mmHg and mean $bc_{micro}pO_2$ fell from 33.8 ± 11.4 to 13.3 ± 9.5 mmHg.

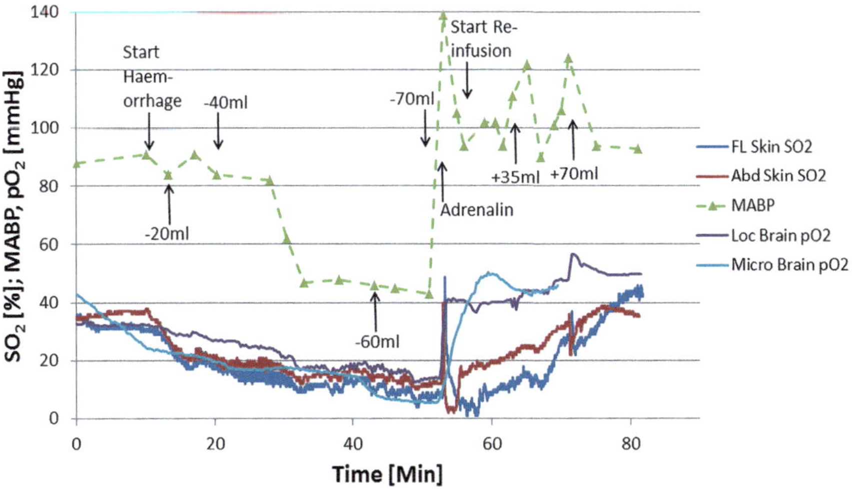

Fig. 8.1 Changes in MABP, $bc_{loc}pO_2$, $bc_{micro}pO_2$, abS_sO_2 and flS_sO_2 during haemorrhage, adrenalin infusion and reinfusion of drawn blood in one experiment

Table 8.1 Mean MABP, $bcpO_2$, abS_sO_2 and flS_sO_2 values during haemorrhage. Absence of an S.D. value indicates that only one experiment involved 90 and 100 ml haemorrhage

Haemorrhage (ml)	Mean MABP (mmHg) ± S.D.	Mean bc_{loc}-pO_2 (mmHg) ± S.D.	Mean bc_{micro}-pO_2 (mmHg) ± S.D.	Mean abS_sO_2 (%) ± S.D.	Mean flS_sO_2 (%) ± S.D.
0	75.5 ± 1.0	30.1 ± 3.1	33.8 ± 11.4	47.4 ± 8.1	43.6 ± 10.9
20	68.0 ± 3.9	28.2 ± 4.6	30.8 ± 7.9	39.4 ± 10.7	37.7 ± 10.1
30	63.3 ± 15.7	24.7 ± 4.0	25.8 ± 4.3	30.9 ± 9.0	35.0 ± 12.5
40	60.7 ± 19.7	23.6 ± 6.4	26.1 ± 5.9	31.2 ± 10.6	32.5 ± 14.1
50	49.0 ± 3.5	17.1 ± 4.2	24.5 ± 7.3	23.3 ± 11.1	29.1 ± 18.0
55	45.0 ± 4.4	15.9 ± 2.2	18.9 ± 4.5	24.1 ± 10.1	26.2 ± 21.8
60	44.0 ± 2.6	17.4 ± 3.4	20.3 ± 15.5	19.3 ± 5.3	18.3 ± 11.9
65	42.0 ± 2.6	13.1 ± 2.5	13.3 ± 9.5	21.9 ± 8.5	23.3 ± 19.8
70	43.0 ± 0.0	16.0 ± 3.3	18.4 ± 18.1	15.4 ± 4.9	14.9 ± 11.2
80	38.5 ± 4.9	12.8 ± 6.6	22.9 ± 9.4	19.8 ± 3.5	25.5 ± 10.8
90	39.0 ± –	17.4 ± –	26.7 ± –	15.9 ± –	14.3 ± –
100	39.0 ± –	14.1 ± –	25.2 ± –	13.9 ± –	5.7 ± –

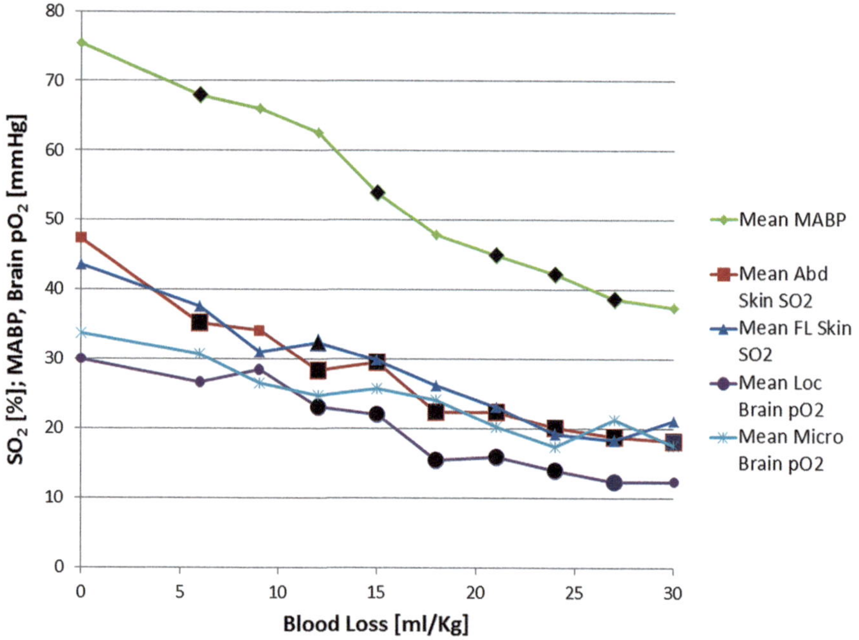

Fig. 8.2 Mean MABP, $bc_{loc}pO_2$, $bc_{micro}pO_2$, abS_sO_2 and flS_sO_2 during haemorrhage in all experiments. The *large data points* indicate statistically significant differences (p<0.05) from pre-haemorrhage values

abS_sO_2 and flS_sO_2 values fell from 47.4 ± 8.1 % and 43.6 ± 10.9 %, respectively, to 21.9 ± 8.5 % and 23.8 ± 19.8 %. All changes, with the exception of mean $bc_{micro}pO_2$, were statistically significant (p<0.05). Figure 8.2 shows the changes in the MABP, $bc_{loc}pO_2$, $bc_{micro}pO_2$, abS_sO_2 and flS_sO_2 in relation to blood loss per kg body weight.

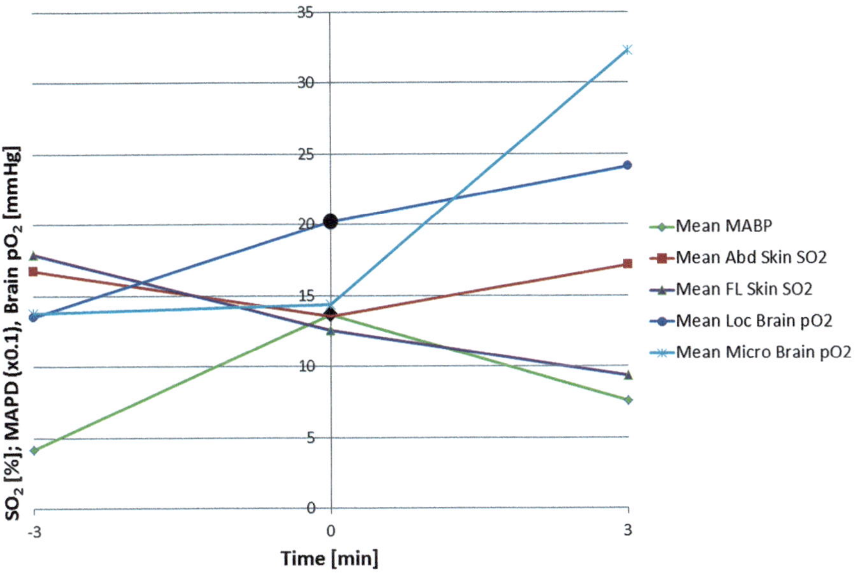

Fig. 8.3 Mean MABP, $bc_{loc}pO_2$, $bc_{micro}pO_2$, abS_sO_2 and flS_sO_2 3 min before, immediately after and 3 min after administration of adrenalin in all experiments. The *large data points* indicate statistically significant differences ($p<0.05$) from pre-adrenalin values. Note that MAPD values have been divided by a factor of 10 to enable comparison with the oxygenation parameters

The results of adrenalin infusion are shown in Fig. 8.3. It can be seen that there was a mean transient increase in MABP to 137±2.6 mmHg ($p<0.005$), falling to 75.7±16.3 mmHg ($p<.05$) within 3 min and $bc_{loc}pO_2$ increased to 20.2±6.0 mmHg ($p<0.05$). $bc_{micro}pO_2$ also increased after 3 min together with a further increase in $bc_{loc}pO_2$; however, there were initial decreases in both abS_sO_2 and flS_sO_2 followed by a return of abS_sO_2 to its pre-adrenalin value but a continued fall in flS_sO_2. None of the latter changes were statistically significant. The changes in MABP, $bc_{loc}pO_2$, $bc_{micro}pO_2$, ab_sSO_2 and fl_sSO_2 during reinfusion are shown in Fig. 8.4. It can be seen that only abS_sO_2 displayed significant differences from pre-haemorrhage values during the intermediate stages of reinfusion. On completion of reinfusion the mean MABP was 99±26.8 mmHg, $bc_{loc}pO_2$ was 32.4±13.1 mmHg, $bc_{micro}pO_2$ was 48.6±11.5 mmHg, abS_sO_2 was 42.1±10.1 % and flS_sO_2 was 34.6±15.7 %. None of the differences between these values and those prior to haemorrhage were significant.

8.4 Discussion

Clearly this was a pilot series of experiments setting out to determine whether it was feasible and practical to monitor both brain and peripheral oxygenation during haemorrhage using techniques that could readily be developed into robust, portable

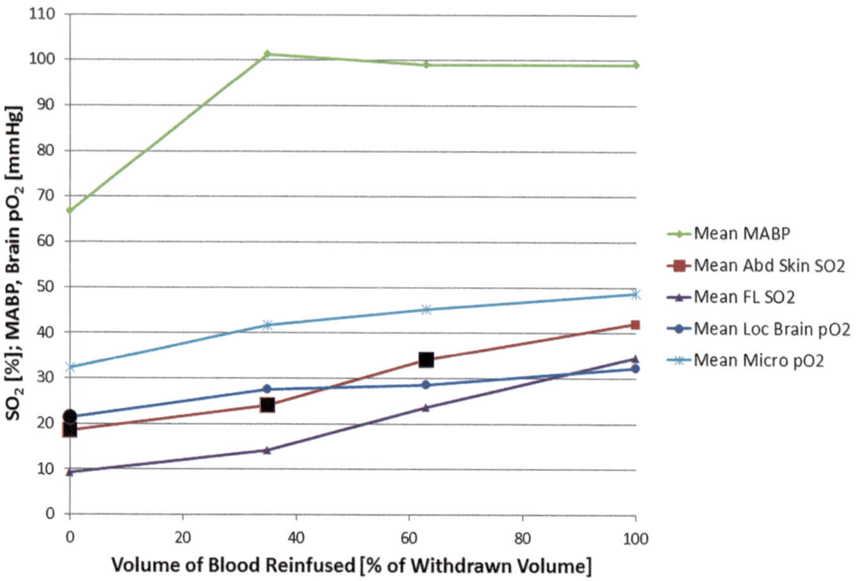

Fig. 8.4 Mean MABP, $bc_{loc}pO_2$, $bc_{micro}pO_2$, abS_sO_2 and flS_sO_2 during reinfusion in all experiments. The *large data points* indicate statistically significant differences ($p<0.05$) from pre-haemorrhage values

and easy-to-use instruments. It was observed in individual experiments that changes in all tissue oxygenation parameters were evident long before any significant fall in blood pressure – although this is not clear in the summary data presented here which are steady-state plots relating to blood volume loss, not time. The parameter displaying the most consistent significant changes during haemorrhage was abS_sO_2 followed by $bc_{loc}pO_2$ and then MABP. $bc_{micro}pO_2$ and flS_sO_2 showed the least consistent changes, but that may be due to the low number of recordings (for technical reasons neither were recorded during the first experiment). Once significant hypotension occurred, changes in peripheral (skin) oxygenation reflected the extent of centralization in order to preserve brain oxygenation. This was particularly evident (anecdotally, as the changes were not significant) after infusion of adrenalin when flS_sO_2 decreased, whereas $bc_{loc}pO_2$, $bc_{micro}pO_2$ and abS_sO_2 all increased. During reinfusion of autologous blood, only abS_sO_2 remained significantly below pre-haemorrhage values until full replacement was complete.

8.5 Conclusions

The simultaneous monitoring of brain and skin oxygenation complement each other in providing immediate information regarding haemodynamic status during the onset of haemorrhagic shock.

The equipment is uncomplicated to use and could be readily adapted for use in emergency situations at scenes of traumatic incidents.

Acknowledgments The authors are most grateful to Joanna Kubin for her expert technical assistance.

References

1. Beilman GJ, Blondet JJ (2009) Near-infrared spectroscopy-derived tissue oxygen saturation in battlefield injuries: a case series report. World J Emerg Surg 4:25
2. Cancio LC, Batchinsky AI, Mansfield JR et al (2006) Hyperspectral imaging: a new approach to the diagnosis of hemorrhagic shock. J Trauma 60(5):1087–1095
3. Miyake M, Grinberg OY, Hou H, Steffen RP, Elkadi H, Swartz HM (2003) The effect of RSR13, a synthetic allosteric modifier of hemoglobin, on brain tissue pO_2 (measured by EPR oximetry) following severe hemorrhagic shock in rats. Adv Exp Med Biol 530:319–329
4. Maxwell TM, Lim RC Jr, Fuchs R, Hunt TK (1973) Continuous monitoring of tissue gas tensions and pH in hemorrhagic shock. Am J Surg 126(2):249–254
5. Clavijo-Alvarez JA, Sims CA, Pinsky MR, Puyana JC (1973) Monitoring skeletal muscle and subcutaneous tissue acid–base status and oxygenation during hemorrhagic shock and resuscitation. Shock 24(2):270–275
6. Haisjackl M, Hasibeder W, Sparr H, Klaunzer S, Putz G, Kroesen G (1989) The use of a transcutaneous pO_2/pCO_2 combination electrode during volume therapy in a child in shock. Anaesthesist 38(8):431–433
7. Martini L, Giavaresi G, Fini M, Faenza S, Petrini F, Giardino R (2003) Hemodynamic indices versus gastric tonometric measurements for prognosis of hemorrhagic shock: a porcine model. Comp Med 53(2):178–185
8. Cavus E, Meybohm P, Dörges V et al (2008) Regional and local brain oxygenation during hemorrhagic shock: a prospective experimental study on the effects of small-volume resuscitation with norepinephrine. J Trauma 64(3):641–648, discussion 648–649
9. Manley GT, Hemphill JC, Morabito D et al (2000) Cerebral oxygenation during hemorrhagic shock: perils of hyperventilation and the therapeutic potential of hypoventilation. J Trauma 48(6):1025–1032
10. Song D, Olano M, Wilson DF et al (1995) Comparison of the efficacy of blood and polyethylene glycol-hemoglobin in recovery of newborn piglets from hemorrhagic hypotension: effect on blood pressure, cortical oxygen and extracellular dopamine in the brain. Transfusion 35(7):552–558
11. Wilson DF, Vinogradov SA, Schears GJ, Esipova TV, Pastuszko A (2012) Monitoring cardiopulmonary function and progression toward shock: oxygen micro-sensor for peripheral tissue. Adv Exp Med Biol 737:221–227
12. Caddick J, Raine C, Harrison D, Erdmann M (2006) Lightguide spectrophotometry to monitor free TRAM flaps. Adv Exp Med Biol 578:351–356
13. Harrison DK, Hawthorn IE (2005) Amputation level viability in critical limb ischaemia: setting new standards. Adv Exp Med Biol 566:325–331

Chapter 9
Hemispheric Differences of Motor Execution: A Near-Infrared Spectroscopy Study

Ingo Helmich, Robert Rein, Nico Niermann, and Hedda Lausberg

Abstract Distal movements of the limbs are predominantly controlled by the contralateral hemisphere. However, functional neuroimaging studies do not unequivocally demonstrate a lateralization of the cerebral activation during hand movements. While some studies show a predominant activation of the contralateral hemisphere, other studies provide evidence for a symmetrically distributed bihemispheric activation. However, the divergent results may also be due to methodological shortcomings. Therefore, the present study using functional near-infrared spectroscopy examines cerebral activation in both hemispheres during motor actions of the right and left hands. Twenty participants performed a flexion/extension task with the right- or left-hand thumb. Cerebral oxygenation changes were recorded from 48 channels over the primary motor, pre-motor, supplementary motor, primary somatosensory cortex, subcentral area, and the supramarginal gyrus of each hemisphere. A consistent increase of cerebral oxygenation was found for oxygenated and for total hemoglobin in the hemisphere contralateral to the moving hand, regardless of the laterality. These findings are in line with previous data from localization [1–3] and brain imaging studies [4–6]. The present data support the proposition that there is no hemispheric specialization for simple distal motor tasks. Both hemispheres are equally activated during movement of the contralateral upper limb.

9.1 Introduction

The principal anatomic organization of the human motor system is characterized by its contralateral control of distal movements [1–3]. Thus, the anatomical organization of the neural pathways suggests that in functional neuroimaging, simple motor

I. Helmich (✉) • R. Rein • N. Niermann • H. Lausberg
Department of Neurology, Psychosomatic Medicine and Psychiatry, Institute of Health Promotion and Clinical Movement Science, German Sport University, Cologne, Germany
e-mail: in.helmich@gmail.com

S. Van Huffel et al. (eds.), *Oxygen Transport to Tissue XXXV*, Advances in Experimental Medicine and Biology 789, DOI 10.1007/978-1-4614-7411-1_9,

tasks should be accompanied by a stronger activation of the contralateral hemisphere than of the ipsilateral hemisphere. However, functional neuroimaging studies using near-infrared spectroscopy (NIRS) do not unequivocally demonstrate a lateralization of the cerebral activation during simple motor tasks [7–11]. While some studies show a predominant activation of the contralateral hemisphere [7, 9], other studies provide evidence for bihemispheric distributed activation patterns [8, 10, 11]. Two studies concluded that the left hemisphere is specialized during motor actions by left lateralized hemodynamic responses [8, 10], whereas Wriessnegger et al. (2008) reported opposite results of right hemispheric lateralized activations [11]. However, the divergent results may also be due to methodological shortcomings. Variables that have been shown to lead to stronger left hemispheric brain activation patterns are complexity [12–15] and sequential motor execution [6, 16] mostly observed during finger opposition tasks. Stronger bilateral responses were observed after proximal motor actions compared to distal execution [17]. Further, the often bilateral organization of neuronal receptive fields in the somatosensory area [18, 19] is also a possible factor diminishing laterality by the sensory stimulus during finger-to-thumb opposition tasks [20]. In this regard we examine the question if there exists a hemispheric specialization of simple distal motor execution by using NIRS with an improved design excluding potential factors leading to different patterns of cerebral activation in the two hemispheres such as complexity [12–15], sequentiality [6, 16], proximity [17], and sensory stimulation [20]. Therefore, we hypothesize that simple distal motor execution leads to increased cerebral oxygenation in motor-related cortices of the contralateral hemisphere for either hand.

9.2 Methods

9.2.1 Sample

Twenty participants (ten females, ten males; mean age 25 years; range 19–33) took part in the study after written informed consent was obtained. None of the participants had a known history of any neurological or psychiatric disorder. The study was approved by the local Ethics Committee of the German Sports University Cologne.

9.2.2 Procedures

A randomized block design was developed using Presentation software consisting of two conditions: (i) seven repetitive self-paced flexion/extension movements with the right-hand thumb (motor rh) and (ii) seven flexion/extension movements with the left-hand thumb (motor lh). Each block lasted 20 s. The time interval between the blocks was 10 s. Participants were instructed to not move any other parts of the body than their right- or left-hand thumbs. During the experiment, the investigator verified that the participants did not move other parts of the body.

9.2.3 *Optical Imaging*

Cerebral oxygenation changes were recorded using a near-infrared optical tomographic imaging device (DYNOT Imaging System, NIRx; wavelengths 760 nm, 830 nm; sampling rate 1.81 Hz). Methodology and underlying physiology are explained in detail elsewhere [21, 22]. A total of 32 optodes were placed in two 4×4 grids above each hemisphere resulting in 48 channels of measurement. The DYNOT system allows every optode to emit and detect optical changes. Optodes were placed with an interoptode distance of 2.5 cm around C3 and C4 according to the 10–20 system [23] symmetrically above each hemisphere. Coordinates of optode positions were collected for spatial registration of NIRS channels into the standard brain from the Montreal Neurological Institute (MNI space) according to Singh et al. (2005) [24] using a 3-D digitizing system (zebris 3D Measuring Systems, zebris Medical GmbH). Optodes covered identical regions above both hemispheres including primary motor, pre-motor, supplementary motor cortex, primary somatosensory cortex, subcentral area, and the supramarginal gyrus. Optodes were mounted with a customized plastic hard shell system on the participant's head to gain placement stability and to avoid movement artifacts.

9.2.4 *Data Analysis*

Data were analyzed using NIRS-SPM software [25, 26]; 48 channels were converted to hemoglobin concentration changes according to Cope et al. [21]. Because fNIRS data can be affected by movement artifacts, each channel of individual participants was visually inspected, and movement artifacts were corrected using the artifact correction toolbox of the NIRS analysis package [27]. Data were baseline corrected to 10 s before stimulus onset and low-pass filtered according to the precoloring method [28]. Unknown global trends were removed by the wavelet-minimum description length detrending algorithm [26]. To assess the vascular response, the stimulation was modeled as a boxcar function convolved with a canonical hemodynamic response function. Data were then fed into a general linear model estimation to obtain beta values for each condition and each species of hemoglobin (oxy (HbO_2)-, deoxy (Hb)-, and total (HbT) hemoglobin). Statistical significance of the intra-condition differences of HbO_2, Hb, and HbT was analyzed using two-sided, one-sample t-tests. To restrict alpha inflation due to multiple statistical comparisons, significant results are reported that exceed a conservative threshold of $p < 0.01$. To investigate which of the two hemispheres is more involved in the task, we calculated analyzes of variance (ANOVAs) with repeated measurements with within-subjects variables (*hemisphere, condition*) using SPSS (IBM SPSS Statistics, Version 20). Multiple post hoc pairwise comparisons were corrected using Bonferroni corrections.

9.3 Results

Flexion/extension with the right-hand thumb resulted in significant ΔHbO_2 increases in 14 channels ($t(19)>2.09$, $p<0.01$) located over the left hemisphere and in six channels over the right hemisphere. Flexion/extension with the left-hand thumb resulted in significant ΔHbO_2 increases in 11 channels over the left hemisphere and 12 channels over the right hemisphere. The 2×2 ANOVA on ΔHbO_2 values showed a significant interaction effect of *hemisphere*condition* ($F(1, 433)=27.379$, $p<0.001$). Post hoc pairwise comparisons (Fig. 9.1) revealed that during motor rh, increased ΔHbO_2 occurred in the left hemisphere with a significant difference from ΔHbO_2 of the right hemisphere ($p<0.01$). Corresponding higher oxygenation occurred in the right hemisphere with a significant difference from the left hemisphere during motor lh ($p<0.01$). Post hoc pairwise comparisons further showed higher oxygenation in the left hemisphere during motor rh with a significant difference from motor lh ($p<0.01$), whereas significantly higher oxygenation in the right

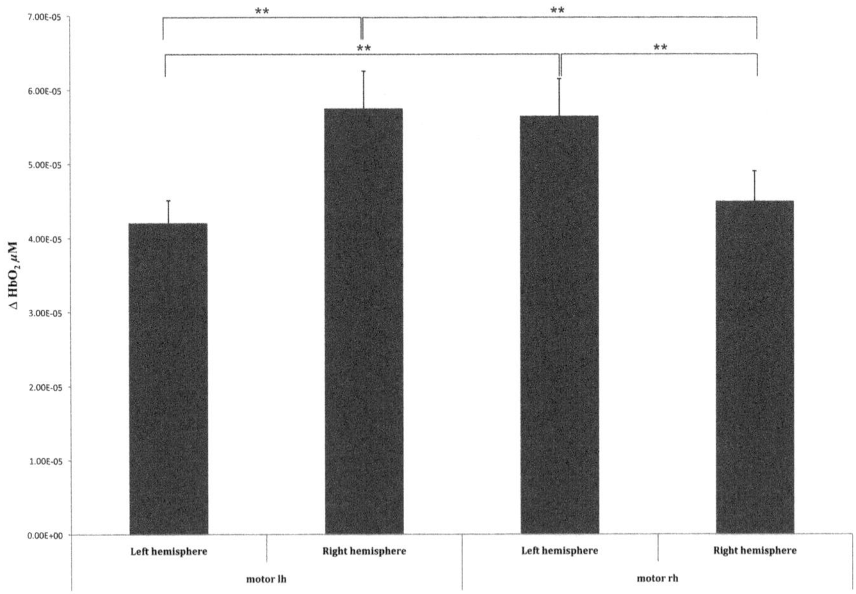

Fig. 9.1 ΔHbO_2 of the right and left hemisphere during flexion/extension with the right-hand thumb (motor rh) and left-hand thumb (motor lh) (**$p<0.01$)

hemisphere occurs during motor lh than during motor rh ($p<0.01$). ΔHbT showed similar results to ΔHbO_2 with eight significant channels above the left hemisphere and six significant channels above the right hemisphere during motor rh. During motor lh, ΔHbT was significantly increased in ten channels above the left hemisphere and in 11 channels above the right hemisphere. The 2×2 ANOVA on the ΔHbT values also showed a significant interaction effect of *hemisphere*condition* ($F(1, 432)=5.617$, $p<0.05$). For ΔHb there were no significant decreases for either hand.

9.4 Discussion

Flexion/extension of the thumb of either hand induced greater increases in ΔHbO_2 and ΔHbT in the contralateral hemisphere, differing significantly from oxygenation changes in the ipsilateral hemisphere. These findings are in line with data from previous brain imaging studies [4–6, 16, 17]. We observed no hemispheric specialization for simple distal motor actions based on equal results for the right- and left-hand thumb movements. Previous studies using NIRS investigating hemispheric differences of hand movements showed divergent results using either finger tapping or opposition tasks [7–11]. Possible variables that might have caused different patterns of cerebral activation in the two hemispheres as complexity, sequentiality, proximity, and sensory stimulation have been excluded in this study. Although the results of this study provide comprehensible results for ΔHbO_2 and ΔHbT, it has to be mentioned that no significant changes for ΔHb have been detected. Therefore, activation patterns in this study are based on ΔHbO_2 and ΔHbT. ΔHbT has been shown to correlate to regional cerebral blood flow as monitored by positron emission tomography [29]. *Conclusion:* We conclude that there is no hemispheric specialization regarding the execution of simple distal motor action. Both hemispheres are equally activated during movement of the contralateral upper limb.

References

1. Penfield W, Boldrey E (1937) Somatic motor and sensory representation in the cerebral cortex of man as studied by electrical stimulation. Brain 60(4):389–443
2. Penfield W, Rasmussen T (1950) The cerebral cortex of man. A clinical study of localization of function. The Macmillan, New York
3. Roland PE, Zilles K (1996) Functions and structures of the motor cortices in humans. Curr Opin Neurobiol 6(6):773–781
4. Sabatini U, Chollet F, Rascol O et al (1993) Effect of side and rate of stimulation on cerebral blood flow changes in motor areas during finger movements in humans. J Cereb Blood Flow Metab 13(4):639–645
5. Rao SM, Binder JR, Bandettini PA et al (1993) Functional magnetic resonance imaging of complex human movements. Neurology 43(11):2311–2318
6. Pulvermuller F, Lutzenberger W, Preissl H, Birbaumer N (1995) Motor programming in both hemispheres: an EEG study of the human brain. Neurosci Lett 190(1):5–8

7. Franceschini MA, Fantini S, Thompson JH, Culver JP, Boas DA (2003) Hemodynamic evoked response of the sensorimotor cortex measured noninvasively with near-infrared optical imaging. Psychophysiology 40(4):548–560
8. Plichta MM, Herrmann MJ, Ehlis AC, Baehne CG, Richter MM, Fallgatter AJ (2006) Event-related visual versus blocked motor task: detection of specific cortical activation patterns with functional near-infrared spectroscopy. Neuropsychobiology 53(2):77–82
9. Sato T, Ito M, Suto T et al (2007) Time courses of brain activation and their implications for function: a multichannel near-infrared spectroscopy study during finger tapping. Neurosci Res 58(3):297–304
10. Suto T, Ito M, Uehara T, Ida I, Fukuda M, Mikuni M (2002) Temporal characteristics of cerebral blood volume change in motor and somatosensory cortices revealed by multichannel near-infrared spectroscopy. Int Congr Ser 1232:383–388
11. Wriessnegger SC, Kurzmann J, Neuper C (2008) Spatio-temporal differences in brain oxygenation between movement execution and imagery: a multichannel near-infrared spectroscopy study. Int J Psychophysiol 67(1):54–63
12. Lausberg H, Cruz RF, Kita S, Zaidel E, Ptito A (2003) Pantomime to visual presentation of objects: left hand dyspraxia in patients with complete callosotomy. Brain 126(Pt 2):343–360
13. Haaland KY, Harrington DL (1996) Hemispheric asymmetry of movement. Curr Opin Neurobiol 6(6):796–800
14. Grafton ST, Hazeltine E, Ivry RB (2002) Motor sequence learning with the nondominant left hand. A PET functional imaging study. Exp Brain Res 146(3):369–378
15. Haaland KY, Harrington DL, Knight RT (2000) Neural representations of skilled movement. Brain 123(Pt 11):2306–2313
16. Solodkin A, Hlustik P, Noll DC, Small SL (2001) Lateralization of motor circuits and handedness during finger movements. Eur J Neurol 8(5):425–434
17. Colebatch JG, Deiber MP, Passingham RE, Friston KJ, Frackowiak RS (1991) Regional cerebral blood flow during voluntary arm and hand movements in human subjects. J Neurophysiol 65(6):1392–1401
18. Zhu Z, Disbrow EA, Zumer JM, McGonigle DJ, Nagarajan SS (2007) Spatiotemporal integration of tactile information in human somatosensory cortex. BMC Neurosci 8:21
19. Disbrow E, Roberts T, Krubitzer L (2000) Somatotopic organization of cortical fields in the lateral sulcus of *Homo sapiens*: evidence for SII and PV. J Comp Neurol 418(1):1–21
20. Khorrami MS, Faro SH, Seshadri A et al (2011) Functional MRI of sensory motor cortex: comparison between finger-to-thumb and hand squeeze tasks. J Neuroimaging 21(3): 236–240
21. Cope M, Delpy DT, Reynolds EO, Wray S, Wyatt J, van der Zee P (1988) Methods of quantitating cerebral near infrared spectroscopy data. Adv Exp Med Biol 222:183–189
22. Obrig H, Villringer A (2003) Beyond the visible–imaging the human brain with light. J Cereb Blood Flow Metab 23(1):1–18
23. Jasper HH (1958) Report to the committee on methods and clinical examination in electroencephalography. Electroencephalogr Clin Neurophysiol 10:371–375
24. Singh AK, Okamoto M, Dan H, Jurcak V, Dan I (2005) Spatial registration of multichannel multi-subject fNIRS data to MNI space without MRI. Neuroimage 27(4):842–851
25. Ye JC, Tak S, Jang KE, Jung J, Jang J (2009) NIRS-SPM: statistical parametric mapping for near-infrared spectroscopy. Neuroimage 44(2):428–447
26. Jang KE, Tak S, Jung J, Jang J, Jeong Y, Ye JC (2009) Wavelet minimum description length detrending for near-infrared spectroscopy. J Biomed Opt 14(3):034004
27. Fekete T, Rubin D, Carlson JM, Mujica-Parodi LR (2011) The NIRS Analysis Package: noise reduction and statistical inference. PLoS One 6(9):e24322
28. Worsley KJ, Friston KJ (1995) Analysis of fMRI time-series revisited–again. Neuroimage 2(3):173–181
29. Villringer A, Chance B (1997) Non-invasive optical spectroscopy and imaging of human brain function. Trends Neurosci 20(10):435–442

Chapter 10
Acute Stress Exposure Preceding Global Brain Ischemia Accelerates Decreased Doublecortin Expression in the Rat Retrosplenial Cortex

Nobuo Kutsuna, Takashi Eriguchi, Hideki Oshima, Takeshi Suma, Kaoru Sakatani, Atsuo Yoshino, and Yoichi Katayama

Abstract **Background:** Psychological distress is a risk factor of stroke in humans and worsens the behavioral and neurological outcomes. In rats, acute stress exposure preceding ischemic events attenuates learning and memory. The retrosplenial cortex (RS) plays an important role in these functions, and global brain ischemia (GBI) or acute stress exposure can induce a decrease in expression of the immature neuronal marker, doublecortin (DCX), in the RS. However, little is known about the DCX expression in the RS after stress exposure prior to GBI. **Methods:** Eighteen male Sprague–Dawley rats were used. Acute stress exposure was applied as the forced swim paradigm and GBI was induced by bilateral common carotid arterial occlusion for 10 min. The rats were divided into three groups: GBI model preconditioned by stress ($n=6$, Group P), GBI model preconditioned by non-stress ($n=6$, Group G), and controls ($n=6$, Group C). We performed immunohistochemistry to observe and analyze the DCX-expressing cells and Fluoro-Jade B (FJB) staining to detect cell death in the RS after GBI in each group. **Results:** The total number of DCX-expressing cells was 1,032, 1,219, and 1,904 in Group P, Group G, and Group C, respectively. The mean number of DCX-expressing cells per unit area was significantly lower in Group P and Group G than in Group C ($P<0.001$). Moreover, the number was significantly lower in Group P than in Group G ($P<0.05$). In each group, no FJB positive cells were observed. **Conclusion:** DCX plays an important role in various cytoskeletal changes. Preconditioning by acute stress exposure accelerated the decrease in DCX expression in the RS after GBI.

N. Kutsuna (✉) • T. Eriguchi • H. Oshima • T. Suma • A. Yoshino • Y. Katayama
Division of Neurosurgery, Department of Neurological Surgery, Nihon University School of Medicine, 30-1 Oyaguchikamicho, Itabashi-ku, 173-8610 Tokyo, Japan
e-mail: nkutsuna@yahoo.co.jp

K. Sakatani
Division of Optical Brain Engineering, Department of Neurological Surgery, Nihon University School of Medicine, Tokyo, Japan

S. Van Huffel et al. (eds.), *Oxygen Transport to Tissue XXXV*, Advances in Experimental Medicine and Biology 789, DOI 10.1007/978-1-4614-7411-1_10,

10.1 Introduction

Psychological distress is a risk factor for stroke in humans [1], and acute stress exposure prior to brain ischemia worsens the behavioral and neurological outcomes in rats [2]. Acute stress preconditioning attenuates learning and memory function especially in ischemic rats [2]. However, little is known about the cellular mechanisms involved between acute stress and brain ischemia.

The retrosplenial cortex (RS), the dorsal cingulate cortex, plays an important role in these functions. The RS is associated with memory and visual spatial functions in rats [3, 4] and primates [5]. Memory [6, 7] and learning [8, 9] represent a complex process based on functional and structural changes at the molecular, synaptic, neuronal, and circuitry levels.

Doublecortin (DCX)-expressing cells play a key role in the plasticity of the brain. DCX is involved in various cytoskeletal changes [10, 11]. DCX-expressing cells exist within the cerebral neocortex and allocortex of the adult brain and are apparently associated with a structural plasticity [12–14].

Global brain ischemia (GBI) induces a decrease in DCX expression in the rat RS, though the mechanisms between GBI and the decreased DCX were not elucidated [15]. Moreover, acute stress exposure induces a similar decrease in the rat RS [16]. These findings suggest a decrease in plasticity potential in the RS after GBI or acute stress exposure.

We propose that preconditioning by acute stress can accelerate the decrease in DCX-expressing cells after GBI, and the resultant findings may help to elucidate cellular mechanisms involved between acute stress and brain ischemia.

10.2 Materials and Methods

10.2.1 Stress Preconditioning and GBI Model

Eighteen male Sprague–Dawley rats (body weight, 250–300 g) were used in the present study. Acute stress exposure was applied as the forced swim paradigm (day 1, 15 min; day 2, 5 min) and the rats with immobile behavior were employed [17]. Transient GBI was induced by bilateral common carotid arterial occlusion for 10 min (day 3) [18]. The rats were divided into three groups: stress exposure preceding GBI ($n=6$, Group P), non-stress exposure preceding GBI ($n=6$, Group G), and controls ($n=6$, Group C). The animals were purchased from Charles River Laboratories (Saitama, Japan) and bred at the Animal Housing Facility of Nihon University. The colony was maintained at 22–23 °C on a 12-h light/dark cycle (lights on at 08:00).

At 7 days after GBI, the rats were transcardially perfusion fixed with lactated Ringer's solution, followed by perfusion of 4 % paraformaldehyde. Coronal serial brain sections were cut in the frontal plane on a vibratome (50 μm). We identified the RS by referring to the rat atlas of Paxinos and Watson [19].

All experimental procedures were conducted in accordance with the Guidelines for the Care and Use of Mammals in Neuroscience and Behavioral Research (National Research Council, National Academy Press, Washington, DC, 2003) and approved by the Animal Care and Use Committee of Nihon University.

10.2.2 Immunohistochemistry and FJB Staining

We performed immunostaining and fluorescence immunostaining with DCX to observe the cells with a plasticity potential. The primary antibodies used in this study were polyclonal goat anti-DCX antibodies (dilution 1:1000, Santa Cruz Biotechnology (California, USA)). We counted the right and left hemispheres separately on two serial sections, for a total of four regions per animal, to examine the differences in expression related to acute stress preconditioning.

We performed Fluoro-Jade B (FJB) staining to investigate the cause of any decrease in DCX-expressing cells. The FJB staining procedure was as reported previously [20].

10.2.3 Measurement and Analysis

A Biozero (BZ-8000; Keyence, Osaka, Japan) and a BZ Analyzer (Keyence) were used to prepare the microphotographs, and a VH Analyzer (Keyence) was used to count the numbers of immunostain-positive cells. We used Neurolucida (Version 3; MicroBrightField, USA) to analyze the area and cell counts without double counting of positive cells and then estimated the cell counts per unit area in the control and GBI model with/without acute stress preconditioning. The statistics software SPSS Statistics 17.0 (Japan IBM, Tokyo, Japan) was used for data analysis. We performed a one-way ANOVA followed by Tukey's tests, and $P < 0.05$ was considered as statistically significant.

10.3 Results

10.3.1 Number of DCX-Expressing Cells

Immunohistochemically DCX-expressing cells in coronal sections were observed in the RS of each case (Fig. 10.1a). Their shape gave the appearance of non-pyramidal cells, they had multiple neurites, and their diameter was 10–25 μm, indicating that they were interneurons. DCX-expressing cells were observed in the subgranular zone (SGZ) within the hippocampal dentate gyrus of each case, indicating the occurrence of neurogenesis in the SGZ (Fig. 10.1b).

The total number of DCX-expressing cells was 1,032, 1,219, and 1,904 cells in Group P, Group G, and Group C, respectively. The mean number of DCX-expressing

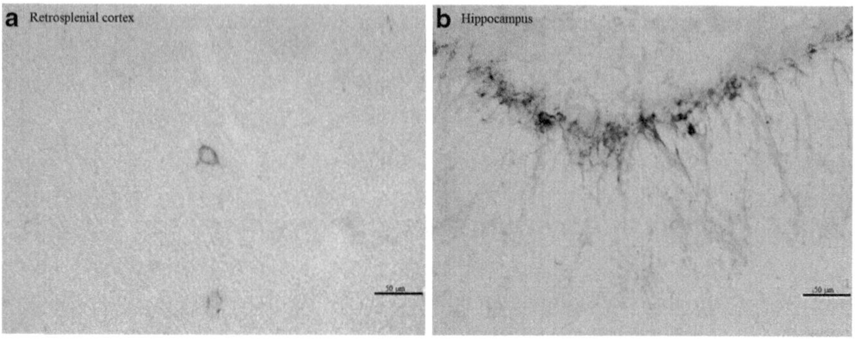

Fig. 10.1 (**a**) Doublecortin (DCX)-expressing cells in the retrosplenial cortex (*RS*) (diaminobenzidine reaction; scale bar, 50 μm; bregma posterior, 3.30 mm). This image was obtained from the RS of a global brain ischemia (*GBI*) model rat after acute stress exposure (Group P; layer, III). Similarly, DCX-expressing cells were present in the RS of the GBI model without acute stress exposure and controls, respectively. (**b**) DCX-expressing cells in the subgranular zone (*SGZ*) of the hippocampal dentate gyrus (diaminobenzidine reaction; scale bar, 50 μm). This image was obtained from the hippocampus of Group P. The cells demonstrate neurogenesis in the SGZ of the hippocampus

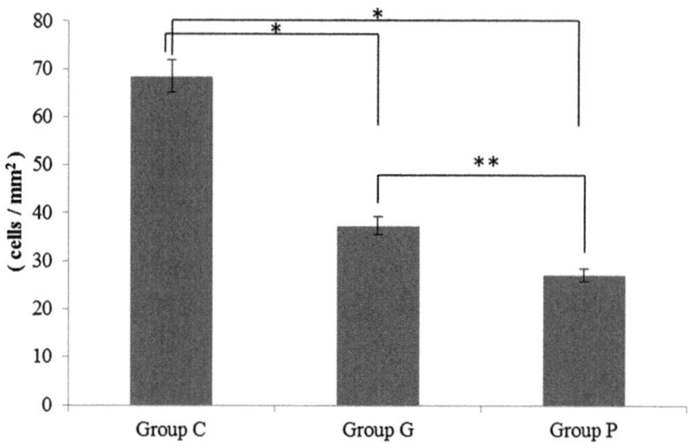

Fig. 10.2 In the global brain ischemia (*GBI*) model, regardless of the preceding acute stress exposure, significantly less doublecortin (*DCX*) expression was observed in the retrosplenial cortex (*RS*) than in the controls (Group C) per unit area ($n = 24$ samples/subregion, right and left hemispheres of two sections from each rat). Acute stress exposure preceding GBI (Group P) revealed a significantly greater decrease in DCX expression in the RS as compared to the normal condition preceding GBI (Group G)

cells per unit area (mean ± SD) was 27 ± 7.57 cells/mm^2 in Group P, 37 ± 9.68 cells/mm^2 in Group G, and 68 ± 18.16 cells/mm^2 in Group C. The mean number of DCX-expressing neurons was significantly lower in Group P and Group G than in Group C (ANOVA followed by Tukey-type test, $P < 0.001$). Moreover, the mean number was significantly lower in Group P than in Group G (ANOVA followed by Tukey-type test, $P < 0.05$) (Fig. 10.2).

10.3.2 FJB Staining

FJB staining is employed for the identification of neuronal cell death to elucidate the mechanism of decreased DCX. As regards the RS, there were no positive cells in Group P, Group G, or Group C, while FJB positive neurons were present in the hippocampal CA1 in Group P and Group G. This result for the RS did not appear to be appropriate for the observed decreased number of DCX-expressing cells.

10.4 Discussion and Conclusion

Preconditioning by acute stress exposure could accelerate the decreased levels of DCX expression in the RS after GBI. The present data showed that the DCX-expressing cells in the RS were significantly decreased after GBI regardless of preconditioning by acute stress exposure in comparison with the controls, as reported previously [15]. Furthermore, acute stress preconditioning through forced swim accelerated the decrease in DCX expression in the RS after GBI as compared to normal preconditioning.

DCX is a microtubule-binding protein and contributes to various cytoskeletal changes involved in the extension of axons and dendrites and new synapse formation in mature neurons [10, 11]. The present results indicate that acute stress preconditioning before GBI accelerated the decrease in plasticity potential after GBI in the RS, though it was not clear whether the actual plasticity decreased in the present study.

DCX-expressing cells in the cortex of adult animals are GABAergic interneurons [13]. The DCX-expressing cells in the present study appeared to be GABAergic interneurons based on their diameter, configuration with bipolar or multipolar formation, and distribution as reported for the RS [16]. It is inferred therefore that interneurons with a plastic potential in the RS may be influenced by acute stress preconditioning.

No FJB positive cells after GBI were observed regardless of acute stress preconditioning, indicating that cells did not fall into cell death after GBI in the present study. Taken together, the decreased DCX expression in the interneurons and findings of FJB staining imply that GABAergic interneurons with a plastic potential may undergo a decrease in their plastic potential without neuronal cell death [21].

Preconditioning by acute stress exposure accelerated the decrease in DCX expression in the RS after GBI. However, the molecular mechanisms between GBI, stress exposure, and the decreased DCX were not elucidated. As described above, the RS plays an important role in memory and space learning which requires complex plastic processes. Alterations in the RS may be associated with more attenuated learning and memory due to acute stress exposure preceding GBI and indicate the involvement of a cellular mechanism between stress exposure and GBI.

Acknowledgments The authors thank Professor Shin Aizawa (Division of Anatomical Science, Department of Functional Morphology, Nihon University School of Medicine) for excellent technical support. This work was supported in part by a Grant-in-Aid for Scientific Research (C-20591725) and by the Strategic Research Program for Brain Science (MEXT); a grant from the Ministry of Education, Culture, Sports, Science and Technology of Japan (C-18591614); a Nihon University School of Medicine Alumni Association 60th Anniversary Fund Research Grant (2012); and a Grant for the Promotion of Industry-University Collaboration at Nihon University.

References

1. Guiraud V, Armor MB, Mas JL, Touzé E (2010) Triggers of ischemic stroke: a systematic review. Stroke 41(11):2669–2677
2. Caso JR, Moro MA, Lorenzo P, Lizasoain I, Leza JC (2007) Involvement of IL-1beta in acute stress-induced worsening of cerebral ischaemia in rats. Eur Neuropsychopharmacol 17(9): 600–607
3. Garden DL, Massey PV, Caruana DA et al (2009) Anterior thalamic lesions stop synaptic plasticity in retrosplenial cortex slices: expanding the pathology of diencephalic amnesia. Brain 132(Pt 7):1847–1857
4. Cooper BG, Mizumori SJ (1999) Retrosplenial cortex inactivation selectively impairs navigation in darkness. Neuroreport 10(3):625–630
5. Maguire EA (2001) The retrosplenial contribution to human navigation: a review of lesion and neuroimaging findings. Scand J Psychol 42(3):225–238
6. Casoli T, Spagna C, Fattoretti P, Gesuita R, Bertoni-Freddari C (1996) Neuronal plasticity in aging: a quantitative immunohistochemical study of GAP-43 distribution in discrete regions of the rat brain. Brain Res 714(1–2):111–117
7. Tanaka J, Horiike Y, Matsuzaki M, Miyazaki T, Ellis-Davies GC, Kasai H (2008) Protein synthesis and neurotrophin-dependent structural plasticity of single dendritic spines. Science 319(5870):1683–1687
8. Holahan MR, Rekart JL, Sandoval J, Routtenberg A (2006) Spatial learning induces presynaptic structural remodeling in the hippocampal mossy fiber system of two rat strains. Hippocampus 16(6):560–570
9. Vazdarjanova A, Ramirez-Amaya V, Insel N, et al (2006) Spatial exploration induces ARC, a plasticity-related immediate-early gene, only in calcium/calmodulin-dependent protein kinase II-positive principal excitatory and inhibitory neurons of the rat forebrain. J Comp Neurol 498: 317–329
10. Francis F, Koulakoff A, Boucher D et al (1999) Doublecortin is a developmentally regulated, microtubule-associated protein expressed in migrating and differentiating neurons. Neuron 23(2):247–256
11. Gleeson JG, Lin PT, Flanagan LA, Walsh CA (1999) Doublecortin is a microtubule-associated protein and is expressed widely by migrating neurons. Neuron 23(2):257–271
12. Xiong K, Luo DW, Patrylo PR, et al (2008) Doublecortin-expressing cells are present in layer II across the adult guinea pig cerebral cortex: partial colocalization with mature interneuron markers. Exp Neurol 211(1):271–282
13. Cai Y, Xiong K, Chu Y, et al (2009) Doublecortin expression in adult cat and primate cerebral cortex relates to immature neurons that develop into GABAergic subgroups. Exp Neurol 216(2):342–356
14. Luzzati F, Bonfanti L, Fasolo A, Peretto P (2009) DCX and PSA-NCAM expression identifies a population of neurons preferentially distributed in associative areas of different pallial derivatives and vertebrate species. Cereb Cortex 19(5):1028–1041
15. Kutsuna N, Murata Y, Eriguchi T et al (2013) DCX-expressing neurons decrease in the retrosplenial cortex after global brain ischemia. Adv Exp Med Biol 765:115–121

16. Kutsuna N, Suma T, Takada Y et al (2012) Decrease in doublecortin expression without neuronal cell death in rat retrosplenial cortex after stress exposure. Neuroreport 23(4): 211–215
17. Porsolt RD, Anton G, Blavey N, Jalfre M (2012) Behavioural despair in rats: a new model sensitive to antidepressant treatments. Eur J Pharmacol 47(4):379–391
18. Yagita Y, Kitagawa K, Ohtsuki T et al (2001) Neurogenesis by progenitor cells in the ischemic adult rat hippocampus. Stroke 32(8):1890–1896
19. Paxinos G, Watson C (2007) The rat brain in stereotaxic coordinates, 6th edn. Academic, USA
20. Schmued LC, Hopkins KJ (2000) Fluoro-Jade B: a high affinity fluorescent marker for the localization of neuronal degeneration. Brain Res 874(2):123–130
21. Friocourt G, Liu JS, Antypa M, Rakic S, Walsh CA, Parnavelas JG (2007) Both doublecortin and doublecortin-like kinase play a role in cortical interneuron migration. J Neurosci 27(14): 3875–3883

Chapter 11
Effects of Transcranial Direct Current Stimulation of the Motor Cortex on Prefrontal Cortex Activation During a Neuromuscular Fatigue Task: An fNIRS Study

Makii Muthalib, Benjamin Kan, Kazunori Nosaka, and Stephane Perrey

Abstract This study investigated whether manipulation of motor cortex excitability by transcranial direct current stimulation (tDCS) modulates neuromuscular fatigue and functional near-infrared spectroscopy (fNIRS)-derived prefrontal cortex (PFC) activation. Fifteen healthy men (27.7 ± 8.4 years) underwent anodal (2 mA, 10 min) and sham (2 mA, first 30 s only) tDCS delivered to the scalp over the right motor cortex. Subjects initially performed a baseline sustained submaximal (30 % maximal voluntary isometric contraction, MVC) isometric contraction task (SSIT) of the left elbow flexors until task failure, which was followed 50 min later by either an anodal or sham treatment condition, then a subsequent posttreatment SSIT. Endurance time (ET), torque integral (TI), and fNIRS-derived contralateral PFC oxygenated (O_2Hb) and deoxygenated (HHb) hemoglobin concentration changes were determined at task failure. Results indicated that during the baseline and post-treatment SSIT, there were no significant differences in TI and ET, and increases in fNIRS-derived PFC activation at task failure were observed similarly regardless of the tDCS conditions. This suggests that the PFC neuronal activation to maintain muscle force production was not modulated by anodal tDCS.

M. Muthalib (✉)
Movement to Health (M2H) Laboratory, Euromov, Montpellier-1 University, Montpellier, France

School of Exercise and Health Sciences, Edith Cowan University, Perth, Australia

Movement Neuroscience Program, Queensland University of Technology, Brisbane, Australia
e-mail: makii.muthalib@gmail.com; makii.muthalib@univ-montp1.fr

B. Kan • K. Nosaka
School of Exercise and Health Sciences, Edith Cowan University, Perth, Australia

S. Perrey
Movement to Health (M2H) Laboratory, Euromov, Montpellier-1 University, Montpellier, France

S. Van Huffel et al. (eds.), *Oxygen Transport to Tissue XXXV*, Advances in Experimental Medicine and Biology 789, DOI 10.1007/978-1-4614-7411-1_11,

11.1 Introduction

Transcranial direct current stimulation (tDCS) is a noninvasive and minimally discomforting electrical brain stimulation technique that is used to modulate cortical excitability and behavioral responses [1, 2]. tDCS has been used for the treatment of a variety of neurological or neuropsychiatric disorders. tDCS applies low-intensity (1 ~ 2 mA) controlled direct currents to the scalp overlying a target region of the brain via surface electrodes. The tDCS effects on cortical excitability are dependent on current polarity, intensity, and duration of the stimulation, such that anodal stimulation over the motor cortex increases corticospinal excitability, whereas cathodal stimulation decreases corticospinal excitability [2].

Neuromuscular fatigue is represented by a decrease in the ability of muscle fibers to generate force, which can be attributed to both central and peripheral mechanisms. Peripheral fatigue is characterized by changes at or distal to the neuromuscular junction, whereas central fatigue refers to a progressive reduction in voluntary neuronal activation of muscle during exercise. Central fatigue can be further differentiated into changes in the activity of spinal motoneurons (spinal fatigue) due to intrinsic cellular properties (e.g., discharge rates) and to reduced motor cortical neuronal drive (supraspinal fatigue), and both these central fatigue levels can be modulated by afferent inputs [3].

Cogiamanian et al. [4] were the first to investigate whether manipulation of motor cortex excitability by anodal tDCS (1.5 mA, 10 min) modulates neuromuscular fatigue in healthy subjects during a subsequent sustained submaximal isometric contraction task (SSIT) at 35 % of maximal voluntary contraction (MVC) of the elbow flexors until task failure. Their results indicated that anodal tDCS improved endurance time (ET) during posttreatment SSIT by ~ 20 % compared to sham tDCS (i.e., baseline SSIT compared to posttreatment SSIT), suggesting a decrease in residual neuromuscular fatigue that was modulated by anodal tDCS via central (supraspinal) neuroplastic mechanisms.

Previous neuroimaging studies have reported that, during neuromuscular fatigue of unilateral small muscle group exercise, the prefrontal cortex (PFC) increases neuronal activation to reinforce muscle force for the continuation of the performance [5–9]. This increase in cortical neuronal activation during neuromuscular fatigue is implied by an enhanced cortical oxygenation and blood flow by means of a neurovascular coupling mechanism [10]. Noninvasive functional near-infrared spectroscopy (fNIRS) neuroimaging measures several physiological parameters related to cortical oxygenation and blood flow and includes measurements of concentration changes in oxygenated (O_2Hb) and deoxygenated (HHb) hemoglobin. Therefore, the aim of this study was to investigate the fNIRS-derived PFC activation responses of anodal tDCS modulation of motor cortex excitability during a subsequent SSIT of the elbow flexors until task failure.

11.2 Methods

11.2.1 Subjects

Fifteen healthy men (mean ± SD age, height, and weight: 27.7 ± 8.4 years, 176.4 ± 7.4 cm, 72.7 ± 8.7 kg, respectively) participated in the study. All subjects had no known health problems (e.g., metabolic or neuromuscular disorders) or any upper extremity muscle or joint injuries. The study conformed to the recommendations of the local Human Research Ethics Committee in accordance with the Declaration of Helsinki.

11.2.2 Experimental Setup

An Eldith tDCS system (Neuroconn, Ilmenau, Germany) was used to deliver constant direct currents (2 mA) through a pair of 0.3-cm-thick square (24 cm^2) sponge electrodes soaked in saline solution (140 mM NaCL). The anode electrode was placed on the scalp overlying the right motor cortical representation of the left arm, and the cathode was placed over the right shoulder based on the electrode setup method described by Cogiamanian et al. [4].

Subjects were seated on a preacher arm curl bench, securing the shoulder angle at 45 ° flexion with a supinated forearm position and an elbow joint angle of 90 °. The elbow joint was aligned with the axis of rotation of an isokinetic dynamometer (Cybex6000, Lumex Inc., Ronkonkoma, USA). Torque signals were collected onto a data acquisition system (PowerLab, ADInstruments, Bella Vista, Australia) at a sampling rate of 200 Hz, and real-time visual feedback of torque signals was displayed on the computer monitor.

A NIRO-200 oximeter (Hamamatsu Photonics K.K., Hamamatsu, Japan) was used to measure PFC O_2Hb and HHb concentration changes using the modified Beer-Lambert Law. The probe unit has two silicon photodiodes as photodetectors on one side and three laser-emitting diodes (775, 810, and 850 nm) on the other side separated from each other by a distance of 4 cm. A probe unit was firmly attached over the skin of the forehead (in a region corresponding to Fp2 according to the 10–20 International EEG system) contralateral to the left exercising arm with a double-sided adhesive tape. The probe unit, in turn, was covered with a black band. The NIRO-200 provides estimates of concentration changes (from an arbitrary baseline of zero) in O_2Hb and HHb at a sampling rate of 6 Hz. The quantification of O_2Hb and HHb concentration changes, expressed in ΔμM, was obtained by including an age-dependent constant differential path length factor ($5.13 + 0.07 \times age^{0.81}$) [11].

11.2.3 Protocol

Subjects underwent anodal or sham tDCS treatment conditions separated by 1–2 weeks and the order of intervention was randomized among the participants. Subjects initially performed three 3-s MVCs with a 45-s rest between contractions, and MVC was determined by taking the mean peak torque of the three MVC trials. Following a 3-min rest, subjects performed a baseline SSIT of the left elbow flexors at 30%MVC until task failure, which was followed 50 min later by either anodal or sham treatment conditions. The sham intervention was identical to the tDCS intervention, except that the current was programmed to return to zero for the remaining time period after 30 s of stimulation. A posttreatment SSIT was performed ~ 10 min following the treatment conditions.

11.2.4 Statistical Analysis

ET at task failure was determined by the time point in which participants were unable to maintain greater than 90 % of the target torque output for more than 2 s, in spite of continual verbal encouragement. Torque integral (TI) was determined as the area under the torque curves until task failure. PFC O_2Hb and HHb concentration changes were determined at task failure (average of 5 s around this time point). ET, TI, and PFC O_2Hb and HHb changes during the baseline and posttreatment SSIT were compared between the anodal and sham tDCS sessions by a two-way repeated measures ANOVA. An individual paired Student's *t*-test was used to determine if PFC O_2Hb and HHb changes were different from pre-exercise levels for both anodal and sham tDCS conditions. Significance was set at $P \leq 0.05$. Data are presented as mean ± SD.

11.3 Results

Figure 11.1 shows typical changes in torque and fNIRS-derived PFC activation (i.e., increase in O_2Hb and decrease in HHb) during an SSIT until task failure. Group results indicated that during the baseline SSIT, there were no significant differences between anodal or sham tDCS conditions for ET, TI, or PFC O_2Hb and HHb changes. Similarly, during the posttreatment SSIT, there was no significant difference between the anodal and sham tDCS conditions for ET (333 ± 119 s vs. 353 ± 146 s, respectively) and TI (5,907 ± 2,557 vs. 6,020 ± 2,597 $Nm \cdot s^{-1}$, respectively). Although right PFC O_2Hb increased and HHb decreased significantly ($P<0.05$) relative to pre-exercise levels during the posttreatment SSIT for both anodal (11.3 ± 4.0 and −0.9 ± 2.2 μM, respectively) and sham (10.1 ± 4.2 and −1.1 ± 1.8 μM, respectively) tDCS conditions, there was no significant difference between the conditions.

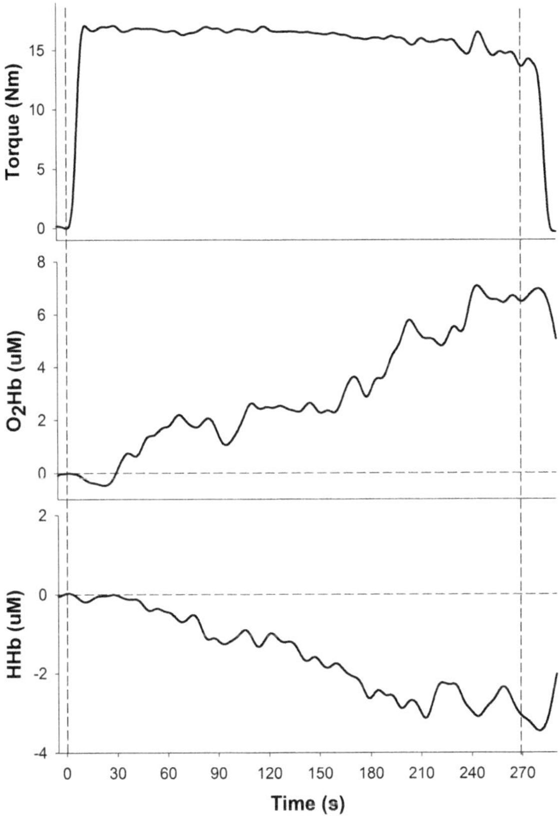

Fig. 11.1 Typical single-subject recordings of the changes in torque and fNIRS-derived PFC activation (i.e., increase in O_2Hb and decrease in HHb) during a 30 % MVC SSIT of the elbow flexors until task failure (*second dashed line*) following 10-min anodal tDCS condition. Endurance time at task failure was determined by the time point in which participants were unable to maintain greater than 90 % of the target torque output for more than 2 s

11.4 Discussion

To the best of our knowledge, this is the first study to determine PFC activation during a tDCS-modulated neuromuscular fatigue task. The present study found no significant effects of anodal tDCS on ET, TI, and fNIRS-derived PFC activation during a subsequent SSIT of the elbow flexors. It is important to note that there was no significant difference in ET results within and between anodal and sham tDCS conditions (i.e., baseline and posttreatment SSIT), which is in contrast to Cogiamanian et al. [4] who reported an ~40 % reduction in ET in sham tDCS and a smaller reduction of ~20 % with anodal tDCS during posttreatment SSIT compared to baseline SSIT. Thus, in the present study, it can be suggested that no residual neuromuscular fatigue was induced by the baseline SSIT on the subsequent posttreatment SSIT.

One reason for the lack of difference between anodal and sham tDCS conditions for the dependent variables could be due to a nonsignificant effect of our tDCS electrode setup to induce changes in cortical excitability. In the present study, no assessment of corticospinal excitability changes was included such as motor-evoked potential (MEP) changes upon transcranial magnetic stimulation (TMS). However, in a subsequent study that utilized TMS (unpublished), no significant changes in MEP were observed after a similar tDCS electrode setup and protocol to that of the present study. Thus, in the present study, it appears that no significant differences in PFC activation during the posttreatment SSIT between anodal and sham tDCS conditions could indicate no modulation of cortical excitability by anodal tDCS and no effect on central (supraspinal) fatigue mechanisms and PFC activation. In future studies, it would be important to use a tDCS electrode setup that has consistently been shown to increase motor cortex excitability (e.g., anode over the motor cortex and cathode over the contralateral frontopolar cortex) [2].

Another possibility for our findings could be that our relatively strict determination of task failure during the 30 % MVC SSIT protocol (i.e., > 90 % of target torque for > 2 s) may not have allowed fatigue to develop to a level that necessitated tDCS modulation. tDCS is known to alter firing rates of motor cortical neurons due to a shift of resting membrane potentials, such that it is assumed that anodal tDCS can increase motor unit recruitment due to the depolarization of resting membrane potentials of motor cortex neurons and corticospinal excitability [2]. Rupp and Perrey [9] reported that fNIRS-derived PFC activation increased (i.e., O_2Hb increase and HHb decrease) from pre-exercise levels to exhaustion (ET = ~460 s) during a unilateral SSIT at 40 % MVC of the ankle extensors. They concluded that centrally mediated decrease in neuronal activation, as reflected by fNIRS-derived PFC activation measurements, was likely not implicated in neuromuscular fatigue during an SSIT of a small muscle group. In the present study, it could be suggested that maximal voluntary neuronal activation was possible at task failure during our SSIT protocol, such that anodal tDCS was unlikely to further increase voluntary activation, suggesting a ceiling effect at corticospinal levels. This then would also explain the similar fNIRS-derived PFC activation levels at task failure between anodal and sham tDCS conditions.

11.5 Conclusion

In conclusion, this study showed that the 10-min anodal tDCS treatment at 2 mA did not affect neuromuscular fatigue and fNIRS-derived PFC activation compared to a sham tDCS treatment. This could be due in part to our tDCS electrode setup to induce an increase in motor cortex excitability, a ceiling effect of tDCS to modulate motor cortex excitability and/or to an absence of supraspinal fatigue mechanisms during the SSIT.

Acknowledgments The authors would like to thank Prof. Marco Ferrari and Prof. Samuele Marcora for their valuable comments on the manuscript.

References

1. Nitsche MA, Liebetanz D, Lang N, Antal A, Tergau F, Paulus W (2003) Safety criteria for transcranial direct current stimulation (tDCS) in humans. Clin Neurophysiol 114(11): 2220–2222
2. Nitsche MA, Paulus W (2000) Excitability changes induced in the human motor cortex by weak transcranial direct current stimulation. J Physiol 527(Pt 3):633–639
3. Gandevia SC (2001) Spinal and supraspinal factors in human muscle fatigue. Physiol Rev 81(4):1725–1789
4. Cogiamanian F, Marceglia S, Ardolino G, Barbieri S, Priori A (2007) Improved isometric force endurance after transcranial direct current stimulation over the human motor cortical areas. Eur J Neurosci 26(1):242–249
5. Mottola L, Crisostomi S, Ferrari M, Quaresima V (2007) Relationship between handgrip sustained submaximal exercise and prefrontal cortex oxygenation. Adv Exp Med Biol 578: 305–309
6. Muthalib M, Ferrari M, Quaresima V, Nosaka K (2012) Frontal cortex activation during electrical muscle stimulation as revealed by functional near-infrared spectroscopy. Adv Exp Med Biol 737:45–49
7. Liu JZ, Shan ZY, Zhang LD, Sahgal V, Brown RW, Yue GH (2003) Human brain activation during sustained and intermittent submaximal fatigue muscle contractions: an FMRI study. J Neurophysiol 90(1):300–312
8. Millet GY, Muthalib M, Jubeau M, Laursen PB, Nosaka K (2012) Severe hypoxia affects exercise performance independently of afferent feedback and peripheral fatigue. J Appl Physiol 112(8):1335–1344
9. Rupp T, Perrey S (2009) Effect of severe hypoxia on prefrontal cortex and muscle oxygenation responses at rest and during exhaustive exercise. Adv Exp Med Biol 645:329–334
10. Perrey S (2008) Non-invasive NIR, spectroscopy of human brain function during exercise. Methods 45(5):289–299
11. Duncan A, Meek JH, Clemence M et al (1995) Optical pathlength measurements on adult head, calf and forearm and the head of the newborn infant using phase resolved optical spectroscopy. Phys Med Biol 40(2):295–304

Chapter 12
The Effect of Inner Speech on Arterial CO_2 and Cerebral Hemodynamics and Oxygenation: A Functional NIRS Study

Felix Scholkmann, Martin Wolf, and Ursula Wolf

Abstract The aim of the present study was (i) to investigate the effect of inner speech on cerebral hemodynamics and oxygenation, and (ii) to analyze if these changes could be the result of alternations of the arterial carbon dioxide pressure ($PaCO_2$). To this end, in seven adult volunteers, we measured changes of cerebral absolute [O_2Hb], [HHb], [tHb] concentrations and tissue oxygen saturation (StO_2) (over the left and right anterior prefrontal cortex (PFC)), as well as changes in end-tidal CO_2 ($P_{ET}CO_2$), a reliable and accurate estimate of $PaCO_2$. Each subject performed three different tasks (inner recitation of hexameter (IRH) or prose (IRP) verses) and a control task (mental arithmetic (MA)) on different days according to a randomized crossover design. Statistical analysis was applied to the differences between pre-baseline, two tasks, and four post-baseline periods. The two brain hemispheres and three tasks were tested separately. During the tasks, we found (i) $P_{ET}CO_2$ decreased significantly ($p<0.05$) during the IRH (~3 mmHg) and MA (~0.5 mmHg) task. (ii) [O_2Hb] and StO_2 decreased significantly during IRH (~1.5 μM; ~2 %), IRP (~1 μM; ~1.5 %), and MA (~1 μM; ~1.5 %) tasks. During the post-baseline period, [O_2Hb] and [tHb] of the left PFC decreased significantly after the IRP and MA task (~1 μM and ~2 μM, respectively). In conclusion, the

F. Scholkmann
Biomedical Optics Research Laboratory, Division of Neonatology, University Hospital Zurich, Zurich 8091, Switzerland

Institute of Complementary Medicine, University of Bern, Imhoof-Pavillon, Inselspital, 3010 Bern, Switzerland

M. Wolf
Biomedical Optics Research Laboratory, Division of Neonatology, University Hospital Zurich, Zurich 8091, Switzerland

U. Wolf (✉)
Institute of Complementary Medicine, University of Bern, Imhoof-Pavillon, Inselspital, 3010 Bern, Switzerland
e-mail: ursula.wolf@kikom.unibe.ch

S. Van Huffel et al. (eds.), *Oxygen Transport to Tissue XXXV*, Advances in Experimental Medicine and Biology 789, DOI 10.1007/978-1-4614-7411-1_12,

study showed that inner speech affects $PaCO_2$, probably due to changes in respiration. Although a decrease in $PaCO_2$ is causing cerebral vasoconstriction and could potentially explain the decreases of $[O_2Hb]$ and StO_2 during inner speech, the changes in $PaCO_2$ were significantly different between the three tasks (no change in $PaCO_2$ for MA) but led to very similar changes in $[O_2Hb]$ and StO_2. Thus, the cerebral changes cannot solely be explained by $PaCO_2$.

12.1 Introduction

In previous studies, we showed that guided rhythmic speech exercises in the context of arts speech therapy (AST) cause changes in heart rate variability [1, 2], cardiorespiratory interactions [3], as well as hemodynamics and oxygenation in the brain and muscle [4–6]. In particular, we demonstrated that during speech exercises, a decrease in cerebral hemodynamics and oxygenation occurred. We hypothesized that this effect might be the result of a decrease in the partial pressure of carbon dioxide in the arterial blood ($PaCO_2$) during speaking [5, 6]. This hypothesis was confirmed in a subsequent study [4]: we found significant changes in end-tidal CO_2 ($P_{ET}CO_2$), a reliable and accurate estimate of $PaCO_2$ [7], during all recitation tasks and even during the control task (mental arithmetic). We concluded that changes in breathing (hyperventilation) during the tasks are mainly to account for the measured changes in hemodynamics and oxygenation mediated by hypocapnia. To further investigate the effect of $PaCO_2$ variations on hemodynamics and oxygenation and in order to avoid a CO_2 reaction, the aim of the present study was to investigate the impact of inner speech tasks on these parameters.

12.2 Material and Methods

Seven healthy subjects (four men, three women, mean age 34.6 ± 9.3 years) participated in this study. The study was carried out as a controlled and randomized crossover trial. The design of the study was in accordance with the Declaration of Helsinki; the approval was obtained from the Ethical Committee of the Canton of Zurich. The participants were German/Swiss German native speakers who had no previous knowledge of AST and were asked not to eat and consume any stimulants (such as caffeine or other ingredients in energy drinks) for at least 2 h before the start of the measurements. Each subject was measured while performing three different tasks, i.e., inner recitation (i.e., reciting without voicing aloud) of hexameter (IRH) or prose (IRP) verses and a control task (mental arithmetic (MA)). Each task was performed on a separate day to avoid potential carry-over effects, and each measurement lasted 38 min (8 min pre-baseline, 5 min task, 5 min recovery, 5 min task, and 15 min recovery). During the measurements, the subjects sat opposite a speech therapist who recited the respective text verse by verse or asked the subjects to perform the MA task. The subject repeated the texts with inner speech.

The following physiological parameters were measured: (i) heart rate (device: Medilog AR12 Plus, Schiller AG, Baar, Switzerland; sampling rate, 4,096 Hz, 16 bit); (ii) absolute concentrations of oxyhemoglobin ([O_2Hb]), deoxyhemoglobin [HHb], total hemoglobin ([tHb]), and tissue oxygen saturation (StO_2) (device: OxiplexTS, ISS Inc., Champaign, USA; sampling rate, 50 Hz); and (iii) $P_{ET}CO_2$ (device: Nellcor N1000 gas analyzer, Nellcor. Inc, Hayward, USA; sampling rate, 50 Hz; measurement range, 0–60 mmHg). The NIRS sensors were placed on the left and right side of the forehead over the left and right anterior prefrontal cortex (PFC) and the $P_{ET}CO_2$ probe directly below the right nostril of the subject. The placement of the sensors is illustrated in Fig. 12.1c.

Movement artifacts in [O_2Hb], [HHb], [tHb], and StO_2 signals were removed using the method presented in [8]. Thereby, care was taken to ensure that no artificial new trends were introduced to the signals. Measurements with too many artifacts were excluded from further analysis. The $P_{ET}CO_2$ signal was calculated by using the raw CO_2 waveform signal, detecting the local maxima of every respiratory cycle and determining the envelope over these local maxima. Each time series was segmented into intervals with a length of 3 min each (see Fig. 12.2). For further analysis, all signals were downsampled to 5 Hz and the NIRS-derived signals were low-pass filtered using a moving average filter (window length, 10 s).

The measured changes of [O_2Hb], [HHb], [tHb], StO_2, and $P_{ET}CO_2$ were then tested on their statistical significance by calculating the median values for each segment, normalizing every median value by subtracting the median value from the first interval (to remove the intra-individual variance of the starting values) and applying the Wilcoxon signed-rank test to test for the null hypothesis that the median values for each interval have a distribution with a zero median. Whether the changes in the left and right PFC are statistically different or not was tested with a Wilcoxon rank sum test. This test was also used to test for group differences. All calculations were performed using Matlab (MathWorks, Natick, Massachusetts, USA).

12.3 Results

Figure 12.1a, b shows the measured changes in [O_2Hb], [HHb], [tHb], StO_2, and $P_{ET}CO_2$ for the right and left PFC and the three different tasks. *During the tasks* (i.e., *intervals 2 and/or 4*), StO_2 and [O_2Hb] decreased significantly ($p < 0.05$) in the right PFC during IRH (~1.5 μM; ~1.5 %), IRP (~1 μM; ~1.5 %), and MA (~1 μM; ~1.5 %) tasks. The left PFC showed a less consistent pattern of decreases: while StO_2 and [O_2Hb] decreased in all three tasks, the decreases were only significant for StO_2 during IRP (~2 %) and for [O_2Hb] during IRP (~1.5 μM) as well as IRH (~1.5 μM). A significant increase of [HHb] took place in the right PFC during MA (~0.5 μM) and IRP (~0.2 μM). [tHb] decreased significantly in the right PFC

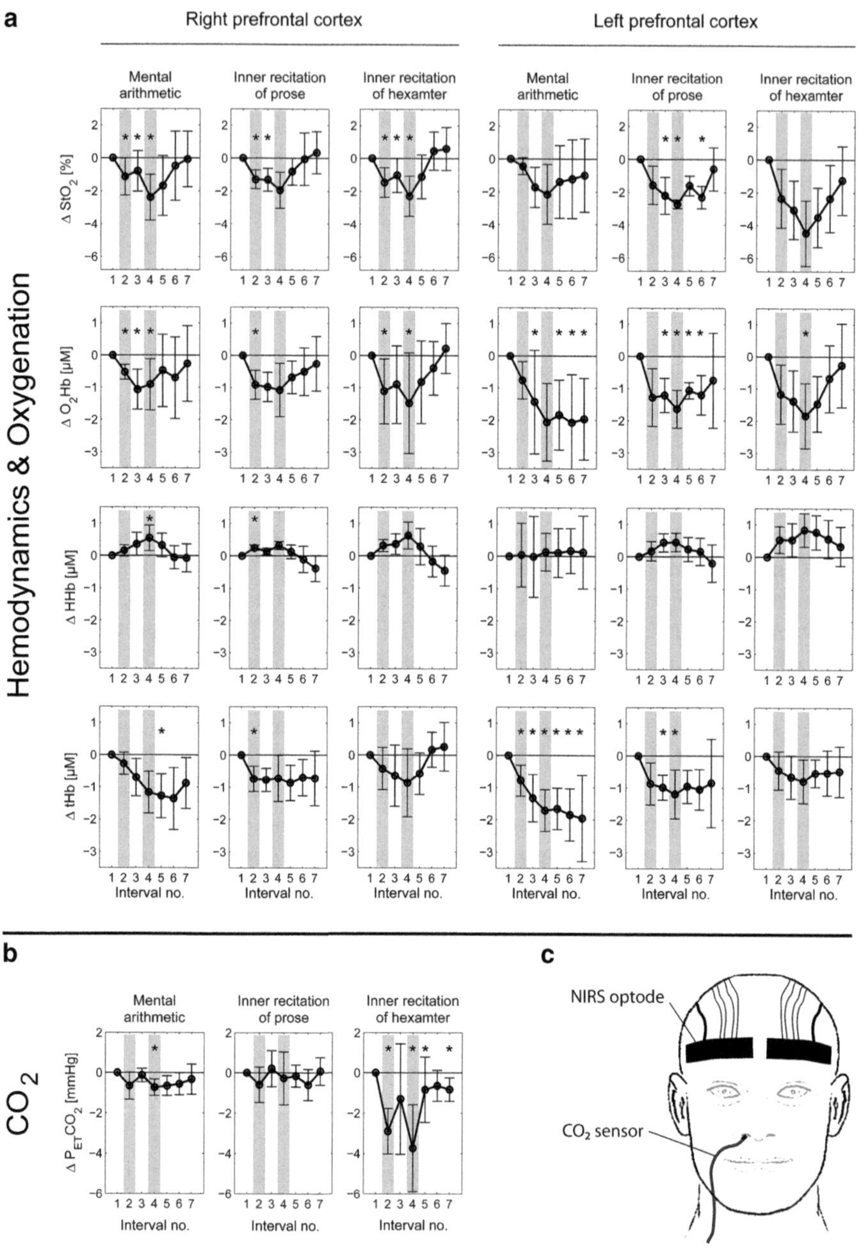

Fig. 12.1 Changes in StO_2, $[O_2Hb]$, [HHb], [tHb] (subfigures **a**), and $P_{ET}CO_2$ (subfigures **b**) over the course of the experiments. Each interval refers to a time span of 3 min. The *shaded areas* indicate the periods when the tasks were performed. All data were shown as median values (*black circles*) ± median absolute deviation (*MAD*). Significant changes are marked with an *asterisk*: $p < 0.05$ (*). Subfigure **c** illustrates the placement of the NIRS probes and the CO_2 sensor on the head of the subject

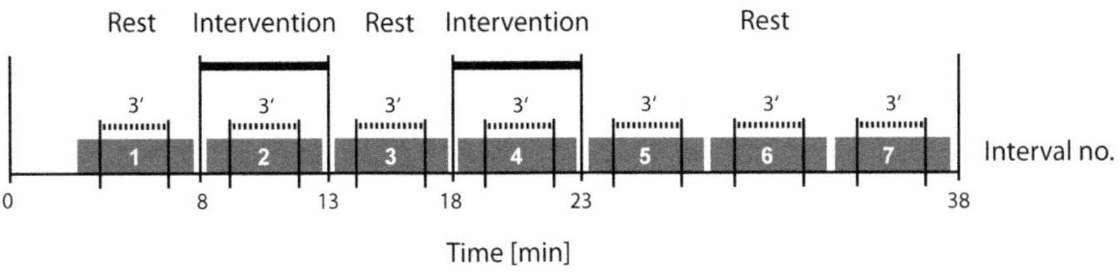

Fig. 12.2 Experimental paradigm

during IRP (~0.5 μM) and in the left PFC during MA (~1.5 μM) as well as IRP (~1 μM). *During the post-baseline period* (i.e., *intervals no 5–7*), [O_2Hb] and [tHb] of the left PFC decreased significantly after the IRP and MA task (both ~2 μM). $P_{ET}CO_2$ decreased significantly during IRH (~3 mmHg) and MA (~0.5 mmHg). The differences in [O_2Hb], [HHb], [tHb], and StO_2 changes between the right and left PFC were not statistically significant. The comparison of the [O_2Hb], [HHb], [tHb], StO_2, and PETCO$_2$ changes with respect to the three different tasks showed that the changes in PETCO$_2$ during the IRH task differed significantly from changes during IPR and MA.

12.4 Discussion

As already indicated in [4], in order to explain the results obtained from speech studies, one should be aware that the measured changes of NIRS-derived hemodynamic and oxygenation signals are the result of at least two major physiological effects. One the one hand, increased neuronal activity leads to an increase in the cerebral metabolic rate of O_2 ($CMRO_2$) which is accompanied by an increase of the cerebral blood flow (CBF) and thus volume (CBV) (neurovascular coupling) [9]. This effect in characteristic changes of the NIRS-derived signals: [O_2Hb] ↑, [HHb] ↓, [tHb] ↑, and StO_2 ↑. On the other hand, changes in $PaCO_2$ have a strong effect on cerebral hemodynamics and oxygenation, i.e., an increase of the frequency and/or volume of breathing (hyperventilation) causes a decrease in $PaCO_2$ (hypercapnia) which leads to a reduction in CBF by cerebral vasoconstriction [10]. This effect is also associated with characteristic changes of the NIRS-derived signals: [O_2Hb] ↓, [HHb] ↑, [tHb] ↓, and StO_2 ↓. The measured changes of the NIRS-derived signals are a combination of both these effects.

The observed significant decrease of $P_{ET}CO_2$ as well as StO_2, [O_2Hb], and [tHb] during all three tasks indicates that the neurovascular coupling seems to be overruled by a hyperventilation-induced hypocapnia which causes a cerebral vasoconstriction. However, it is not clear why the changes in $P_{ET}CO_2$ were significantly different between the three tasks (no change in $P_{ET}CO_2$ for MA) but led to very similar changes in [O_2Hb] and StO_2. This is unexpected since $PaCO_2$ and CBF are almost linearly correlated in the normal physiological range [11]. Thus, the hemodynamic and oxygenation changes cannot solely be explained by $PaCO_2$. Differences in brain activity related to the specific type of task might also explain the results obtained. It is known that at least two factors are influencing mainly the activity of the PFC: stress [12, 13] and specific types of cognitive activity (particularly memory retrieval and multitasking) [14]. Since the three different types of task in our experiment could be associated with different amounts of evoked stress as well as memory retrieval and multitasking, the ratio of neurovascular coupling/CO_2-mediated effects might differ which would explain the variability of the obtained data.

For further research, it would be interesting to investigate (i) what role silent articulation during the inner speech tasks plays on cerebral hemodynamic and oxygenation changes and (ii) how the effects depend on population-characterizing parameters (i.e., age, gender, type of personality). Additionally, (iii) one should consider to place the NIRS optode over the left inferior frontal gyrus since it was shown that this region is associated with inner speech [15].

In conclusion, the study showed that inner speech effects cerebral hemodynamics and oxygenation primarily by changes in $PaCO_2$ caused by variations in respiration and secondarily by increased neuronal activity of the PFC.

Acknowledgments We thank all subjects and the arts speech therapist Andrea Klapproth for their participation in this study, Rachel Folkes for proofreading of the manuscript, and the numerous participants of the ISOTT conferences 2010, 2011, and 2012 for their stimulating discussions about CO_2 and cerebral hemodynamics/oxygenation.

References

1. Bettermann H, von Bonin D, Frühwirth M, Cysarz D, Moser M (2002) Effects of speech therapy with poetry on heart rate rhythmicity and cardiorespiratory coordination. Int J Cardiol 84(1):77–88
2. von Bonin D, Frühwirth M, Heuser P, Moser M (2001) Effects of speech therapy with poetry on heart rate variability and well-being. Forsch Komplementarmed Klass Naturheilkd 8(3):144–160
3. Cysarz D, von Bonin D, Lackner H, Heusser P, Moser M, Bettermann H (2004) Oscillations of heart rate and respiration synchronize during poetry recitation. Am J Physiol Heart Circ Physiol 287(2):H579–H587
4. Scholkmann F, Gerber U, Wolf M, Wolf U (2012) End-tidal CO2: an important parameter for a correct interpretation in functional brain studies using speech tasks. Neuroimage 66:71–79
5. Wolf M, von Bonin D, Wolf U (2011) Speech therapy changes blood circulation and oxygenation in the brain and muscle: a near-infrared spectrophotometry study. Adv Exp Med Biol 701:21–25
6. Wolf U, Scholkmann F, Rosenberger R, Wolf M, Nelle M (2011) Changes in hemodynamics and tissue oxygenation saturation in the brain and skeletal muscle induced by speech therapy – a near-infrared spectroscopy study. Sci World J 11:1206–1215
7. Weinger MB, Brimm JE (1987) End-tidal carbon dioxide as a measure of arterial carbon dioxide during intermittent mandatory ventilation. J Clin Monit 3(2):73–79
8. Scholkmann F, Spichtig S, Muehlemann T (2010) How to detect and reduce movement artifacts in near-infrared imaging using moving standard deviation and spline interpolation. Physiol Meas 31(5):649–662

9. Buxton RB (2012) Dynamic models of BOLD contrast. Neuroimage 62(2):953–961
10. Szabo K, Lako E, Juhasz T, Rosengarten B, Csiba L, Olah L (2011) Hypocapnia induced vasoconstriction significantly inhibits the neurovascular coupling in humans. J Neurol Sci 309(1–2):58–62
11. Grubb RL, Raichle ME, Eichling JO, Ter-Pogossian MM (1974) The effects of changes in $PaCO_2$ on cerebral blood volume, blood flow, and vascular mean transit time. Stroke 5(5): 630–639
12. Sullivan RM, Gratton A (2002) Prefrontal cortical regulation of hypothalamic-pituitary-adrenal function in the rat and implications for psychopathology: side matters. Psychoneuroendocrinology 27(1–2):99–114
13. Buijs RM, Van Eden CG (2000) The integration of stress by the hypothalamus, amygdala and prefrontal cortex: balance between the autonomic nervous system and the neuroendocrine system. Prog Brain Res 126:117–132
14. Koechlin E, Hyafil A (2007) Anterior prefrontal function and the limits of human decision-making. Science 318(5850):594–598
15. Simons CJP, Tracy DK, Sanghera KK et al (2010) Functional magnetic resonance imaging of inner speech in schizophrenia. Biol Psychiatry 67(3):232–237

Chapter 13
Investigation of Frontal Lobe Activation with fNIRS and Systemic Changes During Video Gaming

Ilias Tachtsidis and Antonis Papaioannou

Abstract Frontal lobe activation caused by tasks such as videogames can be investigated using multichannel near-infrared spectroscopy (fNIRS), sometimes called optical topography. The aims of this study are to investigate the effects of video gaming (fighting and puzzle games) in the brain and the systemic physiology and to determine whether systemic responses during the gaming task are associated with the measurement of localised cerebral haemodynamic changes as measured by fNIRS. We used a continuous-wave 8-channel fNIRS system to measure the changes in concentration of oxy-haemoglobin (HbO_2) and deoxy-haemoglobin (HHb) and changes in total haemoglobin ($\Delta tHb = \Delta HbO_2 + \Delta HHb$) over the frontal lobe in 30 healthy volunteers. The Portapres system was used to measure mean blood pressure (MBP) and heart rate (HR), and a laser Doppler was employed to measure the changes in scalp blood flow (or flux). Even though we observed significant changes in systemic variables during gaming, in particular in scalp flow, we also managed to see localised activation patterns over the frontal polar (FP1) region. However, in some channels over the frontal lobe, we also observed significant correlations between the HbO_2 and systemic variables.

13.1 Introduction

Multichannel functional near-infrared spectroscopy (fNIRS), or optical topography (OT), is often employed to detect brain functional activation. fNIRS measures the changes in brain tissue concentrations of oxy-haemoglobin (HbO_2) and deoxy-haemoglobin (HHb) that occur secondary to the brain electrical activity changes due

I. Tachtsidis (✉) • A. Papaioannou
Department of Medical Physics & Bioengineering, University College London, Malet Place Engineering Building, London, UK
e-mail: i.tachtsidis@ucl.ac.uk

S. Van Huffel et al. (eds.), *Oxygen Transport to Tissue XXXV*, Advances in Experimental Medicine and Biology 789, DOI 10.1007/978-1-4614-7411-1_13,

to the activation task. The fNIRS haemodynamic changes should occur at specific locations that overlay the cortical activated areas and should be closely coupled to the task-related timing periods. This assumes that the functional haemodynamic task-related changes are occurring on top of an unchanged global systemic and brain resting state. However, in certain functional experiments, these assumptions are not accurate and can lead to false positives in fNIRS [1]. We have previously used fNIRS to monitor the frontal and prefrontal cortex during anagram-solving tasks and observed significant task-related changes in mean blood pressure (MBP), heart rate (HR) and scalp blood flow (flux) that correlated with the fNIRS signals [2, 3]. In a recent study using multichannel fNIRS to produce maps of the haemodynamic response during anagram solving while simultaneously monitoring the systemic physiology, we observed a number of fNIRS channels to be highly correlated with activation-related systemic changes leading to false-positive cortical activation locations [1]. In our latest study utilising both fNIRS and functional magnetic resonance imaging (fMRI) and angiography during frontal activation tasks, we observed a significant correlation between the changes in HbO_2 and the systemic activation response of the deep scalp veins [4].

Several earlier studies used fNIRS to investigate the effect of videogames over the frontal and prefrontal lobe [5, 6]; however, none of these studies investigated the systemic changes during this type of activation task. The main aims of this study are to determine whether there are significant systemic changes during video gaming and if these changes are significantly associated with the fNIRS haemodynamic measurements.

13.2 Methods

We used two commercially available videogames for the Game Boy Advance SP (Nintendo Corp. Japan), a 'fighting' game (Final Fight One, Capcom) and a 'puzzle' game (Polarium Advance, Nintendo). The former is an arcade game where the player can choose a hero and fight against different enemies in order to complete specific missions. The latter is a very simple puzzle game, where the player has to flip black or white tiles on a square board in order to create horizontal rows of one colour and erase all the tiles in a single stroke to clear the board.

We studied two groups of healthy young volunteers, most of whom had some previous experience in video gaming. The first group ($n=17$, mean age 24 years) did the puzzle game and the second group ($n=13$, mean age 24 years) did the fighting game. These studies were approved by the Research Ethics Committee of UCL.

In order to become familiar with the experimental environment, the volunteers were given the game to practise for about 10 min. Following that, each subject sat in front of a desk on which a computer monitor was placed to alert the subjects via a visual stimulus when to rest and when to start playing the game. The protocol involved a single block of 5 min playing the game continuously (activation period) with a 2-min period of rest before and after the activation block.

A continuous-wave (CW) 8-channel fNIRS system, the Oxymon Mk III (Artinis Medical Systems BV, The Netherlands), was used. This system measures the changes in light attenuation at two wavelengths, 764 nm and 858 nm, and utilises the modified Beer-Lambert law with an age-dependent differential path length factor (DPF) to resolve the concentration changes in oxy (HbO_2)- and deoxy (HHb)-haemoglobin and calculate the changes in total haemoglobin (tHb) which is the summation of ΔHbO_2 and ΔHHb. The optode (source-detector fibre) configuration used in this study was the 8-channel split, which allows eight channel recordings with an inter-optode distance of 40 mm. The optode template was placed on the volunteer's forehead, using the international 10/20 system of electrode placement [7], such that (i) FP1 region was covered between light-emitting fibres (Tx2, Tx4a) and light-receiving fibre (Rx2), corresponding to channels 7 and 8, and (ii) FP2 region was covered between light-emitting fibres (Tx2, Tx3b) and light-receiving fibre (Rx1), corresponding to channels 5 and 6 (Fig. 13.1).

The Portapres system (Finapres Medical Systems) was employed, with the inflatable cuff placed on the index finger of the left hand, in order to measure the mean blood pressure (MBP) and the heart rate (HR). The laser Doppler (Moor Instruments) was used to measure the scalp blood flow (flux) with the laser probe placed on the forehead.

In order to locate the activation channels, the activation period was split in three separate epochs each one having a duration of 10 s. The first epoch was immediately after the beginning of the stimuli (120–129 s); the second was in the middle of the activation period (270–279 s) and the third was at the end of the activation (411–420 s). For each epoch, we calculate a mean value for all measurements and subtracted that from a 10-s mean calculated at the beginning of the rest period (1–10 s). The difference was then compared to zero using a Student's *t*-test to assess the significance ($p \leq 0.05$). We then defined activation as a significant increase in HbO_2, a significant decrease or no change in HHb and a significant increase in tHb [3]. Further, we estimated the correlation between the fNIRS and systemic signals to assess the relation of the brain haemodynamic and systemic signals. No correlation was defined as $0.25 > r > -0.25$.

13.3 Results

Figure 13.1 presents a summary of the percentage of subjects that demonstrated activation for each fNIRS channel. For the fighting game: in the first epoch 62 % of the subjects show activation over channel 8; in the second epoch, 46 % of the subjects show activation (equally) over both channel 5 and 8, while during the third epoch 39 % of the subjects show activation (equally) over both channels 5 and 8. For the puzzle game: in the first and second epochs, 65 % and 71 % of the subjects, respectively, demonstrated activation over channel 7, while in the third epoch, 65 % of the subjects demonstrated activation over channel 5 and only 31 % of the subjects show activation over channel 7.

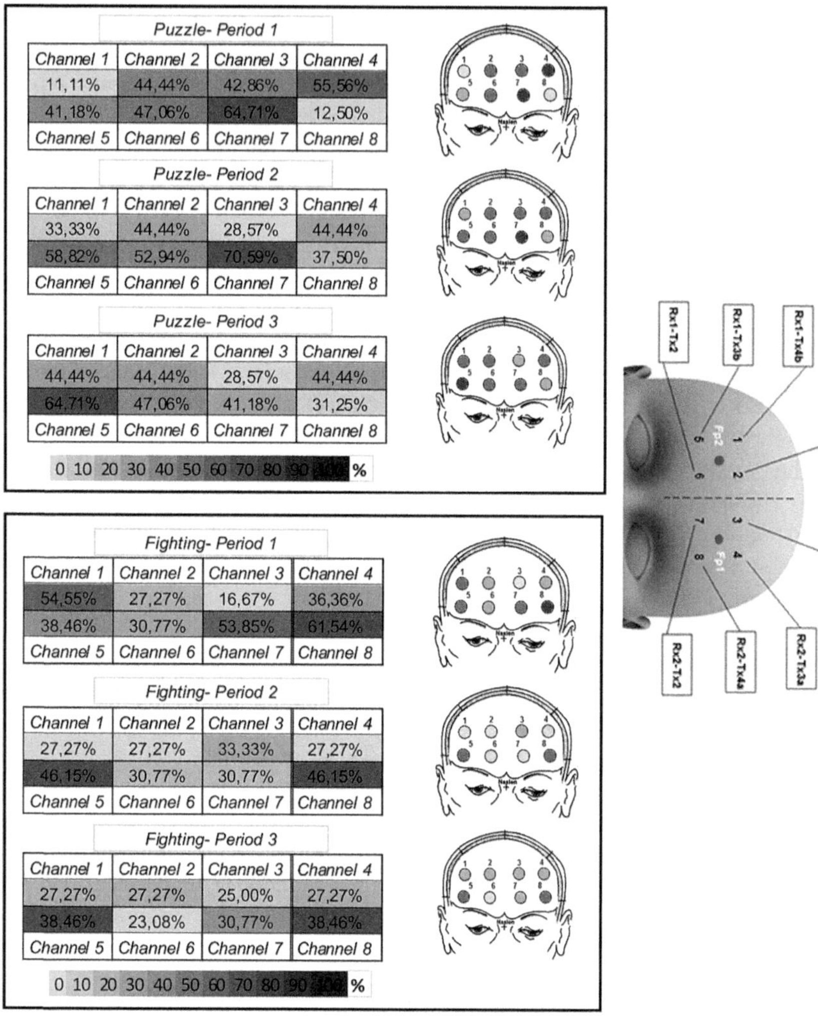

Puzzle- Period 1

Channel 1	Channel 2	Channel 3	Channel 4
11,11%	44,44%	42,86%	55,56%
41,18%	47,06%	64,71%	12,50%
Channel 5	Channel 6	Channel 7	Channel 8

Puzzle- Period 2

Channel 1	Channel 2	Channel 3	Channel 4
33,33%	44,44%	28,57%	44,44%
58,82%	52,94%	70,59%	37,50%
Channel 5	Channel 6	Channel 7	Channel 8

Puzzle- Period 3

Channel 1	Channel 2	Channel 3	Channel 4
44,44%	44,44%	28,57%	44,44%
64,71%	47,06%	41,18%	31,25%
Channel 5	Channel 6	Channel 7	Channel 8

Fighting- Period 1

Channel 1	Channel 2	Channel 3	Channel 4
54,55%	27,27%	16,67%	36,36%
38,46%	30,77%	53,85%	61,54%
Channel 5	Channel 6	Channel 7	Channel 8

Fighting- Period 2

Channel 1	Channel 2	Channel 3	Channel 4
27,27%	27,27%	33,33%	27,27%
46,15%	30,77%	30,77%	46,15%
Channel 5	Channel 6	Channel 7	Channel 8

Fighting- Period 3

Channel 1	Channel 2	Channel 3	Channel 4
27,27%	27,27%	25,00%	27,27%
38,46%	23,08%	30,77%	38,46%
Channel 5	Channel 6	Channel 7	Channel 8

Fig. 13.1 Group analysis shows the percentage of subjects that demonstrated activation in specific channels and epochs

Analysis of the systemic data showed that synchronous with the cerebral haemodynamic changes, significant systemic responses occurred as well. Individual analysis showed that during the fighting game, 77 % of the subjects showed a significant change in at least one of the systemic variables, with MBP and HR showing a consistent increase during the second and third epochs, while flux demonstrated an increase during all epochs. During the puzzle game, 88 % of the subjects showed at least one significant change in all systemic variables; however, MBP and HR varied between subjects and epochs, while flux showed a consistent increase across all epochs. Group analysis of the systemic data for the fighting game across subjects

Table 13.1 Mean (± standard deviation) values for monitored physiological variables during three phases of the experiment. Statistically significant values are highlighted $*p \leq 0.05$

	Puzzle			Fighting		
	1st epoch	2nd epoch	3rd epoch	1st epoch	2nd epoch	3rd epoch
Δ[MBP] (mmHg)	3.7(±9.6)	5.3±(12.4)	5.3±(11.3)	0.3(±7.3)	7.4(±14.1)*	11.8(±23.5)*
Δ[HR] (beats/min)	5.5(±19.8)	9.2±(22.5)	9.9±(25)	2.5(±5.9)	4.5(±8.1)*	9.9(±8.7)*
Δ[Flux] (no units)	4.4(±5.5)*	7.9±(9.2)*	7±(9.1)*	7.5(±11.3)*	12.5(±19)*	11.4(±17)*

Table 13.2 Correlation analysis with number representing the percentage of fNIRS channels

	MBP/HbO_2	MBP/HHb	HR/HbO_2	HR/HHb	$Flux/HbO_2$	Flux/HHb
	Puzzle game					
$r \geq 0.25$	57	26	41	22	70	26
$0.25 > r > -0.25$	34	36	47	53	24	41
$r \leq -0.25$	9	38	12	25	6	33
	Fighting game					
$r \geq 0.25$	40	20	42	14	46	9
$0.25 > r > -0.25$	42	44	41	52	46	55
$r \leq -0.25$	18	36	17	34	8	36

shows a significant positive increase for MBP and HR for the second and third epochs and a significant positive increase for the flux signal across all epochs (Table 13.1). Group analysis of the systemic data for the puzzle game across subjects shows no significant increases for MBP and HR and a significant positive increase for the flux signal across all epochs (Table 13.1).

Correlation analysis revealed significant correlations between the HbO_2 signal and systemic variables (Table 13.2). For the fighting game task, a significant correlation was seen at 40 % of channels between HbO_2 and MBP, 42 % of channels between HbO_2 and HR and 46 % of channels between HbO_2 and flux. For the puzzle game task, a significant correlation was seen at 57 % of channels between HbO_2 and MBP, 41 % of channels between HbO_2 and HR and 70 % of channels between HbO_2 and flux (Table 13.2).

13.4 Discussion

We found significant localised changes in HbO_2 and HHb measured over the frontal lobe during the gaming tasks. In addition, during the fighting game, our group analysis revealed significant changes in MBP and HR occurring at the middle and later periods of the activation block. We observed during both the fighting and puzzle games significant changes in the flux signal throughout the activation period. Correlation analysis between the HbO_2 and HHb with the systemic measurements revealed some individual and channel variability with most of the HbO_2 measurements correlating positively with flux in both the fighting and the puzzles game. There was a variation in the results among the volunteers between the two games. More haemodynamic activation patterns were seen during the puzzle game studies in the third epoch of the activation block compared to the other epochs, whereas in fighting game studies, these patterns were seen in the first period, just after the start of the activation block. In addition, during the fighting game studies, the percentage of activation patterns gradually decreases over time with the third epoch demonstrating the lowest percentage of volunteers showing activation.

Previous studies reported that playing videogames significantly increased systolic and diastolic pressure, heart rate and oxygen consumption in adolescents [8], in particular during the fighting game, which is a very dynamic game (the characters are moving, jumping and punching) that can possibly cause more areas in the brain to be activated (such as the motor cortex). The action involved in such a game can also cause emotional reactions in the subject. In a study exploring if playing a game that contains violence causes any sympathetic or parasympathetic reactions, the investigators concluded that a violent game causes different autonomic responses and affects heart rate and heart rate variability, which is a measure of stress reactivity [9]. In our studies, the high correlation coefficients found between fNIRS channels and systemic variables give an indication that some changes in fNIRS signals are due to changes in systemic variables and not due to haemodynamic changes originating from specific regions of the frontal lobe. Therefore, during analysis of brain activation during tasks such as video gaming, the contribution of the systemic changes should be taken into consideration.

Acknowledgments The authors would like to thank the Wellcome Trust (088429/Z/09/Z).

References

1. Tachtsidis I, Leung TS, Chopra A, Koh PH, Reid CB, Elwell CE (2009) False positives in functional near-infrared topography. Adv Exp Med Biol 645:307–314
2. Tachtsidis I, Leung TS, Devoto L, Delpy DT, Elwell CE (2008) Measurement of frontal lobe functional activation and related systemic effects: a near-infrared spectroscopy investigation. Adv Exp Med Biol 614:397–403
3. Tachtsidis I, Leung TS, Tisdall MM et al (2008) Investigation of frontal cortex, motor cortex and systemic haemodynamic changes during anagram solving. Adv Exp Med Biol 614:21–28
4. Kirilina E, Jelzow A, Heine A et al (2012) The physiological origin of task-evoked systemic artefacts in functional near infrared spectroscopy. Neuroimage 61:70–81
5. Nagamitsu S, Nagano M, Yamashita Y, Takashima S, Matsuishi T (2006) Prefrontal cerebral blood volume patterns while playing video games–a near-infrared spectroscopy study. Brain Dev 28:315–321
6. Matsuda G, Hiraki K (2006) Sustained decrease in oxygenated hemoglobin during video games in the dorsal prefrontal cortex: a NIRS study of children. Neuroimage 29(3):706–711

7. Homan RW, Herman J, Purdy P (1987) Cerebral location of international 10–20 system electrode placement. Electroencephalogr Clin Neurophysiol 66(4):376–382
8. Segal KR, Dietz WH (1991) Physiologic responses to playing a video game. Am J Dis Child 145(9):1034–1036
9. Ivarsson M, Anderson M, Akerstedt T, Lindblad F (2009) Playing a violent television game affects heart rate variability. Acta Paediatr 98(1):166–172

Chapter 14
Effect of Valsalva Maneuver-Induced Hemodynamic Changes on Brain Near-Infrared Spectroscopy Measurements

Atsuhiro Tsubaki, Sho Kojima, Adriane Akemi Furusawa, and Hideaki Onishi

Abstract Near-infrared spectroscopy (NIRS) is widely used to measure human brain activation on the basis of cerebral hemodynamic response. However, a limitation of NIRS is that systemic changes influence the measured signals. The purpose of this study was to clarify the relationship between NIRS signals and blood pressure during the Valsalva maneuver. Nine healthy volunteers performed a 20-s Valsalva maneuver to change their blood pressure. Changes in oxyhemoglobin (O_2Hb) concentration were measured with 34 channels with an inter-optode distance of 30 mm for deep-penetration measurements (deepO_2Hb) and 9 channels with an inter-optode distance of 15 mm for shallow-penetration measurements (shallowO_2Hb). The difference value (diffO_2Hb) between deepO_2Hb and shallowO_2Hb was calculated. Mean arterial pressure (MAP) was recorded by volume clamping the finger pulse, and skin blood flow changes were measured at the forehead. Pearson's correlation coefficients between deepO_2Hb and MAP, shallowO_2Hb and MAP, and diffO_2Hb and MAP were 0.893 ($P<0.01$), 0.963 ($P<0.01$), and 0.831 ($P<0.01$), respectively. The results suggest that regional and systemic changes in the cardiovascular state strongly influence NIRS signals.

A. Tsubaki (✉) • H. Onishi
Institute for Human Movement and Medical Sciences, Niigata University of Health and Welfare, 1398 Shimami-cho, Kita-ku, Niigata-shi, Niigata 950-3198, Japan

Department of Physical Therapy, Niigata University of Health and Welfare, 1398 Shimami-cho, Kita-ku, Niigata-shi, Niigata 950-3198, Japan
e-mail: tsubaki@nuhw.ac.jp

S. Kojima
Institute for Human Movement and Medical Sciences, Niigata University of Health and Welfare, 1398 Shimami-cho, Kita-ku, Niigata-shi, Niigata 950-3198, Japan

Graduate School of Health and Welfare, Niigata University of Health and Welfare, 1398 Shimami-cho, Kita-ku, Niigata-shi, Niigata 950-3198, Japan

A.A. Furusawa
Department of Physical Therapy, Niigata University of Health and Welfare, 1398 Shimami-cho, Kita-ku, Niigata-shi, Niigata 950-3198, Japan

S. Van Huffel et al. (eds.), *Oxygen Transport to Tissue XXXV*, Advances in Experimental Medicine and Biology 789, DOI 10.1007/978-1-4614-7411-1_14,

14.1 Introduction

Near-infrared spectroscopy (NIRS) is widely used to reveal brain activation patterns. This technique can noninvasively measure real-time hemodynamic changes related to cortical neural activity [1–3]. NIRS is advantageous in that it can measure neural activation during dynamic movements and has allowed researchers to visualize cortical activation patterns of human gait [4], balance control [5], and locomotor recovery after stroke [6–8].

In order to detect cortical activation, NIRS measures the concentrations of oxyhemoglobin and deoxyhemoglobin in tissue based on their differential absorption at multiple wavelengths by using the modified Beer-Lambert law [3]. Near-infrared beams are transmitted through the scalp and skull, and measured signals might have task-related cardiovascular responses expressed in the perfusion of extracranial layers [9]. It was suggested that physiological signals arising from cardiac, respiratory, and blood pressure modulations interfere with measurement of the hemodynamic response to brain stimulation [10, 11]. For example, visual stimulation causes blood pressure fluctuations that have confounding effects on brain NIRS [9].

Many techniques were established to resolve this problem. Morren [12] reported the use of independent component analysis to specifically detect neuronal tissue signal compared with other physiological signals. Kohno et al. [13] also used independent component analysis to remove the skin blood flow (SBF) artifact during a locomotor task on a treadmill. Saager and Berger [14] used two-detector paradigms in which measurements were made at two different distance detector fibers from a single source fiber. By clarifying the effects of blood pressure changes on the relationship between oxyhemoglobin (O_2Hb) concentration changes measured by two different inter-optode distances, we can use further analysis to investigate brain activation by using NIRS.

The purpose of this study was to clarify the relationship between changes in O_2Hb concentration and blood pressure by using two-detector measurement during the Valsalva maneuver (VM), which is a blood pressure fluctuation task.

14.2 Methods

Nine healthy volunteers (mean ± SD; age, 21.4 ± 1.0 years; height, 164.8 ± 10.2 cm; weight, 56.0 ± 8.3 kg, five females) participated in this study. All subjects were free of any known cardiovascular or respiratory diseases and were not taking any medications. Each subject received verbal and written explanations of the study objectives, measurement techniques, and risks and benefits associated with the investigation. The study was approved by the Ethics Committee of Niigata University of Health and Welfare (17157-100203) and conformed to the standards set by the Declaration of Helsinki.

Subjects were seated on a chair in a quiet room and were instructed to perform the VM by plugging their nose, closing their mouth, and attempting to expire air for

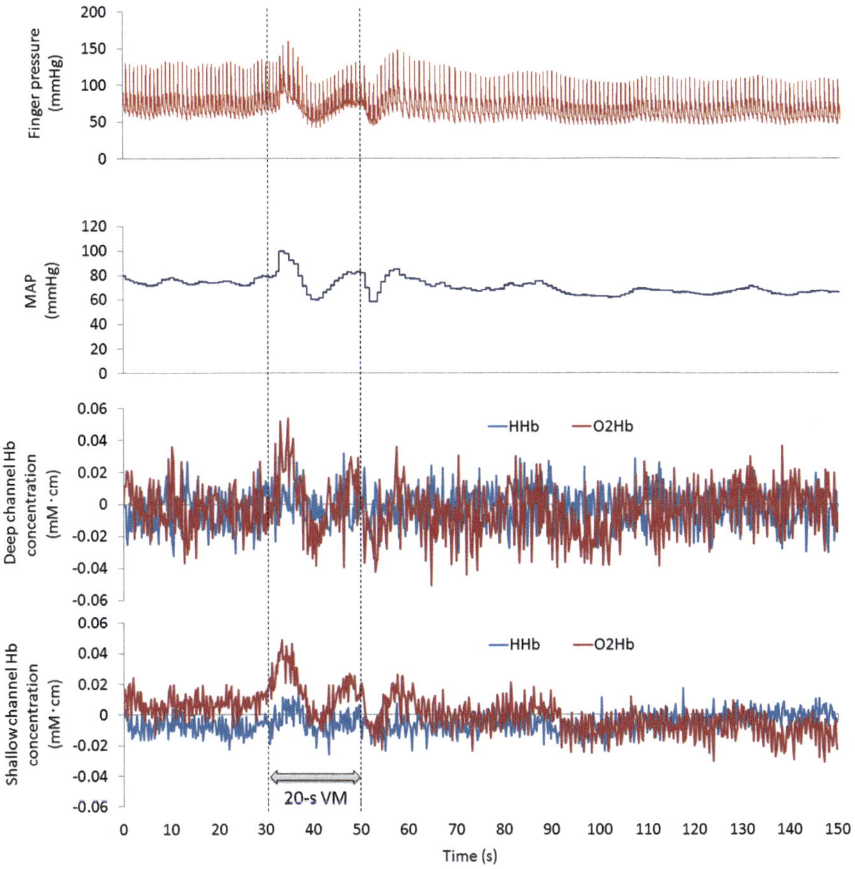

Fig. 14.1 Changes in pulse pressure, MAP, deep-channel hemoglobin (Hb) concentration, and shallow-channel Hb concentration during the 20-s Valsalva maneuver

20 s. The task was preceded by an initial resting phase of 30 s in order to measure baseline conditions. After VM, a 100-s resting phase was used to ensure that physiological parameters recovered to the initial resting state (Fig. 14.1).

A multichannel NIRS imaging system (OMM-3000; Shimadzu Co., Kyoto, Japan) with three wavelengths (780, 805, and 830 nm) was used to take measurements. NIRS optodes were set in a holder at a distance of 30 mm and stabilized by a crepe bandage around the head. The NIRS array covered the left temporoparietal scalp (Fig. 14.2). The Cz position of the international 10–20 system was used to ensure consistent optode placement among all subjects. Changes in O_2Hb concentration were calculated by using the modified Beer-Lambert law on line at a sampling rate of 190 ms. Thirty-four channels with an inter-optode distance of 30 mm were used for deep-penetration measurements (deepO_2Hb), and 9 channels with an inter-optode distance of 15 mm were used for shallow-penetration measurements

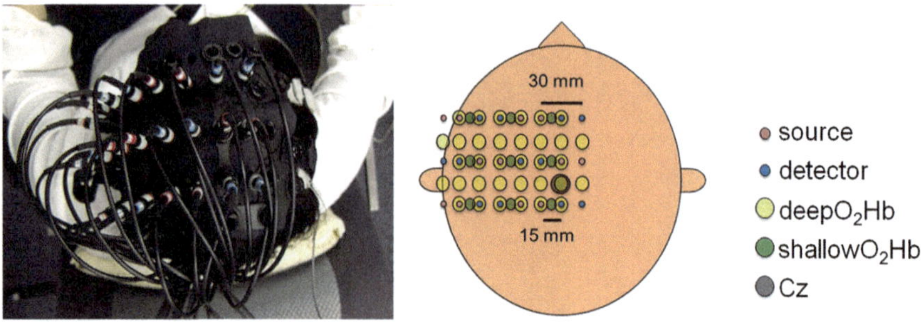

Fig. 14.2 NIRS optode placement and schematic locations of deepO_2Hb and shallowO_2Hb channels

(shallowO_2Hb) using 12 light source fibers and 12 detection fibers. The measured values were used to calculate the difference (diffO_2Hb) between deepO_2Hb and shallowO_2Hb.

Beat-to-beat systolic blood pressure, mean arterial pressure (MAP), and heart rate were recorded by volume clamping the finger pulse with a finger photoplethysmograph (Finometer; Finapres Medical Systems, Amsterdam, The Netherlands) on the left middle finger. A height reference sensor was used to remove confounds related to arm position. SBF changes were measured at the forehead by using a laser Doppler blood flow meter (Omegaflow FLO-CI; Omegawave Inc., Osaka, Japan). These analog data were converted into digital data with an A/D converter (PowerLab; AD Instruments, Australia) at a 1,000-Hz sampling rate.

We calculated diffO_2Hb from the resting phase average. Then, the diffO_2Hb values for deepO_2Hb and shallowO_2Hb and the grand average of deepO_2Hb, shallowO_2Hb, and diffO_2Hb were calculated. MAP and SBF data were down sampled adapting a sampling rate of NIRS measurement. The relationships between deepO_2Hb, shallowO_2Hb, diffO_2Hb, MAP, and SBF during the VM phase were assessed with Pearson's correlation coefficients with significance set at $P<0.05$.

14.3 Results

During the 20-s VM, MAP increased by 20 mmHg 5 s into the task. Then, MAP decreased by 25 mmHg. MAP was lowest 12 s after VM began, after which time it increased slightly. The changes in deepO_2Hb, shallowO_2Hb, and diffO_2Hb were similar to that observed for MAP (Fig. 14.3).

MAP was strongly correlated with shallowO_2Hb ($r=0.963$, $P<0.01$), deepO_2Hb ($r=0.893$, $P<0.01$), and diffO_2Hb ($r=0.831$, $P<0.01$). SBF was correlated with shallowO_2Hb ($r=0.500$, $P<0.01$) and MAP ($r=0.550$, $P<0.01$). The highest correlation coefficient was found between shallowO_2Hb and MAP.

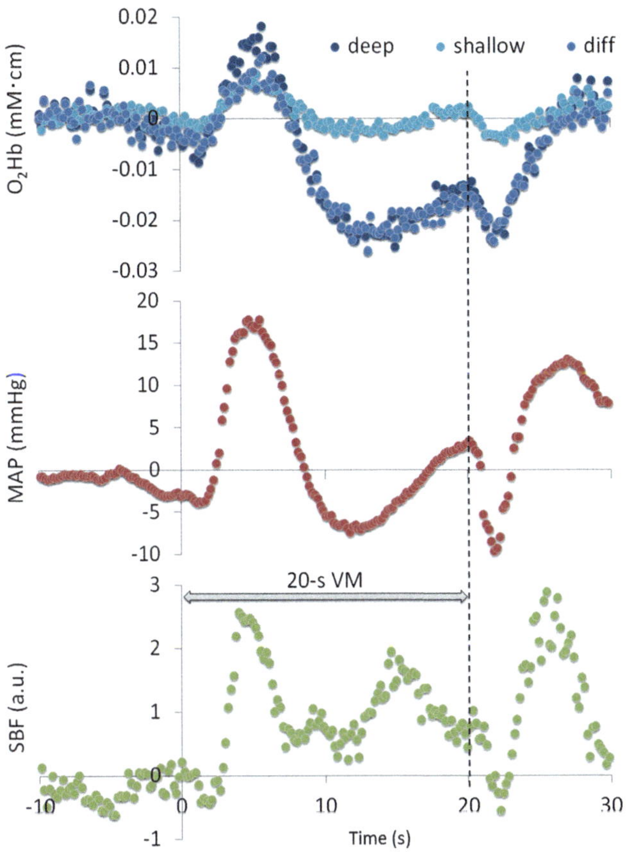

Fig. 14.3 Changes in deepO_2Hb, shallowO_2Hb, diffO_2Hb, MAP, and SBF during the 20-s VM

14.4 Discussion

The main finding of this study was that O_2Hb measured by NIRS and MAP are strongly correlated and that shallowO_2Hb and MAP showed the highest correlation coefficient during VM.

The two different inter-optode distances used in this study, 15 and 30 mm, are acceptable for measuring shallow and deep penetration, respectively. Saager and Berger [14] employed near detector probes placed 5–15 mm from the source probe to measure scalp oxygenation and far detector probes placed 30–40 mm from the source probe to measure brain tissue oxygenation. Minati et al. [9] also used two inter-optode distances to divide shallow and deep penetration during a visual stimulation task. Similarly, Saager et al. [15] utilized two detectors at 5 and 32 mm and corrected NIRS based on their signals. Gagnon et al. [16] placed optodes at 10 and 30 mm over the left motor region of each subject.

The Pearson's correlation coefficients suggest that blood pressure fluctuations might affect O_2Hb changes measured by NIRS. Changes in arterial blood pressure can be identified as phases I to IV [17, 18]. In our study, MAP changes could also be identified in this way. The trend observed for deepO_2Hb, shallowO_2Hb, and diffO_2Hb was quite similar to that noted for MAP changes. Furthermore, we observed a strong relationship between MAP and deepO_2Hb, shallowO_2Hb, and diffO_2Hb. Minati et al. [9] established that the correlation coefficient between shallowO_2Hb from the superficial occipital cortex with an inter-optode distance of 10 mm and MAP was 0.6 during visual stimulation. Kirilina et al. [18] showed that the physiological origin of the systemic artifact in NIRS signals is a task-evoked sympathetic arterial vasoconstriction. The VM can elicit more drastic blood pressure changes compared to visual stimulation, and SBF may also fluctuate along with blood pressure changes. DeepO_2Hb also had a strong relationship with MAP, suggesting that blood pressure changes influenced the intracranial layer. During VM, middle cerebral artery blood velocity changes assessed by transcranial Doppler ultrasound were similar to those noted for MAP [19]. Therefore, O_2Hb in deep penetration might also correlate with MAP.

Respiratory-induced elevations in the partial pressures of arterial carbon dioxide can alter vasodilatation and increase cerebral blood flow [20]. In a future study, we would like to confirm the relationship between the partial pressures of arterial carbon dioxide and O_2Hb during VM.

In conclusion, regional and systemic cardiovascular state changes strongly influence NIRS signals. When we employ tasks that induce blood pressure changes, monitoring blood pressure is necessary to detect and interpret physiological signals.

Acknowledgments This work was supported by a Grant-in-Aid for Exploratory Research from the Niigata University of Health and Welfare.

References

1. Tamura M, Hoshi Y, Okada F (1997) Localized near-infrared spectroscopy and functional optical imaging of brain activity. Philos Trans R Soc Lond B Biol Sci 352(1354):737–742
2. Obrig H, Wenzel R, Kohl M et al (2000) Near-infrared spectroscopy: does it function in functional activation studies of the adult brain? Int J Psychophysiol 35(2–3):125–142
3. Boas DA, Gaudette T, Strangman G, Cheng X, Marota JJ, Mandeville JB (2001) The accuracy of near infrared spectroscopy and imaging during focal changes in cerebral hemodynamics. Neuroimage 13(1):76–90
4. Miyai I, Tanabe HC, Sase I et al (2001) Cortical mapping of gait in humans: a near-infrared spectroscopic topography study. Neuroimage 14(5):1186–1192
5. Mihara M, Miyai I, Hatakenaka M, Kubota K, Sakoda S (2008) Role of the prefrontal cortex in human balance control. Neuroimage 43(2):329–336
6. Miyai I, Yagura H, Oda I et al (2002) Premotor cortex is involved in restoration of gait in stroke. Ann Neurol 52(2):188–194
7. Miyai I, Yagura H, Hatakenaka M, Oda I, Konishi I, Kubota K (2003) Longitudinal optical imaging study for locomotor recovery after stroke. Stroke 34(12):2866–2870

8. Miyai I, Suzuki M, Hatakenaka M, Kubota K (2006) Effect of body weight support on cortical activation during gait in patients with stroke. Exp Brain Res 169(1):85–91
9. Minati L, Kress IU, Visani E, Medford N, Critchley HD (2011) Intra- and extra-cranial effects of transient blood pressure changes on brain near-infrared spectroscopy (NIRS) measurements. J Neurosci Methods 197(2):283–288
10. Franceschini MA, Fantini S, Thompson JH, Culver JP, Boas DA (2003) Hemodynamic evoked response of the sensorimotor cortex measured noninvasively with near-infrared optical imaging. Psychophysiology 40(4):548–560
11. Franceschini MA, Joseph DK, Huppert TJ, Diamond SG, Boas DA (2006) Diffuse optical imaging of the whole head. J Biomed Opt 11(5):054007
12. Morren G, Wolf U, Lemmerling P et al (2004) Detection of fast neuronal signals in the motor cortex from functional near infrared spectroscopy measurements using independent component analysis. Med Biol Eng Comput 42(1):92–99
13. Kohno S, Miyai I, Seiyama A et al (2007) Removal of the skin blood flow artifact in functional near-infrared spectroscopic imaging data through independent component analysis. J Biomed Opt 12(6):062111
14. Saager R, Berger A (2008) Measurement of layer-like hemodynamic trends in scalp and cortex: implications for physiological baseline suppression in functional near-infrared spectroscopy. J Biomed Opt 13(3):034017
15. Saager RB, Telleri NL, Berger AJ (2011) Two-detector Corrected Near Infrared Spectroscopy (C-NIRS) detects hemodynamic activation responses more robustly than single-detector NIRS. Neuroimage 55(4):1679–1685
16. Gagnon L, Cooper RJ, Yücel MA, Perdue KL, Greve DN, Boas DA (2012) Short separation channel location impacts the performance of short channel regression in NIRS. Neuroimage 59(3):2518–2528
17. Tiecks FP, Lam AM, Matta BF, Strebel S, Douville C, Newell DW (1995) Effects of the valsalva maneuver on cerebral circulation in healthy adults. A transcranial Doppler Study. Stroke 26(8):1386–1392
18. Kirilina E, Jelzow A, Heine A et al (2012) The physiological origin of task-evoked systemic artefacts in functional near infrared spectroscopy. Neuroimage 61:70–81
19. Pott F, van Lieshout JJ, Ide K, Madsen P, Secher NH (2000) Middle cerebral artery blood velocity during a valsalva maneuver in the standing position. J Appl Physiol 88(5):1545–1550
20. Ainslie PN, Ogoh S (2010) Regulation of cerebral blood flow in mammals during chronic hypoxia: a matter of balance. Exp Physiol 95(2):251–262

Chapter 15
Effect of Maternal use of Labetalol on the Cerebral Autoregulation in Premature Infants

Alexander Caicedo, Liesbeth Thewissen, Gunnar Naulaers, Petra Lemmers, Frank van Bel, and Sabine Van Huffel

Abstract Hypertensive disorders of pregnancy (HDP) are normally treated to avoid maternal complications. In this study we aimed to investigate if there was an effect of maternal HDP treatment on the cerebral autoregulation of the neonates by analysing measurements of mean arterial blood pressure (MABP) and $rScO_2$ by means of correlation, coherence, and transfer function analysis. We found that these infants presented higher values of transfer function gain, which indicates impaired cerebral autoregulation, with a decreasing trend towards normality. We hypothesised that this trend was due to a vasodilation effect of the maternal use of labetalol due to accumulation, which disappeared by the third day after birth. Therefore, we investigated the values of pulse pressure in order to find evidence for a vasodilatory effect. We found that lower values of pulse pressure were present in these infants when compared with a control population, which, together with increased transfer function gain values, suggests an effect of the drug on the cerebral autoregulation.

A. Caicedo (✉) • S. Van Huffel
Department of Electrical Engineering, ESAT/SCD, KU Leuven, Leuven, Belgium

iMinds Future Health Department, Leuven, Belgium
e-mail: acaicedo@esat.kuleuven.be

L. Thewissen • G. Naulaers
Neonatal Intensive Care Unit, University Hospitals Leuven, KU Leuven, Leuven, Belgium

P. Lemmers • F. van Bel
Department of Neonatology, University Medical Center, Wilhelmina Children's Hospital, Utrecht, The Netherlands

S. Van Huffel et al. (eds.), *Oxygen Transport to Tissue XXXV*, Advances in Experimental Medicine and Biology 789, DOI 10.1007/978-1-4614-7411-1_15,

15.1 Introduction

Hypertension is the most common medical disorder encountered during pregnancy and is estimated to occur in about 6–8 % of pregnancies [1]. Hypertensive disorders of pregnancy (HDP) should be treated in order to prevent maternal complications and improve fetal maturity by permitting prolongation of pregnancy and minimising fetal exposure to possible adverse effects of antihypertensive treatment. There are several advantages and disadvantages of HDP treatment [1]. In pregnancy longer than 34 weeks, induction of labour in the occurrence of hypertension and pre-eclampsia is generally considered the best treatment to improve maternal and neonatal outcome. However, there is considerable morbidity in late preterms explained by the mode of delivery and gestational age [2]. Labetalol is a selective α-1- (peripheral vasodilation) and non-selective β-receptor agonist (preventing reflex tachycardia and maintaining cardiac output), which is often used in HDP treatment. Due to its lipophilic properties, it easily passes the placental barrier, which is in essence a lipid membrane. Hypotension, bradycardia and hypoglycaemia are possible neonatal side effects but may also occur in (preterm) infants regardless of labetalol exposure. Labetalol's half-life in adults is approximately 6 h but accumulation occurs. However, half-life after maternal use in a preterm baby with clinical signs of β-blockage was 24 h [3]. Conflicting evidence exists for specific neonatal side effects described after the use of labetalol for maternal hypertension. Nevertheless oral and intravenous labetalol is used as a first- or second-line treatment in HDP due to its highly effective antihypertensive properties and because it has a better profile than hydralazine and other β-blockers [4, 5]. Scarce information on neonatal cerebral haemodynamics in gestational hypertension and pre-eclampsia is available. However, not much is known about the true influence of maternal use of labetalol on the neonatal haemodynamic parameters (bradycardia, hypotension) in the brain, mainly since cerebral fetal circulation is subject to changes due to brain sparing in severe pre-eclampsia. We aimed to investigate labetalol-induced effects on neonatal cerebral autoregulation mechanisms during the first 3 days of life.

15.2 Methods

15.2.1 Data

The study was performed in 56 infants from the Wilhelmina's Children's Hospital Utrecht (the Netherlands), with a gestational age of 29 (24.7–31.9) weeks and a birth weight of 960 (540–1,585) grams. In all infants peripheral oxygen saturation (SaO_2) was measured continuously by pulse oximetry and mean arterial blood pressure (MABP) by an indwelling arterial catheter. With NIRS, regional oxygen saturation ($rScO_2$) was continuously and noninvasively recorded using the INVOS4100 (Somanetics). MABP, SaO_2 and NIRS signals were simultaneously measured during the first 3 days of life. From the 56 infants, 16 correspond to control subjects, and 40 correspond to the group of mothers who were treated for HDP. From the HDP group, 21 neonates correspond to mothers treated with labetalol (HDP+Lab), and 19 correspond to mothers with other treatment (HDP-Lab). The subjects from the three groups were matched for gestational age, birth weight and gender.

15.2.2 Signal Analysis

Artefacts shorter than 30 s were removed and corrected by interpolation using robust least-squares support vector machines for function estimation [6]. Artefacts longer than 30 s were truncated. Remaining artefacts, if any, were removed manually. Hence, a single continuous measurement was replaced by a set of continuous artefact-free segments. The resulting signals were filtered with a mean average filter and then downsampled to 1 Hz in order to obtain a common sampling frequency.

15.2.3 Mathematical Tools

Cerebral autoregulation was assessed by means of correlation, coherence and transfer function analysis between MABP and $rScO_2$. The correlation, coherence and transfer function scores were calculated using a time-sliding window of length 15 min and overlapping time of 1 min. Coherence and transfer function analysis were performed using the Welch method for the calculation of the respective cross-power and auto-power spectral densities. This method involves a further segmentation of the signals into 5-min epochs with an overlapping of 4.5 min. The average of the coefficients in the frequency ranges 0.003–0.02 Hz (very low frequency range: VLF), 0.02–0.05 Hz (low frequency range: LF) and 0.05–0.1 Hz (high frequency range: HF) was calculated [7] for further analysis. The transfer function was calculated with the following formula:

$$H(f) = \frac{G_{io}(f)}{G_{ii}(f)}$$

where $G_{io}(f)$ represents the input–output cross-power spectrum and $G_{ii}(f)$ represents the input auto-power spectrum. In addition, the pulse pressure was calculated as the difference between the systolic and diastolic blood pressure measurements. Figure 15.1 shows a representative set of measurements for a control subject.

15.2.4 Statistical Analysis

To assess whether the scores and pulse pressure values were different between the populations, the non-parametric Kruskal-Wallis test was used due to the lack of

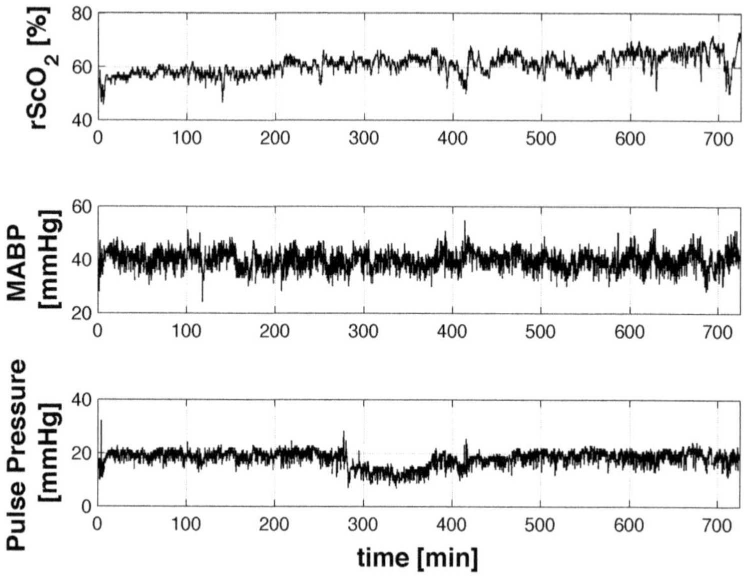

Fig. 15.1 Measurements of $rScO_2$, MABP and pulse pressure for one of the subjects during the first day of life

normality in the data distributions. The statistical analysis was performed using the statistics toolbox from MATLAB. All reported p-values were two-tailed, and a nominal p-value < 0.05 was considered as statistically significant.

15.3 Results

When comparing the correlation, coherence and gain values for the three different populations, taking the complete measurements for the first 3 days of life per group, no statistically significant differences between the scores were found. However, when the analysis was performed in a day-by-day basis, the HDP+Lab presented higher values of gain during the first day of life, in the VLF and LF bands, when compared with the gain values for the Control and the HDP-Lab groups ($p < 0.05$). In addition, the gain values during the first day of life, in the VLF and LF bands, for the HDP+Lab population were higher than the values in the second and third days ($p < 0.05$). This behaviour indicates a progression towards normality. Figure 15.2 shows the gain for a representative subject from the HDP+Lab group. Correlation, coherence and phase were still not statistically significant.

Pulse pressure values were lower for the HDP+Lab group when compared with the other groups. In addition, for the three populations the pulse pressure values were lower during the first day of life and presented an increased profile in the second and third days. Figure 15.3 shows the median values of pulse pressure for the

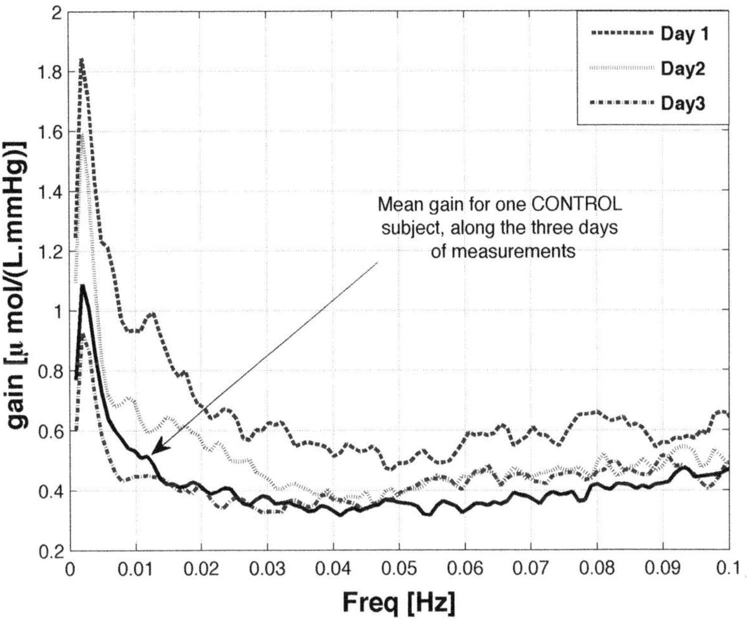

Fig. 15.2 Gain values for a representative subject from the HDP+Lab group. Gain values during the first, the second and third days are shown together with the values from a control subject for comparison

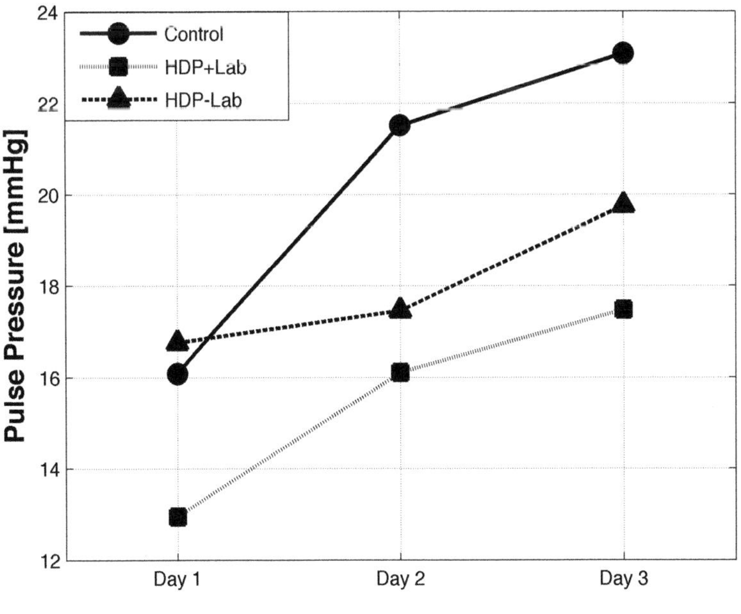

Fig. 15.3 Median values of pulse pressure for the three different groups, Control, HDP+Lab and HDP-Lab, during the first 3 days of life

Table 15.1 Pulse pressure values for the different populations during the first 3 days of life (all values are given in mmHg)

	Day 1	Day 2	Day 3
Control	16.08 (5.97–27.54)	21.50 (10.25–28.47)	23.07 (15.73–32.64)
HDP+Lab	12.94 (9.32–20.17)	16.10 (8.24–25.11)	17.47 (7.67–27.11)
HDP-Lab	16.76 (9.05–23.18)	17.45 (9.87–27.34)	19.76 (10.04–27.87)

three different populations and its evolution during the first 3 days of life. In Table 15.1 the median values of pulse pressure and its range of variation (minimum–maximum values) are indicated.

15.4 Discussion and Conclusion

In this study we found that the maternal use of labetalol induces high values of the transfer function gain between MABP and $rScO_2$. These high values are concomitants with low values of pulse pressure. In [8] low values of pulse pressure indicate the presence of vasodilation due to the use of drugs for hypertension treatment. However, labetalol is a non-selective β-antagonist; its action also reduces cardiac contractility which is reflected as a reduction in pulse pressure. When comparing pulse pressure values from the HDP+Lab and HDP-Lab groups, a stronger reduction is observed in the HDP+Lab group; this may be caused by a combination of vasodilation and reduction in cardiac contractility. We hypothesised that vasodilation was present in this group due to the concomitant reduction in pulse pressure and increase in transfer function gain values. Vasodilation reduces the effect of the myogenic mechanism responsible for cerebral autoregulation, which increases the transfer function gain. This vasodilation may be produced by the accumulation of labetalol due to its maternal use [9]. Interestingly, a trend towards normality can be seen in the gain and pulse pressure values. The gain values return to normality by the end of the third day, while the pulse pressure values were still lower than normal by then. This may be due to a decrease in labetalol. This vasodilatory effect might also be caused by brain sparing. In [10] higher mean values in cerebral blood flow velocity were reported in children with evidence of brain sparing. However, no differences were found in the amount of neonates that presented brain sparing between the HDP+Lab and HDP-Lab groups.

Correlation, coherence and transfer function phase did not present differences between the studied populations. This may be due to the fact that impaired cerebral autoregulation is reflected not only by a coupled dynamics between MABP and cerebral blood flow but also by the strength of this relation. This strength is assessed by the gain of the transfer function but not by the correlation and coherence scores. Moreover, in a recent study carried out in our group [11], we found that among the aforementioned methods, transfer function analysis is the most robust method for cerebral autoregulation assessment. This result is in agreement with the results provided by [12].

In conclusion, there is evidence indicating that labetalol induces changes in the cerebral autoregulation mechanism of the neonates possibly due to its accumulation. This accumulation of labetalol might produce vasodilation, which leads to high values of gain, impairing the myogenic mechanism of cerebral autoregulation. Further studies are necessary to evaluate whether this phenomenon also has an effect on the later neurological outcome of the patients.

Acknowledgments The authors would like to thank the Research Council of KUL, GOA MaNet, PFV/10/002 (OPTEC), IDO 08/013 Autism, several PhD/postdoc and fellow grants; the Flemish Government, FWO (PhD/postdoc grants and projects – G.0427.10N (Integrated EEG-fMRI), G.0108.11 (Compressed Sensing) and G.0869.12N (Tumour imaging)); IWT (TBM070713-Accelero, TBM080658-MRI (EEG-fMRI), TBM110697-NeoGuard, PhD grants), IBBT; Flanders Care (Demonstratieproject Tele-Rehab III (2012–2014)); the Belgian Federal Science Policy Office (IUAP P7/(DYSCO, 'Dynamical systems, control and optimization', 2012–2017); the ESA AO-PGPF-01, PRODEX (CardioControl) C4000103224; and EU (RECAP 209G within INTERREG IVB NWE programme), EU HIP Trial FP7-HEALTH/2007–2013 (n° 260777) and EU ITN Transact 2012.

References

1. Moser M, Brown CM, Rose CH, Garovic VD (2012) Hypertension in pregnancy: is it time for a new approach to treatment? J Hypertens 30(6):1092–1100
2. Langenveld J, Ravelli AC, van Kaam AH et al (2011) Neonatal outcome of pregnancies complicated by hypertensive disorders between 34 and 37 weeks of gestation: a 7-year retrospective analysis of a national registry. Am J Obstet Gynecol 205(6):540.e1–540.e7
3. Haraldsson A, Geven W (1989) Half-life of maternal labetalol in a premature infant. Pharm Weekbl Sci 11(6):229–231
4. Magee LA, Cham C, Waterman EJ, Ohlsson A, von Dadelszen P (2003) Hydralazine for treatment of severe hypertension in pregnancy: meta-analysis. BMJ 327(7421):955–960
5. Magee LA, Elran E, Bull SB, Logan A, Koren G (2000) Risks and benefits of beta-receptor blockers for pregnancy hypertension: overview of the randomized trials. Eur J Obstet Gynecol Reprod Biol 88(1):15–26
6. Caicedo A, Van Huffel S (2010) Weighted LS-SVM for function estimation applied to artifact removal in biosignal processing. In: Proceedings of the 32nd annual international conference of the IEEE Engineering in Medicine and Biology Society (EMBC 2010), Buenos Aires, Argentina, 31 Aug-4 Sept, pp 988–991
7. Wong F, Leung T, Austin T et al (2008) Impaired autoregulation in preterm infants identified by using spatially resolved spectroscopy. Pediatrics 121(3):604–611
8. Karpanou EA, Vyssoulis GP, Stefanadis CI, Cokkinos DV (2006) Differential pulse pressure response to various antihypertensive drug families. J Hum Hypertens 20(10):765–771
9. Purkayastha S, Saxena A, Eubank WL, Hoxha B, Raven PB (2013) Alpha-1 adrenergic receptor control of the cerebral vasculature in humans at rest and during exercise. Exp Physiol 98(2):451–461
10. Scherjon SA, Oosting H, Kok JH, Zondervan HA (1994) Effect of fetal brainsparing on the early neonatal cerebral circulation. Arch Dis Child Fetal Neonatal Ed 71(1):F11–F15
11. Caicedo A, Naulaers G, Lemmers P, van Bel F, Wolf M, Van Huffel S (2012) Detection of cerebral autoregulation by near-infrared spectroscopy in neonates: performance analysis of measurement methods. J Biomed Opt 17(11):117003
12. Hahn GH, Heiring C, Pryds O, Greisen G (2011) Applicability of near-infrared spectroscopy to measure cerebral autoregulation noninvasively in neonates: a validation study in piglets. Pediatr Res 70(2):166–170

Chapter 16
Brain Tissue Oxygen Saturation Increases During the Night in Adolescents

Andreas Jaakko Metz, Fiona Pugin, Reto Huber, Peter Achermann, and Martin Wolf

Abstract How does the oxygen metabolism change during sleep? We aimed to measure the change in brain tissue oxygen saturation (StO_2) before and after sleep with near-infrared spectroscopy (NIRS) using an in-house developed sensor.

P. Achermann and M. Wolf shared senior authorship.

A.J. Metz (✉)
Biomedical Optics Research Laboratory, Division of Neonatology, University Hospital Zurich, Frauenklinikstrasse 10, 8091 Zurich, Switzerland

Institute for Biomedical Engineering, ETH Zurich, Zurich, Switzerland

Member of the PhD program imMed, University Zurich, Zurich, Switzerland
e-mail: andreas.metz@usz.ch

F. Pugin
Child Development Center, University Children's Hospital Zurich, Zurich, Switzerland

Member of the PhD program imMed, University Zurich, Zurich, Switzerland

R. Huber
Child Development Center, University Children's Hospital Zurich, Zurich, Switzerland

Zurich Center for Integrative Human Physiology, University Zurich, Winterthurerstrasse 190, 8057 Zurich, Switzerland

P. Achermann
Institute for Pharmacology and Toxicology, Chronobiology and Sleep Research, University of Zurich, Winterthurerstrasse 190, 8057 Zurich, Switzerland

Zurich Center for Integrative Human Physiology, University Zurich, Winterthurerstrasse 190, 8057 Zurich, Switzerland

M. Wolf
Biomedical Optics Research Laboratory, Division of Neonatology, University Hospital Zurich, Frauenklinikstrasse 10, 8091 Zurich, Switzerland

Institute for Biomedical Engineering, ETH Zurich, Zurich, Switzerland

Zurich Center for Integrative Human Physiology, University Zurich, Winterthurerstrasse 190, 8057 Zurich, Switzerland

S. Van Huffel et al. (eds.), *Oxygen Transport to Tissue XXXV*, Advances in Experimental Medicine and Biology 789, DOI 10.1007/978-1-4614-7411-1_16,

According to the synaptic homeostasis hypothesis [1], synaptic downscaling during sleep would result in reduced energy consumption. Thus, this reduced energy demands should be reflected in the oxygen metabolism and StO_2. Thirteen nights of 7 male subjects (age 11–16 years, one subject contributed only one night, all others two) were included in the analysis. We performed NIRS measurements throughout the entire night. The NIRS sensor was placed close to electrode position Fp1 (international 10/20 system), over the left frontal cortex. Absolute StO_2 and total haemoglobin (tHb) were calculated from the NIRS measurements using a self-calibrating method [2]. StO_2 and tHb during the awake period prior to sleep and after awakening were compared. The subjects were instructed to lie in bed in the same position before and after sleep. Values of the two nights were averaged for each subject. Furthermore, a linear regression line was fit through the all-night StO_2 recordings. We found an increase in StO_2 by 4.32 ± 1.76 % (mean ± SD, paired t-test $p<0.001$, $n=7$) in the morning compared to evening, while tHb did not change (1.02 ± 6.81 μM $p=0.704$, $n=7$). Since the tHb remained at a similar level after sleep, this increase in StO_2 indicates that in the morning more oxygenated blood and less deoxygenated blood was present in the brain compared to the evening. The slope of the regression line was 0.37 ± 0.13 % h^{-1} leading to a similar increase of StO_2 in the course of sleep. This may be interpreted as a reduced oxygen consumption or energy metabolism after sleep.

16.1 Introduction

How does the oxygen metabolism change in the course of sleep? While different studies investigated specific events during sleep, e.g., sleep apnoea [3], differences between sleep stages [4] or wakefulness and sleep [5], the temporal evolution of the oxygen metabolism in the course of sleep is not clear. According to the synaptic homeostasis hypothesis [1], a popular hypothesis about the function of sleep, oxygen demand should be reduced during sleep, and hence brain metabolism should be decreased. A PET study [6] showed that global cerebral blood flow (CBF) after sleep was reduced compared to measurement before sleep onset. In an ultrasound Doppler study in six healthy young males, blood flow velocity in the middle cerebral artery decreased by 6.6 % between pre- and post-sleep measurements [7]. Global CBF measured with PET could be a marker for oxygen consumption [8], and in that sense the reduced global CBF may indicate a reduced energy consumption. However, this relation is not yet established and in order to obtain absolute CBF values, arterial blood sampling is required. This makes PET rather impractical for sleep studies, in particular in children, which will also not be able to sleep in one position throughout the entire night. Near-infrared spectroscopy (NIRS) provides a non-invasive way to indirectly measure the oxygen metabolism which is related to the cerebral tissue oxygen saturation via the link of cytochrome C oxidase [9]. Because of its non-invasive nature and the small size of the sensor, the subjects are free to move during sleep and the discomfort is reduced to a minimum. Hence, NIRS may be an interesting tool to investigate the oxygen metabolism in sleep. The aim of the current study was to investigate sleep-related changes in StO_2 by comparing StO_2 before and after sleep. The synaptic homeostasis hypothesis [1] proposes a reduced energy consumption after sleep compared to before sleep, which should be reflected in StO_2.

16.2 Methods

16.2.1 Subjects

Data of seven healthy subjects (age 11–16 years, mean 13.6 years, all male) were recorded and analysed. Each subject spent two nights in the sleep laboratory at the Children's Hospital Zurich, separated by 3 weeks. Time in bed was approximately 8.5 h. Before the recordings in the sleep laboratory, the subjects had to keep a regular sleep–wake rhythm for at least three nights. The subjects had to fill in a sleep questionnaire and wore an activity monitor at the wrist of the nondominant arm. This allowed us to check compliance with the instructions of keeping a regular sleep–wake rhythm. The last three days before the night, they were not allowed to consume caffeine-containing products. The study was approved by the ethical committee of the Canton of Zurich, and informed consent was obtained from the legal representatives.

16.2.2 Protocol

The near-infrared spectroscopy (NIRS) sensor was placed at the left forehead, close to the position of electrode Fp1, international 10/20 system [10]. Additionally to the NIRS measurement, high-density EEG recordings (128 electrode EEG net, Electrical Geodesics, Inc.) were performed during the night. EEG data are not presented here. In addition to the continuous measurements during the night, we measured 2 min prior to light out and 2 min after awakening. We instructed the subject to lie on his back, looking at the ceiling, and not to move. With an accelerometer (ADXL330, Analog Devices) built into the sensor, we were able to exclude errors resulting from wrong head positions or head movements.

16.2.3 NIRS Measurement

NIRS measurements were performed with an in-house built NIRS device, the *OxyPrem*, which is similar to previous wireless sensors [11]. It measures light attenuation at 760, 805 and 870 nm, at distances 1.5 and 2.5 cm. With this sensor we were able to calculate StO_2 for two different regions using the multi-distance method [2].

Region one is covering an area of approximately 3 cm^2 and was closer to electrode F3 (just below the hairline for most subjects). Region two was covering the same area close to electrode Fp1. Only data of region one are reported.

16.2.4 Post Processing

To calculate StO_2 we use the simplified diffusion constant and did not account for water in tissue, as described in assumption A4 in [12]. This approach is based on the diffusion equation for a semi-infinite medium and a point source and implies the assumption $r(3\mu_a\mu_s') >> 1$. Here r denotes the distance between source and detector; μ_a and μ_s' are the absorption and the reduced scattering coefficients, respectively.

For the scattering coefficients of the brain, we used the values published by Matcher et al. [13] and thus were able to obtain total (tHb), oxygenated (O_2Hb) and deoxygenated haemoglobin (HHb). The relative change in tHb may be an indicator for the change in blood volume. The relation between StO_2 and tHb is given by $StO_2 = O_2Hb/tHB$ and $tHb = O_2Hb + HHb$. To exclude errors in StO_2 resulting from unintended movement of the subject, the accelerometer data were checked. By visual inspection, only those parts were included in the analysis with values around −0.3 and −0.7 g in the y- and z-axis of the accelerometer, respectively. In this position the subject was lying on the back. The constant g represents the earth's gravity ($\approx$9.81 ms^{-2}). The included parts per measurement were averaged to obtain one value in StO_2 and tHb for statistical analysis. To estimate the change of StO_2 over the night, we calculated a linear regression and investigated the slope (%/h; see Fig. 16.2). Sleep stages were visually scored according to standard criteria.

16.2.5 Statistics

For statistical analysis we averaged the two nights per subject, leading to seven evening–morning StO_2 and tHb pairs. These were compared by paired t-tests. All processing was performed by MATLAB® (R2009b and R2011b, the Mathworks®, Natick, Massachusetts, USA).

16.3 Results

We found a significant increase in StO_2 of 4.32 ± 1.76 % (mean ± SD, $p < 0.001$, $n = 7$) in the morning compared to the evening. Mean StO_2 prior to sleep was 69.43 ± 2.02 % and 73.76 ± 2.36 % after awakening. The mean change in tHb was 1.02 ± 6.81 μM ($p = 0.704$, $n = 7$), which was not significant. Individual data are

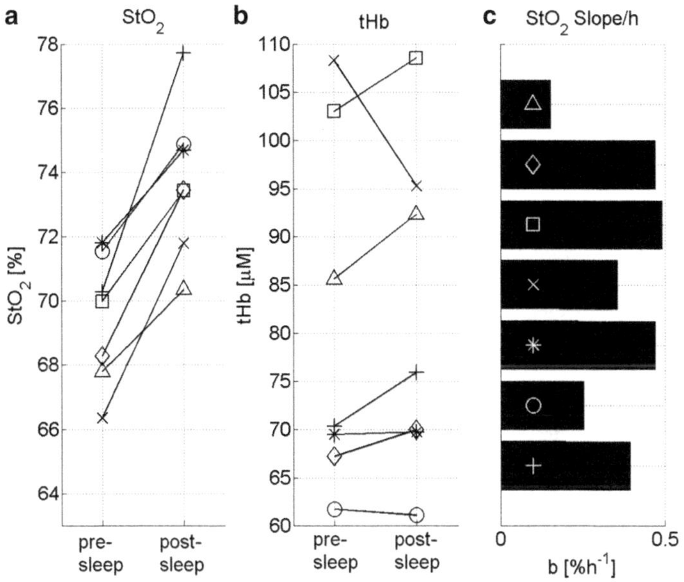

Fig. 16.1 (**a**) Brain tissue oxygen saturation (StO_2) increased in the morning after sleep (post-sleep), compared to the evening before sleep (pre-sleep). The values of the two nights were averaged except for subject (+) which only contributed with one night. (**b**) Change of total haemoglobin (tHb) from evening to morning. On average no change was observed. (**c**) Slope (b) of the linear regression for StO_2 for the whole night (see Fig. 16.2). A positive slope was observed in all subjects and all nights. The *different symbols* indicate the seven subjects

shown in Fig. 16.1. Furthermore, the O_2Hb increased by 4.27 ± 4.46 μM ($p<0.05$, $n=7$) and HHb decreased by 3.25 ± 3.01 μM ($p<0.05$, $n=7$).

The mean linear increase over the night was 0.37 ± 0.13 % h^{-1} (Figs. 16.1 and 16.2). The percentage increase indicates an absolute rather than a relative increase with respect to the baseline value.

16.4 Discussions and Conclusion

In adolescent subjects we observed an increase in StO_2 in wakefulness post-sleep compared to wakefulness pre-sleep. This increase was confirmed when fitting a regression line through all-night StO_2 measurements. Since tHb did not change overnight, this increase in StO_2 may indicate that in the morning more oxygenated and less deoxygenated blood was present in the brain compared to the evening. The oxygen metabolic rate can be expressed as a function of CBF, arterial oxygen saturation (SaO_2) and cerebral tissue oxygen saturation [14], and therefore our finding might be interpreted as a reduced oxygen consumption and thus energy

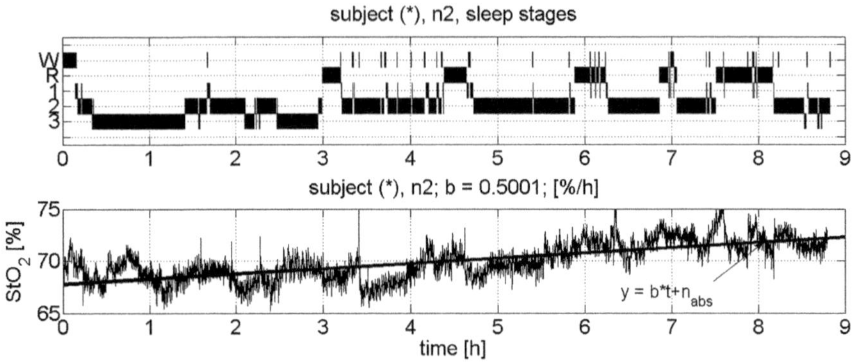

Fig. 16.2 *Top*: Visually scored sleep stages for one night of subject (*) (*W* waking, *R* REM sleep, *1–3* non-REM sleep stages N1–N3). *Bottom*: Corresponding time course of StO_2. A linear regression line was fitted through the data. The slope (b) of the increase during the night is indicated at the *top of the panel*. Please note that the b value here is given for the individual night, while the b values in Fig. 16.1 are averaged over two nights

metabolism (linked by the oxidative phosphorylation, [15], Chap. 19) after sleep. This interpretation requires SaO_2 and the CBF to be constant. Since tHb overall remains constant, the cerebral blood volume (CBV) remains constant and hence does the CBF, which is related to the CBV [16]. Since we did not measure the SaO_2, we cannot be sure whether this assumption holds; however, the subjects were healthy and no sleep apnoeas were observed during the night, which makes the assumption more plausible. A reduced energy metabolism would be in line with the synaptic homeostasis hypothesis [1], which states that synaptic downscaling during the night would lead to a reduced energy demand of the brain. Alternatively, the increased StO_2 could be linked to circadian effects (independent of sleep), e.g., cortisol rhythm which exerts its wake-promoting effect in the early morning hours [17]. As can be seen in Fig. 16.2, the assumed linear trend describes the StO_2 increase during sleep fairly accurate regarding the whole-night changes. But on a shorter timescale, fluctuations were evident. At this point we speculate that these fluctuations are movement induced on the one hand and related to changes in sleep on the other.

In summary, in adolescents we found an increase in cerebral tissue oxygen saturation in the course of sleep, which may represent a reduced oxygen consumption of the brain and therefore a lower energy metabolism. Thus, our data are in line with the synaptic homeostasis hypothesis [1].

Acknowledgments This work was supported by the University Research Priority Program "Integrative Human Physiology", University of Zurich, Switzerland. The authors would like to thank Urs Bachofner and Madlaina Stauffer for their great support with data collection. Furthermore, Andreas Metz would like to thank Raphael Zimmermann for fruitful discussions.

References

1. Tononi G, Cirelli C (2006) Sleep function and synaptic homeostasis. Sleep Med Rev 10(1):49–62
2. Hueber DM, Fantini S, Cerussi AE et al (1999) New optical probe designs for absolute (self-calibrating) NIR tissue hemoglobin measurements. Proc SPIE 3597:618–631
3. Pizza F, Biallas M, Wolf M, Werth E, Bassetti CL (2010) Nocturnal cerebral hemodynamics in snorers and in patients with obstructive sleep apnea: a near-infrared spectroscopy study. Sleep 33(2):205–210
4. Näsi T, Virtanen J, Noponen T, Toppila J, Salmi T, Ilmoniemi RJ (2011) Spontaneous hemodynamic oscillations during human sleep and sleep stage transitions characterized with near-infrared spectroscopy. PLoS One 6(10):e25415
5. Spielman AJ, Zhang G, Yang CM et al (2000) Intracerebral hemodynamics probed by near infrared spectroscopy in the transition between wakefulness and sleep. Brain Res 866(1–2): 313–325
6. Braun AR, Balkin TJ, Wesenten NJ et al (1997) Regional cerebral blood flow throughout the sleep-wake cycle. An H2(15)O PET study. Brain 120(Pt 7):1173–1197
7. Kuboyama T, Hori A, Sato T et al (1997) Changes in cerebral blood flow velocity in healthy young men during overnight sleep and while awake. Electroencephalogr Clin Neurophysiol 102(2):125–131
8. Maquet P (2000) Functional neuroimaging of normal human sleep by positron emission tomography. J Sleep Res 9(3):207–231
9. Tachtsidis I, Tisdall MM, Leung TS et al (2009) Relationship between brain tissue haemodynamics, oxygenation and metabolism in the healthy human adult brain during hyperoxia and hypercapnea. Adv Exp Med Biol 645:315–320
10. Jasper HH (1958) Formal discussion: dendrites. Electroencephalogr Clin Neurophysiol Suppl 35(Supp 10):42–50
11. Muehlemann T, Haensse D, Wolf M (2008) Wireless miniaturized in-vivo near infrared imaging. Opt Express 16(14):10323–10330
12. Metz AJ, Biallas M, Jenny C, Muehlemann T, Wolf M (2013) The effect of basic assumptions on the tissue oxygen saturation value of near infrared spectroscopy. Adv Exp Med Biol 765: 169–175
13. Matcher SJ, Cope M, Delpy DT (1997) In vivo measurements of the wavelength dependence of tissue-scattering coefficients between 760 and 900 nm measured with time-resolved spectroscopy. Appl Opt 36(1):386–396
14. Elwell CE, Henty JR, Leung TS et al (2005) Measurement of CMRO2 in neonates undergoing intensive care using near infrared spectroscopy. Adv Exp Med Biol 566:263–268
15. Nelson DL, Lehninger AL, Cox MM (2008) Lehninger principles of biochemistry. WH Freeman Limited, New York
16. Grubb RL Jr, Raichle ME, Eichling JO, Ter-Pogossian MM (1974) The effects of changes in PaCO2 on cerebral blood volume, blood flow, and vascular mean transit time. Stroke 5(5): 630–639
17. Cajochen C, Chellappa S, Schmidt C (2010) What keeps us awake? The role of clocks and hourglasses, light, and melatonin. Int Rev Neurobiol 93:57–90

Chapter 17
Changes of Cerebral Oxygen Metabolism and Hemodynamics During ECPR with Hypothermia Measured by Near-Infrared Spectroscopy: A Pilot Study

Tsukasa Yagi, Ken Nagao, Kaoru Sakatani, Tsuyoshi Kawamorita, Taketomo Soga, Kimio Kikushima, Kazuhiro Watanabe, Eizo Tachibana, Yoshiteru Tominaga, Katsushige Tada, Ishii Mitsuru, Nobutaka Chiba, Kei Nishikawa, Masakazu Matsuzaki, Harumi Hirose, Atsuo Yoshino, and Atsushi Hirayama

Abstract (Background) The 2010 CPR Guidelines recommend that extracorporeal cardiopulmonary resuscitation (ECPR) using an emergency cardiopulmonary bypass (CPB) should be considered for patients with cardiac arrest. However, it is not yet clear whether this therapy can improve cerebral circulation and oxygenation in these patients. To clarify this issue, we evaluated changes of cerebral blood oxygenation (CBO) during ECPR using near-infrared spectroscopy (NIRS). (Methods) We employed NIRS to measure CBO in the bilateral frontal lobe in patients transported to the emergency room (ER) after out-of-hospital cardiac arrest between November 2009 and June 2011. (Results) Fifteen patients met the above criteria.

T. Yagi, M.D. (✉)
Department of Cardiology, Cardiopulmonary Resuscitation and Emergency Cardiovascular Care, Surugadai Nihon University Hospital, 1-8-13 Kanda Surugadai, Chiyoda-ku, Tokyo 101-8309, Japan

Department of Emergency and Critical Care Medicine, Surugadai Nihon University Hospital, Tokyo, Japan
e-mail: ygt0108@livedoor.com

Department of Cardiology, Kawaguchi Municipal Medical Center, Kawaguchi, Japan

K. Nagao
Department of Cardiology, Cardiopulmonary Resuscitation and Emergency Cardiovascular Care, Surugadai Nihon University Hospital, 1-8-13 Kanda Surugadai, Chiyoda-ku, Tokyo 101-8309, Japan

K. Sakatani
Department of Electrical and Electronics Engineering, Nihon University College of Engineering, Koriyama, Japan

T. Kawamorita • Y. Tominaga • K. Tada • I. Mitsuru • N. Chiba
Department of Emergency and Critical Care Medicine, Surugadai Nihon University Hospital, Tokyo, Japan

S. Van Huffel et al. (eds.), *Oxygen Transport to Tissue XXXV*, Advances in Experimental Medicine and Biology 789, DOI 10.1007/978-1-4614-7411-1_17,

The tissue oxygenation index (TOI) on arrival at the ER was 36.5 %. This increased to 67.8 % during ECPR ($P<0.001$). The one patient whose TOI subsequently decreased had a favorable neurological outcome. (Conclusion) Increase of TOI during ECPR might reflect an improvement in cerebral blood flow, while decrease of TOI after ECPR might reflect oxygen utilization by the brain tissue as a result of neuronal cell survival. NIRS may be useful for monitoring cerebral hemodynamics and oxygen metabolism during CPR.

17.1 Introduction

Cardiac arrest is a major public health problem worldwide. Despite decades of efforts to promote cardiopulmonary resuscitation (CPR) science and education, the survival rate for out-of-hospital cardiac arrest remains low [1–3]. In Japan, the SOS-KANTO study showed that a favorable neurological outcome at 30 days was extremely rare in patients with out-of-hospital cardiac arrest who arrived at the emergency hospital in cardiac arrest [3–6]. The 2010 CPR Guidelines indicated that organized post-cardiac arrest care with an emphasis on multidisciplinary programs that focus on optimizing hemodynamic, neurologic, and metabolic function (including therapeutic hypothermia (TH)) may improve survival to hospital discharge among victims who achieve return of spontaneous circulation (ROSC) following cardiac arrest either in- or out-of-hospital [2]. TH is one intervention that has been shown to improve outcome for comatose adult victims of witnessed out-of-hospital

T. Soga • K. Kikushima • K. Watanabe • K. Nishikawa • M. Matsuzaki • H. Hirose
Department of Cardiology, Cardiopulmonary Resuscitation and Emergency Cardiovascular Care, Surugadai Nihon University Hospital, 1-8-13 Kanda Surugadai, Chiyoda-ku, Tokyo 101-8309, Japan

Department of Emergency and Critical Care Medicine, Surugadai Nihon University Hospital, Tokyo, Japan

E. Tachibana
Department of Cardiology, Cardiopulmonary Resuscitation and Emergency Cardiovascular Care, Surugadai Nihon University Hospital, 1-8-13 Kanda Surugadai, Chiyoda-ku, Tokyo 101-8309, Japan

Department of Cardiology, Kawaguchi Municipal Medical Center, Kawaguchi, Japan

A. Yoshino
Department of Emergency and Critical Care Medicine, Surugadai Nihon University Hospital, Tokyo, Japan

Division of Neurosurgery, Department of Neurological Surgery, Nihon University School of Medicine, Tokyo, Japan

A. Hirayama
Division of Cardiology, Department of Internal Medicine, Nihon University School of Medicine, Tokyo, Japan

cardiac arrest when the presenting rhythm was ventricular fibrillation (VF) [7, 8]. The 2010 CPR Guidelines recommended that comatose adult patients with ROSC after out-of-hospital VF cardiac arrest should be cooled to 32–34 °C for 12–24 h (Class I) [2] and that extracorporeal CPR (ECPR) using an emergency cardiopulmonary bypass (CPB) should be considered for patients with inhospital cardiac arrest when the duration of the no-flow arrest was brief, and the condition leading to the cardiac arrest was reversible or amenable to heart transplantation or revascularization (Class IIb) [2]. Since 1996, we have performed ECPR using emergency CPB with TH, followed by percutaneous coronary intervention (PCI) if necessary, on patients who arrive at the emergency room (ER) in cardiac arrest [9, 10]. Our preliminary study indicated that early attainment of a core temperature of 34 °C during cardiac arrest had neurological benefits for patients with out-of-hospital cardiac arrest who underwent CPB and PCI [10]. However, it is not yet clear whether this therapy can improve cerebral circulation and oxygenation in these patients. To clarify this issue, we evaluated changes of cerebral blood oxygenation (CBO) during ECPR using near-infrared spectroscopy (NIRS). NIRS, an optical technique, is an attractive tool for this purpose because it allows noninvasive, continuous measurements of CBO changes with high time resolution and is easy to use [11, 12].

17.2 Methods

17.2.1 Patients

Between November 2009 and June 2011, we employed NIRS (NIRO-200NX, Hamamatsu Photonics, Japan) to measure CBO in the bilateral frontal lobe in patients transported to the ER after out of hospital cardiac arrest. The patients were included in a prospective observational study. They were enrolled in this study when they met the following criteria: aged 18–74 years, cardiac arrest witnessed by bystanders, presumed cardiac etiology of cardiac arrest according to the Utstein style guidelines [13], defibrillation using automated external defibrillator by bystander and/or emergency medical personnel, and persistent cardiac arrest on arrival at the ER [10]. Exclusion criteria were a tympanic-membrane temperature below 30 °C on arrival at the ER, successful ROSC within 10 min of arrival at the ER with conventional ALS; non-cardiac etiology of cardiac arrest; or pregnancy. Patients were also excluded if their families refused to give informed consent for participation in this study [10].

17.2.2 Procedures

Our treatment protocol of extracorporeal CPR for induction of hypothermia with PCI is shown in Fig. 17.1 [10]. On arrival at the ER, the attending physicians

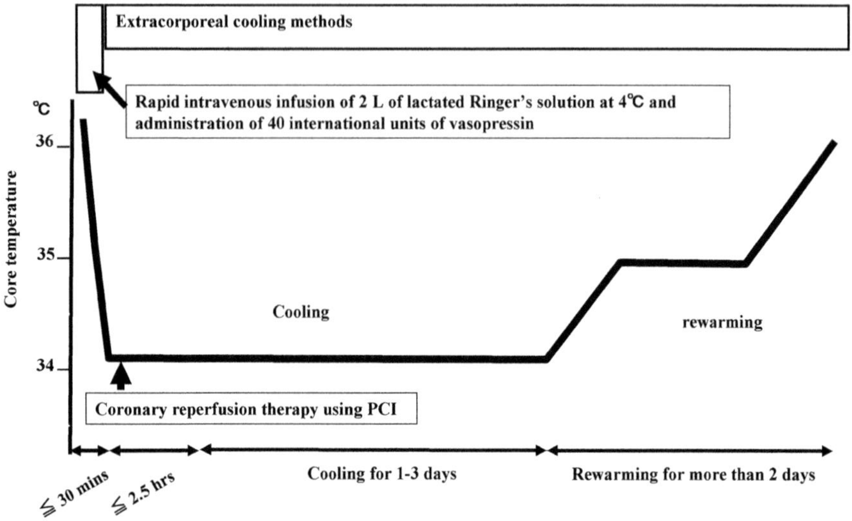

Fig. 17.1 Protocol of extracorporeal cardiopulmonary resuscitation (*CPR*) for induction of hypothermia with percutaneous coronary intervention (*PCI*). On arrival of the patient at the emergency room (*ER*), extracorporeal CPR using emergency cardiopulmonary bypass plus intra-aortic balloon pumping was immediately performed. Subsequently, emergency coronary angiography with PCI was performed in cases of suspected acute coronary syndrome

assessed as soon as possible whether a patient was eligible for this study after conventional ALS, including rapid intravenous infusion of 2 L of lactated Ringer's solution at 4 °C and the administration of 40 international units (IU) of vasopressin, and employed NIRS to measure CBO in the bilateral frontal lobe in patients. CPB plus intra-aortic balloon pumping was initiated when ROSC could not be achieved within 10 min of arrival. After implementation of CPB plus intra-aortic balloon pumping, emergency coronary angiography was performed during cardiac arrest in cases of suspected acute coronary syndrome (ACS). Subsequently, coronary reperfusion therapy using PCI during extracorporeal CPR was performed immediately.

17.2.3 Statistical Analysis

All analyses were performed using the SPSS software package (version 16.0 J SPSS, Chicago, IL, USA). Continuous variables are expressed as mean ± SD. Differences in the mean levels of tissue oxygenation index (TOI) between arrival at the ER, administration of 40 IU of vasopressin, and implementation of CPB were tested by Mann–Whitney *U* test for unpaired values with two-tailed *P* values of <0.05.

17.3 Results

During the study period, 15 patients met the above criteria. Characteristics of these patients are presented in Table 17.1. The mean age was 57.0 ± 12.6 years. The proportion of male patients was 93.3 %. The mean time interval from collapse to the implementation of CPB was 51.6 ± 16.6 min, and the mean time interval from arrival at the ER to implementation of CPB was 17.9 ± 7.7 min. The proportion of patients due to ACS was 66.6 %.

The TOI on arrival at the ER was 36.5 ± 7.0 % (Fig. 17.2). This increased to 42.3 ± 6.9 % following administration of 40 IU of vasopressin ($P < 0.001$) and to 67.8 ± 7.9 % following implementation of CPB ($P < 0.001$). Moreover, oxyhemoglobin (Oxy-Hb) increased by 14.1 ± 6.3 μM, and deoxyhemoglobin (Deoxy-Hb) decreased by 16.8 ± 8.4 μM. After the implementation of CPB, the one patient whose TOI subsequently decreased had a favorable neurological outcome at 30 days after cardiac arrest (Fig. 17.3).

17.4 Discussion

This study shows that ECPR with TH can improve cerebral circulation and oxygenation in these patients. In cardiac arrest patients, clinically relevant recovery depends strongly on the restoration of cerebral function, which in turn depends on two major

Table 17.1 Baseline characteristics of the study populations

Characteristics	Patients ($n = 15$)
Age (years)	57.0 ± 12.6
Male sex (no. (%))	14 (93.3)
Prehospital treatment (no. (%))	
Defibrillations	14 (93.3)
Administration of intravenous epinephrine	7 (46.7)
Initial cardiac rhythm	
VF/pulseless VT	14 (93.3)
PEA	1 (6.7)
Asystole	0
Time interval (min)	
From collapse to implementation of CPB	51.6 ± 16.6
From arrival at the ER to implementation of CPB	17.9 ± 7.7
Cause of cardiac arrest	
Acute coronary syndrome	10 (66.6)
Cardiomyopathy	2 (13.3)
Others	3 (20.0)

VF/pulseless VT ventricular fibrillation/pulseless ventricular tachycardia, *PEA* pulseless electrical activity, *CPB* cardiopulmonary bypass, *ER* emergency room

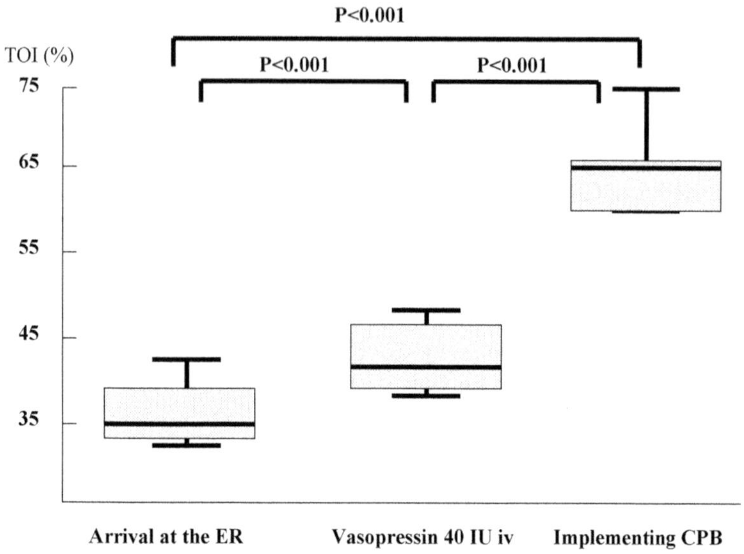

Fig. 17.2 The value of tissue oxygenation index (*TOI*) on arrival at the emergency room (*ER*) was 36.5 ± 7.0 %. This increased to 42.3 ± 6.9 % following administration of 40 IY of vasopressin ($P<0.001$) and to 67.8 ± 7.9 % following implementation of cardiopulmonary bypass (CPB) ($P<0.001$)

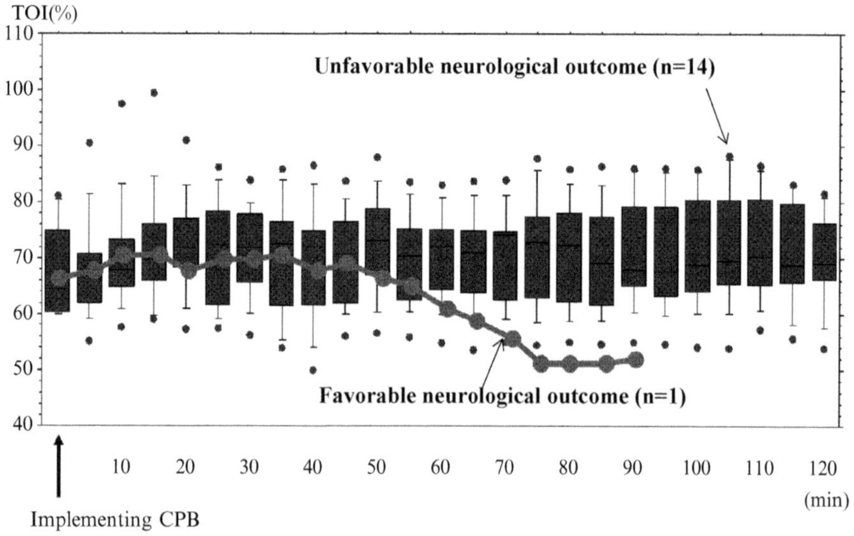

Fig. 17.3 After the implementation of cardiopulmonary bypass (*CPB*), the one patient whose tissue oxygenation index (*TOI*) subsequently decreased had a favorable neurological outcome at 30 days after cardiac arrest

factors, i.e., prevention or alleviation of global ischemia during the "no-flow time" and "low-flow (CPR) time" and minimization of post-reperfusion injury of the brain [14]. We speculated that the TOI on arrival at the ER may reflect the degree of global ischemia that persisted during the prehospital period. Increase of TOI by the implementation of CPB might reflect an improvement in cerebral blood flow, while decrease of TOI after the implementation of CPB might reflect oxygen utilization of the brain tissue as a result of neuronal cell survival. However, it should be noted that changes of TOI could be caused by other factors such as cerebral metabolic rate of oxygen ($CMRO_2$), arteriovenous (AV) volume ratio, and systemic oxygen saturation. Further studies are necessary to clarify these issues.

17.4.1 Study Limitations

There are several limitations to our study. First, it was not a multicenter study for resuscitation after out-of-hospital cardiac arrest. Second, our findings should be considered preliminary because of small sample size: there were only 15 patients in the present study, and there was only one patient with a favorable neurological outcome at 30 days after cardiac arrest.

17.5 Conclusions

We suggest that NIRS may be useful for monitoring cerebral hemodynamics and oxygen metabolism during ECPR.

References

1. Hazinski MF, Nolan JP, Billi JE et al (2010) International consensus on cardiopulmonary resuscitation and emergency cardiovascular care science with treatment recommendations. Circ J 122:S250–S605
2. Field JM, Hazinski MF, Sayre MR et al (2010) American Heart Association guidelines for cardiopulmonary resuscitation and emergency cardiovascular care. Circ J 122:S639–S946
3. SOS-KANTO Study Group (2007) Cardiopulmonary resuscitation by bystanders with chest compression only (SOS-KANTO): an observational study. Lancet 369(9565):920–926
4. SOS-KANTO Committee (2005) Incidence of ventricular fibrillation in patients with out-of-hospital cardiac arrest in Japan: survey of survivors after out-of-hospital cardiac arrest in Kanto area (SOSKANTO). Circ J 69(10):1157–1162
5. SOS-KANTO Study Group (2009) Comparison of arterial blood gases of laryngeal mask airway and bag-valve-mask ventilation in out-of-hospital cardiac arrests. Circ J 73(3):490–496
6. SOS-KANTO Study Group (2011) Atropine sulfate for patients with out-of-hospital cardiac arrest due to asystole and pulseless electrical activity. Circ J 75(3):580–588

7. The Hypothermia After Cardiac Arrest Study Group (2002) Mild therapeutic hypothermia to improve the neurologic outcomes after cardiac arrest. N Engl J Med 346:549–556
8. Berrard SA, Gray TW, Buist MD et al (2002) Treatment of comatose survivors of out-of-hospital cardiac arrest with induced hypothermia. N Engl J Med 346(8):557–563
9. Nagao K, Hayashi N, Kanmatsuse K et al (2000) Cardiopulmonary cerebral resuscitation using emergency cardiopulmonary bypass, coronary reperfusion therapy and mild hypothermia in patients with cardiac arrest outside the hospital. J Am Coll Cardiol 36(3):776–783
10. Nagao K, Kikushima K, Watanabe K et al (2010) Early induction of hypothermia during cardiac arrest improves neurological outcomes in patients with out-of-hospital cardiac arrest who undergo emergency cardiopulmonary bypass and percutaneous coronary intervention. Circ J 74(1):77–85
11. Samra SK, Dorjie P, Zelenock GB, Stanley JC (1996) Cerebral oximetry in patients undergoing carotid endarterectomy under regional anesthesia. Stroke 27(1):49–55
12. Nakamura S, Kano T, Sakatani K et al (2009) Optical topography can predict occurrence of watershed infarction during carotid endarterectomy: technical case report. Surg Neurol 71(5):540–542
13. Cummins RO, Chamberlain DA, Abramson NS et al (1991) Recommended guidelines for uniform reporting of data from out-of-hospital cardiac arrest: the Utstein style – a statement for health professionals from a task force of the American Heart Association, the European Resuscitation Council, the Heart and Stroke Foundation of Canada, and the Australian Resuscitation Council. Circulation 84(2):960–975
14. Ito N, Nanto S, Nagao K, Hatanaka T, Nishiyama K, Kai T (2012) Regional cerebral oxygen saturation on hospital arrival is a potential novel predictor of neurological outcomes at hospital discharge in patients with out-of-hospital cardiac arrest. Resuscitation 83(1):46–50

Part III
Muscle Oxygenation

Chapter 18
Analysis of NIRS-Based Muscle Oxygenation Parameters by Inclusion of Adipose Tissue Thickness

Svenja Grieger, Dmitri Geraskin, André Steimers, and Matthias Kohl-Bareis

Abstract The assessment of muscle oxygenation by non-invasive near-infrared spectroscopy generally assumes a homogeneous medium, and this is flawed for large adipose tissue layers underneath the skin. Here we summarize the influence of the adipose tissue thickness on the oxygenation data, show that the adipose layer can be measured by NIRS and indicate a possible correction algorithm. Spectroscopic evidence suggests the usefulness of this algorithm, however, not in all subjects.

18.1 Introduction

One of the main applications of near-infrared spectroscopy (NIRS) in tissue optics is the measurement of muscle oxygenation which is especially valuable for exercise physiology [1, 2]. As most commercial NIRS monitors or lab set-ups are based on a limited number of light sources and detectors, the underlying assumptions are homogeneous optical properties (absorption, scattering) of the medium. This assumption certainly does not hold when there is lipid tissue covering the muscle. This adipose lipid tissue thickness (ATT) extends from < 1 mm up to values > 15 mm, depending on subject and muscle.

Our experimental set-up consists of a broadband spectroscopy system that records six reflectance spectra at a mean distance of 35 mm from the light source. Light delivering fibres were positioned on the site of different muscles of healthy subjects.

The collection of spectra in the range from around 600 to 1,000 nm allows a detailed analysis of both the spectral lipid signal and haemoglobin concentrations of its oxygenated and deoxygenated forms (oxyHb, deoxyHb) [3, 4]. Here we show, first, the dependence of these haemoglobin parameters on the lipid thickness when

S. Grieger • D. Geraskin • A. Steimers • M. Kohl-Bareis (✉)
RheinAhrCampus Remagen, University of Applied Sciences Koblenz, Remagen, Germany
e-mail: kohl-bareis@rheinahrcampus.de

S. Van Huffel et al. (eds.), *Oxygen Transport to Tissue XXXV*, Advances in Experimental Medicine and Biology 789, DOI 10.1007/978-1-4614-7411-1_18,

analysing the data according to the spatially resolved approach, where the slope of attenuation A with respect to source-detector distance ρ is converted into spectra of the tissue absorption μ_a,

$$\mu_a(\lambda) = \frac{1}{\mu_s'(\lambda)} \cdot \frac{1}{3} \cdot \left(\ln(10) \frac{\partial A(\lambda)}{\partial \rho} - \frac{2}{\rho} \right)^2, \tag{18.1}$$

assuming values of the transport scattering coefficient μ_s'. μ_a spectra are then converted to haemoglobin concentrations and subsequently to oxygen saturation $SO_2 = oxyHb/(oxyHb + deoxyHb)$ [5, 6].

Second, it is demonstrated that ATT can be measured by NIRS and, third, a correction term for Eq. 18.1 is proposed and tested.

18.2 Influence of Adipose Tissue Thickness on Oxygenation Data

In 20 healthy subjects at rest, haemoglobin parameters were obtained according to Eq. 18.1 from five muscle groups, and the values are plotted in Fig. 18.1 as a function of ATT. For most muscle groups there is a clear decrease in the total haemoglobin concentrations and an increase in SO_2 with increasing ATT for all sites. This trend is different in quality as expressed by the correlation coefficient and in the slope for the different muscle groups. The correlation of total haemoglobin tHb and ATT is strongest for values derived from vastus lateralis and vastus medialis ($r = 0.93$ and 0.79, respectively) while weaker for the forearm, possibly explained by a smaller muscle thickness. Similarly, the correlation of SO_2 and ATT_{US} is best for the large muscle groups of the thigh and lowest for the forearm.

18.3 Measurement of Adipose Tissue Thickness by NIRS

Tissue attenuation spectra contain information of the lipid due to its absorption peak close to 920 nm. We tested the correlation of this optical lipid signal with a measurement of the ATT by ultrasound, by magnetic resonance imaging and by a mechanical caliper [3]. In Fig. 18.2 this is shown for a total of 240 data points from six body sites of 20 subjects (both left and right hemispheres), where the amplitude of the optical lipid signal (OLS) is plotted versus ultrasound derived ATT data. When fitting an exponential curve of the form $OLS = p_1 \cdot (1 - \exp(-p_2 \cdot ATT))$, the fit parameters $p_1 = 7.961 \cdot 10^{-3} OD \cdot nm^{-2}$ and $p_2 = 0.0966\ mm^{-1}$ were obtained, and analysing the deviation (residuum) of the experimental data as shown in Fig. 18.2, the error of the optically derived ATT can be estimated as ± 0.76 mm. This error contains not only uncertainties in the optical data but of the ultrasound measurements as well.

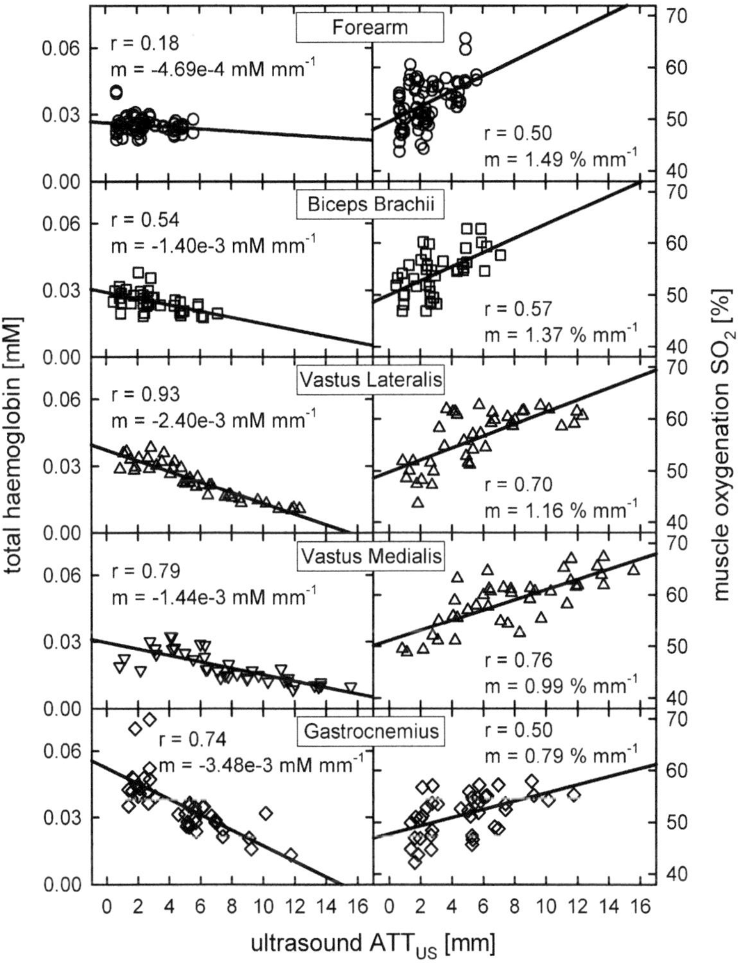

Fig. 18.1 Concentration of total haemoglobin tHb = oxyHb + deoxyHb (*left*) and oxygen saturation SO_2 (*right*) as a function of the ultrasound-measured adipose tissue thickness ATT_{US} ($n = 40$ for each site). Values are separately plotted for the different muscle groups, with the correlation coefficient r and the slope m stated

18.4 Correction Algorithm Including ATT

All steps of the conventional SRS algorithm for a calculation of muscle oxygenation are summarized on the left side of Fig. 18.3. Key is Eq. 18.1 to derive the absorption coefficient μ_a, and this is used, e.g. in the commercial NIRS monitor Hamamatsu NIRO-300 (Hamamatsu Photonics, Japan) for three wavelengths. This equation, derived from the diffusion theory for the transport of light in tissue, represents a simple non-linear relationship between $\partial A/\partial \rho$ and μ_a. Here we propose to replace

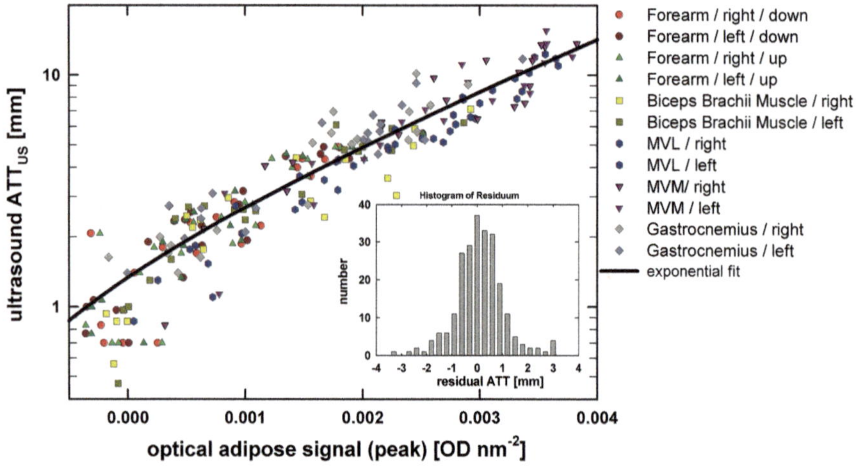

Fig. 18.2 Correlation of optical lipid signal derived from attenuation spectra at 920 nm for 240 measurements (6 muscle sites, 20 subjects, left and right). The *line* represents an exponential fit giving a mean absolute residuum of 0.76 mm. *Inset*: histogram of residuum

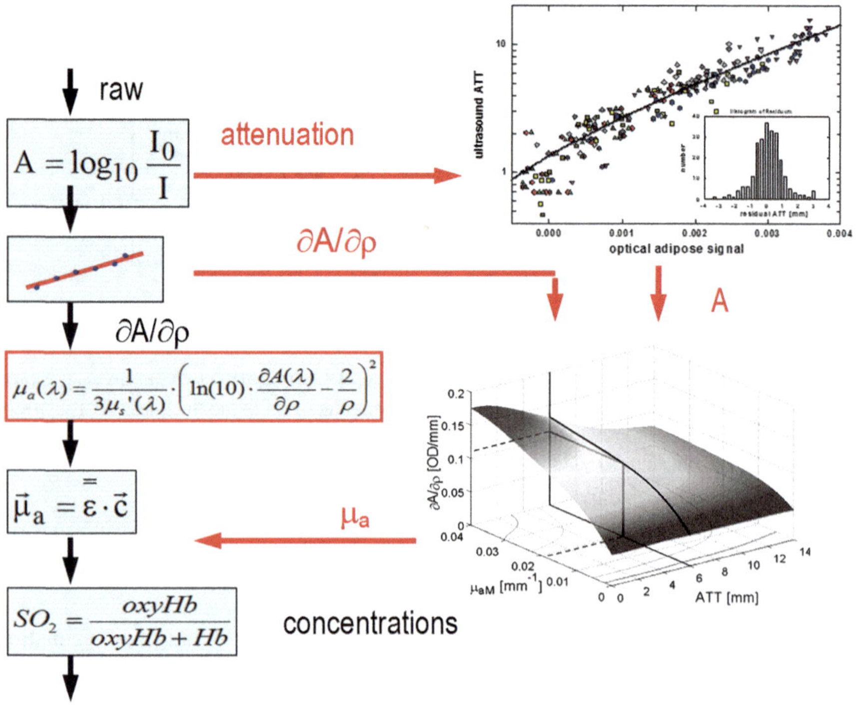

Fig. 18.3 Proposed algorithm for the inclusion of the lipid layer (ATT) in the SRS approach: the analytical equation for a homogenous medium (*left*) is bypassed by Monte Carlo-derived relationships of ∂A/∂ρ versus μ_a

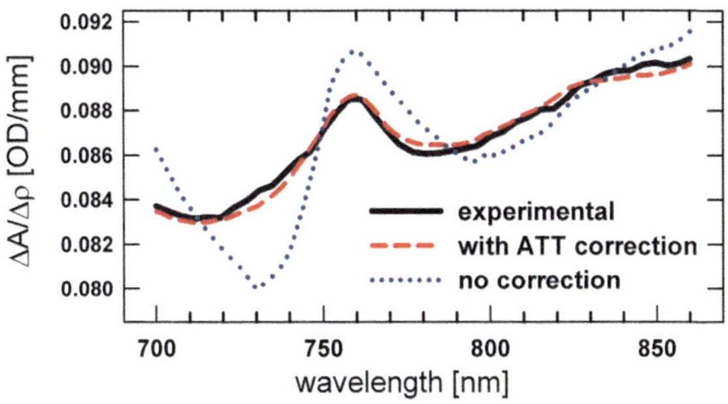

Fig. 18.4 Spectrum of $\Delta A/\Delta\rho$ measured on the vastus lateralis muscle of a volunteer with an ATT of 6.2 mm (ultrasound). When no correction of the ATT was included, the description of residuum of the fitted spectrum is much larger than with the inclusion of the Monte Carlo-based correction. The mean residuum is reduced by a about a factor of 4 with the correction

this analytical function by similar functions derived from Monte Carlo simulations [7] for layered media, where the lipid is represented by one layer covering a second layer, representing muscle. In these simulations both layers can be attributed with different absorption and scattering properties and the $\partial A/\partial\rho$ values can be derived as a function of all these properties. Look-up tables were generated for a number of ATT values and a variety of muscle and lipid absorption coefficients, and the experimental data of $\partial A/\partial\rho$ were converted into absorption coefficients. In this way the assumption of a homogeneous medium is not needed and the ATT included; however, assumptions of the absorption and scattering properties of the lipid layer have to be included.

There are a limited number of objective measures to judge an algorithm, mainly based on physiological interpretation of the calculated haemoglobin values. The broadband SRS approach used here, however, allows the quality of the model assumptions to be assessed by the spectroscopic residuum, by recalculating the spectrum from the fitted haemoglobin concentrations. This is shown in Fig. 18.4 where an experimental $\Delta A/\Delta\rho$ spectrum collected on the site of vastus lateralis of a subject is shown. When no correction term for ATT is used, the description of the experimental spectrum is poor. Inclusion of ATT and correction by the Monte Carlo-based model considerably improves the spectral fit. For this subject this improvement was obtained for all spectra during a long exercise bout.

When analysing data from nine subjects (ATT up to about 15 mm) during an incremental cycling exercise, the results were mixed when judged by the residuum. While in some of the data the magnitude of the residuum was considerably reduced, there was no significant change in others. Therefore, the residuum cannot be taken as convincing spectroscopic evidence for the validity of the model. However, a poor fit points at wrong model assumptions or experimental errors, at least in a part of the subjects.

While the broadband approach allows the residuum analysis, its set-up is more complex than systems with a few wavelengths only. Therefore, data were collected on the site of the calf muscle and the vastus lateralis of 20 subjects with the commercial monitor Hamamatsu NIRO-200NX. This system employs three wavelengths only, and therefore, the residuum analysis is limited. Within errors, no improvement of the residuum was observed when analysing data with the correction algorithm.

18.5 Discussion

In conclusion, we have shown that the adipose tissue thickness has a strong influence on the oxygenation parameters. In many studies this problem is currently circumvented and avoided by selecting subjects with a low ATT. This is clearly unsatisfactory, especially when interrogating tissue of older subjects or even female subjects who tend to have a larger ATT.

It is demonstrated that the thickness of the lipid layer can be derived from the same data analysed for haemoglobin. This might be a valuable tool in its own right. Here, however, it is treated as just one step for the improvement of muscle haemoglobin data.

There are many wrong assumptions in the standard SRS algorithm which treats the tissue as a homogeneous, infinite half-space. We proposed to improve this model by including a two-layer model, with the lower one representing the muscle of unknown absorption coefficient. For the upper layer, representing lipid, we assumed scattering and absorption properties from the literature. In the residuum analysis, however, we found no unbiased argument that this model is truly improved.

References

1. Quaresima V, Lepanto R, Ferrari M (2003) The use of near infrared spectroscopy in sports medicine. J Sports Med Phys Fitness 43(1):1–13
2. Wolf M, Ferrari M, Quaresima V (2007) Progress of near-infrared spectroscopy and topography for brain and muscle clinical applications. J Biomed Opt 12(6):062104
3. Geraskin D, Boeth H, Kohl-Bareis M (2009) Optical measurement of adipose tissue thickness and comparison with ultrasound, magnetic resonance imaging, and callipers. J Biomed Opt 14(4):044017
4. Geraskin D, Platen P, Kohl-Bareis M, Kohl-Bareis M (2007) Algorithms for muscle oxygenation monitoring corrected for adipose tissue thickness. Proc. SPIE 6629, Diffuse Optical Imaging of Tissue, 66290P: doi:10.1117/12.727854
5. Matcher SJ, Kirkpatrick P, Nahid K, Delpy DT (1995) Absolute quantification methods in tissue near infrared spectroscopy. Proc SPIE 2389:486–495
6. Suzuki S, Takasaki T, Ozaki T, Kobayashi Y (1999) A tissue oxygenation monitor using NIR spatially resolved spectroscopy. Proc SPIE 3597:582–592
7. Wang L, Jacques SL, Zheng L (1995) MCML–Monte Carlo modeling of light transport in multi-layered tissues. Comput Methods Programs Biomed 47(2):131–146

Chapter 19
Statistical Treatment of Oxygenation-Related Data in Muscle Tissue

Louis Hoofd and Hans Degens

Abstract Muscle oxygenation is determined not only by the flow and oxygen content of the supplying blood but also by the density of the capillary network, the heterogeneity of the distribution of the capillaries and the properties and distribution of the muscle fibres. The distribution of the capillaries is adequately analysed by the method of capillary domains, which also allows to link capillaries to individual fibres. Thus, capillarisation can be linked to cell properties like fibre cross-sectional surface area and perimeter, and oxygen consumption of the individual muscle fibres. However, in order to meaningfully characterise tissue properties, such linkage has to be done for groups of cells. Since most of the data are not normally distributed – domains are lognormally distributed, but how fibre cross-sectional areas are distributed is unknown – a dedicated statistical analysis is required, particularly since none of the variables is independent.

19.1 Introduction

Tissue oxygen levels can vary considerably over short distances, as shown by direct measurements with microelectrodes [1] and backed up by theory [2, 3]. Several factors play a role in the transport of oxygen to the tissue, such as blood flow and oxygen content of the blood, but some are characteristics of the tissue itself. For instance, both capillary density and the heterogeneity of capillary spacing are important [4, 5],

L. Hoofd (✉)
BaLoH Software, Prins Willem Alexanderstraat 8, 6576, BM Ooij, The Netherlands
e-mail: louis@hoofd.info

H. Degens
Institute for Biomedical Research into Human Movement and Health,
Manchester Metropolitan University, Manchester, UK
e-mail: h.degens@mmu.ac.uk

S. Van Huffel et al. (eds.), *Oxygen Transport to Tissue XXXV*, Advances in Experimental Medicine and Biology 789, DOI 10.1007/978-1-4614-7411-1_19,

while the heterogeneity in capillary blood distribution is of less importance – the overall blood supply of course is [6, 7]. Theoretical models have investigated the role of various tissue factors on tissue oxygenation [4]. It is unresolved, however, if local heterogeneity in oxygen consumption is important since only limited modelling has been possible so far [8]. There is thus a requirement to develop a method to analyse how capillary distribution and its relation to local consumption – for muscle, the situation of fibres with differing oxidative capacity – affect local tissue oxygenation.

In order to make such an analysis meaningful, it should be done for a sufficiently large piece of tissue. For both capillary distribution and capillary–fibre relations, the method of capillary domains [9] has in general proven to be well applicable [10]. Since capillary spacing, both in terms of capillary domain areas and intercapillary distances, has a lognormal distribution [11], adequate statistics can be done, only leaving open the question in how far domains at the border of a tissue cross section should be left out [12].

Relations between capillaries and fibres can be determined through the overlaps between domains and fibres in a cross section of tissue [13, 14]. However, correctness of statistics on such results has never adequately been investigated.

19.2 Methods

A frozen cross section of the deep, oxidative part of the rat plantaris muscle was selected to illustrate the statistical treatment explained here. The muscle cross section was stained for capillaries with alkaline phosphatase, and fibre types were distinguished with myosin ATPase after pre-incubation at pH4.55 [15]. Capillary coordinates and fibre contours were read into a computer file using a digitising tablet as laid down earlier [5, 14–16]. All further processing was done by the computer program AnaTis (BaLoH Software, http://www.baloh.nl); the program can handle series of tissue slices but for this article a single slab suffices. The different stages are visualised – combined – in Fig. 19.1: the upper left part shows only the capillaries, the lower left part also the domains, the upper right part capillaries and fibres and the right bottom part capillaries, fibres and domains. Here only the oxidative fibres are used; the other fibres are in a different file and can be handled separately.

Determining the number of capillaries 'associated with' a fibre by the overlap method [13] can be done in two ways. Firstly, all capillaries of which part of the domain overlaps the fibre can be counted; this gives a whole number 0, 1, 2, 3, … for each fibre. More sophisticatedly, the domain overlap fractions can be summed, giving the 'net' number of capillaries for that particular fibre. For instance, if a fibre is overlapped by three capillary domains, the overlaps being only 34 % of the first domain area, 33 % with the second and 28 % with the third, the 'net' number of capillaries is $0.34+0.33+0.28=0.95$ (referred to as f), while the whole number count would be 3 (referred to as n).

A typical way to show the relationship between capillaries and fibres of a certain type is to plot and fit the number of capillaries belonging to a fibre to the fibre

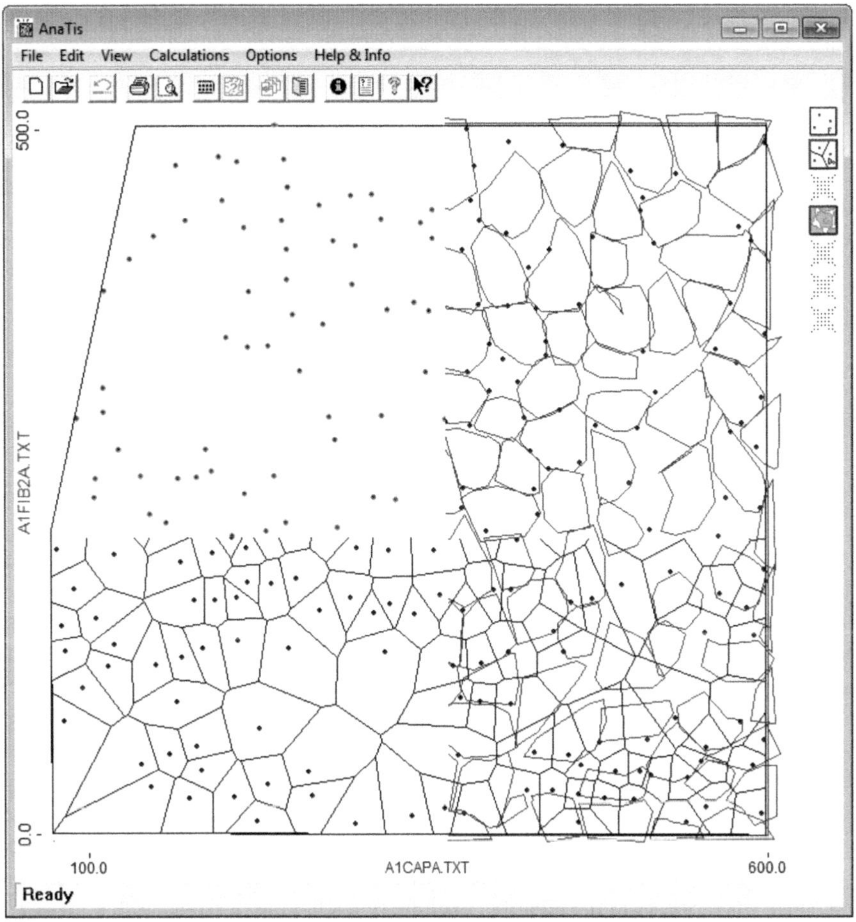

Fig. 19.1 Combined computer plot of tissue cross section showing capillaries (*dots*) domains (*polygons around dots*) and fibres (*contours*)

cross-sectional area, mostly by linear regression. That is done here too. Another possibility is to plot it against fibre perimeter length. However, the statistical treatment is the same as will be the consequences investigated below.

19.3 Results

Figure 19.2 shows plots of the whole number count of capillaries (n) against fibre cross-sectional area A (in the literature, *FCSA*). Each dot represents an individual fibre. For the left panel, lines were calculated in the standard way, where n is the dependent and *FCSA* the independent variable. The lower line shows the relationship

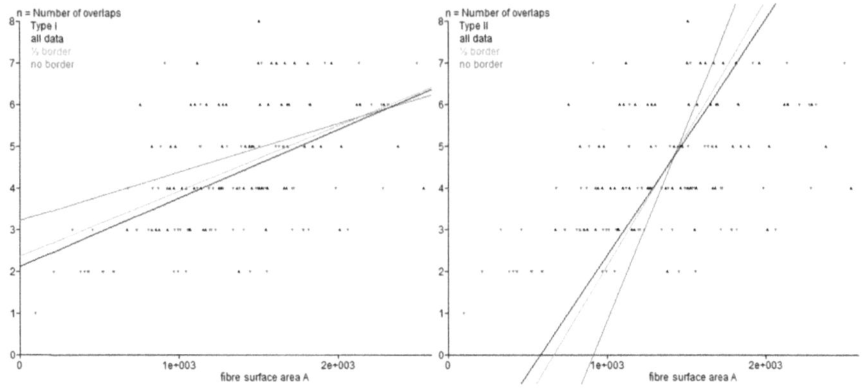

Fig. 19.2 Number of capillaries *n* against fibre cross-sectional area *A* applying two different types of statistics. *Left panel n* as dependent and *right panel n* as independent variable. For explanation see text

when all domains were included, the upper line when fibres overlapping with border domains were excluded and the middle when fibres overlapping with border domains were included for 50 % (analogous to the statistically correct way of domain distribution calculation). In AnaTis these lines and the corresponding data points have different colours. Apart from the statistical problem concerning which line gives the most appropriate representation of the real situation, there is another objection against plotting *n* against *FCSA*; a dependent variable should be continuous, while *n* is only given by whole numbers.

The simplest way to solve this problem is to switch variables in the fit; in other words make *n* independent and *FCSA* dependent. The resulting lines are presented in the right panel and are completely different from the former ones, where the steepest line appears in the data that exclude fibres having overlaps with domains at the border of the region of interest. For example, for the 'in between' lines, the slope now is $b_2 = 6.38 \cdot 10^{-3} \pm 1.08 \cdot 10^{-3}$ μm^{-2} (SE) for *n* as independent variable against $b_1 = 1.56 \cdot 10^{-3} \pm 0.27 \cdot 10^{-3}$ μm^{-2} (SE) for *n* as the dependent variable, a huge difference. The coefficient of correlation (*R*) does not depend on the type of fit and is 0.495; it can be proven that both slopes are related by the equation $b_2 = b_1/R^2$.

The same problem is encountered when forcing the fitted line through the origin; after all, a fibre with zero surface area should have zero capillaries belonging to it. In this case using *n* as dependent variable, the incorrect fit, yields a slope of $3.11 \cdot 10^{-3} \pm 0.10 \cdot 10^{-3}$ μm^{-2} (SE) against $3.45 \cdot 10^{-3} \pm 0.11 \cdot 10^{-3}$ μm^{-2} (SE) for *A* as dependent variable. The difference, however, is much less; also note that the correct slope is higher here.

One might argue that no such problem exists for the fractional number of capillaries *f* (*LCFR*), since this variable is continuous. However, *LCFR* as dependent variable against *FCSA* as independent variable ignores the biological and methodological variation in *FCSA*; at least, the *FCSA* might be different in neighbouring cross sections. For a correct fit, the 'intrinsic' variations of both *LCFR* and *FCSA* should be known, but there is no standard way to determine these. This problem was circumvented by the Deming method [17], assuming both variables to have equivalent variation. For the 'in between' standard fit, with f as dependent variable, the slope is $b_1 = 0.628 \cdot 10^{-3} \pm 0.084 \cdot 10^{-3}$ μm^{-2} (SE); for the Deming fit it is $b_3 = 1.07 \cdot 10^{-3} \pm 0.11 \cdot 10^{-3}$ μm^{-2} (SE). The difference is less than in the whole number situation but still considerable. Again, the coefficient of correlation (R) does not depend on the type of fit and is 0.587; here, both slopes are related by the equation $b_3 = b_1/R$.

Forcing the lines through the origin yields slopes of $0.807 \cdot 10^{-3} \pm 0.028 \cdot 10^{-3}$ μm^{-2} (SE) for the standard approach against $0.856 \cdot 10^{-3} \pm 0.029 \cdot 10^{-3}$ μm^{-2} (SE) for the Deming approach.

19.4 Conclusions

Although the method of capillary domains allows for automated evaluation of distribution of capillaries and their relationship to other tissue fractions, the statistical processing is not so simple. For the capillary distribution, the problem is minor since it was proven to be statistically correct to leave out the domains at two of the four sides of a rectangular tissue cross section. However, often a cross section is not rectangular (see, e.g., the outline Fig. 19.1); in AnaTis this problem is solved by incorporating the border domains for only 50 %.

The same problem is present when correlating fibres to capillaries. However, so far it has not been determined what portion of the fibres with overlaps with border domains should be omitted. A more significant problem, however, is caused by straight-line fits of capillary numbers and fibre sizes. To take n, or the number of domains overlapping a fibre, as dependent variable is statistically incorrect, but is often published as such in the literature, as intuitively one expects the number of capillaries supplying a fibre to be dependent on the size and the metabolic type of a fibre [14, 16] and not the other way around. Because of the large differences in outcome when using the $FCSA$ as a dependent variable, results can only be compared when the same relationships are plotted, i.e., the same variables as dependent and independent. Otherwise, for example, significant differences between results can and will be found where there are none and vice versa. In particular, when differences between groups of animals or muscle or fibre types are to be detected, doing the same statistical analysis is a prerequisite.

Although a strict analysis of statistical correctness has not been done here, the Deming approach seems to be more adequate for relating fractional numbers of capillaries to fibre area (or perimeter length) because neither of these is an exact variable. Analogously, the same holds for fractional number of fibres against domain area which also has been done. If the available statistics program does not support the Deming approach, there are ways to solve this problem, for example, by applying the above-mentioned equation $b_3 = b_1/R$ (or contact the first author). A complete overview of all statistical approaches applied can be found in the manual to AnaTis, available at http://www.baloh.nl.

Correct statistics is often a matter of interpretation, so no definitive judgement is presented here about the validity of each approach, except that the standard statistics of n as dependent variable has to be considered as incorrect. The treatment presented here should be considered as an aid for a decision about the statistics, as well as a classification of the possibilities.

References

1. Kessler M, Höper J, Pohl U (1982) Monitoring of local PO_2 in skeletal muscle of critically ill patients. In: Berk JL (ed) Handbook of critical care, 2nd edn. Little, Brown and Company, Boston, pp 599–609
2. Hoofd L, Turek Z, Olders J (1989) Calculation of oxygen pressures and fluxes in a flat plane perpendicular to any capillary distribution. Adv Exp Med Biol 248:187–196
3. Turek Z, Rakusan K, Olders J, Hoofd L, Kreuzer F (1991) Computed myocardial pO_2 histograms: effects of various geometrical and functional conditions. J Appl Physiol 70(4): 1845–1853
4. Hoofd L, Turek Z (1992) Oxygen pressure histograms calculated in a block of rat heart tissue. Adv Exp Med Biol 317:561–566
5. Degens H, Deveci D, Botto-van Bemden A, Hoofd LJ, Egginton S (2006) Maintenance of heterogeneity of capillary spacing is essential for adequate oxygenation in the soleus muscle of the growing rat. Microcirculation 13(6):467–476
6. Hoofd L, Degens H (2009) The influence of flow redistribution on working rat muscle oxygenation. Adv Exp Med Biol 645:55–60
7. Hoofd L, Turek Z (1994) The influence of flow redistribution on the calculated PO_2 in rat heart tissue. Adv Exp Med Biol 345:275–282
8. Hoofd L, Turek Z, Egginton S (1990) Concentric oxygen diffusion in tissue with heterogeneous permeability and consumption. Adv Exp Med Biol 277:13–20
9. Hoofd L, Turek Z, Kubat K, Ringnalda BE, Kazda S (1985) Variability of intercapillary distance estimated on histological sections of rat heart. Adv Exp Med Biol 191:239–247
10. Al-Shammari AA, Gaffney EA, Egginton S (2012) Re-evaluating the use of voronoi tessellations in the assessment of oxygen supply from capillaries in muscle. Bull Math Biol 74(9):2204–2231
11. van Haelst ACTA, Hoofd L, Turek Z (1985) Lognormal distribution of capillary domains in rat myocardium. J Physiol (London) 366:114P
12. Egginton S, Ross HF (1989) Influence of muscle phenotype on local capillary supply. Adv Exp Med Biol 248:281–291
13. Egginton S, Turek Z, Hoofd L (1987) Morphometric analysis of sparse capillary networks. Adv Exp Med Biol 215:1–12
14. Degens H, Turek Z, Hoofd LJ, Binkhorst RA (1994) Capillary proliferation related to fibre types in hypertrophied aging rat *M. plantaris*. Adv Exp Med Biol 345:669–676
15. Degens H, Turek Z, Hoofd LJ, Van't Hof MA, Binkhorst RA (1992) The relationship between capillarisation and fibre types during compensatory hypertrophy of the plantaris muscle in rat. J Anat 180(Pt3):455–463
16. Wüst RC, Jaspers RT, van Heijst AF et al (1999) Region-specific adaptations in determinants of rat skeletal muscle oxygenation to chronic hypoxia. Am J Physiol Heart Circ Physiol 297(1):H364–H374
17. Deming WE (1943) Statistical adjustment of data. Wiley, New York (Dover Publications edition, 1985)

Chapter 20
O_2 Saturation in the Intercostal Space During Moderate and Heavy Constant-Load Exercise

Takuya Osawa, Ryotaro Kime, Masako Fujioka, Takuya Osada, Norio Murase, and Toshihito Katsumura

Abstract To examine the hypothesis that the relationship between minute ventilation (V_E) and deoxygenation from the intercostal space (IC) would be steady regardless of exercise protocols, if an increase in O_2 consumption of the accessory respiratory muscles with an increase of V_E brings about deoxygenation in IC, we measured the relationship between V_E and O_2 saturation in IC (SO_{2IC}) during a constant-load exercise test (CET), and the relationship was compared with that during a ramp incremental exercise test (RIET). Six male subjects performed RIET. On a different day, the subjects performed a moderate and heavy CET (CET_MOD and CET_HVY, respectively). SO_{2IC} decreased from the start of both CET_MOD and CET_HVY and changed little from 2 min. Moreover, SO_{2IC} was significantly lower during CET_HVY than during CET_MOD. In comparison between RIET and CET_HVY at the similar V_E level, SO_{2IC} was significantly higher during CET_HVY than RIET. These results suggest that the decrease in SO_{2IC} was caused not only by an increase in O_2 consumption in IC with an increase in V_E but also by a decrease in O_2 supply.

T. Osawa (✉)
Department of Sports Sciences, Japan Institute of Sports Sciences, Tokyo, Japan
e-mail: takuya.osawa@jpnsport.go.jp

R. Kime • T. Osada • N. Murase • T. Katsumura
Department of Sports Medicine for Health Promotion, Tokyo Medical University, Tokyo, Japan

M. Fujioka
Faculty of Sport and Health Science, Ritsumeikan University, Shiga, Japan

S. Van Huffel et al. (eds.), *Oxygen Transport to Tissue XXXV*, Advances in Experimental Medicine and Biology 789, DOI 10.1007/978-1-4614-7411-1_20,

20.1 Introduction

During incremental cycling exercise, deoxygenation, measured by near-infrared spectroscopy (NIRS), in the working muscle (the vastus lateralis) was attenuated at high intensity [1–4], and in previous studies, it was thought to be related to metabolism of the respiratory and accessory respiratory muscles [1]. Deoxygenation from the intercostal space (IC) hyperbolically increased until exhaustion. The deoxygenation was thought to be because the recruitment of the accessory respiratory muscles increased the O_2 consumption. However, since the NIRS data reflect the balance of O_2 supply and consumption [5, 6], deoxygenation was not always induced by an increase of O_2 consumption but also by a decrease of O_2 supply.

During constant-load exercise, minute ventilation (V_E) increases gradually from the start of exercise [7]. However, little is known about deoxygenation in IC. If deoxygenation in IC was led by an increase in O_2 consumption, it could be hypothesized that deoxygenation was controlled by the amplitude of V_E, independent of exercise protocols. However, if deoxygenation was different between protocols despite the same V_E, deoxygenation would be influenced not only by the change of O_2 consumption but also by O_2 supply.

The purpose of this study was to examine deoxygenation in IC during constant-load exercise and to compare between deoxygenation during constant-load exercise with that during incremental exercise at the similar V_E level.

20.2 Methods

20.2.1 Subjects

Six active male volunteers participated in this study (age, 22 ± 1 years; height, 170.5 ± 5.7 cm; body mass, 67.3 ± 12.0 kg; mean ± S.D.). Before participation in the experiment, all procedures and any potential risks were explained to each subject and an informed consent document was signed. This study was approved by the local ethics committee, and all work was performed in accordance with the Declaration of Helsinki.

20.2.2 Experiment Design

The subjects performed a ramp incremental exercise test (RIET). The protocol followed 0-W exercise for 1 min and increased at a ramp rate of 20 $W \cdot min^{-1}$ to exhaustion. Ventilatory threshold (VT) calculated by a V-slope method [8] and peak O_2 uptake (VO_{2peak}) were evaluated. On a different day, the subjects performed a moderate (CET_MOD; the intensity was 90 % of VT) and heavy (CET_HVY; the intensity was the middle between VT and VO_{2peak}) constant-load exercise test. The protocol followed a warm-up exercise (the same intensity as CET_MOD) for 3 min and 0-W exercise for 6 min and performed CET_MOD and CET_HVY exercise for 6 min separated by 0-W exercise for 6 min. The warm-up exercise was performed to meet the metabolic balance before CET_MOD and CET_HVY exercise. An electromagnetically braked cycle ergometer (Strength Ergo 8, Fukuda Denshi, Japan) was utilized; the seat and handle height remained constant for each subject, and the pedal frequency was maintained at 60 rpm.

20.2.3 *Measurements*

V_E and O_2 uptake (VO_2) were determined breath by breath by a computerized metabolic cart (AE300S, Minato, Japan). Expiratory flow measurements were performed by a mass flow sensor. V_E was normalized by the peak of V_E during RIET ($\%V_{Epeak-RIET}$). O_2 saturation in IC (SO_{2IC}) was measured with near-infrared spatially resolved spectroscopy (NIRO-200, Hamamatsu photonics, Japan). The optode was placed on the interion between the seventh intercostal space and the anterior axillary line. The source-detector distance was 4 cm.

20.2.4 *Statistical Analyses*

All data represented as means ± S.D. Statistical analyses were performed using the statistical package SPSS for Windows (version 19.0; SPSS, Chicago, IL). SO_{2IC} during CET_MOD and CET HVY were compared by two-way ANOVA, with time and intensity as main effects. The Dunnett post hoc test was performed to examine time changes, and the paired *t*-test was used to compare between CET_MOD and CET_HVY (Fig. 20.1). SO_{2IC} during RIET, CET_MOD, and CET_HVY were compared by the paired *t*-test, at the similar $\%V_{Epeak-RIET}$ (Fig. 20.2). Values of $P<0.05$ were considered significant.

20.3 Results

At VO_{2peak} during RIET, V_E, VO_2, and SO_{2IC} were 126 ± 22 $L \cdot min^{-1}$, 51.7 ± 9.1 $mL \cdot kg^{-1} \cdot min^{-1}$, and 54.6 ± 14.0 %, respectively.

V_E increased progressively as exercise time increased and was significantly higher during CET_HVY than CET_MOD. In contrast, SO_{2IC} decreased from the start of both CET_MOD and CET_HVY and changed little from 2 to 6 min (Fig. 20.1). The amplitude of SO_{2IC} was significantly higher during CET_HVY than CET_MOD. The final values of V_E, $\%V_{Epeak-RIET}$, VO_2, and SO_{2IC} were

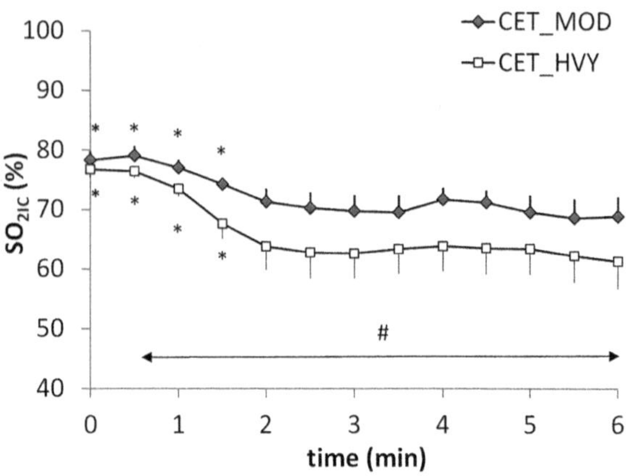

Fig. 20.1 SO_{2IC} kinetics during CET_MOD and CET_HVY. * versus 6 min, # versus MOD, $P<0.05$

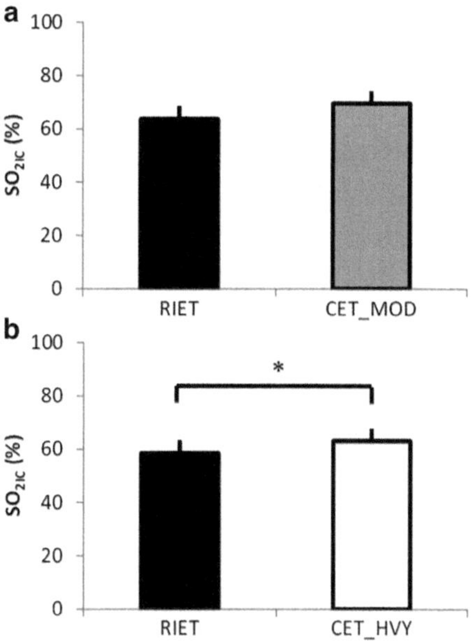

Fig. 20.2 SO_{2IC} at the similar $\%VE_{peak-RIET}$ during RIET and CET_MOD (**a**) and during RIET and CET_HVY (**b**). * RIET versus CET_HVY

59.1 ± 4.6 L·min^{-1}, 46.1 ± 6.9 %, 32.3 ± 6.1 mL·kg^{-1}·min^{-1}, and 68.9 ± 7.3 % during CET_MOD and 103 ± 19 L·min^{-1}, 79.3 ± 11.5 %, 47.9 ± 6.9 mL·kg^{-1}·min^{-1}, and 61.3 ± 10.2 % during CET_HVY, respectively.

In comparison with the similar $\%V_{Epeak-RIET}$ between RIET (48.3 ± 3.3 %) and CET_MOD at the peak of $\%V_{Epeak-RIET}$ (48.4 ± 3.2 %), SO_{2IC} tend to be higher during CET_MOD than RIET, but not significant (Fig. 20.2a). On the other hand, in comparison between RIET (81.0 ± 12.2 %) and CET_MOD at the peak of $\%V_{Epeak-RIET}$ (81.0 ± 11.7 %), SO_{2IC} was significantly higher during CET_HVY than RIET (Fig. 20.2b).

20.4 Discussion

We examined SO_{2IC} kinetics during constant-load exercise and compared SO_{2IC} at similar $\%V_{Epeak-RIET}$ between RIET and CET. The main findings of this study are that SO_{2IC} was little changed over 2 min during both CET_MOD and CET_HVY and that the level of SO_{2IC} was different between RIET and CET_HVY despite similar $\%V_{Epeak-RIET}$. These findings suggest that the decrease of SO_{2IC} is not induced by increased O_2 consumption of the accessory respiratory muscles under the measurement area. In this study, neither blood flow nor O_2 supply in IC was measured, but a decrease in them would mainly result in a decrease in SO_{2IC}.

O_2 saturation is thought to represent the balance between O_2 supply and consumption. We believe that O_2 consumption in the respiratory and accessory respiratory muscles was increased because a rise of V_E led to high work of breathing [9]. However, it would be unlikely that the increase in O_2 consumption caused deoxygenation in these muscles, at least the intercostal muscles. Rather, we assume that SO_{2IC} was more related to O_2 supply changes, which could be effected by sympathetic nerve activity-induced vasoconstriction, which was seen in organs and resting muscles during high-intensity exercise [10, 11]. Blood flow in working muscle was relatively maintained or increased by vasodilation effects such as NO [12], but blood flow and O_2 supply in resting muscles were decreased [11]. Although neither autonomic nerve activity nor blood flow in the resting muscle was measured in this study, previous studies have reported that during high-intensity exercise, the resting muscle deoxygenation increased gradually [13], similar to SO_{2IC} in this study. Hence, in our study where O_2 saturation in IC and the resting muscle were simultaneously monitored, these values were similar during both incremental and decremental exercise (unpublished data). Furthermore, an isocapnic hyperpnea at rest did not deoxygenate IC [14]. Therefore, it was unclear why SO_{2IC} showed different kinetics between protocols, but it was likely that the change was more related to the O_2 supply than to the O_2 consumption.

In summary, the present study demonstrated that SO_{2IC} during CET_HVY did not change despite V_E increasing and that SO_{2IC} during CET_HVY was different from that during RIET in the same V_E. These results imply that a decrease in SO_{2IC} is caused by a decrease of the O_2 supply, rather than an increase of the O_2 consumption.

References

1. Legrand R, Marles A, Prieur F, Lazzari S, Blondel N, Mucci P (2007) Related trends in locomotor and respiratory muscle oxygenation during exercise. Med Sci Sports Exerc 39(1):91–100
2. Osawa T, Kime R, Hamaoka T, Katsumura T, Yamamoto M (2011) Attenuation of muscle deoxygenation precedes EMG threshold in normoxia and hypoxia. Med Sci Sports Exerc 43(8):1406–1413

3. Osawa T, Kime R, Katsumura T, Yamamoto M (2010) Effects of acute hypoxia on the inflection point of muscle oxygenation. Adv Exp Med Biol 662:385–390
4. Wilson JR, Mancini DM, McCully K, Ferraro N, Lanoce V, Chance B (1989) Noninvasive detection of skeletal muscle underperfusion with near-infrared spectroscopy in patients with heart failure. Circulation 80(6):1668–1674
5. Ferrari M, Mottola L, Quaresima V (2004) Principles, techniques, and limitations of near infrared spectroscopy. Can J Appl Physiol 29:463–487
6. Hamaoka T, McCully KK, Quaresima V, Yamamoto K, Chance B (2007) Near-infrared spectroscopy/imaging for monitoring muscle oxygenation and oxidative metabolism in healthy and diseased humans. J Biomed Opt 12(6):062105
7. Kowalchuk JM, Rossiter HB, Ward SA, Whipp BJ (2002) The effect of resistive breathing on leg muscle oxygenation using near-infrared spectroscopy during exercise in men. Exp Physiol 87(5):601–611
8. Beaver WL, Wasserman K, Whipp BJ (1986) A new method for detecting anaerobic threshold by gas exchange. J Appl Physiol 60(6):2020–2027
9. Vogiatzis I, Athanasopoulos D, Habazettl H et al (2009) Intercostal muscle blood flow limitation in athletes during maximal exercise. J Physiol 587(Pt 14):3665–3677
10. Osada T, Katsumura T, Hamaoka T et al (1999) Reduced blood flow in abdominal viscera measured by Doppler ultrasound during one-legged knee extension. J Appl Physiol 86(2):709–719
11. Shiroishi K, Kime R, Osada T, Murase N, Shimomura K, Katsumura T (2010) Decreased muscle oxygenation and increased arterial blood flow in the non-exercising limb during leg exercise. Adv Exp Med Biol 662:379–384
12. Boushel R (2010) Muscle metaboreflex control of the circulation during exercise. Acta Physiol (Oxf) 199(4):367–383
13. Ogata H, Arimitsu T, Matsuura R, Yunoki T, Horiuchi M, Yano T (2007) Relationship between oxygenation in inactive biceps brachii muscle and hyperventilation during leg cycling. Physiol Res 56(1):57–65
14. Guenette JA, Vogiatzis I, Zakynthinos S et al (2008) Human respiratory muscle blood flow measured by near-infrared spectroscopy and indocyanine green. J Appl Physiol 104(4):1202–1210

Chapter 21
Muscle, Prefrontal, and Motor Cortex Oxygenation Profiles During Prolonged Fatiguing Exercise

Thomas Rupp, Marc Jubeau, Guillaume Y. Millet, Bernard Wuyam, Patrick Levy, Samuel Verges, and Stéphane Perrey

Abstract This study aimed to compare changes in skeletal muscle, prefrontal (PFC), and motor (MC) cortex hemodynamics during prolonged (i.e., 4-h) fatiguing whole-body exercise using multichannel near-infrared spectroscopy (NIRS). Ten subjects completed three successive 80-min cycling bouts at 45 % of their maximal power output. After the 4-h cycling, maximal voluntary contraction force of the leg was decreased by ~25 %. Muscle exhibited reproductive deoxygenation patterns during each of the three bouts, whereas intra-bout cerebral hemodynamics were different throughout the protocol. Results demonstrate that specific responses to fatiguing exercise are found between tissues but also between cortical sites involved in cycling, as shown by concomitant PFC hyperoxygenation and MC deoxygenation in the first 80 min of exercise. Further insights are needed to understand the consequences of these changes regarding the integrative control of motor output while fatigue develops over several hours.

T. Rupp (✉) • B. Wuyam • P. Levy • S. Verges
HP2 Laboratory, INSERM U1042, Joseph Fourier University, Grenoble, France
e-mail: trupp.univ@gmail.com

M. Jubeau
HP2 Laboratory, INSERM U1042, Joseph Fourier University, Grenoble, France

Exercise Physiology Laboratory (LPE), Lyon University, Saint-Etienne, France

Motricité Interactions Performance Laboratory (MIP), Nantes University, Nantes, France

G.Y. Millet
HP2 Laboratory, INSERM U1042, Joseph Fourier University, Grenoble, France

Exercise Physiology Laboratory (LPE), Lyon University, Saint-Etienne, France

S. Perrey
Movement to Health Laboratory (M2H), Euromov, Montpellier-I University, Montpellier, France

S. Van Huffel et al. (eds.), *Oxygen Transport to Tissue XXXV*, Advances in Experimental Medicine and Biology 789, DOI 10.1007/978-1-4614-7411-1_21,

21.1 Introduction

Neuromuscular fatigue has been characterized by a decrease in the ability to produce a maximal voluntary contraction force (MVC) or an inability to maintain a certain level of force, potentially originating from multiple sites between the motor planning and the motor production [1]. In the last decade, studies focusing on oxygen transport to tissue with noninvasive techniques have enhanced our understanding of fatigue etiology both at the muscle and brain levels [2, 3].

During progressive maximal cycling, leg muscle tissue saturation index may decrease up to −20 % but often present a plateau in the last part of exercise [4] suggesting that metabolic changes in the muscle may not be the ultimate signal leading to exhaustion. Although cerebral oxygenation is increased during submaximal cycling tasks, critical reductions in prefrontal cortex (PFC) oxygenation occur at the end of short (i.e., <20 min) maximal exertion [5, 6]. PFC is known to project to premotor areas and to be responsible for goal-direct behavior, movement planning, pacing strategies, as well as decision-making [7]. Hence, PFC perturbations may contribute to exercise performance limitation by impairing executive functions (e.g., decision to stop exercising) and central motor drive indirectly, potentially because of an imbalance between reduction in regional cerebral blood flow and increased cerebral metabolic rate and O_2 uptake. Only one study [4] addressed the question of whether PFC and MC oxygenation profiles are consistent during whole-body high-intensity exercise. The authors found significant discrepancies between those sites at exhaustion suggesting that O_2 delivery and/or activation pattern may slightly differ.

Tissue-specific (de)oxygenation perturbations that may occur when fatigue develops over several hours of exercise have never been described. The aim of this study was to compare pattern and magnitude of changes in skeletal muscle, PFC, and MC hemodynamics using multichannel near-infrared spectroscopy (NIRS) during prolonged (i.e., 4-h) fatiguing whole-body exercise.

21.2 Methods

21.2.1 Population and Protocol

Ten healthy trained males gave their written informed consent to participate in the study. Their average (± SD) age, weight, and height were 37 ± 7 year, 73 ± 7 kg, and 180 ± 5 cm, respectively. Subjects performed a 4-h cycling exercise in laboratory at 45 % of their maximal aerobic power output (153 ± 23 W) at a constant pedaling frequency (~80 rpm). Exercise was split in three consecutive 80-min bouts (B1, B2, and B3) separated by 25-min periods of neuromuscular function testing including MVC of the leg extensors on a dedicated ergometer to assess fatigue kinetic. Subjects were instructed to drink no alcoholic or caffeinated beverages and to avoid any physical activity for at least 24 h before testing. The study procedures complied with the Declaration of Helsinki for human experimentation and were approved by the local ethics committee.

21.2.2 Instrumentation

Oxy[HbO_2]-, deoxy[HHb]-, and total[THb]- hemoglobin concentration changes were estimated during exercise over multiple sites using a multichannel continuous wavelengths (780 and 850 nm) NIRS device (Oxymon III, Artinis, the Netherlands). Theoretical and performance details of NIRS have been previously described [8, 9]. Quadriceps muscle oxygenation profiles were assessed from the right *vastus lateralis* using a 4-cm interoptodes distance. Probe holder was secured to the skin using double-sided tape and covered with a black sweatband to shield the optodes from ambient light. PFC oxygenation was assessed between Fp1 and F3 locations according to the international 10–20 EEG system with 3.5-cm interoptodes distance. The probe holder was secured to the skin with double-sided tape and maintained with Velcro headbands. MC hemoglobin concentrations were expressed from the average of a 4-channel square setting (3-cm interoptodes distance) fixed with headbands between Cz and C3 locations.

End-tidal CO_2 ($EtCO_2$) was measured continuously using a respiratory gas monitor connected to the subject via a face mask (Ohmeda RGM, GE Healthcare, Little Chalfont, UK).

Subjects filled a visual analogue scale of perceived exertion (VAS, ranging from 0 = "no difficulty at all" to 100 mm = "I have to stop immediately") at the end of each bout to assess subjective fatigue development during the protocol.

21.2.3 Processing and Statistical Analyses

NIRS parameters were recorded at 25 Hz, filtered with a 2-s width moving Gaussian smoothing algorithm and average over 3-min periods at 20, 40, 60, and 80 min in each bout. Tissues hemodynamics were expressed as relative changes from a stabilized baseline (BL) preceding each exercise bout to assess inter- and intra-bout kinetics throughout the protocol. Single ANOVA with repeated measures was performed to compare MVC and VAS values across the protocol and two-way (bout × time epoch) ANOVA with repeated measures was performed for $EtCO_2$ and each NIRS dependent variable. Post hoc Fisher's LSD tests were applied when appropriate to determine a difference between two mean values. Relationships between changes in tissue oxygenation parameters, between sites and with $EtCO_2$ variations, were determined by Pearson's product correlations. Values are presented as means ± SD in the text and means ± SE in Fig. 21.1. An alpha level of 0.05 was used as the cutoff for significance.

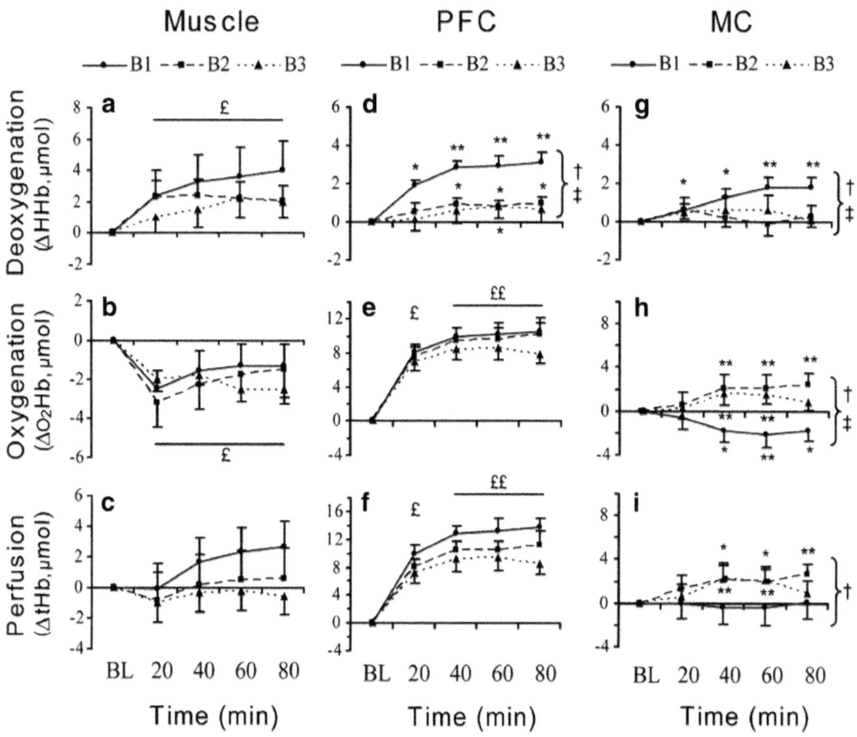

Fig. 21.1 Changes in oxy-([HbO_2]), deoxy-([HHb]), and total hemoglobin ([THb]) from BL during each of the first (B1), second (B2), and third (B3) exercise bouts. Results are shown for the *vastus lateralis* muscle (*left panels*), for the prefrontal cortex (PFC, *middle panels*), and for the motor cortex (MC, *right panels*). Values are mean ± SE. Significant main effect of time: £ versus BL, ££ versus +20 min. Significant main effect of bout: † B2 versus B1, ‡ B3 versus B1. Significant main interaction effect: * versus BL, ** versus +20 min

21.3 Results

21.3.1 Exercise-Induced Fatigue and $EtCO_2$ Changes

MVC was reduced after B1 and further decreased after B2 and B3 (−11 %, −19 %, and −25 % compared to pre-exercise, respectively, $P<0.001$). VAS of perceived exertion increased to 30 ± 16 after B1 and further to 52 ± 22 and 67 ± 25 mm after B2 and B3, respectively ($P<0.001$). $EtCO_2$ increased significantly in each bout between BL and +20 min (+7.1 ± 1.6 mmHg on average) before decreasing until the end of the bouts. Besides, a progressive hypocapnia was seen throughout the protocol as mean bout $EtCO_2$ decreased from 37.4 ± 0.7 mmHg in B1 to 33.3 ± 0.7 and 29.5 ± 0.6 mmHg in B2 and B3, respectively ($P<0.001$).

21.3.2 Exercise-Induced Tissue Oxygenation Profiles

21.3.2.1 Muscle

Muscle did not show any significant bout effect during the protocol, whatever the chromophore (Fig. 21.1a–c). Reproductive typical bout kinetics consisted of a progressive increase in Δ[HHb] and a decrease in Δ[HbO_2], reaching a plateau after ~20 min, without any significant change in Δ[THb] throughout the 80 min.

21.3.2.2 Prefrontal Cortex

As shown in Fig. 21.1e, f, Δ[HbO_2] and Δ[THb] significantly increased during B1 reaching a plateau after ~40 min. No significant main effect of bout was observed throughout the protocol for those chromophores. Δ[HHb] was progressively increased during B1 (Fig. 21.1d, $P<0.05$) from 20 to 80 min and showed a significant main effect of bout throughout the protocol, Δ[HHb] increase during B2 and B3 being significantly reduced compared to B1 ($P<0.05$).

21.3.2.3 Motor Cortex

During B1, MC Δ[HbO_2] was significantly reduced from 20 to 80 min compared to BL (Fig. 21.1h), while Δ[HHb] increased simultaneously (Fig. 21.1g). Δ[THb] was unchanged in B1 compared to BL (Fig. 21.1i). Δ[HbO_2], Δ[HHb], and Δ[THb] showed significant main effect of bout over time as B2 and B3 were different from B1 ($P<0.05$). Both Δ[HbO_2] and Δ[THb] increased from 20 to 80 min in B2 and from 20 to 60 min in B3, while Δ[HHb] did not present any change from BL during B2 and B3. Cerebral and muscle hemodynamics were not correlated during exercise. MC Δ[THb] were not correlated to $EtCO_2$, while CPF Δ[THb] was correlated to $EtCO_2$ changes only at +20 min in B2 ($R^2=0.48$, $P<0.05$) and +20 min in B3 ($R^2=0.45$, $P<0.05$).

21.4 Discussion

MVC and VAS of perceived exertion kinetics indicated increasing levels of objective and subjective neuromuscular fatigue across the 4-h exercise, respectively. Despite progressive fatigue and a tendency towards lower muscle Δ[THb], intra-bout muscle deoxygenation kinetics were reproducible when comparing the three successive bouts. This suggests that the balance between O_2 delivery and O_2 consumption over the repetitive long-duration submaximal exercises was well preserved.

In response to repetitive prolonged submaximal exercise, we observed large increases in PFC Δ[THb] (+11.2±5.8 μMol on average) and Δ[O_2Hb] (+9.6±4.7 μMol on average) compared to what has been described previously in shorter (i.e., 10–20-min) maximal exercises [3–6]. Also, the three- to fourfold increase in Δ[O_2Hb] compared to the Δ[HHb] in PFC argues in favor of an efficient neurovascular coupling throughout the three fatiguing exercise bouts, with no sign of potential mismatch between O_2 utilization and availability for the cortical region.

Very different responses were observed in the MC, in particular during the first 80 min where Δ[THb] did not change and Δ[O_2Hb] significantly decreased in response to exercise. This is the first time that such a different tissue oxygenation pattern between PFC and MC is described during whole-body fatiguing (but non-exhausting) exercise. Our results confirm the relatively specific O_2 delivery and activation pattern of motor-related cortical areas, which has been suggested by others during short maximal exercise [4]. Accordingly, using only PFC location to speculate on potential cerebral limitation of motor performance should be done with caution. Otherwise, differential cortical sensitivity to exercise/fatigue may lead to misinterpretations.

21.4.1 Limitations

To better assess intra-bout kinetics we chose to express data in relation to the respective bout baseline. Unfortunately, it excludes any interpretation on the continuous time course in tissue oxygenation during the 4-h exercise. Besides, even if exercise-induced EtCO_2 changes were not correlated with MC oxygenation kinetics and only poorly correlated with PFC profiles, it remains difficult from the NIRS signal to determine how far exercise-induced systemic changes (e.g., in arterial CO_2, cardiac output) may interfere with the underlying activation-related hemodynamic changes [6]. Our data also raise the question of the exact somatotopic correspondence and function of the investigated areas.

To conclude, our results demonstrate that specific responses to fatiguing exercise are found between tissues and between cortical sites implicated in cycling, as shown by concomitant PFC hyperoxygenation and MC deoxygenation in the first 80 min of exercise. Hence, the pattern and magnitude of hemodynamics perturbations are distinct between muscle, prefrontal, and motor regions throughout a prolonged intermittent exercise. Further insights are needed to understand the consequences of these changes, regarding the integrative control of motor output while fatigue develops over several hours.

Acknowledgments This study was funded by the French National Research Agency (Grant number NT09-653348).

References

1. Enoka RM (2002) Muscle fatigue. In: Neuromechanics of human movement, 3rd edn. Human Kinetics, Champaign
2. Ferrari M, Muthalib M, Quaresima V (2011) The use of near-infrared spectroscopy in understanding skeletal muscle physiology: recent developments. Philos Transact A Math Phys Eng Sci 369(1955):4577–4590
3. Perrey S (2008) Non-invasive NIR, spectroscopy of human brain function during exercise. Methods 45(4):289–299
4. Subudhi AW, Miramon BR, Granger ME, Roach RC (2009) Frontal and motor cortex oxygenation during maximal exercise in normoxia and hypoxia. J Appl Physiol 106(4):1153–1158
5. Rupp T, Perrey S (2008) Prefrontal cortex oxygenation and neuromuscular responses to exhaustive exercise. Eur J Appl Physiol 102(2):153–163
6. Heine M, Subudhi AW, Roach RC (2009) Effect of ventilation on cerebral oxygenation during exercise: insights from canonical correlation. Respir Physiol Neurobiol 166(2):125–128
7. Krawczyk DC (2002) Contributions of the prefrontal cortex to the neural basis of human decision making. Neurosci Biobehav Rev 26(6):631–664
8. Rolfe P (2000) In vivo near-infrared spectroscopy. Annu Rev Biomed Eng 2(1):715–754
9. Elwell CE, Cope M, Edwards AD, Wyatt JS, Delpy DT (1994) Reynolds EO quantification of adult cerebral hemodynamics by near-infrared spectroscopy. J Appl Physiol 77(6):2753–2760

Chapter 22
Aging Affects Spatial Distribution of Leg Muscle Oxygen Saturation During Ramp Cycling Exercise

Shun Takagi, Ryotaro Kime, Norio Murase, Tsubasa Watanabe, Takuya Osada, Masatsugu Niwayama, and Toshihito Katsumura

Abstract We compared muscle oxygen saturation (SmO_2) responses in several leg muscles and within a single muscle during ramp cycling exercise between elderly men ($n = 8$; age, 65 ± 3 years; ELD) and young men ($n = 10$; age, 23 ± 3 years; YNG). SmO_2 was monitored at the distal site of the vastus lateralis (VLd), proximal site of the vastus lateralis (VLp), rectus femoris (RF), vastus medialis (VM), biceps femoris (BF), gastrocnemius lateralis (GL), gastrocnemius medialis (GM), and tibialis anterior (TA) by near-infrared spatial resolved spectroscopy. During submaximal exercise, significantly lower SmO_2 at a given absolute work rate was observed in VLd, RF, BF, GL, and TA but not in VLp, VM, and GM in ELD than in YNG. In contrast, at all measurement sites, SmO_2 at peak exercise was not significantly different between groups. These results indicate that the effects of aging on SmO_2 responses are heterogeneous between leg muscles and also within a single muscle. The lower SmO_2 in older men may have been caused by reduced muscle blood flow or altered blood flow distribution.

22.1 Introduction

Near-infrared spectroscopy (NIRS) has been widely used in measuring muscle oxygenation. Muscle oxygen saturation (SmO_2), which is measured by NIRS, is an indicator of the balance between local O_2 delivery and O_2 consumption. SmO_2 in the

S. Takagi (✉) • R. Kime • N. Murase • T. Watanabe • T. Osada • T. Katsumura
Department of Sports Medicine for Health Promotion, Faculty of Medicine, Tokyo Medical University, 6-1-1 Shinjuku, Shinjuku-ku, Tokyo 160-8402, Japan
e-mail: stakagi@tokyo-med.ac.jp

M. Niwayama
Department of Electrical and Electronic Engineering, Faculty of Engineering, Shizuoka University, Shizuoka 432-8561, Japan

S. Van Huffel et al. (eds.), *Oxygen Transport to Tissue XXXV*, Advances in Experimental Medicine and Biology 789, DOI 10.1007/978-1-4614-7411-1_22,

vastus lateralis (VL) muscle has been founded to decrease during incremental cycling [1], and the decline of SmO_2 was likely attributed to higher O_2 consumption than O_2 delivery.

Aging may reduce O_2 delivery to activating muscle during rest and exercise. A previous study reported that muscle deoxygenation in VL measured by NIRS was enhanced in older compared to younger subjects during cycling exercise due to reduced oxygen supply to activating muscle [2]. However, it has been reported that the O_2 balance is distributed heterogeneously between leg muscles and also within a single skeletal muscle [1, 3]. As aging also alters distribution of blood flow between muscles [4], influence of aging on SmO_2 responses are heterogeneous between muscles and within a single muscle. The purpose of this study was to compare SmO_2 responses between elderly and young subjects in several leg muscles and within a single muscle during cycling exercise.

22.2 Methods

22.2.1 Subjects

Eight elderly men (ELD; age, 65 ± 3 years; height, 166.4 ± 5.4 cm; weight 66.4 ± 5.0 kg; mean ± SD) and ten young men (YNG; age, 23 ± 3 years; height, 174.1 ± 7.3 cm; weight, 66.0 ± 7.0 kg; mean ± SD) participated in the study, which was approved by the Tokyo Medical University Local Research Ethics Committee, Japan. In ELD, one subject was taking a calcium channel blocker, one subject was taking an angiotensin-converting enzyme inhibitor, and one subject was taking a statin and angiotensin II receptor antagonist. All volunteers were informed of the purpose and nature of the study, after which their written informed consent was given.

22.2.2 Experimental Design

The subjects performed 15 W/min (ELD) or 20 W/min (ELD and YNG) ramp bicycle exercise (after a 3-min warm up at 0 or 10 W) until exhaustion (Strength Ergo 8, Fukuda-Denshi, Tokyo, Japan). Pulmonary O_2 uptake (VO_2) was monitored continuously during the experiments to determine peak VO_2 with an online metabolic system (AE300S, Minato Medical Science, Osaka, Japan). Pedal frequency of 50 rpm (for ELD) or 60 rpm (for YNG) was maintained by keeping time with a metronome.

Muscle O_2 saturation (SmO_2) was monitored at the distal site of the VL (VLd), proximal site of the VL (VLp), rectus femoris (RF), vastus medialis (VM), biceps femoris (BF), gastrocnemius lateralis (GL), gastrocnemius medialis (GM), and tibialis anterior (TA) in the left leg by multichannel near-infrared spatial resolved spectroscopy (NIR_{SRS}). VLd was defined as 9–13 cm above the patella (30 % of the length between the patella and the greater trochanter). VLp was defined at a proximal point of 30 % of the length between the patella and the greater trochanter, from the VLd muscle. The SmO_2 values were defined as the SmO_2 averaged over the last 10 s at rest, every 20 W, and exhaustion.

We used a two wavelength (770 and 830 nm) light-emitting diode NIR_{SRS} (Astem Co., Japan). The probe consisted of one light source and two photodiode detectors, and the optode distances were 20 and 30 mm, respectively. In this study, we measured fat layer thickness at each measurement site in the muscles to correct for the light-scattering effects on SmO_2 [5] using an ultrasound device (LogiQ3, GE-Yokokawa Medical Systems, Japan) by placing an ultrasound probe at the same sites as the NIR_{SRS} probes had been placed. Even though an upper limit of fat layer thickness was designated as 1 cm to correct for the light-scattering effects in this study, fat layer thickness was within ~1 cm at each measurement site in all subjects.

22.2.3 *Statistics*

All data are given as means ± standard deviation (SD). Differences in SmO_2 during rest and peak exercise were compared between groups using unpaired *t* tests. To compare changes in SmO_2 during submaximal exercise between groups, a two-way repeated measures analysis of variance was used with age and exercise intensity as factors. Where appropriate, the Bonferroni post hoc test was performed to determine specific significant differences. Because a subject in ELD could not exercise more than 125 W, repeated measures between groups were limited to 20, 40, 60, 80, 100, and 120 W. For all statistical analyses, significance was accepted at $p<0.05$.

22.3 Results

The resting SmO_2 in VLd, BF, GL, and TA was significantly lower in ELD than YNG ($p<0.05$), although it was not significantly different between groups at the other measurement sites (VLp, RF, VM, BF, and GM). There was a significant age × exercise intensity interaction for change in SmO_2 at RF ($p<0.05$), but not at the other measurement sites. Significantly lower SmO_2 at a given absolute work rate was observed in VLd, RF, BF, GL, and TA but not in VLp, VM, and GM. Consequently, there was a significant age × exercise intensity interaction for change in SmO_2 at mean SmO_2 of whole leg muscles (VLd, RF, VM, BF, GL, GM, and TA) ($p<0.05$), and mean SmO_2 was lower in ELD than in YNG during submaximal exercise ($p<0.05$) (Fig. 22.1). In contrast, at all measurement sites, SmO_2 at peak exercise was not significantly different between ELD and YNG. Peak VO_2 (24.4 ± 1.5 vs. 46.4 ± 7.4 mL/kg/min, $p<0.05$) and peak workloads (135 ± 8 vs. 250 ± 30 W, $p<0.05$) were significantly lower in ELD than in YNG.

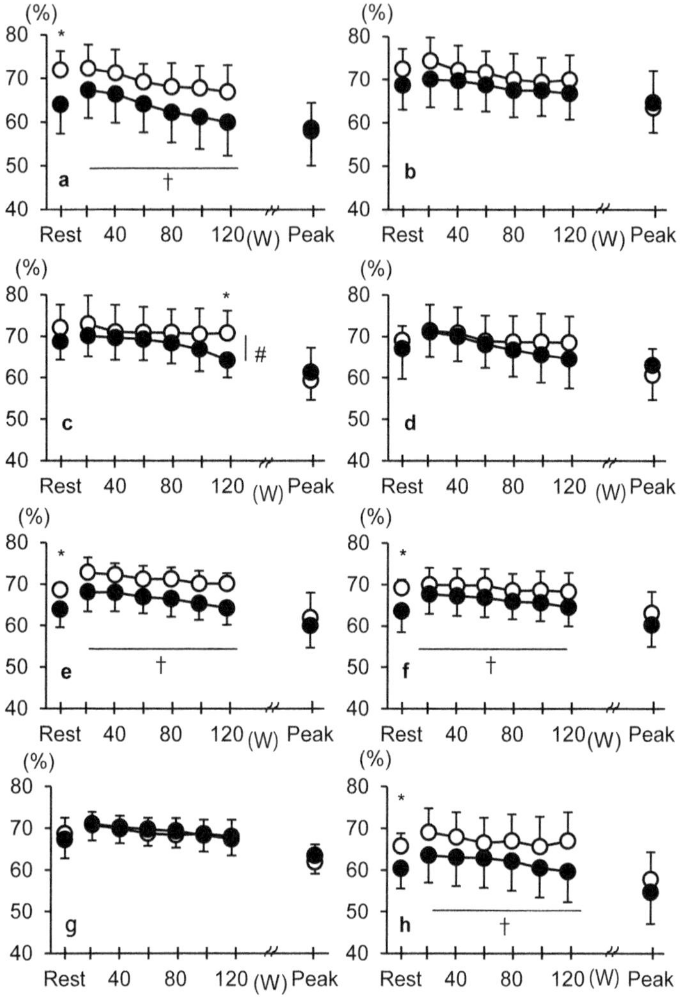

Fig. 22.1 SmO_2 responses in VLd (**a**) VLp (**b**), RF (**c**), VM (**d**), BF (**e**), GL (**f**), GM (**g**), and TA (**h**) muscles during ramp cycling exercise. The *solid circles* show SmO_2 in ELD, and the *open circles* show SmO_2 in YNG. There was a significant difference between ELD and YNG (*, $p<0.05$). There was a significant age × exercise intensity interaction (#, $p<0.05$). There was a main effect of age (†, $p<0.05$)

22.4 Discussion

The results of the present study provided two major findings. First, the SmO_2 responses during rest and submaximal exercise were different between YNG and ELD in some muscles and some parts of muscles. Second, SmO_2 at peak exercise was not affected by aging at any measurement sites. These results suggest that blood flow and metabolic demands are heterogeneous between muscles and also within a single muscle in ELD as well as YNG, and the spatial distribution of leg SmO_2 during submaximal exercise is affected by aging.

The influences of aging on SmO_2 responses were considerably different between muscles during submaximal cycling exercise. Previous studies demonstrated that aging blunts the vasodilation responses in skeletal muscle arterioles and the age-related impairment differs between skeletal muscles [6]. Reduction in blood flow with aging may cause a compensatory increase in O_2 extraction in exercising leg muscles [7]. Therefore, the influence of aging on regional differences in SmO_2 responses may be explained by regional differences in vascular responses. Another possible reason for regional differences in effects of aging on SmO_2 response was differences in the action of the muscles. For example, SmO_2 response in RF, which is one of the knee extensor/hip flexor muscles, seemed to be decreased during relatively high exercise intensity, although it was relatively maintained until around 60 % of peak exercise [8]. Additionally, muscle perfusion and muscle O_2 consumption were presumably affected by differences in muscle fiber composition [9]. In fact, muscle fiber composition was heterogeneous between leg muscles [10] and also within a single muscle [11] in young subjects. Moreover, reduction in muscle fiber area occurred in glycolytic fibers, but not in oxidative muscle fibers with aging [12]. Musch et al. reported that reductions in blood flow mainly occurred in oxidative muscles in older rats during submaximal exercise [4]. Hence, we can speculate that the lower SmO_2 in elderly subjects can be attributed to reduced muscle blood flow or altered blood flow distribution, secondary to attenuation of vasodilation responses and/or alteration of muscle fiber composition. However, further research is needed to clarify the mechanism of regional differences in effects of aging on SmO_2 responses.

Interestingly, peak SmO_2 was not different in any muscles in the present study. In contrast, Costes et al. reported that SmO_2 at peak exercise was lower in older subjects in VL muscle [2]. The disparity was mainly caused by methods of normalizing NIRS signals. In a previous study [2], a cuff ischemia method was applied to normalize the NIRS signal to the maximal oxygenated value and the maximal deoxygenated value. Thus, absolute SmO_2 values at rest and peak exercise cannot be measured by cuff ischemia methods, in contrast to methods of correction of fat layer thickness effects. From our findings, we presume that O_2 extraction in skeletal muscle may be one of the factors in limiting exercise, regardless of age.

In conclusion, the influences of aging on SmO_2 responses are heterogeneous between leg muscles and also within a single muscle. The lower SmO_2 in elderly men may have been caused by reduced muscle blood flow or altered blood flow distribution. Furthermore, regardless of aging, oxygen extraction in skeletal muscle may be one of the factors in limiting peak VO_2, because SmO_2 at peak exercise is similar between ELD and YNG.

Acknowledgments The authors are grateful for revision of this manuscript by Andrea Hope. We also thank Miwako Tanabe, Yuki Takahashi, Chie Hamanaka, Toshimichi Nakazawa, and Tae Minegishi for their helpful technical assistance. This study was supported in part by Grant-in-Aid for JSPS fellows from Japan Society for the Promotion of Science to S.T.

References

1. Kime R, Osada T, Shiroishi K et al (2006) Muscle oxygenation heterogeneity in a single muscle at rest and during bicycle exercise. Jpn J Phys Fit Sport Med 55(Suppl):S19–S22
2. Costes F, Denis C, Roche F, Prieur F, Enjolras F, Barthélémy JC (1999) Age-associated alteration of muscle oxygenation measured by near infrared spectroscopy during exercise. Arch Physiol Biochem 107(2):159–167
3. Kalliokoski KK, Oikonen V, Takala TO, Sipila H, Knuuti J, Nuttila P (2001) Enhanced oxygen extraction and reduced flow heterogeneity in exercising muscle in endurance-trained men. Am J Physiol Endocrinol Metab 280:E1015–E1021
4. Musch TI, Eklund KE, Hageman KS, Poole DC (2004) Altered regional blood flow responses to submaximal exercise in older rats. J Appl Physiol 96(1):81–88
5. Niwayama M, Sone S, Murata H, Yoshida H, Shinohara S (2007) Errors in muscle oxygenation measurement using spatially-resolved NIRS and its correction (Abstract in English). J Jpn Coll Angiol 47(1):17–20
6. Muller-Delp JM, Spier SA, Ramsey MW, Delp MD (2002) Aging impairs endothelium-dependent vasodilation in rat skeletal muscle arterioles. Am J Physiol Heart Circ Physiol 283(4):H1662–H1672
7. Poole JG, Lawrenson L, Kim J, Brown C, Richardson RS (2003) Vascular and metabolic response to cycle exercise in sedentary humans: effect of age. Am J Physiol Heart Circ Physiol 284(4):H1251–H1259
8. Takagi S, Kime R, Niwayama M, Murase N, Katsumura T (2013) Muscle oxygen saturation heterogeneity among leg muscles during ramp exercise. Adv Exp Med Biol 765:273–278
9. Poole DC, Mathieu-Costello O (1996) Relationship between fiber capillarization and mitochondrial volume density in control and trained rat soleus and plantaris muscles. Microcirculation 3(2):175–186
10. Johnson MA, Polgar J, Weightman D, Appleton D (1973) Data on the distribution of fibre types in thirty-six human muscles. An autopsy study. J Neurol Sci 18(1):111–129
11. Lexell J, Henriksson-Larsén K, Sjöström M (1983) Distribution of different fibre types in human skeletal muscles. 2. A study of cross-sections of whole *M. vastus lateralis*. Acta Physiol Scand 117(1):115–122
12. Lexell J (1993) Ageing and human muscle: observations from Sweden. Can J Appl Physiol 18(1):2–18

Chapter 23
Which Is the Best Indicator of Muscle Oxygen Extraction During Exercise Using NIRS?: Evidence that HHb Is Not the Candidate

Ryotaro Kime, Masako Fujioka, Takuya Osawa, Shun Takagi, Masatsugu Niwayama, Yasuhisa Kaneko, Takuya Osada, Norio Murase, and Toshihito Katsumura

Abstract Recently, deoxygenated hemoglobin (HHb) has been used as one of the most popular indicators of muscle O_2 extraction during exercise in the field of exercise physiology. However, HHb may not sufficiently represent muscle O_2 extraction, as total hemoglobin (tHb) is not stable during exercise. The purpose of this study was to measure various muscle oxygenation signals during cycle exercise and clarify which is the best indicator of muscle O_2 extraction during exercise using NIRS. Ten healthy men performed 6-min cycle exercise at both moderate and heavy work rates. Oxygenated hemoglobin (O_2-Hb), HHb, tHb, and muscle tissue oxygen saturation (SmO_2) were measured with near-infrared spatial resolved spectroscopy from the vastus lateralis muscle. Skin blood flow (sBF) was also monitored at a site close to the NIRS probe. During moderate exercise, tHb, O_2-Hb, and SmO_2 displayed progressive increases until the end of exercise. In contrast, HHb remained stable during moderate work rate. sBF remained stable during moderate exercise but showed a progressive decrease at heavy work rate. These results provide evidence that HHb may not sufficiently represent muscle O_2 extraction since tHb is not stable during exercise and HHb is insensitive to exercise-induced hyperaemia.

R. Kime (✉) • S. Takagi • T. Osada • N. Murase • T. Katsumura
Department of Sports Medicine for Health Promotion, Tokyo Medical University, Tokyo, Japan
e-mail: kime@tokyo-med.ac.jp

M. Fujioka
Faculty of Sport and Health Science, Ritsumeikan University, Kyoto, Japan

T. Osawa
Department of Sports Sciences, Japan Institute of Sports Science (JISS), Tokyo, Japan

M. Niwayama
Department of Electronic and Electrical Engineering, Shizuoka University, Shizuoka, Japan

Y. Kaneko
Department of Sports Medicine for Health Promotion, Tokyo Medical University, Tokyo, Japan

Department of Oriental Medicine, Kuretake College of Medical Arts and Sciences, Tokyo, Japan

S. Van Huffel et al. (eds.), *Oxygen Transport to Tissue XXXV*, Advances in Experimental Medicine and Biology 789, DOI 10.1007/978-1-4614-7411-1_23,

23.1 Introduction

Several studies have reported that deoxy-hemoglobin concentration (HHb) reflects the dynamic balance between the delivery and utilization of O_2 and thus HHb has often been used as an indicator of muscle O_2 extraction during exercise [1, 2]. As reported by Quaresima and Ferrari, HHb may represent O_2 extraction when total hemoglobin (tHb) is stable [3]. However, we have reported that tHb progressively increased during cycle exercise at constant work rate [4]. Although the increase of tHB may be caused by exercise-induced hyperemia, we cannot exclude the possibility that tHb changes during exercise may be due to the effects of increased skin blood flow. The purpose of this study was to determine the effects of skin blood flow on tHb changes during cycle exercise at constant work rate and clarify which is the best indicator of muscle O_2 extraction during exercise using NIRS.

23.2 Methods

23.2.1 Subjects

Ten healthy men (age, 27 ± 1 year; height, 169.4 ± 4.1 cm; weight, 65.9 ± 8.4 kg) participated in this study. All subjects were briefed about the experimental protocol, and written informed consent was obtained before the experiment. The institutional review board of Tokyo Medical University approved the research protocol.

23.2.2 Experimental Design

During the first visit, the subjects performed a ramp-incremental cycling exercise test to determine peak pulmonary VO_2 (pVO_2), gas exchange threshold (GET), and work rates for the constant work rate tests. Over the next two visits, moderate and heavy cycling exercise was performed with 7 days between visits. The moderate exercise work rate was conducted at 90 % of GET. The heavy exercise work rate was estimated to require a VO_2 equal to half of the individual difference (Δ) between GET and pVO_2 (i.e., GET + 0.5Δ).

23.2.3 Measurements

Oxygenated hemoglobin (O_2-Hb), HHb, tHb, and muscle tissue oxygen saturation (SmO_2) were measured with near-infrared spatial resolved spectroscopy (NIR_{SRS}) (NIRO-200, Hamamatsu, Japan) from the vastus lateralis muscle. The tHb, O_2-Hb, and HHb responses were evaluated from the minimum values occurring within 1 min after the onset of main exercise to the maximal values occurring within 1 min before the cessation of exercise as functional tHb change (*f*-[tHb]), functional O_2-Hb change (*f*-[O_2-Hb]), and functional HHb change, respectively [4]. A previous study reported that fat layer thickness affects SmO_2 [5]. In contrast, Kek et al. [6] recently reported that SmO_2 can be quantified by the correction of fat layer thickness effects, and the specifications of the NIR_{SRS} were fully described. In this study, we measured fat layer thickness at each measurement site in VL muscle to correct these effects using an ultrasound device (LogiQ3, GE-Yokokawa, Japan) by placing an ultrasound probe on the same sites as the NIR_{SRS} probes had been placed.

Both surface EMG electrodes and the skin blood flow (sBF) probe were attached close to the NIRS probe and monitored during the experiment (ATBF-LN1, Unique Medical, Japan). During all of the tests, pVO_2 and carbon dioxide production (VCO_2) were assessed breath by breath with an online metabolic system (AE-300 Minato, Japan).

23.2.4 Exercise Protocol

The incremental exercise protocol preceded by 1-min unloaded exercise was performed in an upright position. The work rate protocol for the ramp test was 20 W/min and pedal frequency of 60 rpm was maintained by keeping time with a metronome (Strength Ergo 8, Fukuda Denshi, Japan). The constant exercise protocol consisted of 1 min of rest and 1 min of unloaded exercise, followed by two 6-min bouts of each exercise. Before each exercise, 3 min of moderate exercise was performed as a warm-up, followed by 5-min recovery. The seat height and handlebar position on a cycle ergometer were recorded and reproduced on subsequent tests.

23.2.5 Statistics

The changes in recorded variables during the exercise experiments were analyzed by two-way ANOVA for repeated measurements. *f*-[O_2-Hb], *f*-[tHb], and *f*-[HHb] were compared among moderate and heavy work rate using Student's paired *t*-test. Regression and correlation analyses were performed by the least squared residuals method. The level of significance was set at $P<0.05$.

23.3 Results

Typical responses of tHb and O_2-Hb to moderate work rate cycle exercise are illustrated in Fig. 23.1a. Both tHb and O_2-Hb were stable during 1-min unloaded exercise. At the onset of moderate exercise, tHb steeply decreased and reached minimum

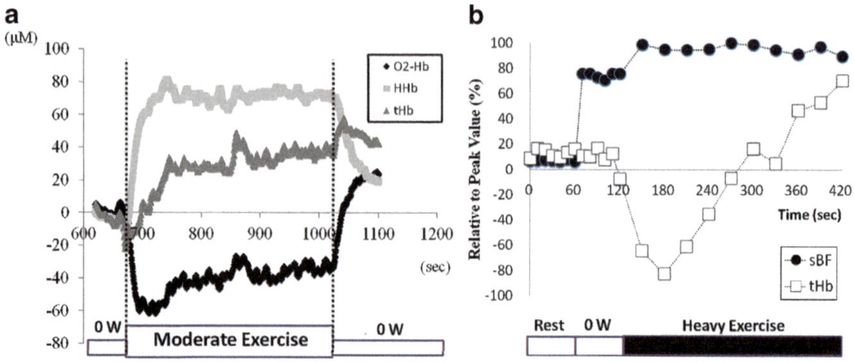

Fig. 23.1 Typical responses of O_2-Hb, HHb, and tHb at moderate work rate (**a**). Typical responses of sBF and tHb at heavy work rate (**b**)

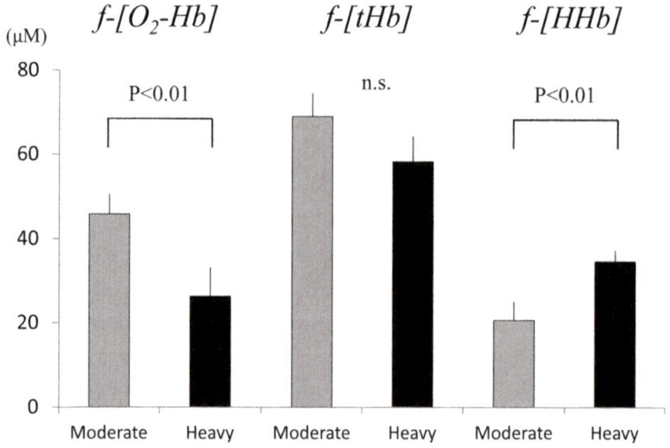

Fig. 23.2 *f*-[O_2-b], *f*-[tHb] and *f*-[HHb] responses at each moderate and heavy work rate

value within 15 s. Thereafter, as the exercise continued, there was a progressive increase of tHb. In addition, after decreasing at the start of exercise, O_2-Hb also progressively increased as the exercise continued, similar to tHb response. At both moderate and heavy work rates, sBF increased steeply at the start of unloaded exercise, even though tHb was stable. During each exercise protocol, sBF was not significantly increased in contrast to tHb, which was gradually increased (Fig. 23.1b). *f*-[O_2-Hb] was significantly higher at moderate than heavy exercise. In contrast, *f*-[tHb] was not significantly different between each work rate. *f*-[HHb] was significantly higher at heavy than moderate exercise (Fig. 23.2). There was significantly positive correlation between *f*-[tHb] and *f*-[O_2-Hb] during exercise ($r=0.729$, $P<0.01$).

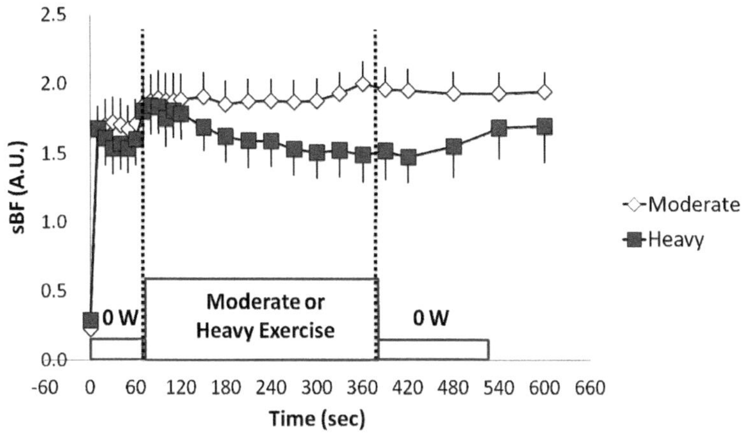

Fig. 23.3 sBF responses at each moderate and heavy work rate

SmO_2 also showed a progressive increase until the end of exercise after an initial drop at the onset of exercise at moderate work rate, but not at heavy work rate. In contrast, HHb remained stable during moderate work rate exercise after an initial increase at the onset of exercise, and HHb at heavy work rate was slightly increased but not significantly changed during the exercise.

sBF increased sharply at the onset of unloaded exercise and was stable during moderate exercise (Fig. 23.3). At heavy work rate, sBF represented a progressive decrease even though tHb was increased (Fig. 23.3).

23.4 Discussion

This study demonstrated that tHb was progressively increased during constant exercise at both moderate and heavy work rates, and the exercise-induced increase of tHb (*f*-[tHb]) was not caused by sBF effects, as tHb and sBF had obviously different kinetics during the constant work rates. These results imply that *f*-[tHb] may be caused by microvascular dilation in muscle tissue. Therefore, HHb may not sufficiently represent muscle O_2 extraction since tHb is not stable during constant exercise and HHb is insensitive to increased O_2 supply to the muscle tissue caused by exercise-induced hyperemia.

We observed that *f*-[tHb] was not significantly different between each work rate. Because higher exercise intensity causes increased adenosine, phosphate, CO_2 and potassium, and decreased blood PO_2, arteriolar dilation may be increased by continued exercise [7]. However, at higher exercise intensities, O_2 supply to the muscle may be restricted by mechanical effects due to increased intramuscular pressure [8–10]. This may explain why blood volume expansion was similar between moderate and heavy work rates. In contrast, *f*-[O_2-Hb] was significantly higher at moderate than heavy exercise. The reason for this may be that muscle O_2 consumption (mVO_2) is greater than muscle O_2 supply at heavy work rate, and the mismatching of dynamic O_2 balance may lead to greater deoxygenation at heavy work rate.

Recently, HHb has been used as one of the most popular indicators of muscle O_2 extraction during exercise in the field of exercise physiology [1, 2]. Basically, however, the increasing rate of HHb during venous occlusion can be used as a mVO_2 index [11]. In other words, the HHb signal may represent a balance between O_2 unloading in muscle tissue and blood outflow from muscle tissue. Therefore, HHb may not demonstrate direct correlation to O_2 delivery, even though both O_2 unloading and blood outflow may (partially or indirectly) affect O_2 delivery.

In conclusion, this study provides evidence that the increased muscle oxygenation during constant work rate cycling exercise may be caused by increased O_2 delivery due to exercise-induced blood volume expansion. Therefore, HHb may not sufficiently represent muscle O_2 extraction since tHb is not stable during constant exercise and HHb is insensitive to increased O_2 delivery to the muscle tissue caused by exercise-induced hyperemia. Since tHb is not stable during exercise, the parameters which have less effects on tHb changes could be a more precise way to evaluate muscle O_2 extraction using NIRS, such as SmO_2 or deoxygenation (HHb minus O_2-Hb).

Acknowledgments The authors are grateful for revision of this manuscript by Andrea Hope. We also thank Mikiko Anjo and Ayaka Sato for their helpful technical assistance. This study was supported in part by Grant-in-Aid for scientific research from the Japan Society for Promotion of Science (24500799) to R. K.

References

1. DeLorey DS, Kowalchuk JM, Paterson DH (2003) Relationship between pulmonary O_2 uptake kinetics and muscle deoxygenation during moderate-intensity exercise. J Appl Physiol 95(1): 113–120
2. Koga S, Poole DC, Fukuoka Y et al (2011) Methodological validation of the dynamic heterogeneity of muscle deoxygenation within the quadriceps during cycle exercise. Am J Physiol Regul Integr Comp Physiol 301(2):R534–R541
3. Quaresima V, Ferrari M (2009) Muscle oxygenation by near-infrared-based tissue oximeters. J Appl Physiol 107(1):371
4. Kime R, Im J, Moser D, Nioka S, Katsumura T, Chance B (2009) Noninvasive determination of exercise-induced vasodilation during bicycle exercise using near infrared spectroscopy. Med Sci Monit 15(3):CR89–CR94
5. Komiyama T, Quaresima V, Shigematsu H, Ferrari M (2001) Comparison of two spatially resolved near-infrared photometers in the detection of tissue oxygen saturation: poor reliability at very low oxygen saturation. Clin Sci (Lond) 101(6):715–718
6. Kek KJ, Kibe R, Niwayama M, Kudo N, Yamamoto K (2008) Optical imaging instrument for muscle oxygenation based on spatially resolved spectroscopy. Opt Express 16(22): 18173–18187
7. Delp MD, Laughlin MH (1998) Regulation of skeletal muscle perfusion during exercise. Acta Physiol Scand 162(3):411–419

8. Saltin B, Rådegran G, Koskolou MD, Roach RC (1998) Skeletal muscle blood flow in humans and its regulation during exercise. Acta Physiol Scand 162(3):421–436
9. Honig CR, Odoroff CL, Frierson JL (1982) Active and passive capillary control in red muscle at rest and in exercise. Am J Physiol 243(2):H196–H206
10. Sadamoto T, Bonde-Petersen F, Suzuki Y (1983) Skeletal muscle tension, flow, pressure, and EMG during sustained isometric contractions in humans. Eur J Appl Physiol 51(3):395–408
11. Homma S, Eda H, Ogasawara S, Kagaya A (1996) Near-infrared estimation of O_2 supply and consumption in forearm muscles working at varying intensity. J Appl Physiol 80(4): 1279–1284

Chapter 24
Tissue Oxygenation During Exercise Measured with NIRS: Reproducibility and Influence of Wavelengths

Erwin Gerz, Dmitri Geraskin, Julia Franke, Petra Platen, André Steimers, and Matthias Kohl-Bareis

Abstract Near-infrared spectroscopy (NIRS) is widely used for the measurement of skeletal muscle oxygenation during exercise as it reflects muscle metabolism, and most studies report a large variability between subjects. Here we assess the data quality of tissue oxygen saturation (SO_2) and oxygenated (oxyHb) and deoxygenated (deoxyHb) haemoglobin concentrations recorded during an incremental cycling protocol in nine healthy volunteers. The protocol was repeated three times on the same day and a fourth session on a different day to estimate the reproducibility of the method with a broadband, spatially resolved spectroscopy (SRS) system. We found that the inter-subject variation in SO_2 (standard deviation ≈ 6 %) was considerably larger than the reproducibility (≈ 1.5 %) both for the same-day and different-day tests. The reproducibility of changes in SO_2 was better than 1 %.

24.1 Introduction

Near-infrared spectroscopy (NIRS) has demonstrated its value in many different areas of exercise physiology, including endurance and strength exercise, both during acute and following chronic exercise training [1, 2]. The objective here is to shed light on the following. It has been noted in a number of studies that the inter-subject variations of the muscle haemoglobin values with its concentrations of

E. Gerz • D. Geraskin • A. Steimers • M. Kohl-Bareis (✉)
RheinAhrCampus, University of Applied Sciences Koblenz, Remagen, Germany
e-mail: kohl-bareis@rheinahrcampus.de

J. Franke • P. Platen
Institute of Sports Medicine, Ruhr University Bochum, Bochum, Germany

Institute of Cardiology and Sports Medicine, German Sport University, Cologne, Germany

S. Van Huffel et al. (eds.), *Oxygen Transport to Tissue XXXV*, Advances in Experimental Medicine and Biology 789, DOI 10.1007/978-1-4614-7411-1_24,

oxygenated and deoxygenated form (oxyHb and deoxyHb, respectively) are large, and therefore, the mean values averaged over subjects are often used. However, limited work has been done so far to show the repeatability or reproducibility of data on the same subject. We address this topic by repeating an incremental cycling exercise three times within a longer exercise trial on 1 day followed by a fourth trial on a different day. Furthermore, we tested whether the modified Lambert-Beer (MLB approach) and spatially resolved spectroscopy (SRS) provided the same haemoglobin concentration changes. Furthermore, the influence of the wavelength is tested.

24.2 Methods

Nine healthy volunteers (five females; mean ± standard deviation (SD) values, age = 29.4 (±7.6) year; body mass = 69.0 (±9.5) kg; height = 178 (±13) cm; $\dot{V}O_2$ max = 53.9 (±9.5) ml·kg^{-1}·min^{-1}; maximal power output P_{max} = 288 (± 77) W) participated in the study. The study was approved by the local Ethics Committee of the German Sport University, Cologne, and all subjects provided their written informed consent. On experimental day 1 (D1), the volunteers performed a maximal incremental exercise protocol on a cycle ergometer (Ergometrics 900, Ergoline GmbH, Germany). Exercise workload was increased stepwise every 3 min (start at P = 30 W, ΔP = 40 W) while maintaining cadence at 70–90 repetitions per min. This protocol was repeated twice separated by rest periods of 40 min, i.e., giving three exercises denoted as D1:E1, E2 and E3. To evaluate the reproducibility of the NIRS signals over a longer time period, the exercise protocol was repeated on a second day (marked as D2:E1) later by 27 (± 18) days.

The spectroscopy system is based on a CCD-camera in combination with a spectrometer with six light detecting fibre bundles of 1 mm diameter arranged in a line separated by Δρ = 2.5 mm at a mean distance <ρ> = 35 mm from the light delivering bundle. Details of the setup are given elsewhere [3]. The attenuation spectra were analysed for the range 700–880 nm according to the spatially resolved spectroscopy (SRS) approach to give haemoglobin concentrations and oxygen saturation [4, 5]. Additionally, changes in haemoglobin were calculated from changes in attenuation by the modified Lambert-Beer (MLB) approach [6] including a differential pathlength factor and its wavelength dependence from the literature [7]. The fibres were placed on the main body of the right vastus lateralis muscle. The adipose tissue thickness (ATT) was measured by ultrasound with mean values of 5.3 (±3.7) mm.

For an assessment of the data quality, first the standard deviation for the different trials was calculated for each subject and each power step and the reproducibility R taken as the mean of these values. When the reproducibility is given as relative value (i.e., in %), it is calculated as the mean over all subjects of the standard deviation of a parameter (e.g., SO_2, ΔSO_2, or deoxyHb, oxyHb) divided by the magnitude of this parameter.

24.3 Results

For each power interval, the time course data were averaged. To avoid transient changes in the NIRS parameters following the power increase, only the last third for each power interval was included, and this analysis was done for all oxygenation parameters both for the SRS and MLB methods. The results are shown in Fig. 24.1 for a representative subject where in (a) and (c), the absolute values of SO_2, oxyHb and deoxyHb based on SRS are shown. For the MLB analysis (e), changes in concentrations are plotted with respect to an arbitrary baseline defined by the reference spectra taken with the integrating sphere preceding the study. In Fig. 24.1b, 24.1d and 24.1f, changes are shown obtained from data of (a), (c) and (e) after subtracting values at P=30 W for each exercise. The low power output condition was used rather than the pre-exercise condition as the variations in the haemoglobin parameters are higher during rest. When comparing the SO_2 and ΔSO_2 versus power traces for the different exercises, the variations are smaller than 2 %. OxyHb (and ΔoxyHb) versus power is markedly different for D1 and D2 and with larger variations than for deoxyHb (and ΔdeoxyHb). The shape of the power dependence of ΔdeoxyHb and to a lesser degree of ΔoxyHb is similar for the SRS and the MLB analysis (compare d and f). However, the absolute magnitude of changes is about a factor of 5 lower for the MLB approach. While there is approximately a linear increase in deoxyHb

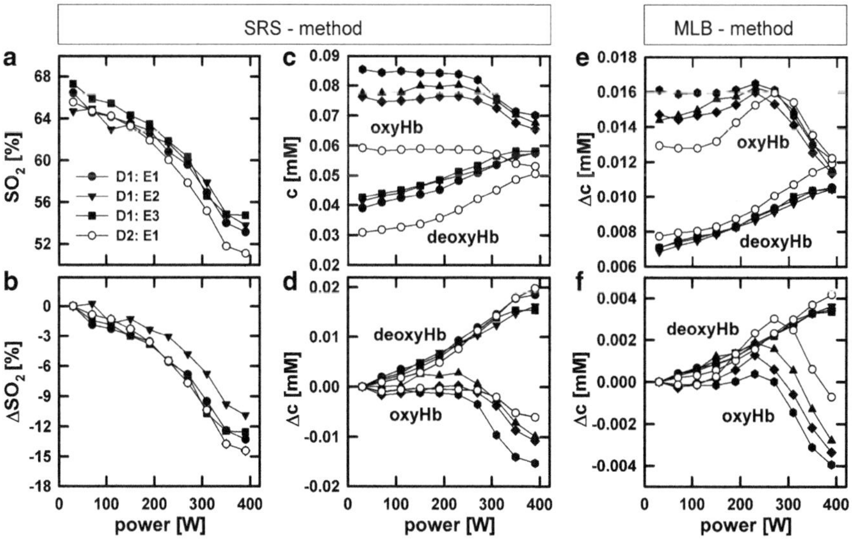

Fig. 24.1 Variations of oxygenation parameters for a representative subject as a function of exercise power. SRS analysis for 700–880 nm: (**a**) SO_2, (**b**) ΔSO_2, (**c**) oxyHb and deoxyHb and (**d**) ΔoxyHb and ΔdeoxyHb. MLB-method: (**e**) ΔoxyHb and ΔdeoxyHb (reference: start of measurement) and (**f**) ΔoxyHb and ΔdeoxyHb. The Δ-values of (**b**), (**d**) and (**f**) are referenced to the first power step (P=30 W) of each trial. The *solid symbols* represent the three exercise trials of day 1 (D1:E1, E2 and E3), while the *open symbols* represent data of day 2 (D2:E1)

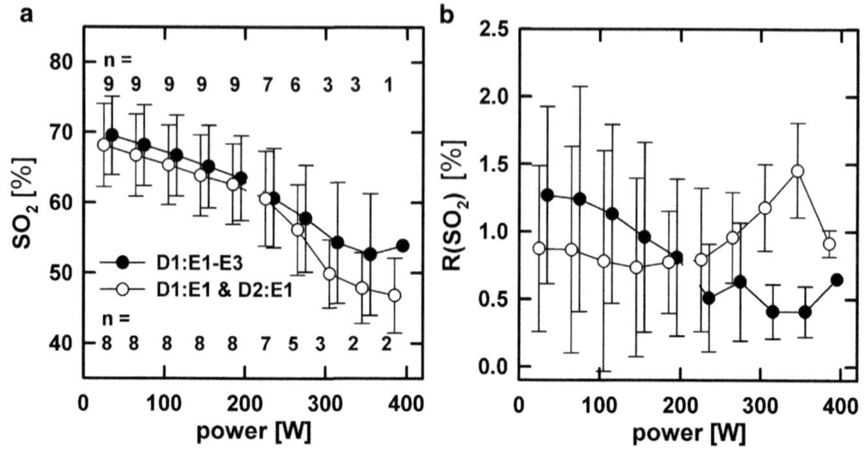

Fig. 24.2 (**a**) Mean (± SD) oxygenation parameters calculated for the SRS analysis of SO_2 as a function of power, with the error bars representing the inter-subject variability. In (**b**), the reproducibility R calculated as the standard deviation (error bars) averaged over all subjects is shown. The number of subjects (n) for each power is stated in (**a**). *Solid symbols*: D1:E1–E3 (n ≤ 9), *open symbols*: D1:E1 and D2:E1 (n ≤ 8)

(and ΔdeoxyHb) with exercise power output, oxyHb (and ΔoxyHb) has a biphasic power dependence with an initial increase (corresponding to a rise in total haemoglobin) followed by a decrease for higher exercise power outputs.

For all subjects, the mean (± SD) values of the oxygenation parameters were calculated for the different exercises and from this the mean and the reproducibility R (see Fig. 24.2). The values for P ≤ 190 W and >190 W cannot directly be compared as a different number of subjects (n) exceeded this power. SO_2 dropped from about 70 % to 50 % with increasing power output, with the initial decrease linearly correlated with power. For same-day and different-day tests, the parameters are very similar. The reproducibility R (Fig. 24.2b) is between 0.5 % and 1.5 % for SO_2 when same-day and different-day repetitions are considered and decreases with increasing power for the same-day test.

Table 24.1 summarizes the reproducibility when averaging over P = 30–190 W, with values given for same-day (D1:E1–E3) and different-day (D1:E1 and D2:E1) comparison. The reproducibility of SO_2 is about 1.5 %. The reproducibility of deoxyHb is somewhat better than for oxyHb for the same-day trials and comparable for the different-day trials. The relative reproducibility R of ΔSO_2 (i.e., the mean of the standard deviation divided by the magnitude of ΔSO_2) is about 12 % and 5.3 % for D1 and D2. Relative R is about 11 % for ΔdeoxyHb and significantly worse (factors 5–6) for ΔoxyHb. The reproducibility of ΔdeoxyHb is considerably better with SRS compared with MLB.

To evaluate the exercise power dependence of R, the values of the highest power steps for each subject were averaged giving a normalized value of the reproducibility R = 0.85 % for SO_2, indicating that R is lower for high exercise power than for low power (R = 1.3 %).

Table 24.1 Summary of the reproducibility R of oxygenation parameters averaged for the workload from 30 to 190 W. The analysis is separated for the tests at the same day (D1:E1–E3, n=9) and for different days (D1:E1 and D2:E1, n=8). For SRS, values are given both for absolute values and changes with respect to the power level for P=30 W

Parameter	Same-day trial	Different-day trial
SRS analysis		
SO_2	1.3 % (absolute)	1.6 % (absolute)
deoxyHb	2.9 %	7.9 %
oxyHb	4.8 %	7.2 %
ΔSO_2	12 %	5.3 %
ΔdeoxyHb	11 %	11 %
ΔoxyHb	66 %	51 %
MLB analysis		
ΔdeoxyHb	27 %	47 %
ΔoxyHb	56 %	48 %

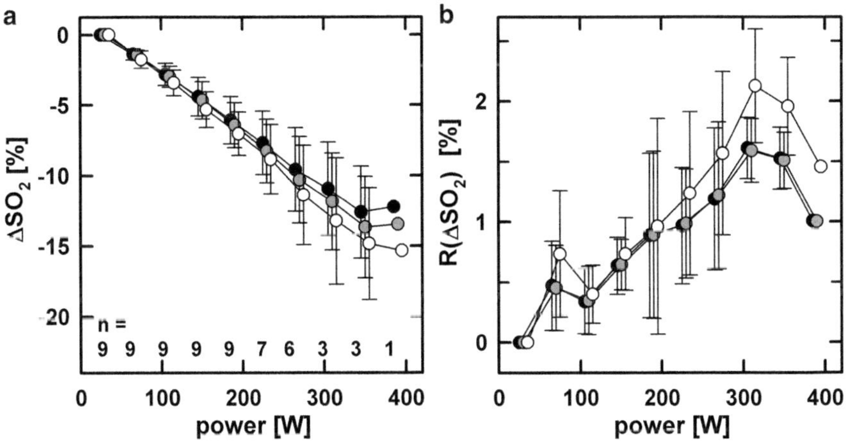

Fig. 24.3 (**a**) Mean (± SD) ΔSO_2 values calculated from SRS analysis and (**b**) reproducibility R comparing results for different wavelength ranges: *solid symbols*, λ=700–880 nm; *grey symbols*, λ=740–860 nm; *open symbols*, 4-λ analysis (775, 810, 850 and 905 nm). Exercises D1:E1–E3 were analysed. To highlight variations due to the wavelength ranges data at all exercise, power steps are shown even when the number of subjects n (compare **a**) is low

The influence of the wavelength range was evaluated by comparing the oxygenation parameters based on (1) the broadband range from 700 to 880 nm, (2) the broadband range from 740 to 860 nm and (3) the four wavelengths $\lambda_{1\text{-}4}$=775, 810, 850 and 905 nm (band width 5 nm each) which correspond to those used in the commercial system NIRO-300 (Hamamatsu Photonics, Japan). The mean (± SD) ΔSO_2 values and the reproducibility R are presented in Fig. 24.3 for these ranges when the same-day exercises were evaluated. As the focus here is on variations due to different wavelengths, the data are shown for all power steps even when the number of subjects is low for the higher exercise power outputs. Up to P = 190 W (n = 9), there is no significant influence of the different wavelength ranges both on the oxygenation parameters and R. The deviations increase at the higher exercise power output (P > 190 W). For P = 390 W, the difference is about 3 % in ΔSO_2 for an overall signal ΔSO_2 of 13 %. However, these data are from a single subject. Similarly, the effect of the wavelength ranges was small for the different-day exercises (data not shown).

24.4 Conclusion

We report here for the first time about a detailed analysis of reproducibility of NIRS data and a comparison of SRS and MLB methodology on data collected during maximal incremental exercise in a group of healthy, recreational active athletes. A well-established exercise protocol was repeated three times on the same day (D1:E1–E3) giving an overall length of about 4 h, which is much longer than most similar NIRS studies. In addition, on a separate day (D2), all subjects repeated the same test to investigate possible variations over an extended period between testing.

When the SRS analysis was used the reproducibility for the subjects, R is about 1.5 % for the same-day and the different-day tests of SO_2 (Table 24.1). The reproducibility of the haemoglobin concentrations is somewhat poorer (relative variations, <5 % for same-day, <8 % for different-day) indicating larger variations in blood volume. However, this is still within the biological variability with respect to the measurement of physiological variables during exercise and is accepted in the exercise science literature. Remarkable is that the reproducibility both for individual subjects and as a mean was found to be smaller for a high workload. This can be interpreted that at lower exercise workload blood volume (and likely blood flow) is more variable due to the metabolic adjustment during the initiation of exercise.

The initial purpose of the study was to verify whether three identical exercises spread over 4 h would result in fatigue concurrent with a faster decrease in SO_2 with exercise power output. This was not observed in our data, as the resulting decrease in SO_2 reached approximately the same level of saturation at the end of each exercise bout. Our expectation and hypothesis was to find large variations in the oxygenation parameters when repeating the exercise, and we were surprised by the high repeatability. From our work, we can infer that even small signals of a few percent in SO_2 can contain information. When comparing MLB and SRS analysis, we found a considerably poorer reproducibility for MLB when changes in haemoglobin are considered rather than absolute values.

Our results indicate that the haemoglobin parameters are comparable for three wavelength ranges analysed, and this applies to the mean values of ΔSO_2, ΔoxyHb and ΔdeoxyHb as a function of exercise workload as well as to the inter-subject variability. Within errors, the reproducibility was the same. From our data, it can be concluded that in terms of reproducibility, the broadband approach has no advantage when the standard SRS methodology is employed and only haemoglobin components are calculated.

References

1. Quaresima V, Lepanto R, Ferrari M (2003) The use of near infrared spectroscopy in sports medicine. J Sports Med Phys Fitness 43(1):1–13
2. Wolf M, Ferrari M, Quaresima V (2007) Progress of near-infrared spectroscopy and topography for brain and muscle clinical applications. J Biomed Opt 12(6):062104
3. Geraskin D, Boeth H, Kohl-Bareis M (2009) Optical measurement of adipose tissue thickness and comparison with ultrasound, magnetic resonance imaging, and callipers. J Biomed Opt 14(4):044017
4. Matcher SJ, Kirkpatrick P, Nahid K, Cope M, Delpy DT (1995) Absolute quantification methods in tissue near infrared spectroscopy. Proc SPIE 2389:486–495
5. Suzuki S, Takasaki S, Ozaki T, Kobayashi Y (1999) A tissue oxygenation monitor using NIR spatially resolved spectroscopy. Proc SPIE 3597:582–592
6. Cope M, Delpy DT (1988) A system for long term measurement of cerebral blood and tissue oxygenation in newborn infants by near infrared transillumination. Med Biol Eng Comput 26(3):289–294
7. Essenpreis M, Elwell CE, Cope M, van der Zee P, Arridge SR, Delpy DT (1993) Spectral dependence of temporal point spread functions in human tissues. Appl Opt 32(4):418–425

Chapter 25
Using Portable NIRS to Compare Arm and Leg Muscle Oxygenation During Roller Skiing in Biathletes: A Case Study

Catherine M. Hesford, Stewart Laing, and Chris E. Cooper

Abstract Portable near-infrared spectroscopy (NIRS) has been shown to be a useful and reliable tool for monitoring muscle oxygenation and blood volume changes during dynamic exercise in elite athletes. The wearable nature of such technology permits the measurement of specific muscles/muscle groups during realistic sport-specific exercise tasks in an outdoor environment. The aim of this case study was to observe the effect on arm and leg muscle oxygenation of roller skiing over a typical outdoor racing course. Such information is required by coaches in order to ascertain whether an athlete is using the correct technique at different stages of the course. Two wearable NIRS devices (PortaMon, Artinis Medical Systems) were used to compare muscle tissue oxygen saturation (TSI%) and total haemoglobin (tHb) changes in the quadriceps muscle group (*vastus lateralis*) and a muscle of the upper arm (*triceps*) during roller skiing. During the flat section, quadriceps ΔTSI remained steady in both subjects, whereas triceps ΔTSI showed a reduction (−10 %). During the steep uphill section of the course, arm and leg TSI decreased equally in one subject (ΔTSI = −10 %), whereas there was a difference between the two muscle groups in the other subject ($\Delta TSI_{quadriceps} = -2$ %; $\Delta TSI_{triceps} = -7$ %). A difference was also seen between subjects during the downhill section of the course. This study presents the first example of the use of portable NIRS to assess oxygenation and blood volume changes in multiple muscle groups during roller skiing in a realistic, outdoor setting.

C.M. Hesford (✉) • C.E. Cooper
Centre for Sports and Exercise Science, School of Biological Sciences, University of Essex, Colchester, UK
e-mail: chesfo@essex.ac.uk

S. Laing
Department of Sport and Exercise Sciences, University of Bangor, Bangor, UK

S. Van Huffel et al. (eds.), *Oxygen Transport to Tissue XXXV*, Advances in Experimental Medicine and Biology 789, DOI 10.1007/978-1-4614-7411-1_25,

25.1 Introduction

The recent development of reliable and robust portable near-infrared spectrometers allows for the collection of data relating to local muscle oxygenation and blood volume changes during exercise in a realistic performance setting. Muscle oxygenation and haemoglobin concentration changes have been measured in the *vastus lateralis* muscles during field running in moderately trained athletes [1, 2]. Research with elite athletes during short-track speed skating race simulation highlighted an asymmetry between the oxygenation changes in right and left leg quadriceps [3]. Differences in muscle oxygenation were also seen when skating over different race distances and between male and female subjects [4]. The use of multiple NIRS devices on one subject has proven useful in the speed skating studies, as it permitted simultaneous comparison of a range of local muscle sites, offering a more detailed account of local muscle physiology than could be provided by 'global' physiological measures such as pulmonary VO_2 or blood lactate concentration.

In the sport of biathlon, there are two distinct disciplines: cross-country skiing and shooting. Cross-country skiing requires excellent levels of aerobic endurance; during the summer training season, athletes use roller skiing as a primary training exercise. The technique employed for roller skiing is largely identical to on-snow cross-country skiing. Cross-country skiing requires a whole-body action, with forward propulsion being provided not only by the legs but also by the arms in the 'poling' technique, where the athlete uses long skiing poles to help with forward motion. The influence of the arms (particularly the triceps muscle) is particularly important when skiing uphill. Previous work which used NIRS to evaluate changes in haemoglobin oxygen saturation during treadmill roller skiing at differing inclines found no effect of increasing incline on quadriceps oxygenation but failed to measure triceps oxygenation and account for the effects of increased poling [5].

Great Britain (GB) biathlon squad coaches highlighted a number of questions relating to technique employed by their athletes during cross-country skiing, most notably they wanted to know if the athletes recovered sufficiently on downhill sections of the skiing course, and also wanted to know whether all athletes fully utilised both arms and legs to provide forward propulsion when skiing uphill. Portable NIRS may offer the opportunity to measure changes in muscle oxygenation in both the arms and legs during skiing at different gradients and therefore help to provide the coaches with answers to some of their questions. As such, this case study was undertaken to test the idea that portable NIRS could monitor muscle oxygenation and blood volume changes during roller skiing in a realistic setting. It was hypothesised that skiing over different gradients would elicit different levels of oxygenation in the arm and leg muscles, with the steepest uphill gradient resulting in the highest level of desaturation in both the arms and the legs.

25.2 Methods

Subjects were elite male biathletes ($n=2$), both members of the GB squad. The subjects for this case study were chosen as the two most experienced members of the GB biathlon squad. The study was conducted in Ruhpolding, Germany, which is the squad's training base. The roller-skiing course which the subjects completed was chosen because it contained a mixed profile, comprising a flat section, a gentle incline, and downhill section, and a final steep incline. This permitted the subjects to employ the varying roller-skiing techniques that are required for skiing at different gradients (see Fig. 25.1).

Portable NIRS devices (PortaMon, Artinis, the Netherlands) were attached to the right *vastus lateralis* and *triceps* muscles of the subjects. The PortaMon is a dual wavelength spatially resolved spectrophotometer, which simultaneously provides data pertaining to haemoglobin concentration via the Beer-Lambert method, and a quantitative measure of tissue oxygen saturation (TSI%). The DPF value used was 4.

Devices were fixed into place using micropore tape and secured using a navy blue bandage. The thickness of adipose tissue overlying the muscle under investigation was 3.2 mm (*triceps*) and 4.4 mm (*vastus lateralis*) for subject A and 3.6 mm (*triceps*) and 5.2 mm (*vastus lateralis*) for subject B. Both subjects reported that the devices were comfortable to wear. Although the PortaMon devices provide the facility for 'online' real-time wireless data collection, it was decided to make 'offline' measurements in this case, due to difficulties in remaining in wireless range of the skiers at all times. Therefore, all data were stored onto the memory of the NIRS devices and subsequently downloaded for analysis. Data were collected at a frequency of 10 Hz. Video data of each athlete were synchronised with NIRS-detected data to permit analysis of TSI and tHb changes during each section of the racing track.

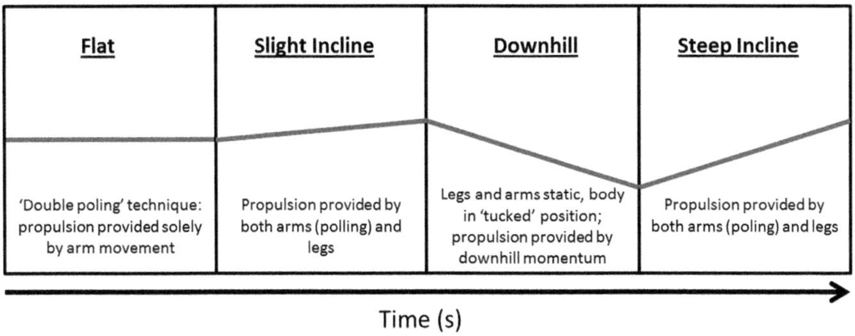

Fig. 25.1 A schematic diagram to show the changing inclines experienced by the athletes on this roller-skiing course, and the principle techniques employed at each stage

25.3 Results

The roller-skiing racing course was completed in 274 s by subject A and 278 s by subject B. Table 25.1 presents the changes in total haemoglobin (tHb) during each phase of the course. It can be seen that in both subjects, the flat section, where propulsion was provided solely by the 'poling' technique elicited a large drop in tHb in the triceps, but not in the quadriceps of each subject. There was a decrease in tHb in both muscle groups during the two phases of uphill skating, and an increase in tHb (suggesting an increase in blood volume) during the downhill section, when the muscles have an opportunity to recover.

Figure 25.2 displays the TSI changes seen in each athlete for the first 30 s of each level of gradient on the course. Both subjects show a similar TSI trend during the flat (double poling) section of the course, with a decrease in triceps TSI%, and a stable quadriceps TSI. In the 'slight incline' stage, there is a reduction in TSI seen in both muscles, and this reduction is greater for subject B than subject A. During the downhill section, when the skier should be in a relaxed position, without any forward propulsive force being applied by arm or legs, subject B shows a rapid resaturation in both triceps and quadriceps, whereas subject A only shows a resaturation in the triceps. The desaturation observed during the steep incline is slightly greater than the desaturations seen during the slight incline in both subjects, although there is only a small change in the quadriceps oxygenation value of subject A during this time.

25.4 Conclusions

The results of this case study suggest that portable NIRS is a viable tool which can be used to assess muscle oxygenation in multiple muscle groups during cross- country (roller) skiing. The two subjects monitored showed largely similar results, although subject A showed smaller changes in quadriceps oxygenation (TSI) than subject B. This may mean that there is a difference between the two athletes in terms of the ability to adequately oxygenate the working muscle (since TSI provides a measure of the dynamic balance between oxygen delivery and utilisation), or it could mean that subject A was not using his leg muscles to the same extent as

Table 25.1 Mean tHb values during each stage of the roller-skiing course in each subject. Values are mean changes (with SD in parentheses) from a baseline set at the start of each stage

	Subject A		*Subject B*	
	Vastus lateralis	Triceps	Vastus lateralis	Triceps
Flat	1.5 (0.5)	−17.5 (2.5)	−9.3 (0.9)	−23.0 (3.0)
Slight incline	−10.0 (1.0)	−8.9 (2.2)	−10.1 (1.9)	−19.2 (2.3)
Downhill	8.4 (1.2)	15.6 (6.6)	14.0 (4.0)	10.0 (6.2)
Steep incline	−9.4 (1.2)	−14.3 (1.7)	−20.6 (2.4)	−19.8 (2.7)

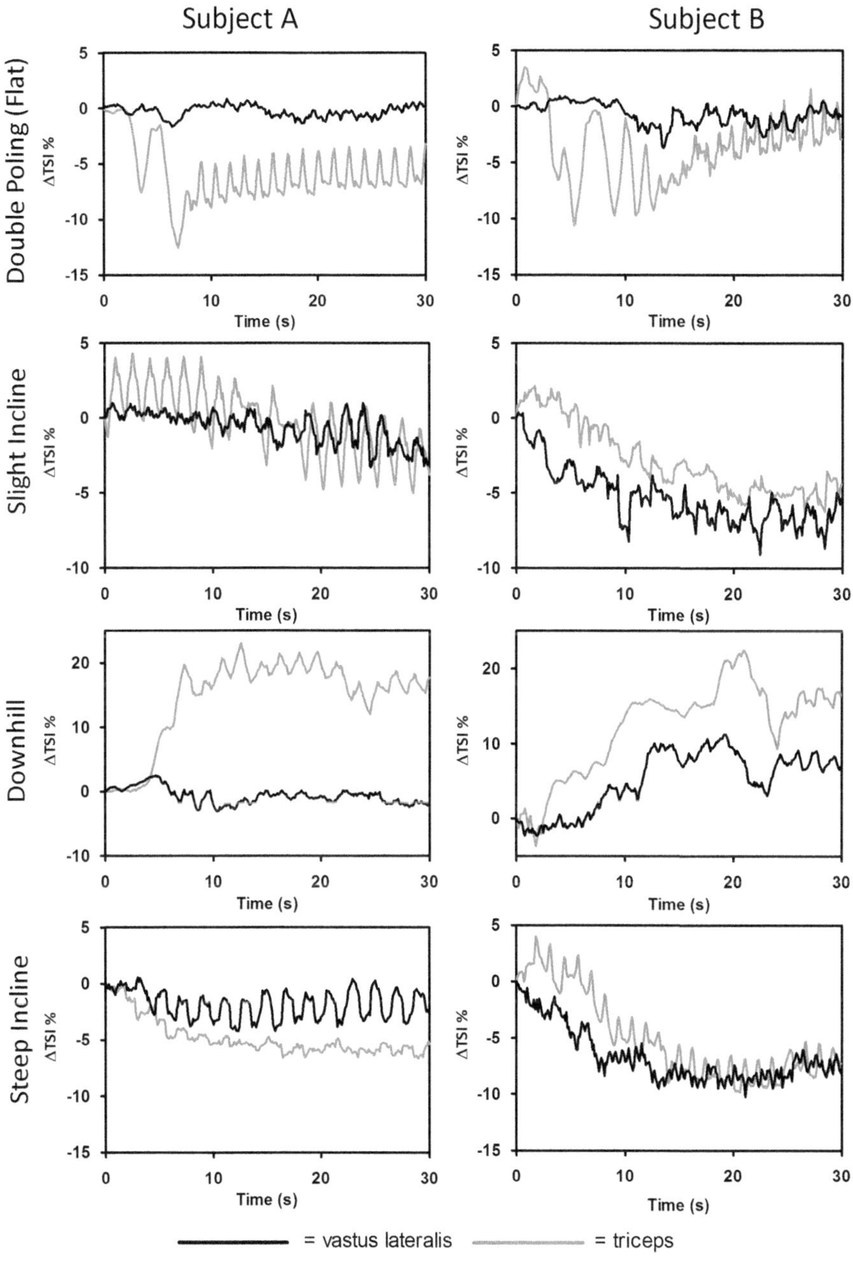

Fig. 25.2 Changes in both vastus lateralis and triceps TSI in each individual during all four stages of the roller skiing course. Data shows changes from a baseline set at the beginning of each stage

subject B, instead gaining propulsion primarily through the poling action. It is proposed that further studies will be carried out (potentially on snow), which can further explore this area, by including a stage when only the legs are used, and examining the oxygenation data during that exercise. It is also interesting to note that subject A does not show a resaturation in the *vastus lateralis* during the downhill stage. This may suggest that the quadriceps muscle group is contracted during the descent, rather than being relaxed, as it appears is the case for subject B (who shows rapid quadriceps resaturation during the downhill phase). This information could be used by the coaches to instruct the skier to alter his technique to capitalise on the downhill skiing phase, by relaxing the leg muscles to allow muscle reoxygenation and temporary recovery.

The results of this study suggest that this application of the NIRS technique could be very useful for both coaches and athletes, as it permits an increased and more detailed understanding of the balance between oxygen delivery and utilisation in specific muscle groups, and provides insight into the parts of a race during which recovery is possible in each muscle group, for each individual.

Acknowledgments The authors wish to thank the athletes and coaches of the Great Britain Biathlon squad for their enthusiastic participation in this work.

References

1. Buchheit M, Bishop D, Haydar B, Nakamura FY, Ahmaidi S (2010) Physiological responses to shuttle repeated-sprint running. Int J Sports Med 31(6):402–409
2. Buchheit M, Cormie P, Abbiss CR, Ahmaidi S, Nosaka KK, Laursen PB (2009) Muscle deoxygenation during repeated sprint running: effect of active vs. passive recovery. Int J Sports Med 30(6):418–425
3. Hesford CM, Laing SJ, Cardinale M, Cooper CE (2012) Asymmetry of quadriceps muscle oxygenation during elite short-track speed skating. Med Sci Sports Exerc 44(3):501–508
4. Hesford CM, Laing S, Cardinale M, Cooper CE (2013) The effect of race distance on muscle oxygenation during short-track speed skating. Med Sci Sports Exerc 45(1):83–92
5. Im J, Nioka S, Chance B, Rundell KW (2001) Muscle oxygen desaturation is related to whole body VO_2 during cross-country ski skating. Int J Sports Med 22(5):356–360

Chapter 26
The Use of Portable NIRS to Measure Muscle Oxygenation and Haemodynamics During a Repeated Sprint Running Test

Ben Jones, Catherine M. Hesford, and Chris E. Cooper

Abstract Portable near-infrared spectroscopy (NIRS) devices were originally developed for use in exercise and sports science by Britton Chance in the 1990s (the RunMan and microRunman series). However, only recently with the development of more robust, and wireless systems, has the routine use in elite sport become possible. As with the medical use of NIRS, finding applications of the technology that are relevant to practitioners is the key issue. One option is to use NIRS to track exercise training-induced adaptations in muscle. Portable NIRS devices enable monitoring during the normal 'field' routine uses to assess fitness, such as repeat sprint shuttle tests. Knowledge about the acute physiological responses to these specific tests has practical applications within team sport training prescription, where development of both central and peripheral determinants of high-intensity intermittent exercise needs to be considered. The purpose of this study was to observe NIRS-detected parameters during a repeat sprint test. We used the PortaMon, a two wavelength spatially resolved NIR spectrometer manufactured by Artinis Inc., to assess NIR changes in the gastrocnemius muscle of both the left and right leg during high-intensity running. Six university standard rugby players were assessed (age 20 ± 1.5 years; height 183 ± 1.0 cm; weight 89.4 ± 5.8 kg; body fat 12.2 ± 3.0 %); the subjects completed nine repeated shuttle runs, which incorporated forward, backward and change of direction movements. Individual sprint time, total time to complete test, blood lactate response (BL), heart rate values (HR) and haemoglobin variables (ΔHHb, ΔtHb, ΔHbO_2 and ΔTSI%) were measured. Total time to complete the test was 260 ± 20 s, final blood lactate was 14.3 ± 2.8 mM, and maximal HR 182 ± 5 bpm. NIRS variables displayed no differences between right and left legs. During the test, the group-averaged data showed a clear decrease in HbO_2 (max. decrease 11.41 ± 4.95 μM), increase in HHb (max. increase 17.65 ± 4.48 μM) and drop in

B. Jones (✉) • C.M. Hesford • C.E. Cooper
Centre for Sports and Exercise Science, University of Essex, Colchester, UK
e-mail: bjonesa@essex.ac.uk

S. Van Huffel et al. (eds.), *Oxygen Transport to Tissue XXXV*, Advances in Experimental Medicine and Biology 789, DOI 10.1007/978-1-4614-7411-1_26,

%TSI (max. drop −24.44±4.63 %). tHb was largely unchanged. However, large interindividual differences were seen for all the NIRS parameters. In conclusion, this observational study suggests that a portable NIRS device is both robust and sensitive enough to detect haemoglobin changes during a high-intensity repeated shuttle run test. It therefore has the possibility to be used to assess exercise training-induced adaptations following a specific training protocol. However, it is at present unclear, given the individual variability, whether NIRS can be used to assess individual performance. We recommend that future studies report individual as well as group data.

26.1 Introduction

This observational study was used to assess usefulness of the PortaMon™ (Artinis Medical Systems BV, the Netherlands) device during a repeat sprint test. There are currently a very limited number of studies that have used portable NIRS in the assessment of haemoglobin variables during repeated high-intensity sprint running. NIRS provides a local measurement for assessing oxygenation in muscle. The data provided by NIRS can establish an insight into muscle reoxygenation and deoxygenation rates following exercise and potentially indicate improved aerobic function as a result of training [1]. This assessment could indicate either an improvement or deterioration in aerobic capacity at a local level. This has importance for the repeat sprint ability required by players for rugby union [2].

It is of importance to understand the physiological responses that may occur within the muscle of an athlete during high-level activity. Sport-specific tests are designed to elicit stresses upon the elite athlete similar to those that may occur during competition. Effective non-invasive monitoring of haemoglobin variables during sport-specific tests could allow the strength and conditioning coach to observe exercise-induced changes within the muscle following a training period [3].

26.2 Methods

Participants were six male university-level rugby players (age 20±1.5 year; height 183±1.0 cm; weight 89.4±5.8 kg; body fat 12.2±3.0 %). Subjects had no history or clinical signs of cardiovascular or pulmonary disease (Par-Q). All subjects gave voluntary written consent to participate in the experiment.

26.2.1 Experimental Overview

Subjects were asked to complete a standardized repeat sprint shuttle test, known as the England Anaerobic shuttle run test or E-set. This test is currently used by a number of professional sports clubs, including the England rugby union squad. The E-set is designed to measure rugby-specific anaerobic performance (in terms of distances and times of work) – the ability to repeat bouts of high-intensity activity with short recovery periods. The E-set is thought to mimic the movement patterns, work: rest ratios, sprint distances and total time requirements of a rugby sevens game. The repeated down-up movement, back pedalling and change of direction elements are considered to be sport-specific to rugby sevens.

Subjects were asked to sprint maximally during each sprint and not to adopt a 'pacing' strategy. Subjects' heart rate (HR) responses were measured pretest and immediately posttest (Polar S610i, Finland). Blood lactate (BL) response was measured pretest, immediately posttest and 3 min posttest (Biosen C line clinic, EKF diagnostic). BL assessment was taken via an earlobe manual lancet sample; 20 μL was collected into capillary tubes, stored in solution and analysed. Haemoglobin variables of the right and left gastrocnemius muscles (near-infrared spectroscopy (NIRS)) were recorded throughout the testing period. All testing was performed outdoors on an all-weather tennis court surface, ambient temperature ranged from 10 °C to 14 °C. Subjects were told not to exercise on the day prior to testing.

26.2.1.1 Near-Infrared Spectroscopy Measurements

The portable NIRS apparatus (PortaMon, Artinis Medical Systems BV, the Netherlands) used in this study was a two-wavelength continuous system, which simultaneously uses the modified Beer-Lambert and spatially resolved spectroscopy (SRS) methods. Myoglobin changes were assumed to be minor compared to haemoglobin. Changes in tissue oxyhaemoglobin (HbO_2), deoxyhaemoglobin (HHb) and total haemoglobin (tHb) were measured using the difference in absorption characteristics of light at 750 and 850 nm. Values for HbO_2, HHb and tHb are reported as a change from baseline (30-s averaging before each test) in micromolar units (μM). To correct for scattering of photons in the tissue, a DPF of 5.0 was used for calculation of absolute concentration changes as previously used by [4]. The tissue haemoglobin saturation index (TSI) was calculated using SRS methods.

NIRS probes were positioned on the medial gastrocnemius, approximately 7 cm from the knee joint and along the vertical axis of the calf. The majority of studies have reported NIRS variables from the vastus lateralis during exercise. However, due to the impact upon the quadriceps during the E-set protocol, NIRS placement was moved to the gastrocnemius as previously described by [5]. A surgical marker was used to mark probe placement for accurate positioning. The device was held securely in place with tape and a black sports covering to prevent contamination from ambient light. It should be noted that motion artifact has previously been reported as an issue during dynamic exercise testing, primarily as a result of the use of non-portable NIRS systems, where an optical fibre attached to a sensor becomes loose and disturbs the sensor [6]. Portable devices such as the PortaMon™ with no loose wires reduce the effect of motion artifact. Attention to secure probe placement and attachment before commencement of testing, followed by a careful device detachment following testing, during which researchers observed an indentation in the skin suggested that any movement artifact due to probe movement was minimal. During all tests, the NIRS system was connected to a personal computer by Bluetooth™ for data acquisition (1 Hz), analog-to-digital conversion and subsequent analysis.

26.3 Results

Tables 26.1 and 26.2 indicate that subjects exercised to 90 % heart rate maximum produced high BL levels, and BL continued to increase for up to 3 min postexercise, suggesting subjects provided volitional maximal effort during testing. Subjects' E-set scores were considerably higher than scores for elite rugby players which are generally <215.

Averaged data (Fig. 26.1) showed clear decreases in TSI during the sprint phase and increases in the recovery period. However, there were clear individual differences. Some individuals showed fast TSI recovery kinetics during rest periods and a clear hyperemic response at the end of the test (Fig. 26.2); others were unable to reoxygenate during recovery periods and TSI did not return to baseline after the end of the final 3-min recovery period following test cessation (Fig. 26.3). The six recovery periods are indicated in (Figs. 26.2 and 26.3) by R1–R6.

Table 26.1 Group E-set physiological responses ($n=6$)

Physiological variables	X (±SD)
E-set score (s)	260±20
Heart rate pretest (bpm)	80±5
Heart rate posttest (bpm)	182±5
Blood lactate pretest (mmol)	1.07±0.20
Blood lactate posttest (mmol)	14.30±2.88

Table 26.2 Individual E-set physiological responses

Subject	Max HR (bpm)	E-set time (s)	BL IMM posttest (mmol)	BL 3 min posttest (mmol)	Max Δ TSI decrease (%)
S1	185	293.6	15.4	16.0	–27.9
S2	180	255.5	10.8	10.7	–22.6
S3	183	273.1	15.2	17.4	–21.2
S4	179	242.3	18.7	19.7	–25.7
S5	190	248.4	11.5	12.4	–18.2
S6	173	244.7	14.0	13.9	–30.8
Mean	182	260.0	14.3	15.0	–24.4
SD	5.8	20.0	2.8	3.3	–4.6

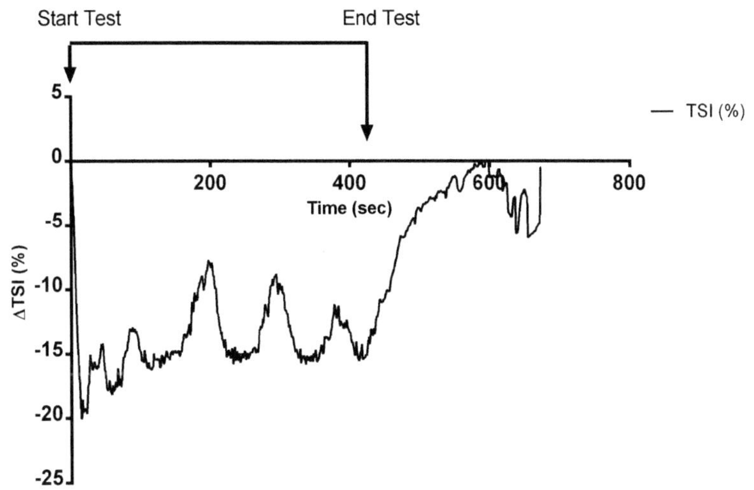

Fig. 26.1 Average group TSI% changes during the E-set test

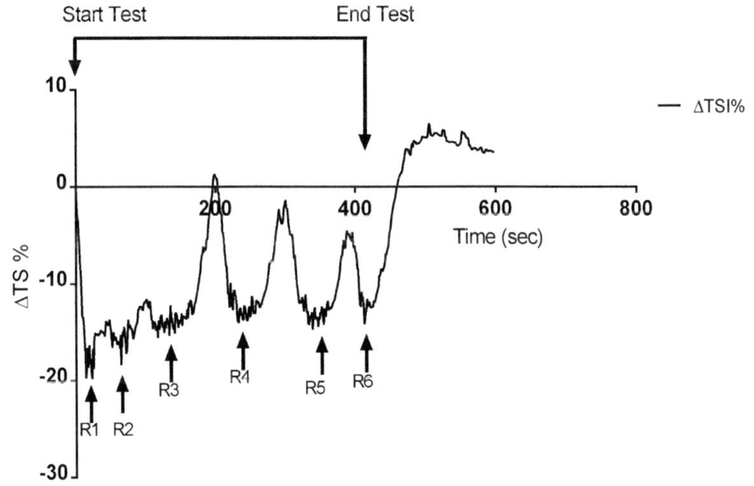

Fig. 26.2 Individual TSI% changes during the E-set test

During the test, the group-averaged data (Fig. 26.4) showed a clear decrease in HbO_2 (max. decrease 11.41 ± 4.95 μM) and increase in HHb (max. increase 17.65 ± 4.48 μM); tHb was largely unchanged until the end of the exercise test, where a hyperemic response was observed.

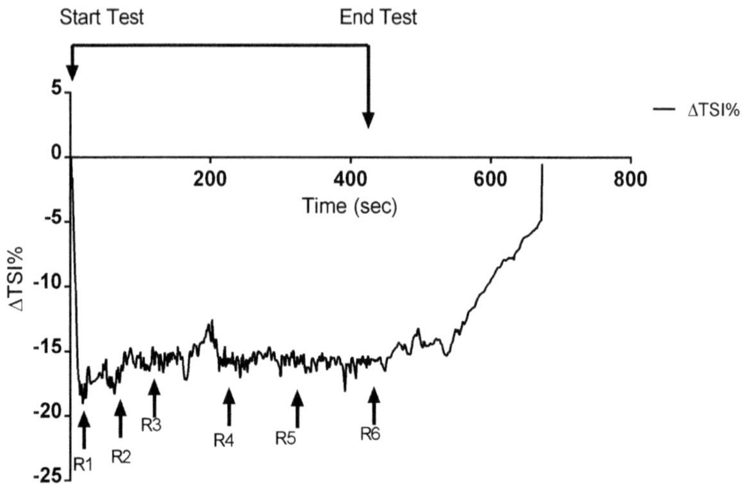

Fig. 26.3 Individual TSI% changes during the E-set test

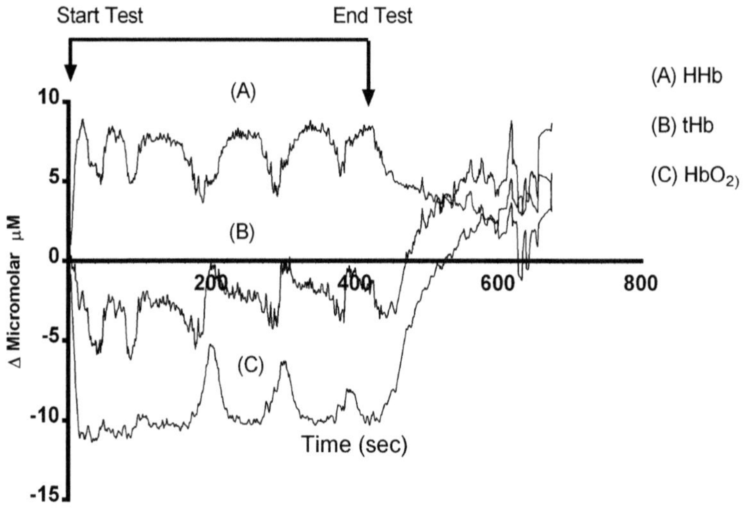

Fig. 26.4 Group haemoglobin variables changes (ΔHHb, ΔtHb, ΔHbO_2)

26.4 Conclusions

The results of this observational study suggest that a portable NIRS device is both robust and sensitive enough to detect haemoglobin changes during a high-intensity repeated shuttle run test. A recent study [1] examining muscle reoxygenation kinetics following various running intensities supports the finding of this study.

The portable NIRS device therefore has the possibility to be used to assess exercise training-induced adaptations following a specific training protocol, although it is unclear at present whether NIRS can be used to assess individual, as opposed to group, performance.

The results largely demonstrate the expected physiological responses seen when moderately trained individuals are exposed to exercise maximally and repeatedly. The average results of the group data demonstrate a clear decrease in HbO_2, an increase in HHb and drop in %TSI during sprint efforts. During rest periods, %TSI largely recovers, HbO_2 increases and HHb decreases. At the end of the test period, a clear and expected hyperemic response occurred as evidenced by an increase in tHb that remained unchanged during the testing period. However, there was considerable individual variability, and we recommend that future training studies report individual as well as group data. The variability seen in the NIRS data may have reflected the variations in fitness levels (evidenced by E-set scores). This variability may decrease in an elite population; this will be the subject of future studies.

References

1. Buchheit M, Ufland P (2011) Effects of endurance training on performance and muscle reoxygenation rate during repeated-sprint running. Eur J Appl Physiol 111(2):293–301
2. Buchheit M, Bishop D, Haydar B, Nakamura FY, Ahmaidi S (2010) Physiological responses to shuttle repeated-sprint running. Int J Sports Med 31(6):402–409
3. Buchheit M, Cormie P, Abbiss CR, Ahmaidi S, Nosaka KK, Laursen PB (2009) Muscle deoxygenation during repeated sprint running: effect of active vs. passive recovery. Int J Sports Med 30(6):418–425
4. Duncan A, Meek JH, Clemence M et al (1995) Optical pathlength measurements on adult head, calf and forearm and the head of the newborn infant using phase resolved optical spectroscopy. Phys Med Biol 40(2):295–304
5. Buchheit M, Ufland P, Haydar B, Laursen PB, Ahmaidi S (2011) Reproducibility and sensitivity of muscle reoxygenation and oxygen uptake recovery kinetics following running exercise in the field. Clin Physiol Funct Imaging 31(5):337–346
6. Ferrari M, Muthalib M, Quaresima V (2011) The use of near-infrared spectroscopy in understanding skeletal muscle physiology: recent developments. Philos Trans A Math Phys Eng Sci 369(1955):4577–4590

Part IV
Tumor Oxygenation

Chapter 27
Amifostine Acts Upon Mitochondria to Stimulate Growth of Bone Marrow and Regulate Cytokines

Wenlong Lv, Mei Zhang, Zhenhuan Zhang, Luqiang Huang, Shanmin Yang, Liangjie Yin, Jinsheng Hong, Deping Han, Chun Chen, Sadasivan Vidyasagar, Paul Okunieff, and Lurong Zhang

Abstract Amifostine is a first-line cytoprotective drug used to prevent radiotherapy-induced or chemotherapy-induced injuries. However, its mechanism of action is not well understood. In this study, freshly harvested bone marrow cells were treated with amifostine and analyzed with a series of mitochondrial indices. In vitro results showed that bone marrow cells treated with amifostine 0.5 h before irradiation (0.5 Gy) experienced several benefits, as compared to vehicle controls, including (1) reduced reactive oxygen species levels, which reduced the production of free radicals; (2) better preservation of mitochondria, as indicated by MitoTracker-positive staining and the increased intensity of staining; (3) reduced apoptosis, as demonstrated by Annexin V staining; and (4) a better proliferation rate, as illustrated by MTT assay. Our in vitro studies showed that amifostine-treated mice exhibited (1) higher ATP production; (2) reduced plasma IL-2 levels, suppressing the immune response triggered by radiotoxicity; and (3) enhanced radiation-induced production of granulocyte colony-stimulating factor. All of these processes benefit recovery from radiation-induced damage.

W. Lv • D. Han • C. Chen
Department of Radiation Oncology, First Affiliated Hospital of Fujian Medical University, Fuzhou, Fujian, China

Department of Radiation Oncology, UF Shands Cancer Center, University of Florida, Gainesville, FL, USA

M. Zhang • Z. Zhang • L. Huang • S. Yang • L. Yin • J. Hong • S. Vidyasagar
P. Okunieff • L. Zhang (✉)
Department of Radiation Oncology, UF Shands Cancer Center, University of Florida, Gainesville, FL, USA
e-mail: lurongzhang@ufl.edu

S. Van Huffel et al. (eds.), *Oxygen Transport to Tissue XXXV*, Advances in Experimental Medicine and Biology 789, DOI 10.1007/978-1-4614-7411-1_27,

27.1 Introduction

Amifostine, (Ethiofos, WR2721, S-2-[3-aminopropylamino] ethylphosphorothioic acid) a low molecular weight compound, is an inactive prodrug that is hydrolyzed in vivo by alkaline phosphatase to the active cytoprotective thiol metabolite; it is subsequently converted into mixed disulfides, which are powerful scavengers for free radicals [1]. When it is intravenously injected 0.5 h before irradiation, it effectively protects normal tissues from radiation-induced damage [2]. While it was originally produced for the US army, its clinical utility for the protection of normal tissue against radiotoxicity and chemotoxicity has been extensively explored. It is now a first-line antioxidant drug [3] that has been used to reduce the incidence of xerostomia in patients undergoing radiotherapy, to decrease the cumulative nephrotoxicity associated with platinum-containing agents, and to protect the bone marrow (BM) from various toxicities [3–5]. However, its mechanism of action is not fully understood. In this study, we explored the effects of amifostine on mitochondrial function, apoptosis, immune modulation, and BM regeneration using BM cells to elucidate its underlying molecular mechanism.

27.2 Methods

Eight-week-old, male, NIH Swiss mice (National Cancer Institute, Frederic, MD, USA) were used for all experiments. Animal protocols were approved by the Animal Ethics Committee at the University of Florida (Gainesville, FL, USA).

For in vitro studies, BM cells were freshly harvested from mouse femurs. Cells (5×10^5/treatment for apoptosis and mitochondrial indices or 8×10^3 cells/well in a U-bottom 96-well microplate for proliferation) were cultured in Dulbecco's modified Eagle's medium (DMEM) containing 10 % heat-inactivated newborn calf serum with 1 % penicillin/streptomycin. BM cells from the same mouse were divided into two treatment groups: (1) saline as vehicle control and (2) incubation with 10 nM of amifostine for 0.5 h. The groups were irradiated with a dose of 0.5 Gy at a dose rate of 0.98 Gy/min via a 137-cesium source (Gammacell-40, Atomic Energy of Canada Limited, Chalk River, Ontario, Canada). After being cultured overnight, cells were assessed with flow cytometry, used according to manufacturer's instructions, for the following indices: (1) reactive oxygen species (ROS) using CM-H2DCFDA (final concentration of 5 μM, Invitrogen Cat # C6827, Grand Island, NY, USA); (2) mitochondrial membrane potential (MMP) using JC-1 (Invitrogen Cat # M34152); (3) mitochondrial level using MitoTracker Red CM-H2XRos (Invitrogen Cat # 7513); and (4) apoptosis using Annexin V/PI staining. BM cell proliferation was assessed with the MTT method (Invitrogen Cat # V13154) [6].

For in vivo studies, NIH Swiss mice (5/group) were pretreated with saline (as vehicle controls) or amifostine (50 mg/kg) 0.5 h before 10-Gy total body irradiation (TBI) and euthanized 1 h later. (1) For analysis of ATP level, freshly prepared BM cells from each group were lysed in 100 μl of ATP assay buffer (BioVision Inc., Milpitas, CA, USA), homogenized in perchloric acid, and centrifuged at 15,000 g for 2 min to pellet insoluble materials. Thereafter, 25 μl of supernatant was added to the 96-well plate with ATP assay buffer to reach a final volume of 50 μl/well; this was incubated at room temperature for 0.5 h and read at an excitation (Ex) of 535 nm and an emission (Em) of 587 nm with a fluorometric microreader. The standard curve was performed on the same plate for calculation of the ATP amount of unknown samples. (2) To determine the effect of amifostine on cytokines, plasma was collected from each group and subjected to enzyme-linked immunosorbent assay (ELISA) with mouse IL-2 and G-CSF kits (R&D Systems, Minneapolis, MN, USA).

For statistical analysis, the saline-treated groups were counted as baseline and expressed as 100 %. The percentage of alteration of the amifostine-treated groups was presented as (mean value of amifostine-treated group/mean value of saline-treated group) × 100 %. An independent Student's *t*-test was used to determine the significance between the treatment and vehicle control groups. A P value of <0.05 was regarded as statistically significant.

27.3 Results

Amifostine is a powerful scavenger for free radicals and a useful antioxidant that likely reduces the level of radiation-induced primary and secondary ROS. Indeed, our results (Fig. 27.1a) showed that ROS levels were reduced in amifostine-treated irradiated cells, as compared to the vehicle controls ($P=0.022$).

The level of detectable MitoTracker represents the number of functional mitochondria [7]. Our results demonstrated that amifostine preserved more mitochondria in the irradiated cells than in the vehicle controls (Fig. 27.1b, $P=0.047$).

The protection of mitochondria is likely to reduce intrinsic apoptosis. Annexin V/PI staining showed that amifostine-treated irradiated cells had a lower percentage of apoptotic cells, as compared to the vehicle controls (Fig. 27.2a, $P=0.003$). MTT results (Fig. 27.2b) demonstrated that amifostine enhanced the proliferation of irradiated cells, as compared to the vehicle controls ($P=0.026$).

Radiation-reduced ATP production was reversed by in vivo treatment with amifostine at a dose of 50 mg/kg (Fig. 27.3, $P=0.005$), indicating that mitochondrial functions are well preserved by amifostine.

IL-2 is a critical immunomodulator that responds to radiation. Amifostine better reduced plasma IL-2 levels in the irradiated cells than in the vehicle controls (Fig. 27.4a, $P=0.022$). Moreover, studies have shown that radiation can trigger the release and production of G-CSF [8]. Our data support this finding (Fig. 27.4b, $P=0.0027$). Notably, this endogenous induction could be further enhanced by amifostine treatment (Fig. 27.4b, $P=0.024$).

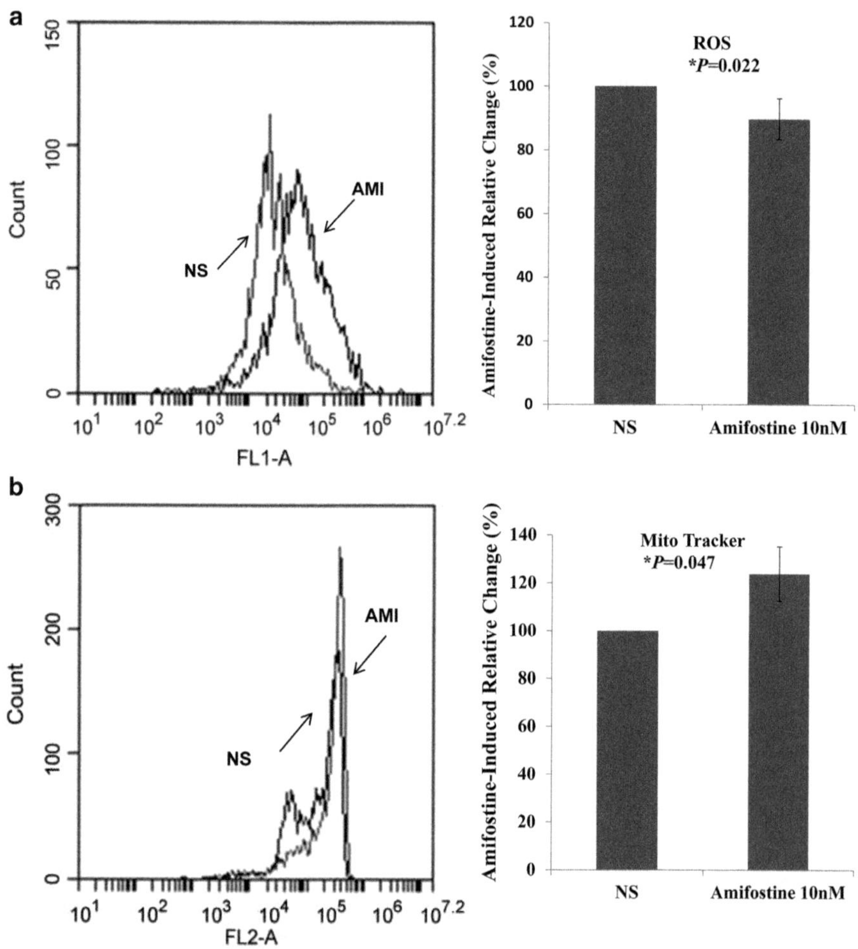

Fig. 27.1 Amifostine protected BM mitochondria from radiation-induced damage. (**a**) ROS was reduced in amifostine-treated irradiated BM cells, as compared to the vehicle controls ($P=0.022$). (**b**): MitoTracker staining showed that amifostine preserved more mitochondria in irradiated BM cells than in the vehicle controls ($P=0.047$)

27.4 Discussion

Our in vitro studies showed that amifostine reduced ROS, preserved mitochondria (Fig. 27.1), reduced apoptosis, and enhanced BM cell proliferation (Fig. 27.2). Our in vivo studies showed that it preserved ATP production (Fig. 27.3), reduced IL-2 production (Fig. 27.4a), and enhanced radiation-induced production of G-CSF (Fig. 27.4b).

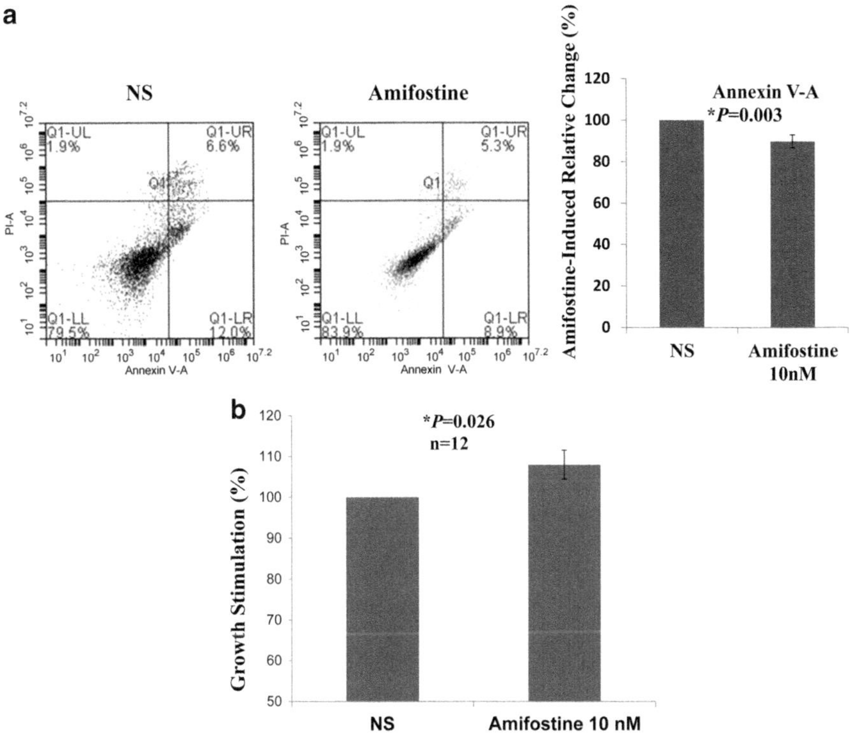

Fig. 27.2 Amifostine protected BM cells from radiation-induced damage. (**a**) Annexin V/PI staining showed that amifostine-treated irradiated BM cells had a lower percentage of apoptotic cells, as compared to the vehicle controls ($P=0.003$). (**b**) MTT results demonstrated that amifostine enhanced the proliferation of irradiated BM cells, as compared to the vehicle controls ($P=0.026$)

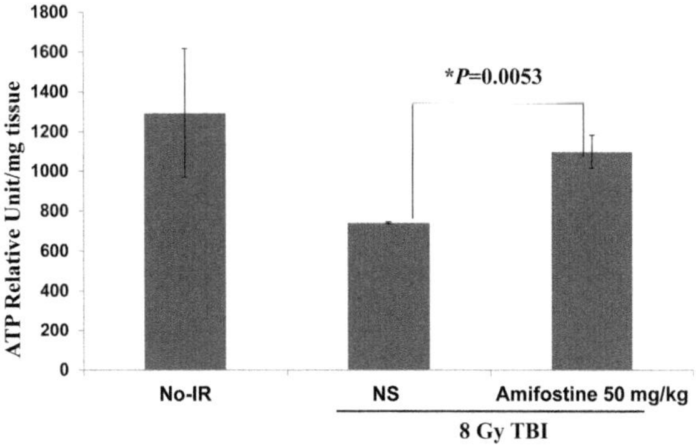

Fig. 27.3 Amifostine preserved ATP levels. The radiation-induced reduction in ATP production was reversed by in vivo treatment with amifostine at a dose of 50 mg/kg ($P=0.005$)

The reduction of ROS levels by amifostine is likely due to its ability to scavenge free radicals. It reduces not only the damage from radiation-generated free radicals but also the secondary wave of free radicals generated by the host's response to radiation [9]. After eluding ROS damage, mitochondrial number and functions are

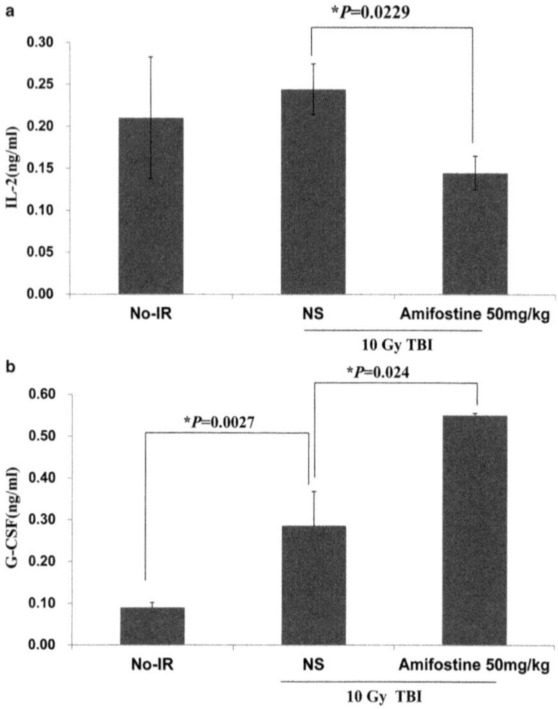

Fig. 27.4 Amifostine regulated key cytokines. (**a**) Amifostine-treated irradiated mice exhibited reduced plasma IL-2 levels, as compared to vehicle control mice ($P = 0.022$). (**b**) Amifostine significantly increased G-CSF production in irradiated mice, as compared to vehicle control mice ($P = 0.024$)

preserved to help generate ATP, thereby supplying energy for cells, tissues, organs, and the whole body to function and interact with the environment. By preventing radiation-induced mitochondrial damage, amifostine also reduces intrinsic apoptosis and allows more BM cells to survive and proliferate.

The process by which amifostine reduces plasma IL-2 levels is complicated; however, less activation of T and B lymphocytes by IL-2 reduces immune overreactions to radiation and helps to maintain homeostasis. In addition, radiation triggers the production of endogenous G-CSF, which is a sign of a host's compensatory reaction. Amifostine could enhance the production of G-CSF, the most essential hematogenic factor, thereby hastening BM recovery after irradiation.

All of these factors account for the mechanism by which amifostine exerts its protective effect on BM, regulates the immune response, and reduces radiation damage.

References

1. Ozcan Arican G (2005) Cytoprotective effects of amifostine and cysteamine on cultured normal and tumor cells treated with paclitaxel in terms of mitotic index and 3H-thymidine labeling index. Cancer Chemother Pharmacol 56(2):221–229
2. Hensley ML, Hagerty KL, Kewalramani T et al (2009) American Society of Clinical Oncology 2008 clinical practice guideline update: use of chemotherapy and radiation therapy protectants. J Clin Oncol 27(1):127–145

3. Kouvaris JR, Kouloulias VE, Vlahos LJ (2007) Amifostine: the first selective-target and broad-spectrum radioprotector. Oncologist 12(6):738–747
4. Culy CR, Spencer CM (2001) Amifostine: an update on its clinical status as a cytoprotectant in patients with cancer receiving chemotherapy or radiotherapy and its potential therapeutic application in myelodysplastic syndrome. Drugs 61(5):641–684
5. Capizzi RL, Scheffler BJ, Schein PS (1993) Amifostine-mediated protection of normal bone marrow from cytotoxic chemotherapy. Cancer 72(11):3495–3501
6. Invitrogen (2009) Amplex red hydrogen peroxide/peroxidase assay kit. http://probes.invitrogen.com/media/pis/mp22188.pdf. Updated 26 June 2009. Accessed 01 May 2013
7. Invitrogen (2008) MitoTracker mitochondrion-selective probes. https://www.ipmc.cnrs.fr/fichiers/recherche/microscopie/Sondes/MitoGreen.pdf. Updated 29 Mar 2013. Accessed 01 May 2013
8. Pape H, Orth K, Heese A et al (2006) G-CSF during large field radiotherapy reduces bone marrow recovery capacity. Eur J Med Res 11(8):322–328
9. Riley PA (1994) Free radicals in biology: oxidative stress and the effects of ionizing radiation. Int J Radiat Biol 65(1):27–33

Chapter 28
Hypoxia, Lactate Accumulation, and Acidosis: Siblings or Accomplices Driving Tumor Progression and Resistance to Therapy?

Arnulf Mayer and Peter Vaupel

Abstract This chapter briefly summarizes the most important processes by which hypoxia, lactate accumulation, and acidosis may influence malignant progression and therapeutic resistance of solid malignant tumors. While these phenomena are often elements of an integrated reaction, they may occur independently of each other under certain circumstances. The latter information may be of interest with regard to possible "targeted" therapeutic interventions.

28.1 Hypoxia

Evidence supporting the existence of hypoxic tissue areas in solid tumors is derived from data originating from a variety of methods [1]. These include invasive measurements of intratumoral oxygen partial pressures using polarographic needle electrodes ("Eppendorf" microsensor) and histological assays based on the immunodetection of so-called endogenous or exogenous hypoxia markers. In addition, different imaging methods have been developed, which, however, at the present time have not been adopted widely in the clinic. The major cause of tumor hypoxia is an enlargement of the intratumoral diffusion distances of oxygen beyond a critical threshold, which is estimated to be equal to approximately 80 μm at the arterial end of the microvessel. This main origin of continuous or "chronic" hypoxia is modified by other factors, including a reduced oxygen-transport capacity of the blood (anemia) and an increased interstitial fluid pressure, which may lead to a flow stop in microvessels. Besides the phenomenon of continuous tumor hypoxia, one also observes intermittent or "acute" hypoxia, which may be caused by fluctuations

A. Mayer (✉) • P. Vaupel
Department of Radiooncology and Radiotherapy, University Medical Center of the Johannes Gutenberg University, Mainz, Germany
e-mail: arnmayer@uni-mainz.de

S. Van Huffel et al. (eds.), *Oxygen Transport to Tissue XXXV*, Advances in Experimental Medicine and Biology 789, DOI 10.1007/978-1-4614-7411-1_28,

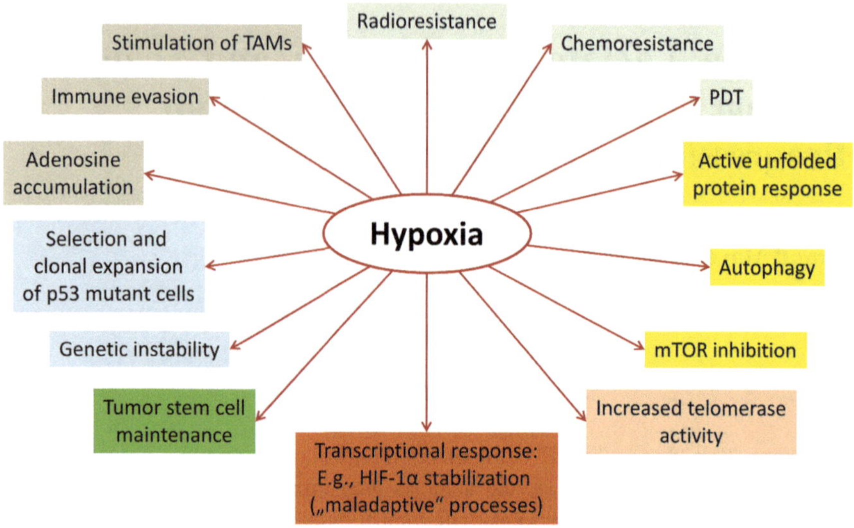

Fig. 28.1 Tumor hypoxia is a central driver of malignant progression and resistance to therapy (selection of mechanisms)

in the flux of erythrocytes or by temporary obstructions of tumor capillaries, e.g., by cell aggregates.

Figure 28.1 shows a synopsis of the various mechanisms by which tumor hypoxia may contribute to a more aggressive phenotype and to an increased resistance to therapy. The discovery that hypoxia is one of the most important factors mediating radioresistance can be traced back to the beginning of the twentieth century. We know today that the mechanism behind this observation is a modification of the free radical chemistry under hypoxic conditions, which has also been shown to be important for some forms of chemotherapy and photodynamic therapy [2]. Since the early 1990s, clinical studies indicated that the pathophysiological significance of hypoxia is clearly not limited to this modification of the radiosensitivity of tumor cells [3]. Hypoxia can lead to an increase of the genetic instability of cancer cells both by inducing mutations and by inhibiting DNA repair [4]. Hypoxia may also act as a selective force favoring the emergence of genetically hypoxia-resistant phenotypes. For example, p53-negative, apoptosis-resistant cell populations may emerge after repeated exposures of cells to hypoxia and reoxygenation [5]. Hypoxia has been shown to be important for the maintenance of the stem cell phenotype, and some types of stem cells have been observed to reside in a "hypoxic niche" in vivo [6]. Furthermore, hypoxia can play an important role in the attenuation of an antitumor immune response. For example, macrophages of the pro-tumorigenic M2 phenotype have been found preferentially in hypoxic tumor areas [7]. Consistent with this finding, other reports have demonstrated that hypoxic tumors contain a higher number of macrophages compared to non-hypoxic tumors. Additionally, higher quantities of intratumoral macrophages have been shown to correlate with a poorer patient

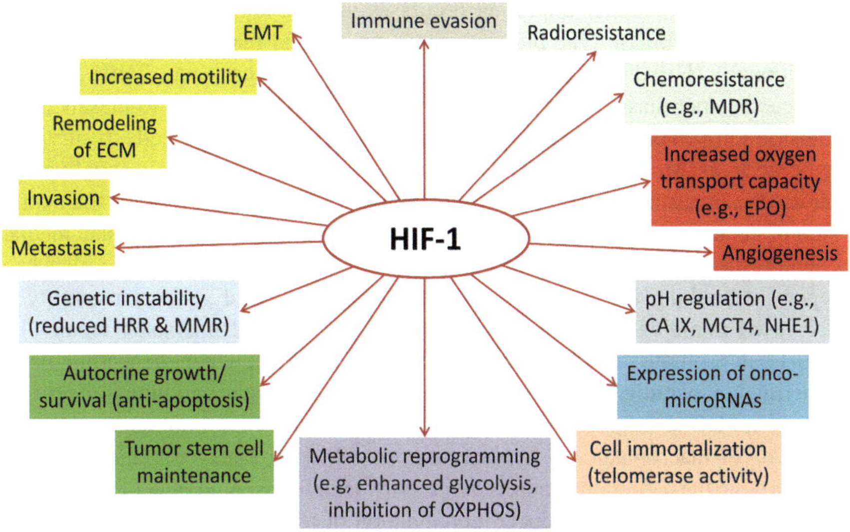

Fig. 28.2 HIF-1 as the central driver of hypoxia-induced transcriptional "maladaptation" in cancer (selection of mechanisms)

prognosis [8]. Under hypoxic conditions, an increased expression of the cytokine CCL28 has been detected, which may lead to intratumoral accumulation of immunosuppressive regulatory T cells which express the cognate receptor CXCR10 [9]. Under hypoxic conditions, adenosine may accumulate in the extracellular space and stimulate adenosine receptors (of the A_{2A} and A_{2B} subtypes) on T cells, thereby leading to an inhibition of antitumor T cell responses [10]. Hypoxia has also been shown to be able to trigger the unfolded protein response and autophagy, which may promote tumor growth and resistance to anticancer therapy [11]. It should be mentioned, however, that both processes can also be antitumorigenic, depending on the specific experimental conditions. Although mTOR inhibition is currently being evaluated as a therapeutic strategy, e.g., in malignant gliomas, hypoxia-mediated suppression of mTOR has recently been shown to prevent irreversible cellular senescence, which may attenuate the efficacy of DNA-damaging agents [12]. Arguably, the overall most significant consequence of hypoxia is a large-scale change of the proteome, which is mediated by the activity of several transcription factors, among which the hypoxia-inducible factor 1 (HIF-1) plays the most important role [13].

More than 800 direct target genes of HIF-1 are known, and a large number of these have been shown to have a direct pathogenic role within the malignant phenotype (see Fig. 28.2, [13]). HIF-1 is a major trigger of proangiogenic cytokines (e.g., VEGF) in tumor cells. Furthermore, HIF-1 can promote vasculogenesis by the recruitment of CXCR4-positive stem cells from the bone marrow via SDF-1. HIF-1 increases the oxygen-transport capacity of the blood by upregulating EPO. Activation of HIF-1 leads to increased cell motility and invasiveness, mediates the ability to remodel the extracellular matrix, and can confer an augmented metastatic potency. These pivotally important processes may be initiated directly by HIF-1, e.g., via the urokinase-type plasminogen activator and matrix metalloproteinases. Additionally, HIF-1 can transactivate transcription factors (e.g., TWIST) which induce the metastasis-promoting cellular program of epithelial-to-mesenchymal transition [14]. HIF-1 may promote radioresistance by allowing cells to survive in hypoxic areas. Moreover, basal HIF-1 expression, but – interestingly – not hypoxia-induced expression of HIF-1, has been demonstrated to play a role for the expression of genes involved in DNA repair [15]. Target genes of HIF-1 can also mediate chemoresistance, e.g., by induction of the MDR-1 gene. There have also been reports describing a role of HIF-1 in mediating increased genetic instability by decreased homologous recombination repair and reduced mismatch repair [16]. HIF-1 can promote the differentiation of TH17 cells [17], which, depending on the experimental paradigm, have been described to both promote and inhibit the growth of tumors. HIF-1-induced SDF-1 may also contribute to the aforementioned accumulation of macrophages in hypoxic tumors [18]. HIF-1 may stimulate proliferation through the induction of autocrine growth factor loops. A number of publications have described an HIF-1-induced upregulation of telomerase and HIF-1 activated genes which are considered to play a role in the stem cell phenotype. Finally, a central mechanism of HIF-1-mediated maladaptation consists of an extensive metabolic reprogramming which leads to a downregulation of mitochondrial oxidative phosphorylation, e.g., via inhibition of the pyruvate dehydrogenase reaction by PDK-1 and promotion of selective autophagy of mitochondria. Simultaneously, HIF-1 mediates the induction of a glycolytic phenotype by increasing glucose influx (e.g., via GLUT-1), upregulation of key enzymes of glycolysis, and by an increase in the efflux of lactate via the monocarboxylate transporter subtype MCT-4 [13, 19].

28.2 Lactate

A substantial part of intratumoral lactate accumulation is the result of HIF-1-mediated metabolic reprogramming. However, comparative analyses of the distribution patterns of hypoxia (as assessed by pimonidazole staining) and locoregional lactate concentrations (analyzed using imaging bioluminescence) have revealed that both parameters are not necessarily co-localized in all cases [20]. Indeed, several HIF-1-independent mechanisms of intratumoral lactate accumulation have been described, e.g., the activation of MYC [21]. Additionally, high lactate levels may also be the consequence of an insufficient waste drainage in poorly vascularized tumor areas. The matter is further complicated by the existence of an intratumoral lactate shuttle between hypoxic (lactate-producing) and normoxic (lactate-consuming) cells [22]. Lactate has been hypothesized to mediate radioresistance by virtue of its antioxidant properties. Lactate also exhibits immunosuppressive properties and promotes cell motility, invasion, and metastasis. Furthermore, lactate may induce angiogenesis, mediate resistance to apoptosis, and may promote a stem cell phenotype. Importantly, lactate can indirectly stabilize HIF-1α and may thus perpetuate the activation of HIF-1 independent of hypoxia [23].

28.3 Acidosis

HIF-1-induced metabolic reprogramming also contributes to the marked extracellular acidosis often found in malignant tumors by upregulating glycolysis. Nevertheless, direct measurements of intratumoral oxygen and pH levels have revealed unequal distributions of both parameters at the microregional level [24, 25], and glycolysis-deficient cells have been shown to retain the ability to acidify the extracellular environment in vivo [26]. Additional pathogenetic mechanisms yielding an intensified tissue acidosis are based on substantial hydrolysis of ATP (derived from breakdown of substrates other than glucose), glutaminolysis, ketogenesis, and CO_2/carbonic acid production [27]. The spectrum of the pathophysiological consequences of intratumoral acidosis includes many processes mentioned for HIF-1 and lactate: acidosis plays a role in mediating radioresistance (e.g., [28]), immune evasion [29], increased cell motility, invasion, metastasis [30, 31], promotion of angiogenesis through VEGF [32], and the stem cell phenotype [33]. Moreover, an acidic extracellular milieu diminishes the effectiveness of basic chemotherapeutic drugs (e.g., doxorubicin, daunorubicin, [34]). Similar to hypoxia and HIF-1, acidosis may contribute to the genetic instability of tumor cells [35] and – similar to hypoxia – is a possible trigger for autophagy [36]. Finally, acidosis has been shown to stabilize HIF-1α independent of hypoxia by nucleolar sequestration of VHL [37].

28.4 Conclusions

Factors of the microenvironment presented in this report trigger an overlapping range of processes which promote tumor growth and mediate resistance to therapy. The broadest spectrum of these processes is initiated by hypoxia and HIF-1, which are also often at the root of lactate accumulation and intratumoral acidosis. With this in mind, the three factors may be regarded as "siblings." However, both of the latter factors may also be triggered independently of hypoxia and, importantly, similar pathogenic processes (e.g., radioresistance) may be initiated by all three factors via entirely independent mechanisms (e.g., modification of the spectrum of free radicals generated by radiation vs. scavenging of free radicals). Therefore, the factors discussed here may also act as "accomplices," depending on the specific triggers for each of them in individual tumors.

References

1. Vaupel P, Mayer A (2007) Hypoxia in cancer: significance and impact on clinical outcome. Cancer Metastasis Rev 26(2):225–239
2. Vaupel P (2009) Physiological mechanisms of treatment resistance. In: Molls M, Vaupel P, Nieder C, Anscher MS (eds) The impact of tumor biology on cancer treatment and multidisciplinary strategies. Springer, Berlin\Heidelberg, pp 273–290

3. Höckel M, Schlenger K, Aral B, Mitze M, Schäffer U, Vaupel P (1996) Association between tumor hypoxia and malignant progression in advanced cancer of the uterine cervix. Cancer Res 56(19):4509–4515
4. Klein TJ, Glazer PM (2010) The tumor microenvironment and DNA repair. Semin Radiat Oncol 20(4):282–287
5. Graeber TG, Osmanian C, Jacks T et al (1996) Hypoxia-mediated selection of cells with diminished apoptotic potential in solid tumours. Nature 379(6560):88–91
6. Suda T, Takubo K, Semenza GL (2011) Metabolic regulation of hematopoietic stem cells in the hypoxic niche. Cell Stem Cell 9(4):298–310
7. Movahedi K, Laoui D, Gysemans C et al (2010) Different tumor microenvironments contain functionally distinct subsets of macrophages derived from Ly6C(high) monocytes. Cancer Res 70(14):5728–5739
8. Lewis C, Murdoch C (2005) Macrophage responses to hypoxia: implications for tumor progression and anti-cancer therapies. Am J Pathol 167(3):627–635
9. Facciabene A, Peng X, Hagemann IS et al (2011) Tumour hypoxia promotes tolerance and angiogenesis via CCL28 and T(reg) cells. Nature 475(7355):226–230
10. Sitkovsky M, Lukashev D (2005) Regulation of immune cells by local-tissue oxygen tension: HIF1 alpha and adenosine receptors. Nat Rev Immunol 5(9):712–721
11. Wouters BG, Koritzinsky M (2008) Hypoxia signalling through mTOR and the unfolded protein response in cancer. Nat Rev Cancer 8(11):851–864
12. Leontieva OV, Natarajan V, Demidenko ZN, Burdelya LG, Gudkov AV, Blagosklonny MV (2012) Hypoxia suppresses conversion from proliferative arrest to cellular senescence. Proc Natl Acad Sci USA 109(33):13314–13318
13. Semenza GL (2012) Hypoxia-inducible factors: mediators of cancer progression and targets for cancer therapy. Trends Pharmacol Sci 33(4):207–214
14. Yang MH, Wu MZ, Chiou SH et al (2008) Direct regulation of TWIST by HIF-1α promotes metastasis. Nat Cell Biol 10(3):295–305
15. Wirthner R, Wrann S, Balamurugan K, Wenger RH, Stiehl DP (2008) Impaired DNA double-strand break repair contributes to chemoresistance in HIF-1α-deficient mouse embryonic fibroblasts. Carcinogenesis 29(12):2306–2316
16. To KK, Sedelnikova OA, Samons M, Bonner WM, Huang LE (2006) The phosphorylation status of PAS-B distinguishes HIF-1α from HIF-2α in NBS1 repression. EMBO J 25(20): 4784–4794
17. Dang EV, Barbi J, Yang HY et al (2011) Control of T(H)17/T(reg) balance by hypoxia-inducible factor 1. Cell 146(5):772–784
18. Palazón A, Aragonés J, Morales-Kastresana A, de Landázuri MO, Melero I (2012) Molecular pathways: hypoxia response in immune cells fighting or promoting cancer. Clin Cancer Res 18(5):1207–1213
19. Semenza GL (2011) Hypoxia-inducible factor 1: regulator of mitochondrial metabolism and mediator of ischemic preconditioning. Biochim Biophys Acta 1813(7):1263–1268
20. Yaromina A, Quennet V, Zips D et al (2009) Co-localisation of hypoxia and perfusion markers with parameters of glucose metabolism in human squamous cell carcinoma (hSCC) xenografts. Int J Radiat Biol 85(11):972–980
21. Osthus RC, Shim H, Kim S et al (2000) Deregulation of glucose transporter 1 and glycolytic gene expression by c-Myc. J Biol Chem 275(29):21797–21800
22. Sonveaux P, Vegran F, Schroeder T et al (2008) Targeting lactate-fueled respiration selectively kills hypoxic tumor cells in mice. J Clin Invest 118(12):3930–3942
23. Hirschhaeuser F, Sattler UG, Mueller-Klieser W (2011) Lactate: a metabolic key player in cancer. Cancer Res 71(22):6921–6925
24. Vaupel PW, Frinak S, Bicher HI (1981) Heterogeneous oxygen partial pressure and pH distribution in C3H mouse mammary adenocarcinoma. Cancer Res 41(5):2008–2013
25. Helmlinger G, Yuan F, Dellian M, Jain RK (1997) Interstitial pH and pO_2 gradients in solid tumors in vivo: high-resolution measurements reveal a lack of correlation. Nat Med 3(2): 177–182

26. Newell K, Franchi A, Pouyssegur J, Tannock I (1993) Studies with glycolysis-deficient cells suggest that production of lactic acid is not the only cause of tumor acidity. Proc Natl Acad Sci USA 90(3):1127–1131
27. Vaupel P (2004) Tumor microenvironmental physiology and its implications for radiation oncology. Semin Radiat Oncol 14(3):198–206
28. Haveman J (1980) The influence of pH on the survival after X-irradiation of cultured malignant cells. Effects of carbonylcyanide-3-chlorophenylhydrazone. Int J Radiat Biol Relat Stud Phys Chem Med 37(2):201–205
29. Lardner A (2001) The effects of extracellular pH on immune function. J Leukoc Biol 69(4):522–530
30. Raghunand N, Gatenby RA, Gillies RJ (2003) Microenvironmental and cellular consequences of altered blood flow in tumours. Br J Radiol 76(1):S11–S22
31. Calorini L, Peppicelli S, Bianchini F (2012) Extracellular acidity as favouring factor of tumor progression and metastatic dissemination. Exp Oncol 34(2):79–84
32. Fukumura D, Xu L, Chen Y, Gohongi T, Seed B, Jain RK (2001) Hypoxia and acidosis independently up-regulate vascular endothelial growth factor transcription in brain tumors in vivo. Cancer Res 61(16):6020–6024
33. Hjelmeland AB, Wu Q, Heddleston JM et al (2011) Acidic stress promotes a glioma stem cell phenotype. Cell Death Differ 18(5):829–840
34. Thews O, Gassner B, Kelleher DK, Schwerdt G, Gekle M (2006) Impact of extracellular acidity on the activity of P-glycoprotein and the cytotoxicity of chemotherapeutic drugs. Neoplasia 8(2):143–152
35. Morita T, Nagaki T, Fukuda I, Okumura K (1992) Clastogenicity of low pH to various cultured mammalian cells. Mutat Res 268(2):297–305
36. Wojtkowiak JW, Rothberg JM, Kumar V et al (2012) Chronic autophagy is a cellular adaptation to tumor acidic pH microenvironments. Cancer Res 72(16):3938–3947
37. Mekhail K, Gunaratnam L, Bonicalzi ME, Lee S (2004) HIF activation by pH-dependent nucleolar sequestration of VHL. Nat Cell Biol 6(7):642–647

Chapter 29
Breast Cancer Detection of Large Size to DCIS by Hypoxia and Angiogenesis Using NIRS

Shoko Nioka, Mitch Shnall, Emily Conant, Shih Chang Wang, Visjna Baksa Reynolds, Boon Chye Ching, Juliana Ho Teng Swan, Pau Choo Chung, Lili Cheng, Darbin Shieh, Yungchi Lin, Chenghung Chung, Sheng Hao Tseng, and Britton Chance

Abstract This investigation aimed to test all tumor-bearing patients who undergo biopsy to see if angiogenesis and hypoxia can detect cancer. We used continuous-wave near-infrared spectroscopy (NIRS) to measure blood hemoglobin concentration to obtain blood volume or total hemoglobin [Hb_{tot}] and oxygen saturation for the angiogenesis and hypoxic biomarkers. The contralateral breast was used as a reference to derive the difference from breast tumor as a difference in total hemoglobin

---

S. Nioka (✉)
Biochemistry and Biophysics, University of Pennsylvania,
Philadelphia, PA, USA

Institute of Computer and Communication Engineering, National Cheng
Kung University, Tainan, Taiwan
e-mail: shoko@nioka.net

M. Shnall • E. Conant • B. Chance
Radiology, University of Pennsylvania,
Philadelphia, PA, USA

S.C. Wang • V.B. Reynolds
Radiology, National University of Singapore, Singapore

B.C. Ching • J.H.T. Swan
Radiology, National Cancer Center Singapore, Singapore

P.C. Chung • Y. Lin • C. Chung
Institute of Computer and Communication Engineering, National Cheng
Kung University, Tainan, Taiwan

L. Cheng
Radiology, National Cheng Kung University, Tainan, Taiwan

D. Shieh
Institute of Oral Medicine, Center for Advanced Optoelectronics Technology,
National Cheng Kung University, Tainan, Taiwan

S.H. Tseng
Opto Physics, National Cheng Kung University, Tainan, Taiwan

S. Van Huffel et al. (eds.), *Oxygen Transport to Tissue XXXV*, Advances in Experimental Medicine and Biology 789, DOI 10.1007/978-1-4614-7411-1_29,

Δ[HB_{tot}] and a difference in deoxygenation Δ([Hb]-[HbO2]). A total of 91 invasive cancers, 26 DCIS, 45 fibroblastomas, 96 benign tumors excluding cysts, and 67 normal breasts were examined from four hospitals. In larger-size tumors, there is significantly higher deoxygenation in invasive and ductal carcinoma in situ (DCIS) than in that of benign tumors, but no significant difference was seen in smaller tumors of ≤ 1 cm. With the two parameters of high total hemoglobin and hypoxia score, the sensitivity and specificity of cancer detection were 60.3 % and 85.3 %, respectively. In summary, smaller-size tumors are difficult to detect with NIRS, whereas DCIS can be detected by the same total hemoglobin and hypoxic score in our study.

29.1 Introduction

Breast cancer is a leading cause of death in women worldwide. There has been a great improvement in breast cancer survival since the mid-1970s, and in the USA during the period from 1999 to 2006, 90 % were expected to survive their disease for at least 5 years. The newly released data show an even higher survival rate (98 %) if the cancer is localized in the breast [1]. The increase in survival rate has been attributed to both screening and improved treatment. However, X-ray mammography, which is the only breast cancer screening test, has a low sensitivity (33 %) [2] and has a high incidence of biopsy. Thus, we are still searching for more reliable ways to diagnose breast cancer through functional and biochemical signatures of cancers.

There is much evidence showing a higher concentration of blood in breast cancers. This has been demonstrated in a good review article in 1999 [3] as well as more recently [4]. These studies used near-infrared spectroscopy (NIRS) to determine the blood concentration and hemoglobin saturation in humans. We have conducted a multicenter clinical trial on breast cancers to determine blood concentration and desaturation of blood indicating both angiogenesis and hypermetabolism as a signature for breast cancer [5]. That clinical study involved 44 cancer cases with more than 100 breast tumors and showed quite high sensitivity and specificity (both over 90 %). This study, on mostly large, palpable cancers, was intended to show that NIRS can detect cancers. However, the results did not represent the true patient population, which includes smaller cancers as well as cancer in situ. Thus, in the present study, we have expanded our investigation to focus on early detection of cancers with a smaller tumor size and ductal carcinoma in situ (DCIS).

29.2 Methods

29.2.1 Patient Recruitment

Forty-eight normal breast studies dating from 1997, when we recruited patients coming for breast screening, together with a few tumor-bearing patients, were included from the Hospital of the University of Pennsylvania (HUP). Between 2001

Table 29.1 Number of patients and tumor types in the four multicenter hospitals

	Total Subjects	DCI/LCI	DCIS	Benign	FA	Normal
HUP	165	50	13	34	8	60
NUS	82	17	8	41	12	4
NCCS	51	12	2	12	25	0
NCKU	27	12	3	9	0	3
Total	325	91	26	96	45	67

and 2006, we also recruited patients on the basis of biopsy for most of the cancers, DCIS, benign, and fibroadenoma, a total of nearly 100 cases (Table 29.1). Between 2007 and 2009, patients were recruited at the National University Hospital of Singapore (NUS) and the National Cancer Center of Singapore on the basis of breast biopsy. Between 2009 and 2010, patients at the Taiwan National Cheng Kung University Hospital (NCKU) with positive breast tumor biopsies were recruited. This gave a total for the 4 hospitals of 91 invasive breast cancers and 26 DCIS as well as 96 benign tumors and 45 fibroadenomas (FA) together with 67 normal cases without tumors. Note: cysts and inflammatory diseases were not included in this investigation as benign tumor cases. Fibroadenoma was treated separately from benign tumors as it demonstrates different optical characteristics [6]. The ages of the four groups of subjects are 55.5 ± 1.6, 50.8 ± 2.6, 51.2 ± 3.3, and 52.5 ± 1.5 (mean ± se) for ICA, DCIS, FA, and benign tumor, respectively. In summary, the tumor-bearing population mostly comprised cancers suspected from X-ray mammography screening and confirmed by biopsy, and, as a result, this population included more of the smaller tumor size ≤ 1 cm and DCIS than the previous study [5].

29.2.2 *Data Acquisition by NIRS*

We used the same or a very similar continuous-wave spectroscopy (CWS) device described previously [5]. This has 8 silicon diode detectors, 3–4 cm equidistant from a central multi-wavelength LED which gives 10 ms light pulses. The NIR LED has three wavelengths: 760, 805, and 850 nm. The 3 cm source-detector distance was used for smaller-sized or shallow tumor-bearing breasts and the 4 cm source-detector distance for larger-sized or deeper tumor-bearing breasts. The sensors detect the light from the LED simultaneously after photon migration through the tumor.

29.2.3 *Data Analysis*

The raw data from the CWS system are three wavelengths of light intensities from a calibration phantom of known absorption and scattering coefficient, light intensities from the tumor-bearing breast, and the contralateral breast as the reference, with the location being a mirror image of the tumor position. A modified Beer-Lambert Law, with a differential path length factor (DPF) of 5, was used to calculate the concentrations of two species of hemoglobin, deoxy- and oxyhemoglobin or [Hb] and [HbO_2] (micromolar or μM), under the assumption that all breasts have the same scattering coefficient. It was also assumed that the only absorbing chromophores in the breast were the two species of hemoglobin, deoxy- and oxyhemoglobin. This assumption works well when we observe the difference in Hb and HbO_2 and Δ[Hb] and Δ[HbO_2], between one breast and the contralateral side, where both should have similar quantities of other chromophores and which would be canceled out.

The parameters from the CWS-NIRS are the sum of Δ[Hb], Δ[HbO_2], Δ[Hb] + Δ[HbO_2], or Δ[Hb_{tot}] and the difference of the two, Δ[Hb]-Δ[HbO_2], or ΔDeoxygenation. The delta Δ stands for the difference between the breast with the tumor and the contralateral breast.

Our tests showed that in normal healthy breasts, there is no difference in the optical characteristics between the left and right breast so that the total hemoglobin and deoxygenation differences are near zero. Therefore, for this investigation we used delta Δ, the difference from the contralateral breast, where any deviation from zero in the total hemoglobin and deoxygenation indicates an abnormality. We tested the parameters Δ[Hb_{tot}] and ΔDeoxygenation for cancers, benign tumors, and healthy breasts. We used Student's *t*-test to determine if there is a difference between cancers and benign tumors. Breast tumors were categorized as invasive cancer, either ductal or lobular (IC), ductal cancer in situ (DCIS), fibroadenoma (FA), or benign tumors. Cysts were excluded from the benign tumors for two reasons: cysts are well defined by ultrasound and the optical characteristics of cysts are very different from other types of benign tumors with no blood volume inside. The NIRS method is sensitive to the tumor size, so our test results will be influenced by the tumor size, and so there is a need to test small and large tumor sizes separately. Smaller-size tumors were defined as ≤ 1 cm and larger tumors as > 1 cm in diameter. Invasive carcinoma has the largest tumor size with 2.1 ± 0.17 cm as mean ± se, DCIS has the smallest tumor size (1.2 ± 0.3 cm), and FA and benign tumor sizes are 1.6 ± 0.3 cm and 1.3 ± 0.2 cm, respectively. In summary, we determined the probability that invasive breast cancer and cancer in situ are different (high Δ[Hb_{tot}] and low oxygen or ΔDeoxygenation) from normal and benign tumors; in order to calculate sensitivity and specificity, we used a two-dimensional nomograph (Δ[Hb_{tot}] and ΔDeoxygenation) to score the cancer diagnostic capability. In this way, we were able to score Δ[Hb_{tot}] and ΔDeoxygenation with the same weight. We considered the 90 % confidence level of normal breast area in the nomogram (Fig. 29.1a), along with high Δ[Hb_{tot}] and low oxygen (ΔDeoxygenation) at the cutting point of 5 μM to calculate sensitivity and specificity. This approach is very similar to that in our previous paper [5].

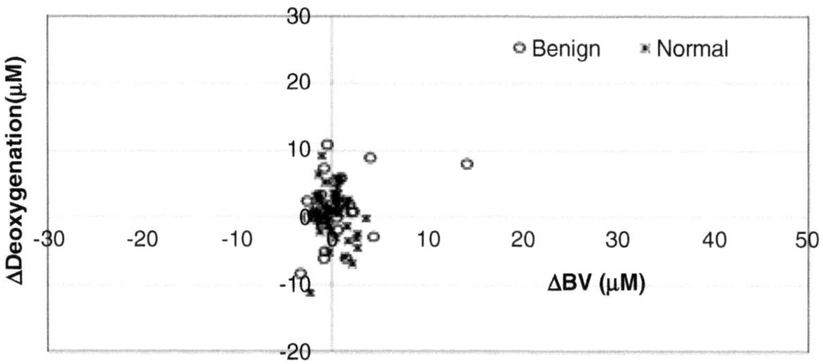

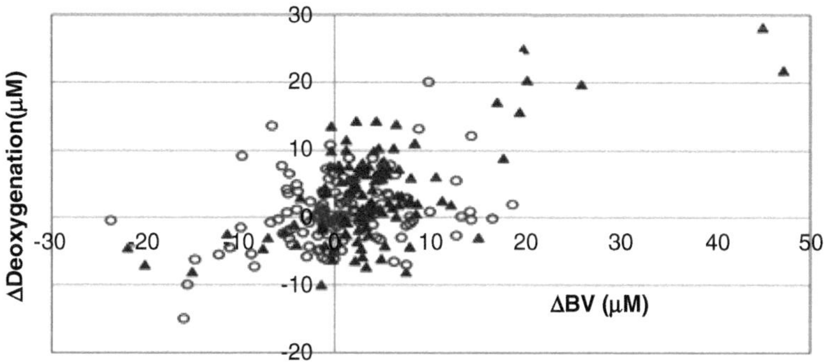

Fig. 29.1 (**a**) Normal breast versus benign tumor in a nomogram of $\Delta[Hb_{tot}]$ or ΔBV and ΔDeoxygenation (control study). Normal breast (*solid circle*) and benign tumor (*star*) show no difference in nomogram distribution. (**b**) Nomogram of breast cancers (IDC/ILC and DCIS, *triangle*) and benign tumors (FA included, *open circle*) is plotted against $\Delta[Hb_{tot}]$ or ΔBV and ΔDeoxygenation. Sensitivity and specificity scores were calculated based on the threshold of high $\Delta[Hb_{tot}]$ and ΔDeoxygenation excluding the lower corner of ≤5 μM of $\Delta[Hb_{tot}]$ and ΔDeoxygenation

29.3 Results

29.3.1 Normal Breast

Figure 29.1a shows the results of the clinical study mostly at the University of Pennsylvania (HUP) mammography screening facility. The averages for normal and benign ΔDeoxygenation were 0.65±0.52 μM (±se) and 0.98±0.86 μM (±se), respectively, and the averages of normal and benign $\Delta[Hb_{tot}]$ were 0.13±0.27 μM (±se) and 0.49±0.61 μM (±se), respectively. The results showed no difference between normal breasts and breasts with benign tumors. It was also shown that there was no difference in the $\Delta[Hb_{tot}]$ and ΔDeoxygenation between left and right breasts amongst normal breasts. Furthermore, the distribution of data shows that the 90 % confidence level lay between 3.8 and 1.4 μM of ΔDeoxygenation and $\Delta[Hb_{tot}]$ for normal breasts. We therefore set the breast cancer cases at the 5 μM line when we used specificity and sensitivity tests later on.

Table 29.2 ΔDeoxygenation (ΔDeoxy) and $\Delta[Hb_{tot}]$ in four breast tumors (mean ± se) > 1 cm (top) and ≤ 1 cm (bottom)

	Tumor > 1 cm								
	n	ΔDeoxy		se	p	$\Delta[Hb_{tot}]$		se	p
IDC/ILC	57	4.62	±	0.71	0.03	4.88	±	0.83	0.11
DCIS	12	5.45	±	2.43	0.01	6.13	±	3.98	0.19
FA	20	0.11	±	0.82	0.07	2.59	±	2.50	0.84
Benign	26	0.31	±	1.23		2.05	±	1.37	
	Tumor ≤ 1 cm								
	n	ΔDeoxy		se	p	$\Delta[Hb_{tot}]$		se	p
IDC/ILC	24	1.72	±	2.40	0.71	0.98	±	1.99	0.75
DCIS	8	2.50	±	2.40	0.64	0.85	±	2.89	0.95
FA	25	−0.25	±	1.12	0.89	−0.04	±	1.49	0.85
Benign	55	2.38	±	0.91		0.32	±	1.08	

Data are in μM, p values are comparison to benign tumor

29.3.2 Invasive Cancer, Cancer In Situ, and Benign Tumors

Table 29.2 summarizes the results of the analysis. There were three times as many larger invasive cancers ($n = 57$) than there were small ones ($n = 24$). For DCIS, there were similar numbers for both the larger- and smaller-size DCIS. There were differences in the tumor $\Delta[Hb_{tot}]$ and ΔDeoxygenation between tumors > 1 cm and smaller ones. In general, larger-size cancers demonstrated higher $\Delta[Hb_{tot}]$ and ΔDeoxygenation than did smaller cancers. As a result, when we compared the cancers to benign tumors, only the larger cancers (invasive and DCIS) had statistically significantly higher ΔDeoxygenation.

29.3.3 Overall Nomogram of Cancer Detection

In order to assess the detectability of breast cancer, we categorized invasive cancers and DCIS as cancers and fibroblastoma as a benign tumor. We used both high $\Delta[Hb_{tot}]$ and high ΔDeoxygenation as equally weighted parameters of cancer detection. All the small- and large-size and unknown-size tumors are included into the nomogram. These four categories of tumors were plotted as $\Delta[Hb_{tot}]$ against ΔDeoxygenation (Fig. 29.1b).

We set a quadrant of higher $\Delta[Hb_{tot}]$ and low oxygen (ΔDeoxygenation) as the criteria for cancer detection. In the high $\Delta[Hb_{tot}]$ and high ΔDeoxygenation quadrant, we drew a marginal line between 5 μM of $\Delta[Hb_{tot}]$ and 5 μM of ΔDeoxygenation and then excluded the lower corner of the quadrant. Any tumor which fell into the quadrant without being in the lower corner was considered positive for cancer detection.

Sensitivity and specificity were analyzed in order to score cancer detectability. Sensitivity was 60.3 % and specificity was 85.3 %. The positive predictive value (PPV) and negative predictive value (NPV) were 76.9 % and 72.6 %, respectively.

29.4 Discussion

This investigation has yielded a sensitivity of 60 % and a specificity of 85 %. This showed lower cancer detectability than was found in our previous study from 2005 where both sensitivity and specificity were over 90 %. We have considered some reasons for the lower scores. First, because most of the cancers presented here were not palpable, we may have missed the location of the cancer since there was no co-registration modality used during the optical measurements. Secondly, in the measurement of smaller-sized masses, data were averaged between the volume of the cancer and the surrounding tissue; i.e., a larger normal tissue catchment volume may have contributed to making the values less as it calculates the average volume which has a larger amount of normal breast tissues nearby. The average $\Delta[Hb_{tot}]$ and ΔDeoxygenation values for the smaller-size cancers were much smaller than those of larger cancers by a factor of more than 2. Thirdly, we do not know if DCIS had developed angiogenesis and hypoxia in the surrounding breast tissue. If it had not developed these conditions, cancer detection would have failed.

We also observed lower than expected average values of $\Delta[Hb_{tot}]$ and ΔDeoxygenation (Table 29.2) in both invasive cancers and DCIS. The data we obtained were much lower than those obtained by others [7–13]. The other seven studies reported 24–130 μM of $\Delta[Hb_{tot}]$ compared to that of normal tissue (17–39 μM). Therefore, the difference of $\Delta[Hb_{tot}]$ between cancer and normal ranges from 7 to over 100 μM ($n=7$) and 30.4 ± 14.3 μM on the mean ± se. In our case, $\Delta[Hb_{tot}]$ was 5 μM or less (2–3 μM in smaller-size cancers). It should be noted that data from other publications used a between-subjects effect, whereas we used a within-subjects effect for the comparison of cancer and normal/benign tumors. Furthermore, they made comparisons to normal breasts whereas in this study we made comparisons to benign tumors.

In this investigation the depth of the tumor was not analyzed as a factor for detection, only changing the source-detector distance from 3 to 4 cm when the tumor depth was greater than 1.5 cm. In the phantom study we conducted before the clinical trial, it was impossible to reach the tumor with a source-detector separation of 3 cm. The non-adjustment of the optimal source-detector separation in each case may have led to the lower sensitivity in this investigation.

The level of detectability of DCIS was not the result of the tumors being smaller in size as the tumor size of DCIS was not very different from invasive cancers. The real issue was whether DCIS developed angiogenesis and hypoxia. Our results (Table 29.2) show there was no difference in $\Delta[Hb_{tot}]$ and ΔDeoxygenation between invasive cancers and DCIS in either smaller- or larger- sized tumors, and DCIS in large masses shows significant hypoxia, as much as in invasive cancers. Thus, our investigation suggests that DCIS has already developed at least hypoxia in the tumor.

Judging from these results, we would need two solutions in order to improve the sensitivity and specificity values. The first solution would need to provide a more accurate localization of the tumor. NIRS devices with an ultrasound sensor in the middle to co-register the location [8] are available, and this may be an ideal way to make sure the tumor is at the center of the catchment volume. The second solution would need to solve the partial volume problem of the cancer, especially when the tumor size is ≤ 1 cm. When the volume of a tumor is ≤ 10 % of the catchment volume of the light migration path in the breast tissue, the parameter value is diluted by a factor of 10 so that it will be impossible to have any contrast even if the cancer has double the values of the surrounding normal tissue. For example, if the total hemoglobin concentrations of cancer and normal tissue are 70 and 40 μM, respectively [8], then a 10 times dilution makes $\Delta[Hb_{tot}]$ only 3 μM. This value looks more like the values obtained in this investigation. Partial volume calculation requires very accurate measurement of tumor size in three dimensions with multiple measurements of depth. Thus, we need to search for a co-registration modality, such as ultrasound-guided localization, for real cancer characterization in the future.

Our previous study from 2005 [5] used a different cancer threshold when we scored a probability of cancer. The threshold used previously was a $\Delta[Hb_{tot}]$ of 12 μM and a ΔDeoxygenation of 13 μM. The difference between these two studies can be attributed to the size of the tumors. The sensitivity for detection of cancer will increase with the size of the tumor. With a larger average mass size, the marginal $\Delta[Hb_{tot}]$ and ΔDeoxygenation can be increased by a factor of 2.

References

1. SEER Cancer Statistics Review 1975–2009 (Vintage 2009 Populations) Based on Nov 2011 SEER data submission, posted to the SEER web site [cited Apr 2012]. Available from http://seer.cancer.gov/csr/1975_2009_pops09/
2. Kriege M, Brekelmans CT, Boetes C et al (2004) Efficacy of MRI and mammography for breast-cancer screening in women with a familial or genetic predisposition. N Engl J Med 351(5):427–437
3. Hawrysz DJ, Sevick-Muraca EM (2000) Developments toward diagnostic breast cancer imaging using near infrared optical measurements and fluorescent contrast agents. Neoplasia 2(5):388–417
4. Fantini S, Sassaroli A (2012) Near-infrared optical mammography for breast cancer detection with intrinsic contrast. Ann Biomed Eng 40(2):398–407
5. Chance B, Nioka S, Zhang J et al (2005) Breast cancer detection based on incremental biochemical and physiological properties of breast cancers: a six-year, two-site study. Acad Radiol 12(8):925–933
6. Nioka S, Chance B (2005) NIR spectroscopic detection of breast cancer. Technol Cancer Res Treat 4(5):497–512
7. Ntziachristos V, Yodh AG, Schnall MD, Chance B (2002) MRI-guided diffuse optical spectroscopy of malignant and benign breast lesions. Neoplasia 4(4):347–354
8. Zhu Q, Hegde PU, Ricci A Jr et al (2010) Early-stage invasive breast cancers: potential role of optical tomography with US localization in assisting diagnosis. Radiology 256(2):367–378

9. Grosenick D, Wabnitz H, Moesta KT, Mucke J, Schlag PM, Rinneberg H (2005) Time-domain scanning optical mammography: II. optical properties and tissue parameters of 87 carcinomas. Phys Med Biol 50(11):2451–2468
10. Fang Q, Selb J, Carp SA et al (2011) Combined optical and X-ray tomosynthesis breast imaging. Radiology 258(1):89–97
11. Intes X (2005) Time-domain optical mammography SoftScan: initial results. Acad Radiol 12(8):934–947
12. Wang J, Pogue BW, Jiang S, Paulsen KD (2010) Nearinfrared tomography of breast cancer hemoglobin, water, lipid, and scattering using combined frequency domain and CW measurements. Opt Lett 35(1):82–84
13. Cerussi A, Shah N, Hsiang D, Durkin A, Butler J, Tromberg BJ (2006) In vivo absorption, scattering, and physiologic properties of 58 malignant breast tumors determined by broadband diffuse optical spectroscopy. J Biomed Opt 11(4):044005

Chapter 30
Impact of Extracellular Acidosis on Intracellular pH Control and Cell Signaling in Tumor Cells

Anne Riemann, Angelika Ihling, Bettina Schneider, Michael Gekle, and Oliver Thews

Abstract Cells in solid tumors generate an extracellular acidosis due to the Warburg effect and tissue hypoxia. Acidosis can affect the functional behavior of tumor cells, causing, e.g., multidrug resistance. In this process ERK1/2 and p38 mitogen-activated protein kinases (MAPK) seem to play a key role. However, the underlying mechanism of MAPK activation by extracellular acidosis remains unclear. Experiments were performed in three tumor and three normal tissue cell lines in which the cells were exposed to an extracellular pH of 6.6 for 3 h. Intracellular pH (pH_i), protein expression and activation, acidosis-induced transactivation, and reactive oxygen species (ROS) formation were measured. Extracellular acidosis resulted in a rapid and sustained decrease of pH_i leading to a reversal of the extra-/intracellular pH gradient. Extracellular acidosis led to p38 phosphorylation in all cell types and to ERK1/2 phosphorylation in three of six cell lines. Furthermore, p38 phosphorylation was also observed during sole intracellular lactacidosis at normal pH_e. Acidosis-enhanced formation of ROS, probably originating from mitochondria, seems to trigger MAPK phosphorylation. Finally, acidosis increased phosphorylation of the transcription factor CREB and resulted in increased transcriptional activity. Thus, an acidic tumor microenvironment can induce a longer-lasting p38 CREB-mediated change in the transcriptional program.

30.1 Introduction

In comparison to normal tissues, many solidly growing tumors show a marked extracellular acidosis with pH (pH_e) values down to 5.5 [1] which results either from an insufficient oxygen supply to the tissue due to a functionally inadequate

A. Riemann • A. Ihling • B. Schneider • M. Gekle • O. Thews (✉)
Julius-Bernstein Institute of Physiology, Martin-Luther-University of Halle-Wittenberg, Magdeburger Str. 6, 06112 Halle (Saale), Germany
e-mail: oliver.thews@medizin.uni-halle.de

S. Van Huffel et al. (eds.), *Oxygen Transport to Tissue XXXV*, Advances in Experimental Medicine and Biology 789, DOI 10.1007/978-1-4614-7411-1_30,

vascularization [1] or from the Warburg effect [2]. Both cases enforce the glycolytic metabolism in tumor cells leading to a pronounced increase in lactic acid formation (resulting in remarkably high lactate concentrations in the tissue).

Previous studies could demonstrate that the extracellular acidosis affects the proliferation of tumors and the malignant behavior or functional properties of the tumor cells, e.g., the functional activity of the drug-transporting p-glycoprotein [3, 4], leading to a reduced chemosensitivity of tumors. These studies also revealed that for this impact of acidosis on drug transport activities, intracellular MAP kinases play a major role. Inhibition of p38 and/or ERK1/2 MAPK prevented the acidosis-induced chemoresistance.

However, the mechanism by which the extracellular pH affects the intracellular MAPK still remains unclear. Also the question whether the extracellular acidosis directly affects the pathways (e.g., by G protein-coupled membrane sensors for H^+ concentration) or acidosis leads to an intracellular acidification which then activates MAPK has to be addressed. Finally, the study should clarify whether other cellular processes besides the p-glycoprotein activity are affected.

30.2 Materials and Methods

30.2.1 Cell Lines

The experiments were performed in three different tumor cell lines (NCI-H358 human bronchioalveolar carcinoma, subline AT1 of the rat R-3327 prostate carcinoma, LS513 human colorectal carcinoma) and in three cell lines from normal tissues (MDCK-C7 normal renal collecting duct epithelium cells of the canine, OK normal epithelial cells from renal proximal tubule of the opossum kidney, CHO immortalized ovary cells of the Chinese hamster). Cells were grown in their adequate medium supplemented with 10 % fetal calf serum (FCS) at 37 °C under a humidified 5 % CO_2 atmosphere and subcultivated twice a week. Control cells were exposed to bicarbonate-HEPES-buffered Ringer solution adjusted to pH 7.4. Extracellular acidosis (pH 6.6) was applied using bicarbonate-MES (morpholinoethanesulfonic acid)-buffered Ringer solution with pH adjusted to 6.6.

30.2.2 Intracellular pH

Cytosolic pH of single cells was determined using the pH-sensitive dye BCECF (2,7-bis-(2-carboxyethyl)-5-(and-6)-carboxyfluorescein, acetoxymethyl ester; Invitrogen, Paisley, UK) as described before [5]. In brief, cells were incubated with Ringer solution containing 5 mM BCECF-AM for 15 min. Then, the cover slips were rinsed with superfusion solution and transferred to the stage of an inverted

Axiovert 100 TV microscope (Zeiss, Oberkochen, Germany). The excitation wavelengths were 450 nm/490 nm; emitted light was measured through a band-pass filter (515–565 nm) every 10 s. After background subtraction, fluorescence intensity ratios were calculated. pH calibration was performed after each experiment by the nigericin (Sigma, St. Louis, USA) technique using a two-point calibration (pH 6.8 and 7.5).

30.2.3 *Western Blot and CRE-SEAP Reporter Gene Assay*

Western blotting was performed according to standard protocols. Cells were lysed (0.1 % Triton X-100 in PBS, protease inhibitor cocktail, 37 mg/l sodium orthovanadate or 150 mM NaCl, 10 mM Tris pH 7.4, 1 % Nonidet P-40, 0.1 % SDS, 1 % sodium deoxycholate, 0.1 % Triton X-100, 1 mM EDTA, protease inhibitor cocktail, 184 mg/l sodium orthovanadate) and cell protein was determined by the BCA method (BC assay reagents from Uptima), separated by SDS-PAGE, and transferred to a nitrocellulose membrane. Subsequently, membranes were incubated with antibodies specific for ERK1/2, p38, phospho-ERK1/2, phospho-p38, CREB, and phospho-CREB (1:1,000, cell signaling). The bound primary antibody was visualized using horseradish peroxidase-conjugated secondary antibodies and the ECL system (Pierce/Thermo Fisher Scientific) with the Molecular Imager ChemiDoc XRS System (Bio-Rad, Munich, Germany). Quantitative analysis was performed with Quantity One software (Bio-Rad).

Transactivation was assessed by the Mercury™ Pathway Profiling reporter gene assay system from Clontech Inc. using secretory alkaline phosphatase (SEAP) under the control of defined *cis*-regulatory response elements (CRE) as reporter, essentially as described earlier [6]. In brief, the cells were transfected with a pCRE-SEAP constructs or empty vectors. SEAP activity in the media was determined with the AttoPhos System from Promega (Mannheim, Germany) and normalized to the transfection control (ß-galactosidase).

30.2.4 *ROS Formation*

Formation of reactive oxygen species (ROS) was assessed with the fluorescent dye CM-H2DCFDA (Molecular Probes, Leiden, Netherlands). Cells were seeded in 24-well plates and incubated for 30 min with dye after the indicated treatments. Subsequently, fluorescence (excitation 485 nm; emission 535 nm) was measured using a multiwell counter (Infinite, Tecan, Berlin, Germany). The increase of fluorescence over the blank value expressed per mg protein was used as a measure for ROS formation.

30.3 Results

In order to analyze the impact of the pH_e on the intracellular compartment, the change of the intracellular pH (pH_i) was measured. In all cell lines (tumors and normal tissues), pH_i decreased with acidic pH_e. At pH_e 7.4 the pH_i was on average −0.24 lower (depending on the cell line). However, at pH_e 6.6 pH_i remained even higher than the extracellular space (+0.18). This phenomenon that the normal pH gradient (with a higher pH in the extracellular space) reversed under acidic condition (pH_i more alkalic than pH_e) was seen in four out of the six cell lines (two tumor and two normal tissue cell lines; AT1, LS513, MDCK-C7, OK). In order to elucidate the cellular mechanisms of intracellular pH homeostasis, H^+ or anion exchangers were inhibited. When AT1 cells were overloaded with lactate (by adding 40 mM lactate to the medium) in combination with an inhibitor of the bicarbonate exchanger (200 μM DIDS), the intracellular pH reached 6.95 ± 0.02 (even at pH_e 7.4), a value which was comparable to that found during extracellular acidification. With this procedure it became possible to acidify solely the intracellular space and to separate effects of extra- and intracellular pH.

When exposing the cells to low pH_e, an activation of the p38 and the ERK1/2 MAP kinases was seen. In all six cell lines, p38 was significantly phosphorylated at pH 6.6 (Fig. 30.1a), whereas ERK1/2 phosphorylation was seen only in AT1, OK, and CHO cells. In the other cell lines, a reduced pH_e had no impact on ERK1/2 (Fig. 30.1a). With the results of the experiments on pH homeostasis, it became possible to study the impact extra- and intracellular pH separately. In AT1 cells extracellular acidification activated both p38 and ERK1/2, whereas sole intracellular

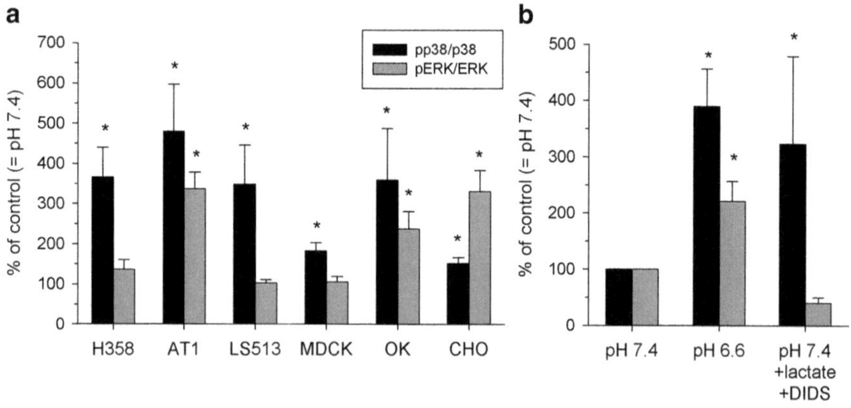

Fig. 30.1 (**a**) ERK1/2 and p38 phosphorylation under acidic conditions (pH = 6.6) in different tumor and normal tissue cell lines. Values are expressed as % of control; $n = 5$–12. (**b**) Impact of intracellular acidification with lactate and DIDS (at extracellular pH 7.4) on MAPK phosphorylation in AT1 cells; n=5–16. (*) $p < 0.05$

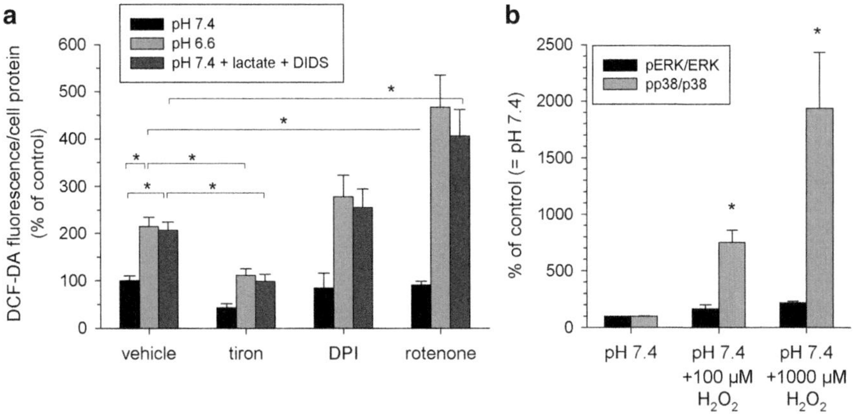

Fig. 30.2 (**a**) ROS formation in AT1 cells measured by DCF-DA fluorescence under control ($pH = 7.4$) and acidic ($pH = 6.6$) conditions and in cells where solely the intracellular space was acidified ($pH = 7.4 + lactate + DIDS$) in combination with application of the ROS scavengers tiron, DPI, or rotenone; $n = 9–12$. (**b**) Impact of ROS (at extracellular pH 7.4) on ERK1/2 and p38 phosphorylation in AT1 cells; $n = 6–7$. (*) $p < 0.05$

acidosis (pH_e 7.4 with lactate + DIDS) led only to p38 phosphorylation (Fig. 30.1b) indicating separate activation mechanisms for both pathways.

Experiments on the mechanism by which changes in H^+ concentration lead to an ERK or p38 phosphorylation led to the finding that acidosis induced a marked increase in ROS formation. When cells were exposed to pH 6.6, ROS production was more than doubled, an effect which relies on the intracellular acidification (Fig. 30.2a). The increased ROS levels could be counteracted by ROS scavenging (tiron) and were even more pronounced when the mitochondrial complex I was inhibited by rotenone indicating that ROS from mitochondria seems to play an important role (Fig. 30.2a). On the other hand ROS were able to phosphorylate p38 (but not ERK1/2) (Fig. 30.2b). When H_2O_2 was added to the medium, p38 activation followed in a dose-dependent manner.

Finally, other signals downstream p38 were analyzed. It turned out that in two of the three tumor cell lines (AT1, LS513), the transcription factor CREB (cAMP response element-binding protein) was phosphorylated by the extracellular acidosis (Fig. 30.3a). Inhibition of p38 signaling pathway with SB203580 reduced acidosis-induced CREB phosphorylation to control level (Fig. 30.3b). Since CREB transactivates CRE-controlled genes, the question arose whether extracellular acidosis is able to change gene expression. By using a CRE-SEAP reporter gene assay, it could be demonstrated that extracellular acidosis significantly induced CRE-dependent gene expression by 200 ± 11 % in AT1 cells.

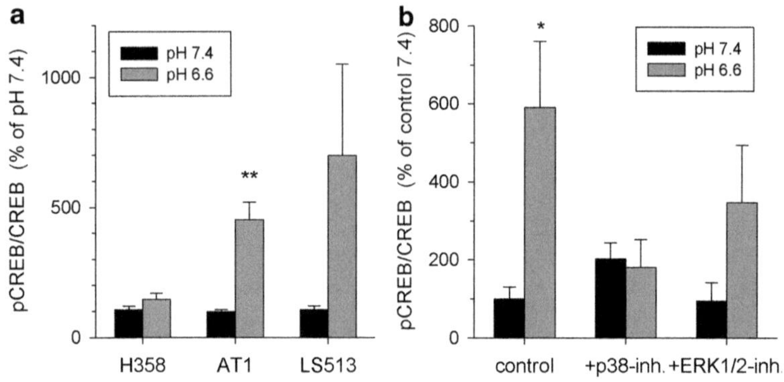

Fig. 30.3 (**a**) Impact of extracellular pH on CREB phosphorylation in different tumor cell lines. Measured values were normalized to pH 7.4 ($n=6$–10). (**b**) CREB phosphorylation in AT1 cells under control (pH = 7.4) and acidic (pH = 6.6) conditions in combination with inhibitors of the p38 (SB203580) and the ERK1/2 (U0126) MAPK. Values are expressed relative to the levels at pH 7.4 in control cells (*) $p<0.05$; $n=3$

30.4 Discussion

Due to forced glycolytic metabolism, solid tumors often show pronounced extracellular acidosis [1]. However, tumor cells (as well as normal tissue cells) possess very effective H^+ transport mechanisms by which intracellular protons are eliminated. Although these mechanisms cannot prevent an acidification of the intracellular space when protons are added extracellularly, the extent of pH change is much smaller. In the AT1 cell line [7], it turned out that the most important ion transporter for maintaining pH homeostasis seems to be a Na^+-dependent HCO_3^-/Cl^- exchanger (NDCBE) [8]. By inhibition of the bicarbonate exchange (together with lactate load), the intracellular space could be acidified solely.

Acidosis led to a phosphorylation of p38 in all cell lines studied; however, ERK1/2 activation was seen only in half of the cell lines. In the AT1 cells (studied in more detail), it seems that p38 is mainly activated by the intracellular pH since the sole intracellular acidification (at normal extracellular pH) activated p38 to the same extent (Fig. 30.1b). In AT1 cells it could also be demonstrated that kinases (MKK3/6) upstream of p38 were activated in a pH-dependent manner [7]. These data are in good accordance with the results from cardiomyocytes where pH_i was essential for hypoxia-induced p38 phosphorylation [9]. Our data therefore reveal that p38 activation is a general (not tumor-specific) mechanism to acidotic stress. ERK1/2 was activated in three of six cell lines (Fig. 30.1b). This behavior (p38 activated by acidosis but not ERK1/2) was also described for cardiomyocytes [9]. At the moment it cannot be decided whether this discrepancy is a cell line-specific feature or whether ERK1/2 phosphorylation is less sensitive to acidosis.

The functional link between acidotic stress signal and p38 activation still remains incompletely understood. Previous results indicated that the involvement of the signaling pathways EGFR, PI3K, PKC, Src family, and PKA is highly unlikely since inhibition of these mechanisms could not prevent acidosis-induced MAPK activation [7]. Also OGR1, described as a membrane H^+ sensor linked to MAPK [10], does not play a role.

However, the cellular level of ROS seems to be involved. On the one hand, extra- and intracellular acidoses were able to significantly increase the cellular ROS production (Fig. 30.2b), and on the other hand, adding H_2O_2 to the medium resulted in a strong activation of p38 while scavenging ROS by tiron suppressed pH-dependent p38 phosphorylation. These results reveal that ROS act as a "signaling molecule" for the pH-dependent activation of MAPK. However, the question of how acidosis induces ROS formation (probably in mitochondria) and how ROS activate p38 has to be addressed in further studies.

Acidosis-induced MAPK activation itself plays an important role for the malignant behavior of tumors. Besides the regulation of multidrug resistance [3, 4], p38 can transactivate CRE-controlled genes. In the present study it was demonstrated that extracellular acidosis phosphorylates CREB which stays in line with previous results on chondrocytes [11]. Using a reporter gene assay, our data clearly show that acidosis controls gene expression. In conclusion, the present study illustrates that tumor acidosis resulting from glycolytic metabolism has a strong impact on the functional behavior of tumor cells. By activating MAPK pH may influence functional properties of tumor cells (e.g., chemoresistance, cell motility) but also gene expression. Taking these findings into account, new therapeutic strategies may be developed by either manipulating tumor pH or interfering with the signaling chains triggered by acidosis.

Acknowledgments This study was supported by Deutsche Krebshilfe (grants 106774/106906), the BMBF (ProNet-T3 Ta-04), and the Wilhelm-Roux program of the Medical School, University Halle-Wittenberg.

References

1. Vaupel P, Kallinowski F, Okunieff P (1989) Blood flow, oxygen and nutrient supply, and metabolic microenvironment of human tumors: a review. Cancer Res 49(23):6449–6465
2. Vander Heiden MG, Cantley LC, Thompson CB (2009) Understanding the Warburg effect: the metabolic requirements of cell proliferation. Science 324(5930):1029–1033
3. Sauvant C, Nowak M, Wirth C, Schneider B, Riemann A, Gekle M, Thews O (2008) Acidosis induces multi-drug resistance in rat prostate cancer cells (AT1) in vitro and in vivo by increasing the activity of the p-glycoprotein via activation of p38. Int J Cancer 123: 2532–2542
4. Thews O, Gassner B, Kelleher DK, Schwerdt G, Gekle M (2006) Impact of extracellular acidity on the activity of p-glycoprotein and the cytotoxicity of chemotherapeutic drugs. Neoplasia 8(2):143–152
5. Gekle M, Golenhofen N, Oberleithner H, Sibernagl S (1996) Rapid activation of Na^+/H^+ exchange by aldosterone in renal epithelial cells requires Ca^{2+} and stimulation of a plasma membrane proton conductance. Proc Natl Acad Sci USA 93(19):10500–10504

6. Pfau A, Grossmann C, Freudinger R, Mildrenberger S, Benesic A, Gekle M (2007) Ca^{2+} but not H_2O_2 modulates GRE-element activation by the human mineralocorticoid receptor in HEK cells. Mol Cell Endocrinol 264(1–2):35–43
7. Riemann A, Schneider B, Ihling A et al (2011) Acidic environment leads to ROS-induced MAPK signaling in cancer cells. PLoS One 6(7):e22445
8. Boron WF, Chen L, Parker MD (2009) Modular structure of sodium-coupled bicarbonate transporters. J Exp Biol 212(Pt 11):1697–1706
9. Zheng M, Reynolds C, Jo SH, Wersto R, Han Q, Xiao RP (2005) Intracellular acidosis-activated p38 MAPK signaling and its essential role in cardiomyocyte hypoxic injury. FASEB J 19(1):109–111
10. Ludwig MG, Vanek M, Guerini D et al (2003) Proton-sensing G-protein-coupled receptors. Nature 425(6953):93–98
11. Goldberg R, Reshef-Bankai E, Coleman R, Green J, Maor G (2006) Chronic acidosis-induced growth retardation is mediated by proton-induced expression of Gs protein. J Bone Miner Res 21(5):703–713

Chapter 31
Tumor Oxygenation: An Appraisal of Past and Present Concepts and a Look into the Future

Arisztid G. B. Kovách Lecture

Peter Vaupel

Abstract Since 1970, the multifactorial pathogenesis of the deficient and heterogeneous oxygenation of transplanted murine tumors and of human cancers (including parameters determining oxygen delivery, e.g., blood flow, diffusion geometry, oxygen transport capacity of the blood) has been investigated in vivo. Hypoxia and/or anoxia was quantitatively assessed and characterized using microtechniques and special preclinical tumor models. Hypoxia subtypes were identified, and critical supply conditions were theoretically analyzed. In the 1980s, first experiments on humans were carried out in cancers of the rectum and of the oral cavity. In the 1990s, the clinical investigations were carried out on cancers of the breast and of the uterine cervix, clearly showing that hypoxia is a hallmark of locally advanced human tumors. In multivariate analysis, hypoxia was found to be an independent, adverse prognostic factor for patient survival due to hypoxia-driven malignant progression and hypoxia-associated resistance to anticancer therapy.

31.1 Introduction

During the directorship of Professor Gerhard Thews, research at the Institute of Physiology, University of Mainz, traditionally focussed on oxygen transport in blood, lung, brain, and heart. Joining his research team in 1970 as a postdoctoral research fellow, I was asked to "investigate the oxygen transport and respiratory gas exchange in other clinically relevant tissues and organs to expand the scope of research of the Institute" (G. Thews). After a careful and time-consuming literature

P. Vaupel (✉)
Department of Radiotherapy and Radiooncology, Klinikum rechts der Isar, Technical University, Ismaninger Strasse 22, 81675 Munich, Germany
e-mail: vaupel@uni-mainz.de

S. Van Huffel et al. (eds.), *Oxygen Transport to Tissue XXXV*, Advances in Experimental Medicine and Biology 789, DOI 10.1007/978-1-4614-7411-1_31,

search, I finally decided to study oxygen transport to the spleen (murine, rabbit, human) and to solid, malignant tumors, since reliable oxygenation data for these tissues were not available at that time, especially in terms of translation of the preclinical data to the clinical setting. In the following, chronology-oriented chapters are presented, and selected data obtained over the last 40 years are described, clearly showing the progress in relevant information.

The oxygenation status and data on the respiratory gas exchange of the spleen have been described earlier [1–3].

31.2 The Past

Since 1970, in vivo investigations have been carried out on isotransplanted rat tumors after the development and implementation of a "tissue-isolated" tumor model in the rat kidney involving a single artery feeding the tumor and a single vein draining the tumor, thus enabling the measurement of total blood flow and of the biologically relevant arteriovenous concentration differences of the substrates and catabolites of interest [4–6]. Key results using this tumor preparation are as follows (for details, see [4–12]): (a) tumor blood flow (TBF) and oxygen availability exhibit pronounced intra-tumor and inter-tumor heterogeneities; (b) tumor oxygenation is distinctly poorer than in normal tissue and shows similar heterogeneities to those found for TBF; (c) increasing oxygen availability through increasing TBF, arterial oxygen content, and hemoglobin concentration (cHb) can increase oxygen uptake and can improve tissue oxygenation; (d) oxygen availability is the major determinant of the oxygen consumption rate of cancers in situ; (e) oxygen consumption rate of cancers in situ is thus a function of TBF and arterial oxygen content; (f) weight-related TBF and tissue oxygenation generally decrease with increasing tumor size (not necessarily applicable to the clinical setting, see below); and (g) contrary to conventional belief, there is no evidence for a general mitochondrial dysfunction, speaking against a principal role of the Warburg effect in its original concept [6, 13].

Modulation of the tumor oxygenation status has been described as a result of therapeutic measures (irradiation [14], localized hyperthermia [15–19], photodynamic therapy [20], normobaric and hyperbaric hyperoxia [6, 21, 22], improvement of perfusion [23, 24], and correction of anemia using erythropoietin [25]).

Between 1977 and 1985, HbO_2 saturation of single red blood cells (RBCs) in tumor microvessels was registered in experimental rat tumors [26, 27], in cancers of the oral cavity and of the rectum, and in primary and metastatic bone tumors [28–30]. In accordance with the studies on experimental murine tumors, the oxygenation status in human cancers was poorer than in the normal tissue, exhibited pronounced intra-tumor and inter-tumor heterogeneities, and was positively correlated with the vascular density. In contrast to the experimental situation, the O_2 status in human tumors showed no size dependency.

In 1985, investigations were started to assess the oxygen status of orthotopically xenografted human breast cancers in immune-deficient rnu/rnu rats. In order to allow measurements of TBF and the relevant arteriovenous concentration differences, a novel "tissue-isolated tumor" model was implemented [31, 32]. Experiments using different tumor histologies showed that – comparable to experimental murine tumors – the oxygen consumption and the median tissue pO_2 both were a function of TBF and the oxygen availability, respectively [33]. Theoretical analysis of the oxygen supply conditions in these xenografted human tumors led to the conclusion that oxygen seems to be the limiting substrate for unlimited proliferation and glucose for tumor cell survival [34, 35].

Using ^{31}P-NMR-spectroscopy studies starting in 1987, correlations between the bioenergetic status, the tissue oxygenation, and the intracellular pH (pH_i) were evaluated. In experimental murine tumors, pH_i was neutral to alkaline whereas the extracellular pH (pH_e) was acidic [36]. Intracellular pH was found to be alkaline as long as the median tissue pO_2 was above 10 mmHg. Below this threshold, pH_i became acidic [37–41] and the gradient between the intracellular pH and extracellular pH flattened.

In 1989 systematic investigations on gynecological patient cancers (cervix, breast, vulva) were initiated. In these patients, the pretherapeutic oxygenation status of primary and recurrent tumors was assessed using the pO_2 histography system [42–46]. Publication of these data had a tremendous impact in defining the role of tumor hypoxia in malignant progression and therapeutic resistance [47–52]. Key findings were as follows: (a) approx. 60 % of pretreatment cervical cancers were hypoxic; (b) cancer oxygenation was distinctly poorer than that of the normal tissues at the site of tumor growth; (c) the extent of hypoxia was independent of clinical size, stage, histology, grade, lymph node status, and various patient demographics; (d) hypoxia was aggravated in anemic patients; (e) hypoxia was less pronounced on transgression of stage IVA cervical cancers into the bladder wall; (f) recurrent tumors were more hypoxic than their primaries; and (g) there was no typical topological distribution of hypoxic areas within tumors (periphery vs. center).

Since 1990, investigations on hypoxia-driven malignant progression followed, based on the observations that in multivariate analysis, hypoxia was found to be a strong, independent, and adverse prognostic factor for overall and disease-free survival in cervical cancer patients [53–58].

In the last 10 years, the recognition of tumor hypoxia as a pivotal factor driving the development of a highly malignant phenotype – in which the HIF system, genetic instability, and clonal selection play a central role – has encouraged attempts to correlate the expression of "endogenous" hypoxia markers (HIFs, GLUT-1, CA IX) with the oxygenation status in identical, non-necrotic tumor microareas. Our results clearly showed that there is no correlation between the protein expression of these markers and pO_2 data measured with O_2 microsensors [59–62]. This supports the hypothesis that the HIF system can be stabilized even under normoxic conditions (e.g., by oncogenic growth factors, certain cytokines, glucose deprivation, acidosis, and gene mutations). From this it can be concluded that HIF-1α and its target genes cannot be considered as strict hypoxia markers, but instead should be considered to be markers of hypoxia-associated malignant progression.

31.3 The Present

Since 2008 our research focus is on the classification and quantification of hypoxia subtypes in xenografted human squamous cell carcinomas of the head and neck. In these experiments, hypoxia subtypes are categorized as follows: (a) continuous (chronic) hypoxia due to diffusion limitations or sustained microvascular flow stop by disturbed Starling forces, (b) intermittent (acute) hypoxia due to temporary obstructions of tumor microvessels or distinct fluctuations of RBC fluxes, and (c) hypoxemic hypoxia due to patient anemia or plasma flow in microvessels only [63]. Using tumor cryosections and (immuno-) fluorescence, detection and quantification of these subtypes showed that chronic hypoxia is the dominating subtype in vital tumor tissue, followed by acute and hypoxemic hypoxia. Analyses using microcirculatory supply units yielded pronounced (tumor size-dependent) intra-tumor heterogeneity and distinct variability between different tumor lines [64–67].

31.4 The Future

The extent of hypoxia subtypes, their respective fractions of total hypoxia, their time frames, and biological and therapeutic consequences will be investigated in the near future. Furthermore, detection and reliable quantification of hypoxia subtypes in the clinical setting are urgently needed, especially for critical judgment of fractionation schedules in radio(chemo-)therapy.

Acknowledgments The cooperation and excellent contribution of all research group members and colleagues to this research at the following institutions over the past 42 years is highly appreciated:

- Dept. Physiology, University of Mainz, Germany (1970–1978)
- Dept. Therapeutic Radiology, Henry Ford Hospital, Detroit, USA (1979)
- Dept. Applied Physiology, University of Mainz, Germany (1980–1986)
- Dept. Radiation Medicine, Harvard Medical School, Boston, USA (1987–1989)
- Dept. Physiology and Pathophysiology, University of Mainz, Germany (1990–2008)
- Dept. Radiooncology, University Medical Center, Mainz, Germany (since 2008)
- Dept. Radiotherapy, Technical University, Munich, Germany (since 2008)

Appendix A Citation Record of Most-Cited Publications of Vaupel et al. (ISI Web of Science, August 21, 2012)

1705 × Vaupel et al (1989) Cancer Res 49: 6449
1018 × Höckel et al (1996) Cancer Res 56:4509
895 × Höckel & Vaupel (2001) J Natl Cancer Inst 93:268
618 × Vaupel et al (1991) Cancer Res 51:3316

518 × Höckel et al (1993) Radiother Oncol 26:45
320 × Vaupel & Mayer (2007) Cancer Metastasis Rev 26:225
315 × Vaupel et al (1981) Cancer Res 41:2008
289 × Höckel et al (1991) Cancer Res 51:6098
235 × Vaupel (2004) Semin Radiat Oncol 14:198
231 × Vaupel et al (2001) Semin Oncol 28 (suppl 8):29
220 × Höckel et al (1999) Cancer Res 59:4525
206 × Vaupel (2004) Oncologist 9 (suppl 5):10
205 × Vaupel et al (2001) Med Oncol 18:243
205 × Kallinowski et al (1990) Int J Radiat Oncol Biol Phys 19:953
204 × Tatum et al (2006) Int J Radiat Biol 82:699
203 × Höckel et al (1996) Semin Radiat Oncol 6:3
168 × Höckel & Vaupel (2001) Semin Oncol 28 (suppl 8)36
164 × Vaupel & Harrison (2002) Oncologist 9 (suppl 5):4
161 × Kallinowski et al (1989) Cancer Res 49:3759
– 150 × Bicher et al (1980) Radiology 137:523

References

1. Vaupel P, Braunbeck W, Thews G (1973) Respiratory gas exchange and pO_2 distribution in splenic tissue. Adv Exp Med Biol 37A:401–406
2. Vaupel P, Manz R, Müller-Klieser W (1979) Respiratory gas exchange in the rat spleen in situ and intrasplenic oxyhemoglobin saturation. Pflügers Arch 379(1):109–111
3. Vaupel P, Frinak S, Müller-Klieser W, Manz R (1981) The end of a postulate: there are no hostile metabolic conditions within the normal spleen. Bibl Anat 20:403–406
4. Vaupel P, Günther H, Grote J, Aumüller G (1971) Atemgaswechsel und Glucosestoffwechsel von Tumoren (DS- Carcinosarkom) in vivo. Z Ges Exp Med 156(4):283–294
5. Schwarz W, Schulz V, Kersten M, Wörz R, Vaupel P (1971) Durchblutung und Sauerstoffverbrauch gewebsisolierter Impftumoren (DS- Carcinosarkom) in vivo. Z Krebsforsch 75(3):161–173
6. Vaupel P (1974) Atemgaswechsel und Glucosestoffwechsel von Implantationstumoren (DS-Carcinosarkom) in vivo. Funktionsanalyse biolog Systeme 1:1–138
7. Vaupel P (1977) Hypoxia in neoplastic tissue. Microvasc Res 13(3):399–408
8. Manz R, Otte J, Thews G, Vaupel P (1983) Relationship between size and oxygenation status of malignant tumors. Adv Exp Med Biol 159:391–398
9. Thews G, Vaupel P (1976) O_2 supply conditions in tumor tissue in vivo. Adv Exp Med Biol 75:537–548
10. Vaupel P, Thews G (1974) PO_2 distribution in tumor tissue of DS-carcinosarcoma. Oncology 30(6):475–484
11. Vaupel P, Frinak S, Bicher HI (1981) Heterogeneous oxygen partial pressure and pH distribution in C3H mouse mammary adenocarcinoma. Cancer Res 41(5):2008–2013
12. Vaupel P, Müller-Klieser W (1986) Cell line and growth site as relevant parameters governing tumor tissue oxygenation. Adv Exp Med Biol 200:633–643
13. Vaupel P, Mayer A (2012) Availability, not respiratory capacity governs oxygen consumption of solid tumors. Int J Biochem Cell Biol 44(9):1477–1481
14. Vaupel P, Frinak S, O'Hara M (1984) Reoxygenation of malignant mammary tumors after a single large dose of irradiation. Adv Exp Med Biol 180:773–782

15. Bicher HI, Hetzel FW, Sandhu TS et al (1980) Effects of hyperthermia on normal and tumor microenvironment. Radiology 137(2):523–530
16. Vaupel P, Frinak S, Müller-Klieser W, Bicher HI (1982) Impact of localized hyperthermia on the cellular microenvironment in solid tumors. Natl Cancer Inst Monogr 61:207–209
17. Vaupel P, Kallinowski F (1987) Physiological effects of hyperthermia. Recent Results Cancer Res 104:71–109
18. Vaupel P, Kallinowski F, Kluge M (1988) Pathophysiology of tumors in hyperthermia. Recent Results Cancer Res 107:65–75
19. Vaupel P, Kelleher DK (2010) Pathophysiological and vascular characteristics of tumors and their importance for hyperthermia: heterogeneity is the key issue. Int J Hyperthermia 26(3):211–223
20. Kelleher DK, Thews O, Scherz A, Salomon Y, Vaupel P (2004) Perfusion, oxygenation status and growth of experimental tumors upon photodynamic therapy with Pd-bacteriopheophorbide. Int J Oncol 24(6):1505–1511
21. Müller-Klieser W, Vaupel P, Manz R (1983) Tumor oxygenation under normobaric and hyperbaric conditions. Br J Radiol 56(668):559–564
22. Thews O, Kelleher DK, Vaupel P (2002) Dynamics of tumor oxygenation and red blood cell flux in response to inspiratory hyperoxia combined with different levels of inspiratory hypercapnia. Radiother Oncol 62(1):77–85
23. Vaupel P, Kelleher DK, Thews O (1998) Modulation of tumor oxygenation. Int J Radiat Oncol Biol Phys 42(4):843–848
24. Jung C, Müller-Klieser W, Vaupel P (1984) Tumor blood flow and O_2 availability during hemodilution. Adv Exp Med Biol 180:281–291
25. Kelleher DK, Matthienssen U, Thews O, Vaupel P (1996) Blood flow, oxygenation, and bioenergetic status of tumors following erythropoietin treatment in normal and anemic rats. Cancer Res 56(20):4728–4734
26. Vaupel P, Grunewald WA, Manz R, Sowa W (1977) Intracapillary HbO_2 saturation in tumor tissue of DS- Carcinosarcomas during normoxia. Adv Exp Med Biol 94:367–375
27. Vaupel P, Manz R, Müller-Klieser W, Grunewald WA (1979) Intracapillary HbO_2 saturation in malignant tumors during normoxia and hyperoxia. Microvasc Res 17(2):181–191
28. Müller-Klieser W, Vaupel P, Manz R, Schmidseder R (1981) Intracapillary oxyhemoglobin saturation in malignant tumors in humans. Int J Radiat Oncol Biol Phys 7(10):1397–1404
29. Wendling P, Manz R, Thews G, Vaupel P (1984) Heterogeneous oxygenation of rectal carcinomas in humans: a critical parameter for preoperative irradiation? Adv Exp Med Biol 180:293–300
30. Strube HD, Vaupel P, Manz R (1983) Inhomogene Oxygenierung bei primär malignen Knochentumoren sowie bei Knochenmetastasen und mögliche Konsequenzen für die Tumortherapie. Z Orthop 121:514
31. Vaupel P, Kallinowski F, Dave S, Gabbert H, Bastert G (1985) Human mammary carcinomas in nude rats–a new approach for investigating oxygen transport and substrate utilization in human tissues. Adv Exp Med Biol 191:737–751
32. Dave S, Kallinowski F, Vaupel P (1985) Blood flow and oxygen supply to human mammary carcinomas transplanted into nude rats. Adv Exp Med Biol 191:753–762
33. Vaupel P, Fortmeyer HP, Runkel S, Kalinowski F (1987) Blood flow, oxygen consumption and tissue oxygenation of human breast cancer xenografts in nude rats. Cancer Res 47(13):3496–3503
34. Vaupel P, Kallinowski F, Groebe K (1988) Evaluation of oxygen diffusion distances in human breast cancer using inherent in vivo data: role of various pathogenetic mechanisms in the development of tumor hypoxia. Adv Exp Med Biol 222:719–726
35. Groebe K, Vaupel P (1988) Evaluation of oxygen diffusion distances in human breast cancer xenografts using tumor-specific in vivo data: role of various mechanisms in the development of tumor hypoxia. Int J Radiat Oncol Biol Phys 15(3):691–697
36. Kallinowski F, Vaupel P (1988) pH distributions in spontaneous and isotransplanted rat tumors. Br J Cancer 58(3):314–321

37. Vaupel P, Okunieff P, Kallinowski F, Neuringer LJ (1989) Correlations between 31P-NMR spectroscopy and tissue O_2 tension measurements in a murine fibrosarcoma. Radiat Res 120(3):477–493
38. Vaupel P, Schaefer C, Okunieff P (1994) Intracellular acidosis in murine fibrosarcomas coincides with ATP depletion, hypoxia and high levels of lactate and total Pi. NMR Biomed 7(3):128–136
39. Vaupel P, Okunieff P, Neuringer LJ (1989) Blood flow, tissue oxygenation, pH distribution, and energy metabolism of murine mammary adenocarcinomas during growth. Adv Exp Med Biol 248:835–845
40. Vaupel P (1996) Is there a critical tissue oxygen tension for bioenergetic status and cellular pH regulation in solid tumors? Experientia 52(5):464–468
41. Vaupel P, Kallinowski F, Okunieff P (1989) Blood flow, oxygen and nutrient supply, and metabolic microenvironment of human tumors: a review. Cancer Res 49(23):6449–6465
42. Vaupel P, Schlenger K, Knoop C, Höckel M (1991) Oxygenation of human tumors: evaluation of tissue oxygen distribution in breast cancers by computerized O_2 tension measurements. Cancer Res 51(12):3316–3322
43. Höckel M, Schlenger K, Knoop C, Vaupel P (1991) Oxygenation of carcinomas of the uterine cervix: evaluation of computerized O_2 tension measurements. Cancer Res 51(22):6098–6102
44. Höckel M, Schlenger K, Höckel S, Aral B, Schäffer U, Vaupel P (1998) Tumor hypoxia in pelvic recurrences of cervical cancer. Int J Cancer 79(4):365–369
45. Vaupel P, Mayer A, Höckel M (2006) Oxygenation status of primary and recurrent squamous cell carcinomas of the vulva. Eur J Gynaecol Oncol 27(2):142–146
46. Vaupel P, Höckel M, Mayer A (2007) Detection and characterization of tumor hypoxia using pO2 histography. Antioxid Redox Signal 9(8):1221–1235
47. Höckel M, Vaupel P (2001) Tumor hypoxia: definitions and current clinical, biologic and molecular aspects. J Natl Cancer Inst 93(4):266–276
48. Höckel M, Schlenger K, Mitze M, Schäffer U, Vaupel P (1996) Hypoxia and radiation response in human tumors. Semin Radiat Oncol 6(1):3–9
49. Vaupel P, Kelleher DK, Höckel M (2001) Oxygen status of malignant tumors: pathogenesis of hypoxia and significance for tumor therapy. Semin Oncol 28(2 Suppl 8):29–35
50. Vaupel P (2004) Tumor microenvironmental physiology and its implications for radiation oncology. Semin Radiat Oncol 14:198–206
51. Vaupel P, Mayer A, Höckel M (2004) Tumor hypoxia and malignant progression. Methods Enzymol 381:335–354
52. Vaupel P, Mayer A (2007) Hypoxia in cancer: significance and impact on clinical outcome. Cancer Metastasis Rev 26(2):225–239
53. Höckel M, Knoop C, Schlenger K et al (1993) Intratumoral pO_2 predicts survival in advanced cancer of the uterine cervix. Radiother Oncol 26(1):45–50
54. Höckel M, Schlenger K, Aral B, Mitze M, Schäffer U, Vaupel P (1996) Association between tumor hypoxia and malignant progression in advanced cancer of the uterine cervix. Cancer Res 56(19):4509–4515
55. Höckel M, Schlenger K, Höckel S, Vaupel P (1999) Hypoxic cervical cancers with low apoptotic index are highly aggressive. Cancer Res 59(18):4525–4528
56. Vaupel P (2008) Hypoxia and aggressive tumor phenotype: implications for therapy and prognosis. Oncologist 13(3):21–26
57. Vaupel P, Harrison L (2004) Tumor hypoxia: causative factors, compensatory mechanisms, and cellular response. Oncologist 9(5):4–9
58. Vaupel P (2004) The role of hypoxia- induced factors in tumor progression. Oncologist 9(5):10–17
59. Mayer A, Wree A, Höckel M, Leo C, Pilch H, Vaupel P (2004) Lack of correlation between expression of HIF-1α protein and oxygenation status in identical tissue areas of squamous cell cancers of the uterine cervix. Cancer Res 64(16):5876–5881
60. Mayer A, Höckel M, Wree A, Vaupel P (2005) Microregional expression of glucose transporter GLUT-1 and oxygenation status: lack of correlation in locally advanced cervical cancers. Clin Cancer Res 11(7):2768–2773

61. Mayer A, Höckel M, Vaupel P (2005) Carbonic anhydrase IX expression and tumor oxygenation status do not correlate at the microregional level in locally advanced cancers of the uterine cervix. Clin Cancer Res 11(20):7220–7225
62. Mayer A, Höckel M, Vaupel P (2006) Endogenous hypoxia markers in locally advanced cancers of the uterine cervix: reality or wishful thinking? Strahlenther Onkol 182(9):501–510
63. Bayer C, Shi K, Astner ST, Maftei CA, Vaupel P (2011) Acute versus chronic hypoxia: why a simplified classification is simply not enough. Int J Radiat Oncol Biol Phys 80(4):965–968
64. Maftei CA, Bayer C, Shi K, Astner ST, Vaupel P (2011) Quantitative assessment of hypoxia subtypes in microcirculatory supply units of malignant tumors using (immuno-) fluorescence techniques. Strahlenther Onkol 187:260–266
65. Maftei AC, Bayer C, Shi K, Astner ST, Vaupel P (2011) Changes in the fraction of total hypoxia and hypoxia subtypes in human squamous cell carcinomas upon fractionated irradiation: evaluation using pattern recognition in microcirculatory supply units. Radiother Oncol 101(1):209–216
66. Maftei AC, Bayer C, Shi K, Vaupel P (2012) Intra- and intertumor heterogeneities in total, and acute hypoxia in xenografted squamous cell carcinomas: detection and quantification using (immuno-) fluorescence techniques. Strahlenther Onkol 188(7):606–615
67. Bayer C, Vaupel P (2012) Acute versus chronic hypoxia in tumors: controversial data concerning time frames and biological consequences. Strahlenther Onkol 188(7):616–627

Chapter 32
In Vivo Metabolic Evaluation of Breast Tumor Mouse Xenografts for Predicting Aggressiveness Using the Hyperpolarized ^{13}C-NMR Technique

He.N. Xu, Stephen Kadlececk, Ben Pullinger, Harrila Profka, Kejia Cai, Hari Hariharan, Rahim Rizi, and Lin Z. Li

Abstract In vivo imaging/spectroscopic biomarkers for solid tumor aggressiveness are needed in the clinic to facilitate cancer diagnosis and treatment strategies. In mouse models of human melanoma and breast cancer, we were able to detect the metabolic differences among tumors of different metastatic potential and between normal and cancer tissues by optical imaging of the mitochondrial redox state of snap-frozen tissue samples. Such metabolic differences indicate that tumors of different aggressiveness have different metabolic homeostasis, which supports that kinetic parameters such as rate constant(s) can also serve as biomarkers for cancer aggressiveness and treatment response. Here we present our preliminary study on the mouse xenografts of the aggressive and indolent human breast cancer cell lines using the hyperpolarized ^{13}C-NMR (HP-NMR) technique. By recording the time courses of ^{13}C-pyruvate tracer and its metabolite signals in vivo, particularly the ^{13}C-lactate signal, the apparent rate constants of both the forward and reverse reactions catalyzed

H.N. Xu
Department of Radiology, University of Pennsylvania, Philadelphia, PA, USA

Britton Chance Laboratory of Redox Imaging, Johnson Research Foundation, Department of Biochemistry and Biophysics, University of Pennsylvania, Philadelphia, PA, USA

S. Kadlececk • B. Pullinger • H. Profka • K. Cai • H. Hariharan • R. Rizi
Department of Radiology, University of Pennsylvania, Philadelphia, PA, USA

L.Z. Li (✉)
Department of Radiology, University of Pennsylvania, Philadelphia, PA, USA

Britton Chance Laboratory of Redox Imaging, Johnson Research Foundation, Department of Biochemistry and Biophysics, University of Pennsylvania, Philadelphia, PA, USA

Abramson Cancer Center, University of Pennsylvania, Philadelphia, PA, USA

Institute of Translational Medicine and Therapeutics, University of Pennsylvania, Philadelphia, PA, USA
e-mail: linli@mail.med.upenn.edu

S. Van Huffel et al. (eds.), *Oxygen Transport to Tissue XXXV*, Advances in Experimental Medicine and Biology 789, DOI 10.1007/978-1-4614-7411-1_32,

by lactate dehydrogenase (LDH) were extracted via the ratiometric modeling of the two-site exchange reaction that we developed. Data from four breast tumors (MCF-7, MDA-MB-468, and MDA-MB-231 medium and large) with different aggressiveness are included. We demonstrate the feasibility to quantify the apparent rate constants of LDH reactions in breast tumor xenografts.

32.1 Introduction

Cancer mortality is mainly the result of tumor metastasis. In vivo imaging of solid tumor aggressiveness or metastatic potential is needed in the clinic to facilitate cancer treatment strategies. Cancer is considered as "a paradigm of genetically-defined metabolic abnormalities" [1]. Several studies have shown tumor abnormalities in glucose metabolism, Krebs cycle, and the mitochondrial electron transport chain [2–5]. Our previous ex vivo studies on animal models demonstrated that the quantitative mitochondrial redox state imaging biomarkers differentiated solid tumor aggressiveness and distinguished normal from malignant tissues and may therefore be potentially useful for clinical diagnosis and treatment of cancer [6–8]. All these results show that the difference in aggressiveness of tumors can be manifested in their intracellular metabolic state including the mitochondrial redox status. Since the in vitro cell cultures or perfused organ models hardly represent the living tissue conditions [9], it is desirable to quantitatively interrogate in vivo metabolism in living tissue by noninvasive techniques.

The HP-NMR technique can produce ^{13}C-NMR signals ~10,000 times stronger than the regular NMR technique and thus has been employed to probe cell metabolism in vivo [10]. A rapidly growing number of HP-NMR experiments have been carried out to investigate the in vivo enzymatic reactions for various purposes, such as tumor grading [11], redox status imaging [12], and monitoring therapeutic effects of anticancer drugs [13]. Usually these studies are achieved by monitoring the time courses of the injected hyperpolarized ^{13}C-enriched substrate such as ^{13}C-pyruvate and its metabolites such as ^{13}C-lactate produced by lactate dehydrogenase (LDH)-catalyzed reaction, followed by extracting the kinetic parameters, such as apparent reaction rate constants. Higher lactate levels in tumors than in normal tissue have been shown by many metabolic studies on prostate cancer patients [14–16] and have been correlated with prostate tumor grades [11]. Increased aerobic glycolysis in cancer – the Warburg effect – has been associated with the observed lactate elevation [17].

Our long-term goal is to investigate whether the in vivo metabolism can provide biomarkers for cancer aggressiveness. We hypothesize that tumor aggressiveness may be predicted based on the difference in LDH reaction rate constants. In the present study, we employed the HP-NMR technique on several breast tumors with different aggressiveness to interrogate the LDH-catalyzed reaction using ^{13}C-NMR signals of pyruvate and lactate.

32.2 Materials and Methods

Breast cancer lines MCF-7, MDA-MB-468, and MDA-MB-231 with increasing order of metastatic potential were cultured and inoculated subcutaneously into athymic nude mice to grow xenografts [8, 18, 19]. During the experiment, tumor-bearing mice were sedated and maintained with 1.5 % isoflurane. A 1.4-cm home-made dual-tuned surface coil was placed over the tumor. The mouse was placed in a 9.4-T wide-bore, vertical bore magnet equipped with a Varian console. Body temperature was monitored via a fluoroptic rectal probe (Luxtron) and maintained in the range of 34–37 °C with heated air. A hyperpolarized [1-^{13}C]-pyruvic acid sample was prepared by the dynamic nuclear polarization (DNP) method using a HyperSense polarizer (Oxford Instruments) [20]. The final concentration of the hyperplolarized tracer was ~75 mM in neutral isotonic solution. This sample (~10 μL/g mouse body weight) was then injected into mice via tail vein catheter over a period of ~10 s. Single-pulse ^{13}C-NMR spectra were collected with a 9 ~ 15 nominal flip angle every 1 or 2 s for about 2 min.

A customized MATLAB® program was used for data analysis. Line broadening of 20 Hz was applied before Fourier transform of the FID signals. The spectral baseline was removed by fitting it as a fourth order polynomial function. A coarse fitting was performed first by summing up the spectra of all time points followed by fitting the sum to Lorentzian functions to obtain an estimate of the peak position and width of the major ^{13}C-labeled pyruvate and lactate. Their peak areas at each time point t were then obtained by fitting the individual NMR spectrum at time t to Lorentzian functions. The time courses of the pyruvate signal $P(t)$ and the converted lactate signal $L(t)$ as well as their ratio $R_{lp}(t)$ ($=L(t)/P(t)$) were obtained. SNR threshold was set at 2.5. Using the ratiometric analysis method for ^{13}C exchange (see Appendix) [21], we modeled the $R_{lp}(t)$ with customized MATLAB® (MathWorks, 2010b) programs on the basis of the following equation:

$$R_{lp}(t) = \left\{ r\left[1 + R_{lp}(t_0)\right] + \left[R_{lp}(t_0) - r\right] e^{-s(t-t_0)} \right\} / \left\{ 1 + R_{lp}(t_0) + \left[r - R_{lp}(t_0)\right] e^{-s(t-t_0)} \right\}$$

where $r=k^+/k^-$ and $s=k^++k^-$, and k^+ is the apparent rate constant for the forward reaction and k^- the rate constant for the backward reaction catalyzed by LDH, i.e.,

$$\text{Pyruvate} \underset{k^-}{\overset{k^+}{\leftrightarrow}} \text{Lactate}$$

32.3 Results and Discussions

The top row of Figure 32.1 shows the time courses of the signals of ^{13}C-pyruvate and its metabolites (e.g., ^{13}C-lactate) in three breast cancer mouse xenografts with different aggressiveness. The signals of ^{13}C-pyruvate peaked at ~4, 15, and 12 s for MCF-7, MDA-MB-468, and MDA-MB-231, respectively, after the injection of the hyperpolarized tracers (t=0 s). The ^{13}C-lactate signals peaked later and surpassed

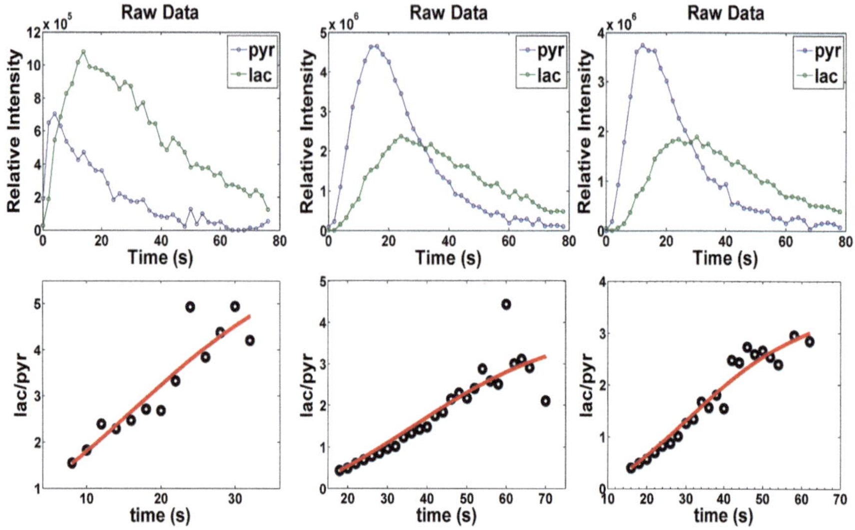

Fig. 32.1 *Upper row*: time courses of the metabolites in mouse xenografts: MCF-7 (*left*), MDA-MB-468 (*middle*), and MDA-MB-231(large) (*right*); *bottom row*: the time courses and their fits of the corresponding tumor's lactate/pyruvate ratios

Table 32.1 Apparent rate constants of individual tumors with different aggressiveness

Tumor	k^+	k^-	k^+/k^-	Size (mm^3)
MCF-7	0.068 ± 0.028	0.011 ± 0.007	6.4 ± 2.3	228
MDA-MB-468	0.040 ± 0.010	0.010 ± 0.004	3.9 ± 0.8	397
MDA-MB-231(large)	0.047 ± 0.007	0.013 ± 0.003	3.5 ± 0.4	1,380
MDA-MB-231(medium)	0.065 ± 0.004	0.014 ± 0.001	4.6 ± 0.3	157

the ^{13}C-pyruvate signals demonstrating the significant conversion of ^{13}C-pyruvate to ^{13}C-lactate by LDH in vivo. The bottom row of Fig. 32.1 shows the time courses of the ^{13}C-lactate/^{13}C-pyruvate ratios and the ratiometric fits. Apparently we can see that the indolent MCF-7 tumor had the fastest pyruvate to lactate conversion. The quantitative results of rate constants are summarized in Table 32.1. From the calculated rate constant ratios, we can see that the indolent MCF-7 tumor had the largest k^+/k^- and the most aggressive MDA-MB-231 (large) had the smallest one.

We have demonstrated the feasibility of obtaining both the apparent forward and backward LDH reaction rate constants in breast tumor mouse xenografts using the ratiometric fitting. Usually the rate constant modeling is based on the time courses of the metabolite signals, which are sensitive to various instrument/protocol parameters. The modeling of the time courses of the metabolite ratios can cancel out some confounding factors such as variations in the degree of spin polarization and the flip angles of radiofrequency pulse, etc.

In further studies with more tumor samples, we hope to investigate the possible tumor size-dependence of the kinetic parameters as obtained in this study, and whether these parameters can be used for the in vivo detection of cancer aggressiveness.

32.4 Conclusions

In this report we presented our preliminary results of using the HP-NMR technique to probe the LDH-catalyzed reaction in three types of breast cancer xenografts with different aggressiveness. Ratiometric modeling was demonstrated to quantify the apparent reaction rate constants. Further experiments are being carried out to investigate the metabolic flux differences among tumors with different aggressiveness.

Acknowledgement This work is supported by an NIH grant R01CA155348 (L.Z. Li). We thank Drs. Jerry D. Glickson and Moses Darpolor for valuable discussions and Ms. Lily Moon for setting up some of the mouse models used in the study.

Appendix A

The MR signals of pyruvate, $P(t)$, and lactate, $L(t)$, can be described by the two-site exchange model using the following differential equations:

$$\frac{dP}{dt} = -\rho P - k^{+}P + k^{-}L \tag{1a}$$

$$\frac{dL}{dt} = -\rho L + k^{+}P - k^{-}L \tag{1b}$$

where we have assumed the relaxation rates ρ for L and P are the same [21]. The solution of Eq. 1a and 1b is

$$\begin{aligned} P = {} & P(0)[k^{-}e^{-\rho t} + k^{+}e^{-(\rho+k^{-}+k^{+})t}]/(k^{-}+k^{+}) \\ & + L(0)[k^{-}e^{-\rho t} - k^{-}e^{-(\rho+k^{-}+k^{+})t}]/(k^{-}+k^{+}) \end{aligned} \tag{2a}$$

$$\begin{aligned} L = {} & P(0)[k^{+}e^{-\rho t} - k^{+}e^{-(\rho+k_L+k_P)t}]/(k^{-}+k^{+}) \\ & + L(0)[k^{+}e^{-\rho t} + k^{-}e^{-(\rho+k^{-}+k^{+})t}]/(k^{-}+k^{+}) \end{aligned} \tag{2b}$$

where $L(0)$ and $P(0)$ are the signal intensities at $t=0$. Let us divide Eq. 2b by Eq. 2a, and we have

$$L(t)/P(t) = \left\{ r\left[1+\frac{L(0)}{P(0)}\right] + \left[\frac{L(0)}{P(0)} - r\right]e^{-st} \right\} / \left\{ 1+\frac{L(0)}{P(0)} + \left[r - \frac{L(0)}{P(0)}\right]e^{-st} \right\} \tag{3}$$

where $r=k^{+}/k^{-}$, $s=k^{+}+k^{-}$. If we also generalize the initial time point to $t=t_0$, Eq. 3 becomes

$$\begin{aligned} R_{lp}(t) = L(t)/P(t) = {} & \left\{ r\left[1+R_{lp}(t_0)\right] + \left[R_{lp}(t_0) - r\right]e^{-s(t-t_0)} \right\} \\ & /\left\{ 1+R_{lp}(t_0) + \left[r - R_{lp}(t_0)\right]e^{-s(t-t_0)} \right\} \end{aligned} \tag{4}$$

References

1. DeBerardinis RJ, Thompson CB (2012) Cellular metabolism and disease: what do metabolic outliers teach us? Cell 148(6):1132–1144
2. Thompson CB (2009) Metabolic enzymes as oncogenes or tumor suppressors. N Engl J Med 360(8):813–815
3. Locasale JW, Cantley LC (2011) Metabolic flux and the regulation of mammalian cell growth. Cell Metab 14(4):443–451
4. Ishikawa K, Takenaga K, Akimoto M et al (2008) ROS-generating mitochondrial DNA mutations can regulate tumor cell metastasis. Science 320(5876):661–664
5. Ishikawa K, Koshikawa N, Takenaga K, Nakada K, Hayashi J (2008) Reversible regulation of metastasis by ROS-generating mtDNA mutations. Mitochondrion 8(4):339–344
6. Li LZ, Zhou R, Xu HN et al (2009) Quantitative magnetic resonance and optical imaging biomarkers of melanoma metastatic potential. Proc Natl Acad Sci USA 106(16):6608–6613
7. Li LZ, Zhou R, Zhong T et al (2007) Predicting melanoma metastatic potential by optical and magnetic resonance imaging. Adv Exp Med Biol 599:67–78
8. Xu HN, Nioka S, Glickson JD, Chance B, Li Z (2010) Quantitative mitochondrial redox imaging of breast cancer metastatic potential. J Biomed Opt 15:036010
9. Veech RL (2006) The determination of the redox states and phosphorylation potential in living tissues and their relationship to metabolic control of disease phenotypes. Biochem Mol Biol Educ 34(3):168–179
10. Kurhanewicz J, Vigneron DB, Brindle K et al (2011) Analysis of cancer metabolism by imaging hyperpolarized nuclei: prospects for translation to clinical research. Neoplasia 13(2):81–97
11. Albers MJ, Bok R, Chen AP et al (2008) Hyperpolarized 13C lactate, pyruvate, and alanine: noninvasive biomarkers for prostate cancer detection and grading. Cancer Res 68(20): 8607–8615
12. Bohndiek SE, Kettunen MI, Hu DE et al (2011) Hyperpolarized [1-^{13}C]-ascorbic and dehydroascorbic acid: vitamin C as a probe for imaging redox status in vivo. J Am Chem Soc 133(30):11795–11801
13. Brindle K (2012) Watching tumours gasp and die with MRI: the promise of hyperpolarised 13C MR spectroscopic imaging. Br J Radiol 85(1014):697–708
14. Swanson MG, Zektzer AS, Tabatabai ZL et al (2006) Quantitative analysis of prostate metabolites using 1H HR-MAS spectroscopy. Magn Reson Med 55(6):1257–1264
15. Tessem M-B, Swanson MG, Keshari KR et al (2008) Evaluation of lactate and alanine as metabolic biomarkers of prostate cancer using 1H HR-MAS spectroscopy of biopsy tissues. Magn Reson Med 60(3):510–516
16. Cornel EB, Smits GA, Oosterhof GO et al (1993) Characterization of human prostate cancer, benign prostatic hyperplasia and normal prostate by in vitro 1H and 31P magnetic resonance spectroscopy. J Urol 150(6):2019–2024
17. Gatenby RA, Gillies RJ (2004) Why do cancers have high aerobic glycolysis? Nat Rev Cancer 4(11):891–899
18. Xu HN, Zheng G, Tchou J et al (2013) Characterizing the metabolic heterogeneity in human breast cancer xenografts by 3D high resolution fluorescence imaging. SpringerPlus 2:7
19. Xu HN, Nioka S, Chance B, Li LZ (2012) 3-D high-resolution mapping of the heterogeneity in mitochondrial redox state of human breast tumor xenografts. Adv Exp Med Biol 737:169–174
20. Pullinger B, Profka H, Ardenkjaer-Larsen JH, Kuzma NN, Kadlecek S, Rizi RR (2012) Metabolism of hyperpolarized [1-^{13}C]pyruvate in the isolated perfused rat lung – an ischemia study. NMR Biomed 25(10):1113–1118
21. Li LZ, Kadlececk S, Xu HN et al (2013) Ratiometric analysis in hyperpolarized NMR (I): test of the two-site exchange model and the quantification of reaction rate constants. NMR Biomed. Epub doi: 10.1002/nbm.2953

Chapter 33
Mapping the Redox State of CHOP-Treated Non-Hodgkin's Lymphoma Xenografts in Mice

He.N. Xu, Tahreem A. Mir, Seung-Cheol Lee, Min Feng, Namisa Farhad, Regine Choe, Jerry D. Glickson, and Lin Z. Li

Abstract Drug treatment may alter the metabolism of cancer cells and may alter the mitochondrial redox state. Using the redox scanner that collects the fluorescence signals from both the oxidized flavoproteins (Fp) and the reduced form of nicotinamide adenine dinucleotide (NADH) in snap-frozen tumor tissues, we investigated the effects of chemotherapy on mouse xenografts of a human diffuse large B-cell lymphoma cell line (DLCL2). The mice in the treatment group were treated with CHOP – cyclophosphamide (C) + hydroxydoxorubicin (H) + Oncovin (O) + prednisone (P) using the following regimen: CHO administration on day 1 followed by

This article is dedicated to the memory of late Dr. Britton Chance who participated in the study with extraordinary scientific enthusiasm at the age of 97.

H.N. Xu • T.A. Mir • M. Feng • N. Farhad
Department of Radiology, University of Pennsylvania, Philadelphia, PA, USA

Britton Chance Laboratory of Redox Imaging, Johnson Research Foundation, Department of Biochemistry and Biophysics, University of Pennsylvania, Philadelphia, PA, USA

S.C. Lee • J.D. Glickson
Department of Radiology, University of Pennsylvania, Philadelphia, PA, USA

R. Choe
Department of Biomedical Engineering, University of Rochester, Rochester, NY, USA

L.Z. Li (✉)
Department of Radiology, University of Pennsylvania, Philadelphia, PA, USA

Britton Chance Laboratory of Redox Imaging, Johnson Research Foundation, Department of Biochemistry and Biophysics, University of Pennsylvania, Philadelphia, PA, USA

Abramson Cancer Center, University of Pennsylvania, Philadelphia, PA, USA

Institute of Translational Medicine and Therapeutics, University of Pennsylvania, Philadelphia, PA, USA
e-mail: linli@mail.med.upenn.edu

S. Van Huffel et al. (eds.), *Oxygen Transport to Tissue XXXV*, Advances in Experimental Medicine and Biology 789, DOI 10.1007/978-1-4614-7411-1_33,

prednisone administration on day 1–5. On day 5 the mitochondrial redox state of the treated group was slightly more reduced than that of the control group ($p = 0.049$), and the Fp content of the treated group was significantly decreased ($p = 0.033$).

33.1 Introduction

Lymphoma is the most common hematological cancer in the USA. It occurs in two forms, Hodgkin's disease (HD) and non-Hodgkin's lymphoma (NHL), the latter being the fifth most common cancer in the USA and far more common than HD. For decades, first line standard therapy of NHL has consisted of 6–8 cycles of CHOP treatment [1, 2], where cyclophosphamide is an alkylating agent that cross-links DNA, hydroxydoxorubicin (doxorubicin or Adriamycin) intercalates between DNA bases and in coordination with iron generates reactive oxygen species (ROS), Oncovin (vincristine) binds to tubulin to prevent cells from duplicating, and prednisone is a corticosteroid. Our previous ^{1}H-MRS study of DLCL2 xenografts in mice (models of diffuse large B-cell lymphoma [2], the most common form of NHL) demonstrated a significant decrease in lactate in the tumor after one cycle of CHOP chemotherapy, where the therapeutic effectiveness was based on tumor volume and decreased proliferation rate measured by Ki67 staining [3].

Optical imaging has much higher spatial resolution than MRS and can detect changes in metabolic heterogeneity following therapeutic intervention. Mitochondrial NADH is intrinsically fluorescent and is the main source of reducing equivalents for the mitochondrial respiratory chain. Mitochondrial metabolism is reflected in the redox state of the NAD system, which is in tight equilibrium with the flavin system. The oxidized flavoproteins (Fp) including FAD are also intrinsically fluorescent. Previously, we showed that fluorescence imaging of the mitochondrial redox state was sensitive to metabolic alterations in tumors [4, 5] and tissue metaplasia [6] and that this method has the potential ability to detect therapeutic effects on cellular metabolism. Here we present preliminary redox scanning results on CHOP-treated DLCL2 xenografts.

33.2 Materials and Methods

As previously described [3], WSU-DLCL2 cells were subcutaneously inoculated in the flanks of 5–7-week-old female nude mice to induce tumor xenografts. Tumor-bearing mice were randomly chosen and divided into the treatment group and the control group. CHOP treatment is as follows: cyclophosphamide, 40 mg/kg i.v., day 1; hydroxydoxorubicin, 3.3 mg/kg i.v., day 1; Oncovin, 0.5 mg/kg i.v., day 1; and prednisone, 0.2 mg/kg p.o., day 1–5. Sham treatment (saline) was performed on the control group. All animal studies were performed in accordance with the guidelines of the University of Pennsylvania Institutional Animal Care and Use Committee (IACUC).

Anesthetized tumor-bearing mice were snap-frozen in liquid N_2 1 day after the last treatment. The frozen tumors (three from the treated group and three from the control group) were quickly removed from the mouse body with a handsaw and embedded in homemade mounting media followed by redox scanning as previously described [5, 7]. The PC-collected scanning data were analyzed with a customized MATLAB® program that uses the reference standards to convert the intensity-based images into concentration-based images and generates redox ratio images.

Each xenograft was scanned with three to five sections spacing 800–1,000 μm with the first section approximately starting at about 1,000–1,500 μm beneath the skin. Three tumors in each group of a total of six tumors were scanned. The redox indices (Fp, NADH, and Fp redox ratio, i.e., Fp/(Fp+NADH)) were averaged for each tissue section. Univariate analysis was conducted on these redox indices using IBM SPSS Statistics (version 20) with tissue depth as the covariate and $p<0.05$ taken as statistically significant. Data were finally reported as mean ± standard error of mean (SEM).

33.3 Results

As shown in Table 33.1, after one cycle of CHOP administration, one tumor in the treated group exhibited volume reduction, while the other two kept growing at a slower rate compared to the control group. No significant changes were found in the tumor volume between the two groups on either day 1 (treated 1,001 ± 50 mm^3 vs. control 1,204 ± 728 mm^3) or day 5 (treated 1,188 ± 504 mm^3 vs. control 1,671 ± 1,025 mm^3) nor is the volume ratio of day 5 to day 1 significantly different between the two groups ($p>0.05$).

The redox scanning results are summarized in Table 33.2. The Fp redox ratio, Fp/(NADH+Fp) of treated group was 0.45 ± 0.02, while that of the control group

Table 33.1 Tumor volume change

Mouse #	V_ratio (day 5/day 1)	Mouse #	V_ratio (day 5/day 1)
Treated		*Control*	
444	1.32	422	1.19
446	1.9	449	2.08
448	0.37	450	0.77

p (unpaired *t*-test) = 0.82

Table 33.2 Redox indices ± SEM averaged across tissue sections within each group

	Ctrl	CHOP-treated	*P*-value
Fp redox ratio	0.53 ± 0.03	0.45 ± 0.02	0.049
Fp (μM)	560 ± 67	349 ± 60	0.033
NADH (μM)	446 ± 48	403 ± 43	0.52

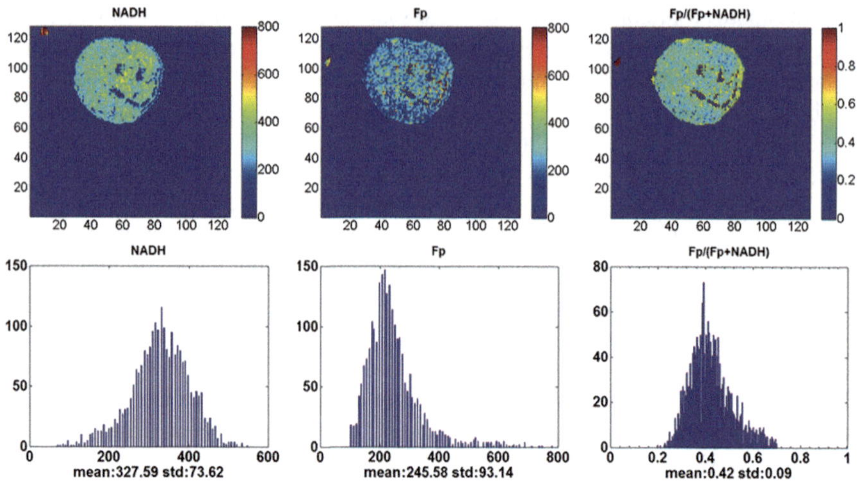

Fig. 33.1 Typical pseudo-color images of the redox indices (*top row*) and their corresponding histograms (*bottom row*) of a tumor in CHOP-treated group (1,500 μm under the skin). From *left to right*: NADH nominal concentration (μm), Fp nominal concentration (μm), and the Fp redox ratio (0–1). The mean value and SD are shown below the x-axes on each histogram. Image resolution: 200 μm

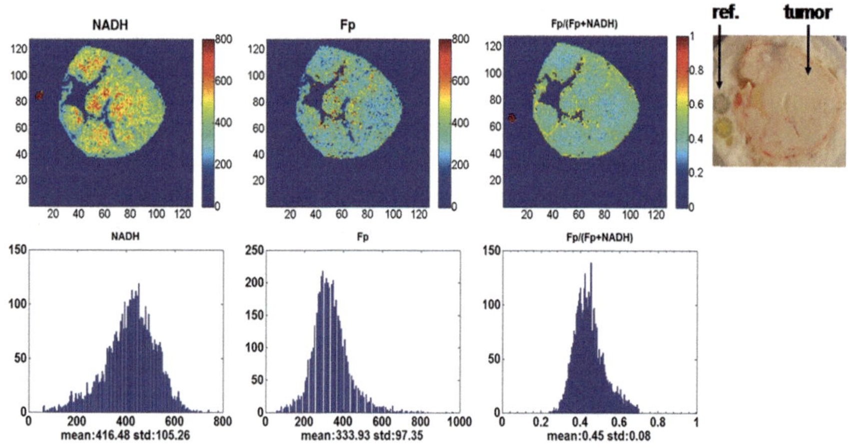

Fig. 33.2 Typical pseudo-color redox images (*top row*) and their corresponding histograms of a tumor in the control group (3,500 μm under the skin). Image resolution: 200 μm

was 0.53 ± 0.03, indicating that the treatment caused the tumors to become slightly more reduced compared to the control group ($p = 0.049$). The Fp content of the treated group was significantly decreased ($p = 0.033$). No significant change was detected in NADH. The typical redox images of the two groups are shown in Figs. 33.1 and 33.2. We did not detect significant changes in the standard deviation of any of the redox indices (Fp redox ratio, Fp, and NADH).

33.4 Discussions

Since mitochondria are targets of many anticancer drugs due to their central roles in cellular energy metabolism and apoptotic signalling pathways [8, 9], they might also provide early biomarkers for therapeutic response. Studies on therapeutic effects on mitochondria isolated from human cancer cells showed that doxorubicin acts on DNA independent of microtubules and does not induce the release of cytochrome c even at very high dosage [10]. Investigations based on cell culture models revealed that short-term (<30 min) treatment with doxorubicin rendered some human cancer cell lines more oxidized with respect to their mitochondrial redox state (more Fp and less NADH) with a large amount of ROS production, whereas long-term (48 h) treatment induced cell cycle arrest and cell death [11]. On the other hand, as an antitubulin agent and one of the vinca alkaloids, vincristine is a cell-cycle-specific drug that inhibits cell growth exclusively during metaphase by inhibiting microtubule dynamics and assembly leading to cell cycle arrest [12–14]. Investigations also showed that a substantial amount of tubulin inherently existed in the mitochondria with a role in apoptosis via interaction with the permeability transition pore, whereas antitubulin agents induce the release of cytochrome c from isolated mitochondria[15]. It was shown that the apoptotic cells have significantly higher Fp redox ratios [16].

All these studies were performed in vitro on isolated mitochondria or cell culture. It is not clear how the tissue mitochondrial redox state is modified by therapy under in vivo condition.

Our previous MRS study showed that successive three cycles of CHOP treatment caused decreased proliferation closely matched by the reduced lactate concentrations [3]. In this study, we examined the therapeutic effect of CHOP on DLCL2 xenografts by mapping the mitochondrial NADH and Fp fluorescent signals and thus the mitochondrial redox state of the tumors across multiple tissue sections. Our preliminary results show that one 5-day cycle of CHOP treatment causes these tumors to become slightly more reduced in their redox state. As there has not been a calibration procedure available at the tissue level to determine the exact mitochondrial redox state [17, 18], we can only speculate that these lymphomas are likely to be mainly in either State 3 (active proliferation) or State 4 (at rest). State 2 under starvation is unlikely due to relatively uniformly strong NADH signals observed in the majority of tumor regions. The decrease in Fp after treatment indicate suppression of mitochondrial metabolism. This is further supported by increased PME/βNTP ratio [3] and decreased βNTP concentration (CHOP 2.0±0.1 vs. sham 3.4±0.1 μmole/g wet weight, unpublished data by SC Lee) after three cycles of CHOP treatment as measured by ^{31}P-MRS on tissue extracts. It appears that CHOP treatment may decrease both glycolysis and mitochondrial metabolism in DLCL2 tumors. Although we are not certain about the exact mechanism, this result is consistent with a significant decrease of cell proliferation index Ki67 staining that was observable after one cycle of CHOP treatment [3].

Our previous metastatic potential studies of melanoma and breast cancer mouse models indicated that aggressive tumors had more oxidized tumor cores, whereas indolent tumors were relatively homogenous and less oxidized [4, 5]. Although the fluorescence intensity of both NADH and Fp depends on cell density, it was shown that cell density changes between the CHOP-treated and the control groups were insignificant [3]. Thus, decreased Fp content and Fp redox ratio in CHOP-treated tumors should not be due to cell density differences, and the treatment appeared to have induced a cellular metabolic change towards less malignancy.

Individual differences in treatment response occur frequently. In the current study, one tumor in the treated group exhibited a large reduction in volume and higher Fp redox ratio compared to the other two. Additional histological investigations are needed to determine whether one cycle of CHOP treatment induced apoptosis in this tumor.

33.5 Conclusions

We report preliminary redox scanning data on CHOP treatment effects on DLCL2 xenografts. One cycle of treatment caused the tumors to become slightly more reduced in their mitochondrial redox state. To our knowledge, this study is the first to explore the therapeutic effect of CHOP on the mitochondrial redox state of lymphoma. We plan to conduct more in-depth studies with a larger sample size and more cycles of treatment.

Acknowledgments Many thanks to Dr. Huaqing Zhao for valuable discussions about statistical analysis. This work was supported by the Center of Magnetic Resonance and Optical Imaging (CMROI) – an NIH-supported research resource P41RR02305 (R. Reddy), the Small Animal Imaging Resources Program (SAIR) 2U24-CA083105 (J.D. Glickson & L. Chodosh), 2R01-CA101700 (J.D. Glickson), and NIH k99/R00-CA126187 (R. Choe).

References

1. Fisher RI, Gaynor ER, Dahlberg S et al (1993) Comparison of a standard regimen (CHOP) with three intensive chemotherapy regimens for advanced non-Hodgkin's lymphoma. N Engl J Med 328(14):1002–1006
2. Mohammad RM, Al-Katib A, Aboukameel A, Doerge DR, Sarkar F (2003) Kucuk O Genistein sensitizes diffuse large cell lymphoma to CHOP (cyclophosphamide, doxorubicin, vincristine, prednisone) chemotherapy. Mol Cancer Ther 2(12):1361–1368
3. Lee S-C, Huang MQ, Nelson DS et al (2008) In vivo MRS markers of response to CHOP chemotherapy in the WSU-DLCL2 human diffuse large B-cell lymphoma xenograft. NMR Biomed 21(7):723–733
4. Li LZ, Zhou R, Xu HN et al (2009) Quantitative magnetic resonance and optical imaging biomarkers of melanoma metastatic potential. Proc Natl Acad Sci USA 106(16):6608–6613
5. Xu HN, Nioka S, Glickson JD et al (2010) Quantitative mitochondrial redox imaging of breast cancer metastatic potential. J Biomed Opt 15(3):036010

6. Xu HN, Nioka S, Chance B, Li LZ (2011) Heterogeneity of mitochondrial redox state in premalignant pancreas in a PTEN null transgenic mouse model. Adv Exp Med Biol 701: 207–213
7. Xu HN, Wu B, Nioka S, Chance B, Li ZL (2009) Quantitative redox scanning of tissue samples using a calibration procedure. J Innov Opt Health Sci 2:375–385
8. Fulda S, Galluzzi L, Kroemer G (2010) Targeting mitochondria for cancer therapy. Nat Rev Drug Discov 9(6):447–464
9. D'Souza GG, Wagle MA, Saxena V, Shah A (2011) Approaches for targeting mitochondria in cancer therapy. Biochim Biophys Acta 1807(6):689–696
10. Andre N, Braguer D, Brasseur G et al (2000) Paclitaxel induces release of cytochrome c from mitochondria isolated from human neuroblastoma cells. Cancer Res 60(19):5349–5353
11. Kuznetsov AV, Margreiter R, Amberger A, Saks V, Grimm M (2011) Changes in mitochondrial redox state, membrane potential and calcium precede mitochondrial dysfunction in doxorubicin-induced cell death. Biochim Biophys Acta 1813(6):1144–1152
12. Lobert S, Vulevic B, Correia JJ (1996) Interaction of vinca alkaloids with tubulin: a comparison of vinblastine, vincristine, and vinorelbine. Biochemistry 35(21):6806–6814
13. Owellen RJ, Hartke CA, Dickerson RM, Hains FO (1976) Inhibition of tubulin-microtubule polymerization by drugs of the Vinca alkaloid class. Cancer Res 36(4):1499–1502
14. Owellen RJ, Owens AH Jr, Donigian DW (1972) The binding of vincristine, vinblastine and colchicine to tubulin. Biochem Biophys Res Commun 47(4):685–691
15. Carre M, Andre N, Carles G et al (2002) Tubulin is an inherent component of mitochondrial membranes that interacts with the voltage-dependent anion channel. J Biol Chem 277(37): 33664–33669
16. Ranji M, Jaggard DL, Chance B (2006) Observation of mitochondrial morphology and biochemistry changes undergoing apoptosis by angularly resolved light scattering and cryoimaging. Biophotonics and immune responses. Proc SPIE 6087:60870K
17. Chance B, Baltscheffsky H (1958) Respiratory enzymes in oxidative phosphorylation. VII. Binding of intramitochondrial reduced pyridine nucleotide. J Biol Chem 233(3):736–739
18. Chance B, Williams GR (1955) A method for the localization of sites for oxidative phosphorylation. Nature 176(4475):250–254

Chapter 34
Maternal Bias in Mouse Radiosensitivity: The Role of the Mitochondrial PTP

Steven Bingrong Zhang, David Maguire, Mei Zhang, Amy Zhang, Lurong Zhang, Steven Swarts, and Paul Okunieff

Abstract This study investigated, at the molecular level, mitochondrial responses to radiation. In three mouse strains, we found the following: (1) mitochondrial response to calcium stress was associated with a strain's susceptibility to γ-radiation; (2) γ-radiation increased this calcium stress response in a dose-responsive manner; (3) the mitochondrial DNA (mtDNA) copy number in the liver of the radiosensitive mouse strain was significantly lower, as compared to that of the radioresistant strain; (4) adenine nucleotide translocase (ANT) mRNA copy numbers were significantly lower in the radiosensitive strain; (5) the F1 offspring (BC/C57M) of radiosensitive females mated with radioresistant males exhibited a significant difference in calcium stress response from that of the radiation-resistant strain, but the reverse cross did not exhibit this difference; and (6) only those mitochondria extracted from the livers of irradiated BC/C57M mice exhibited a heightened calcium stress response. We propose that a genetic change in ANT and a postirradiation change involving either mtDNA-encoded protein replacement or altered mtDNA association fit these data.

34.1 Introduction

As the major site for the creation of reactive oxygen species (ROS) within cells, mitochondria are particularly prone to oxidative stress-induced damage. Moreover, they lack histone protection for the mitochondrial genome and possess only limited repair capability for this maternally inherited mitochondrial DNA (mtDNA). Mitochondria provide over 90 % of cellular energy yet also play a central role in

S.B. Zhang (✉) • D. Maguire • M. Zhang • A. Zhang • L. Zhang • S. Swarts • P. Okunieff
Department of Radiation Oncology, UF Shands Cancer Center, University of Florida, 2033 Mowry Road, Gainesville, FL 32610, USA
e-mail: bszhang@ufl.edu

S. Van Huffel et al. (eds.), *Oxygen Transport to Tissue XXXV*, Advances in Experimental Medicine and Biology 789, DOI 10.1007/978-1-4614-7411-1_34,

apoptotic and necrotic cell death [1]. Although dysfunctions in mitochondrial proteins (whether nucleus encoded or mtDNA encoded) are major contributors to degenerative diseases, cancer, and aging [1–7], the precise molecular mechanisms behind this phenomenon are still unclear. The mitochondrial permeability transition pore (mPTP) is a major contributor to apoptosis and necrosis [8, 9], and increased susceptibility to mPTP opening has been documented in the liver [10] and lymphocyte mitochondria [11] of more mature mice. In addition, Gong et al. found that radiation significantly increased ATP levels and mitochondrial membrane potential in a tumor cell line [12].

However, to our knowledge, no report explores the role of mitochondrial DNA integrity and mPTP opening in association with the oxidative stress susceptibility of a host or cell type. The mPTP is a mitochondrial component that undergoes a sudden increase in permeability to solutes with a molecular mass ≤1,500 Da, leading to organelle swelling and structural modifications. No clear consensus exists on the precise molecular structure of the mPTP, other than that the nuclear-encoded protein ANT is a consistently reported inclusion. The creation of null mutants of ANT casts doubt as to the absolute necessity of that inclusion and instead indicates that ANT acts as a regulator of mPTP function [13]. Other mPTP components may include cyclophilin D, voltage-dependent anion channels, stem subunits of ATP synthase, and members of the proapoptotic and antiapoptotic Bax-Bcl2 protein family. Although mPTP channels may be constructed from a range of subunits, they require an inner membrane anchor and a channel former capable of spanning the inner membrane. Any mPTP channel lacking (or deficient in) a functional ANT is more sensitive to Ca^{2+}-induced opening.

34.2 Research Design and Methods

Lung-tumor susceptible (BALB/c), intermediately resistant (NIH Swiss), and relatively resistant (C57BL/6) mouse strains were exposed to total body irradiation (TBI) and examined for various mitochondrial characteristics, including the effect on mitochondrial oxidative stress sensitivity over time. Approximately 15–20 mice per group (7 weeks old) were exposed to a 137-cesium source for a dose of 5- to 10-Gy TBI. Mice were then returned to normal management. At various intervals, mice were euthanized for liver tissue collection. Tissues were used immediately or frozen at −70 °C for later use.

Relative mtDNA copy number, mPTP opening, and ANT mRNA level were examined. Mitochondrial PTP function was estimated using calcium-induced swelling. The Kirby method was used to extract genomic DNA from samples [14]. Nuclear DNA and mtDNA were analyzed by real-time polymerase chain reaction (RT-PCR) using mouse mtDNA target (12srRNA gene) and nuclear DNA target (18sRNA) primers. ANT2 mRNA quantitative RT-PCR was performed on hepatic RNA extracted using Trizol (Invitrogen, Life Technologies, Grand Island, NY) according to the manufacturer's protocol. Complementary DNA was made via reverse transcription using iSCRIPT (Bio-Rad, Hercules, CA). RT-PCR was performed using RT^2 SYBR Green/Fluorescein qPCR Mastermix (SABiosciences, QIAGEN, Inc., Valencia, CA) with cDNA and target primers on a multicolor RT-PCR detection system (iQ 5, Bio-Rad). Individual PCRs were performed in triplicate on samples. β-actin, a housekeeping gene, and ANT2 assays were used to obtain average cycle threshold (Ct) values for these gene transcripts. Average β-actin Ct values were subtracted from ANT2 Ct values to obtain ΔCt values. ΔΔCt values were then obtained by subtracting experimental sample ΔCt values from the control sample ΔCt. Relative ANT2 expression was then calculated using $2^{-\Delta\Delta Ct}$.

34.3 Results

Significant differences in postirradiation survival were observed between the strains exposed to 7.5-Gy TBI (Fig. 34.1a). BALB/c mice had the highest mortality rate following irradiation, whereas the C57BL/C mice were least sensitive. NIH Swiss mice exhibited an intermediate survival pattern. Of the mitochondrial functional measurements performed on the isolated hepatic mitochondria, a significant strain-related difference was found only in mPTP (Fig. 34.1b and c). Sensitivity (the slope of Ca^{2+}-induced swelling) was greatest in the BALB/c mice and least in the C57BL/6 strain. Radiation increased mPTP swelling in the C57BL/6 (Fig. 34.1d) and NIH Swiss (Fig. 34.1e) strains at the acute stage, indicating radiation-dose dependency. Following irradiation, the mtDNA copy number ratio was considerably higher in the liver of the C57BL/6 strain, as compared to that of the BALB/c strain (Fig. 34.1f). ANT mRNA transcript levels (Fig. 34.1g) were lowest in the BALB/c strain and highest in the C57BL/6 strain, indicating that they are related to strain susceptibility to radiation. Only mitochondria from the F1 offspring (BC/C57M) of radiosensitive BALB/c females mated with radioresistant C57BL/6 males showed a significant difference from the C57BL/6 phenotype response to calcium-induced stress (Fig. 34.1h). The BC/C57M hybrid also showed a marked difference (Fig. 34.1i) in calcium swelling response 2 months after 7-Gy TBI, as compared to the opposite cross (Fig. 34.1j), indicating maternal inheritance bias of this characteristic.

34.4 Discussion

In previous studies, persistent mt/nDNA ratio elevations were reported for both γ-radiation [13, 15–18] and X-radiation [19–23]. Increased levels of gamma polymerase but decreased TFAM were found following irradiation, which might protect damaged cells or allow recombination repair without burdening cells with the need to produce new mitochondrial proteins [18]. More recently, we detected subtle mitochondrial membrane proteomic changes following irradiation of radiosensitive mouse strains [16]. A gel-band shift in the protein profile of mitochondrial

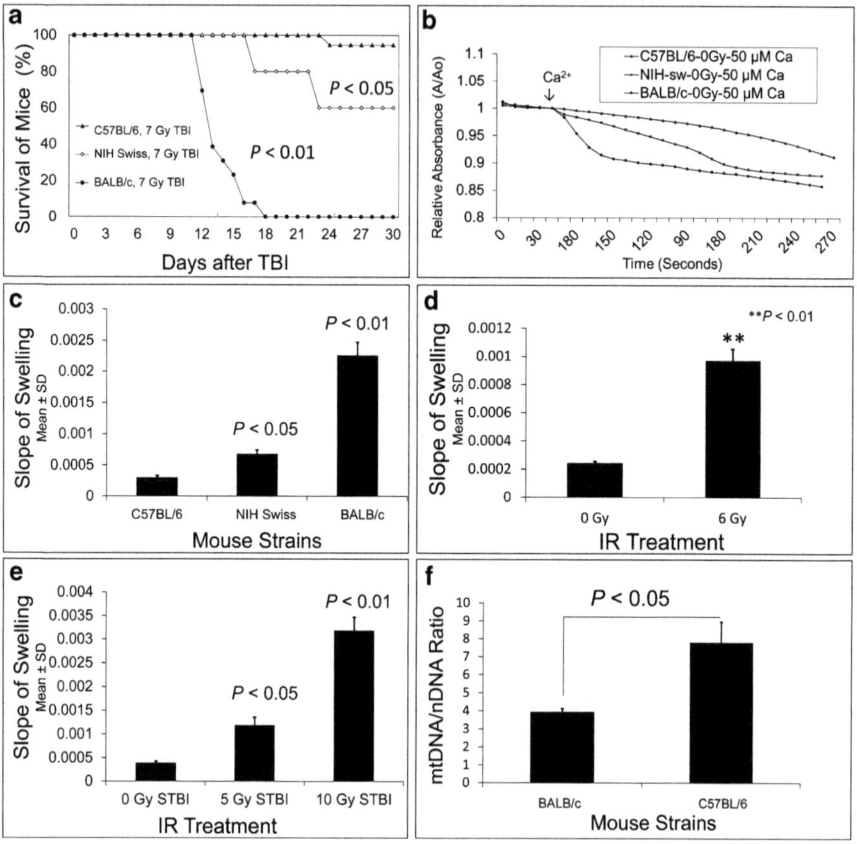

Fig. 34.1 (**a**) Post γ-irradiation survival in BALB/c, NIH Swiss, and C57BL/6 mouse strains ($n=5$/group); (**b**) Ca^{2+} swelling assay in three mouse strains ($n=4$–6/group); (**c**) mPTP activity (slopes of Ca^{2+} swelling) in three mouse strains (C57BL/6 mice were compared with NIH Swiss and BALB/c mice; $n=4$–6/group); (**d**) 9 h after irradiation, mPTP activity in C57BL/6 mice (0 and 6 Gy groups were compared; $n=5$/group); (**e**) 7 days after irradiation, mPTP activity in C57BL/6 strain (0 Gy group was compared with the 5 and 10 Gy groups; $n=5$ mice/group); (**f**) mt/nDNA ratios in tissues of two mouse strains ($n=5$/group); (**g**) relative ANT mRNA levels in three mouse strains (BALB/c mice were compared with NIH Swiss and C57BL/6 mice; $n=5$/group); (**h**) mPTP activity in two mouse strains and F1 hybrids (C57BL/6 mice were compared with CB, BALB/c, and BC mice; $n=5$/group); (**i**) Ca^{2+}–induced swelling of liver mitochondria from F1 hybrids (C57BL/6 father with BALB/c mother) 2 months after irradiation, as compared to nonirradiated hybrids ($n=4$/group); (**j**) Ca^{2+}–induced swelling of hepatic liver mitochondria from F1 hybrids (BALB/c father with C57BL/6 mother) 2 months after irradiation, as compared to nonirradiated hybrids ($n=4$/group)

membrane proteins extracted from mouse livers was consistent with a new protein association producing a higher native MW species.

In this investigation, we found a strong relationship between calcium-induced mPTP opening and mouse strain radiosensitivity that was consistent with ANT gene mutation. We also found that radiation influenced mPTP status, depending on the

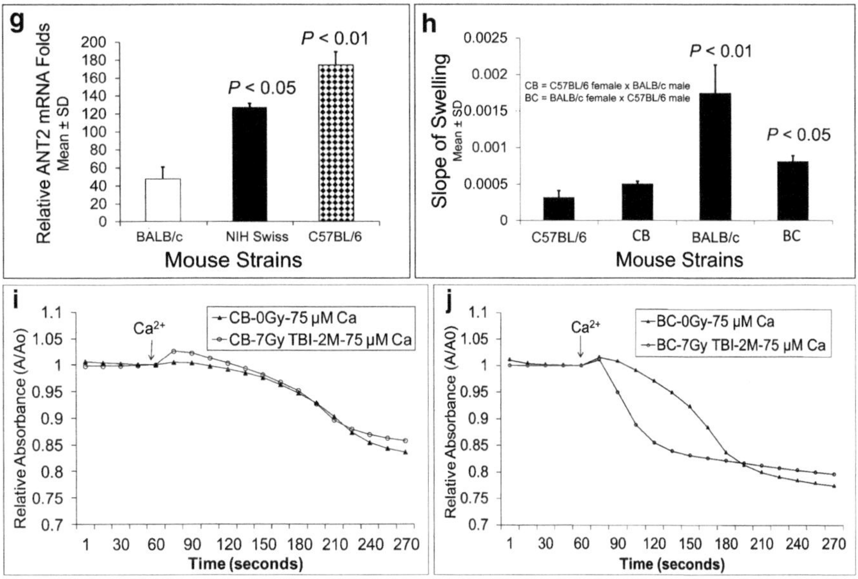

Fig. 34.1 (continued)

strain's genetic susceptibility to radiation. In an attempt to rationalize this latter difference, we considered two plausible mechanisms. Irradiation may lead to the release of a previously ANT-bound mtDNA-encoded protein, thus allowing new associations with a nuclear-encoded protein of higher MW, concurrently leading to an mPTP with increased sensitivity. Less likely, an mtDNA product, such as newly mutated mtDNA, might associate with ANT, thereby inducing mPTP activity and increasing the apparent ANT MW. Mitochondrial DNA has previously been reported to associate with various inner membrane proteins, including ANT [24].

This research seeks to better define the molecular mechanisms within mitochondria responsible for radiation sensitivity during radiotherapy. Our findings may lead to development of new diagnostic methods to identify high-risk groups for radiation hypersensitivity and new treatment modalities. In future work, we plan to further explore mtDNA structural and functional responses to radiation and analyze relevant intracellular signaling pathways in mouse strains over time.

Acknowledgments This project is supported in part by RC1AI0710519, RC2-AI-01075100, RC1-AI0101274 (NIAID/NIH), and Shands Cancer Center startup funds (University of Florida). We thank Kate Casey-Sawicki for helping to prepare this manuscript for publication.

References

1. Wang C, Youle RJ (2009) The role of mitochondria in apoptosis. Annu Rev Genet 43:95–118
2. Park CB, Larsson NG (2011) Mitochondrial DNA mutations in disease and aging. J Cell Biol 193(5):809–818

3. Lee HC, Wei YH (1997) Role of mitochondria in human aging. J Biomed Sci 4(6):319–326
4. Barazzoni R, Short KR, Nair KS (2000) Effects of aging on mitochondrial DNA copy number and cytochrome c oxidase gene expression in rat skeletal muscle, liver, and heart. J Biol Chem 275(5):3343–3347
5. Beal MF (1998) Mitochondrial dysfunction in neurodegenerative diseases. Biochim Biophys Acta 1366(1–2):211–223
6. Lin MT, Beal MF (2006) Mitochondrial dysfunction and oxidative stress in neurodegenerative diseases. Nature 443(7113):787–795
7. Lesnefsky EJ, Lundergan CF, Hodgson JM et al (1996) Increased left ventricular dysfunction in elderly patients despite successful thrombolysis: the GUSTO-I angiographic experience. J Am Coll Cardiol 28(2):331–337
8. Halestrap AP, McStay GP, Clarke SJ (2002) The permeability transition pore complex: another view. Biochimie 84(2–3):153–166
9. Crompton M, Virji S, Doyle V, Johnson N, Ward JM (1999) The mitochondrial permeability transition pore. Biochem Soc Symp 66:167–179
10. Goodell S, Cortopassi G (1998) Analysis of oxygen consumption and mitochondrial permeability with age in mice. Mech Ageing Dev 101(3):245–256
11. Rottenberg H, Wu S (1997) Mitochondrial dysfunction in lymphocytes from old mice: enhanced activation of the permeability transition. Biochem Biophys Res Commun 240(1):68–74
12. Gong B, Chen Q, Almasan A (1998) Ionizing radiation stimulates mitochondrial gene expression and activity. Radiat Res 150:505–512
13. Zhang H, Maguire D, Swarts S et al (2009) Replication of murine mitochondrial DNA following irradiation. Adv Exp Med Biol 645:43–48
14. Kirby KS (1956) A new method for the isolation of ribonucleic acids from mammalian tissues. Biochem J 64(3):405–408
15. Zhang SB, Zhang M, Cao Y et al (2012) Delayed effects of radiation on mitochondrial DNA in radiation-sensitive organs. Adv Exp Med Biol 737:139–145
16. Maguire D, Zhang B, Zhang A, Zhang L, Okunieff P (2013) The role of mitochondrial proteomic analysis in radiological accidents and terrorism. Adv Exp Med Biol 765:139–145
17. Zhang L, Zhang M, Zhang B et al (2012) Radiation-induced elevation of plasma DNA in mice is associated with genomic background. Adv Exp Med Biol 737:147–153
18. Zhang H, Maguire DJ, Zhang M, Zhang L, Okunieff P (2011) Elevated mitochondrial DNA copy number and POL-gamma expression but decreased expression of TFAM in murine intestine following therapeutic dose irradiation. Adv Exp Med Biol 701:201–206
19. Cleaver JE (1992) Replication of nuclear and mitochondrial DNA in X-ray-damaged cells: evidence for a nuclear-specific mechanism that down-regulates replication. Radiat Res 131(3):338–344
20. Gubina NE, Evdokimovskii EV, Ushakova TE (2010) Mitochondrial genetic apparatus functioning in mice spleen cells under radiation-induced apoptosis. Mol Biol (Mosk) 44(6):1027–1035
21. Abdullaev SA, Anishchenko ES, Gaziev AI (2010) Mutant copies of mitochondrial DNA in tissues and plasma of X-rays exposed mice. Radiats Biol Radioecol 50(3):318–328
22. Abdullaev SA, Antipova VN, Gaziev AI (2006) Extracellular mutant mitochondrial DNA content is sharply elevated in the blood plasma of irradiated mice. Mol Biol (Mosk) 43(6):1063–1069
23. Yamamori T, Yasui H, Yamazumi M, Wada Y, Nakamura Y, Nakamura H, Inanami O (2012) Ionizing radiation induces mitochondrial reactive oxygen species production accompanied by upregulation of mitochondrial electron transport chain function and mitochondrial content under control of the cell cycle checkpoint. Free Radic Biol Med 53(2):260–270
24. Bogenhagen DF, Wang Y, Shen EL, Kobayashi R (2003) Protein components of mitochondrial DNA nucleoids in higher eukaryotes. Mol Cell Proteomics 2(11):1205–1216

Chapter 35
Interleukin 11 Protects Bone Marrow Mitochondria from Radiation Damage

Luqiang Huang, Zhenhuan Zhang, Wenlong Lv, Mei Zhang, Shanmin Yang, Liangjie Yin, Jinsheng Hong, Deping Han, Chun Chen, Steve Swarts, Sadasivan Vidyasagar, Paul Okunieff, and Lurong Zhang

Abstract Interleukin 11 (IL-11) is a multifunctional cytokine isolated from bone marrow (BM)-derived stromal cells that promotes hematopoiesis and prolongs the life span of lethally irradiated animals. However, the underlying mechanism for the protective effect of IL-11 on BM is unclear. In this study, we explored the effect of IL-11 on irradiated BM cells. Freshly harvested BM cells were pretreated with 20 ng/ml of recombinant IL-11 for 30 min, irradiated with a dose of 0.5 Gy, cultured for 24 h, and then subjected to several assays. In vitro data showed that, as compared to the vehicle controls, IL-11: (1) reduced the production of reactive oxygen species; (2) reduced the alteration of mitochondrial membrane potential; (3) increased MitoTracker staining, suggesting that the number of mitochondria and their functions were better maintained; and (4) reduced apoptosis of BM cells and enhanced BM cell proliferation. In vivo studies of mice pretreated with saline or 100 μg/kg of

L. Huang
Zhongshan Hospital, Xiamen University, Xiamen, Fujian, P. R. China

Department of Radiation Oncology, UF Shands Cancer Center, University of Florida, Gainesville, FL, USA

Z. Zhang • M. Zhang • S. Yang • L. Yin • S. Swarts • S. Vidyasagar • P. Okunieff • L. Zhang (✉)
Department of Radiation Oncology, UF Shands Cancer Center, University of Florida, Gainesville, FL, USA
e-mail: lurongzhang@ufl.edu

W. Lv
Department of Radiation Oncology, UF Shands Cancer Center, University of Florida, Gainesville, FL, USA

Department of Radiation Oncology, First Affiliated Hospital of Fujian Medical University, Fuzhou, Fujian, China

J. Hong • D. Han • C. Chen
Department of Radiation Oncology, First Affiliated Hospital of Fujian Medical University, Fuzhou, Fujian, China

S. Van Huffel et al. (eds.), *Oxygen Transport to Tissue XXXV*, Advances in Experimental Medicine and Biology 789, DOI 10.1007/978-1-4614-7411-1_35,

IL-11 at 12 and 2 h before 10-Gy total body irradiation (TBI) demonstrated that G-CSF and IL-6 were significantly upregulated, whereas IL-2 and IL-4 were reduced. We found that IL-11 protects mitochondrial functions, acts with G-CSF and IL-6 to stimulate the growth of radiation-damaged BM, and reduces the immune response to radiation injury.

35.1 Introduction

Ionizing radiation damages DNA and other macromolecules either by directly breaking strands or by indirectly disrupting functions via reactive oxygen species (ROS) [1]. After irradiation, the host reacts to repair the damaged DNA, regenerate stem cells, and maintain homeostasis by avoiding an immune overreaction. As part of these responses, interleukin 11 (IL-11), a multifunctional cytokine that belongs to the IL-6 family, is produced by bone marrow (BM)-derived stromal cells [1]. This key multifunctional regulator promotes hematopoiesis [2, 3] and regulates the immune response and bone metabolism.

Studies have shown that IL-11 can reduce radiotoxicity and increase the survival rate of irradiated mice [4]. However, the underlying mechanisms by which IL-11 exerts its protective effect are unknown. To elucidate this phenomenon, we examined the effect of IL-11 on (1) mitochondria, (2) apoptosis, (3) the proliferation of irradiated BM cells, (4) immunoregulators, and (5) growth factors for hematopoiesis.

35.2 Methods

Eight-week-old, male, NIH Swiss mice (National Cancer Institute, Frederic, MD, USA) were used for all experiments. Animal protocols were approved by the Animal Ethics Committee at the University of Florida (Gainesville, FL, USA).

For in vitro studies, BM cells were extracted from mouse femurs and prepared according to standard procedure [5]. Freshly prepared BM cells ($n = 100{,}000$) were divided into two groups and treated with saline (vehicle control) or 20 ng/ml of recombinant IL-11 (rIL-11) for 30 min and then exposed to 0.5 Gy. After being cultured overnight, the cells were assessed with flow cytometry quantization analysis for the following indices: (1) ROS using CM-H2DCFDA (final concentration of 5 μM, Invitrogen Cat # C6827, Grand Island, NY); (2) mitochondrial membrane potential (MMP) using JC-1 (Invitrogen Cat # M34152, 2 μg/100 μl); (3) mitochondrial level using MitoTracker Red CM-H2XRos (Invitrogen Cat # 7513); and (4) apoptosis using Annexin V/PI staining.

For assessment of cell proliferation, freshly prepared BM cells were seeded into U-bottom 96-well plates (3,000 cells/well) in triplicate, treated with saline or 20 ng/ml of rIL-11 for 30 min, and exposed to 0.5 Gy. The cells were further cultured for 72 h and then subjected to standard MTT assay.

For in vivo studies, mice ($n = 5$/group) were pretreated with saline (vehicle control) or 100 μg/kg of rIL-11 at 12 and 2 h before 10-Gy total body irradiation (TBI) with a 137-cesium source (Best Theratronics Ltd, Ottawa, Canada) at a dose rate of 0.9 Gy/min. One hour later, mice were euthanized for analysis of mitochondria and plasma levels of cytokines. For assessment of mouse IL-2, IL-4, IL-6, and G-CSF, mouse plasma was subjected to enzyme-linked immunosorbent assay (ELISA) with kits purchased from R&D Systems (Minneapolis, MN, USA).

The mean and standard deviation were calculated from raw data. The saline-treated groups were counted as baseline and expressed as 100 %. The percentage of relative change of IL-11 treatment in flow cytometry analysis was expressed as: (results of IL-11 treatment group/results of saline treatment group) × 100 %. The ELISA results for cytokines were calculated from the standard curve that was performed simultaneously with the tested plasma samples. An independent Student's *t*-test was used to determine the significance between the treatment and control groups. A P value of < 0.05 was regarded as statistically significant.

35.3 Results

Radiation-induced ROS directly damages DNA and other macromolecules [6]. Figure 35.1a shows that IL-11 administered to BM cells 30 min before irradiation prevented the cells from generating ROS (17.2 % vs. 4.0 %, $P < 0.001$).

MMP is the key factor for electron/proton transport and ATP/ADP/AMP regeneration. Figure 35.1b (JC-1 staining) shows that vehicle control cells could not maintain MMP as effectively as IL-11-treated cells ($P < 0.01$).

MitoTracker probes can passively diffuse across the plasma membrane and accumulate in active mitochondria [7]. More active mitochondria lead to stronger staining. As Fig. 35.1c shows, IL-11-treated BM cells preserved more active mitochondria, as compared to the vehicle controls ($P < 0.01$). Similarly, mice treated with IL-11 had more active mitochondria, as compared to the vehicle controls (Fig. 35.1d). The data illustrate that IL-11 had a beneficial effect on mitochondria in vivo.

Damage to mitochondria can trigger apoptosis. Annexin V/PI staining and flow cytometry demonstrated that cells treated with IL-11 before irradiation exhibited reduced apoptosis, as compared to the vehicle controls (Fig. 35.2a, $P < 0.01$). The MTT assays showed that the cells escaping from apoptosis exhibited increased proliferation (Fig. 35.2b, $P < 0.01$).

To determine if the in vitro effect of IL-11 could be translated in vivo, we explored the responses of immunoregulators and stimulators of BM regeneration in mice treated with rIL-11 before irradiation. ELISA results showed that IL-11 reduced IL-2 and IL-4 (Fig. 35.3a and b, $P < 0.05$ and $P < 0.05$, respectively). Moreover, our results demonstrated that plasma IL-6 and G-CSF were significantly increased by IL-11 (Fig. 35.4a and b).

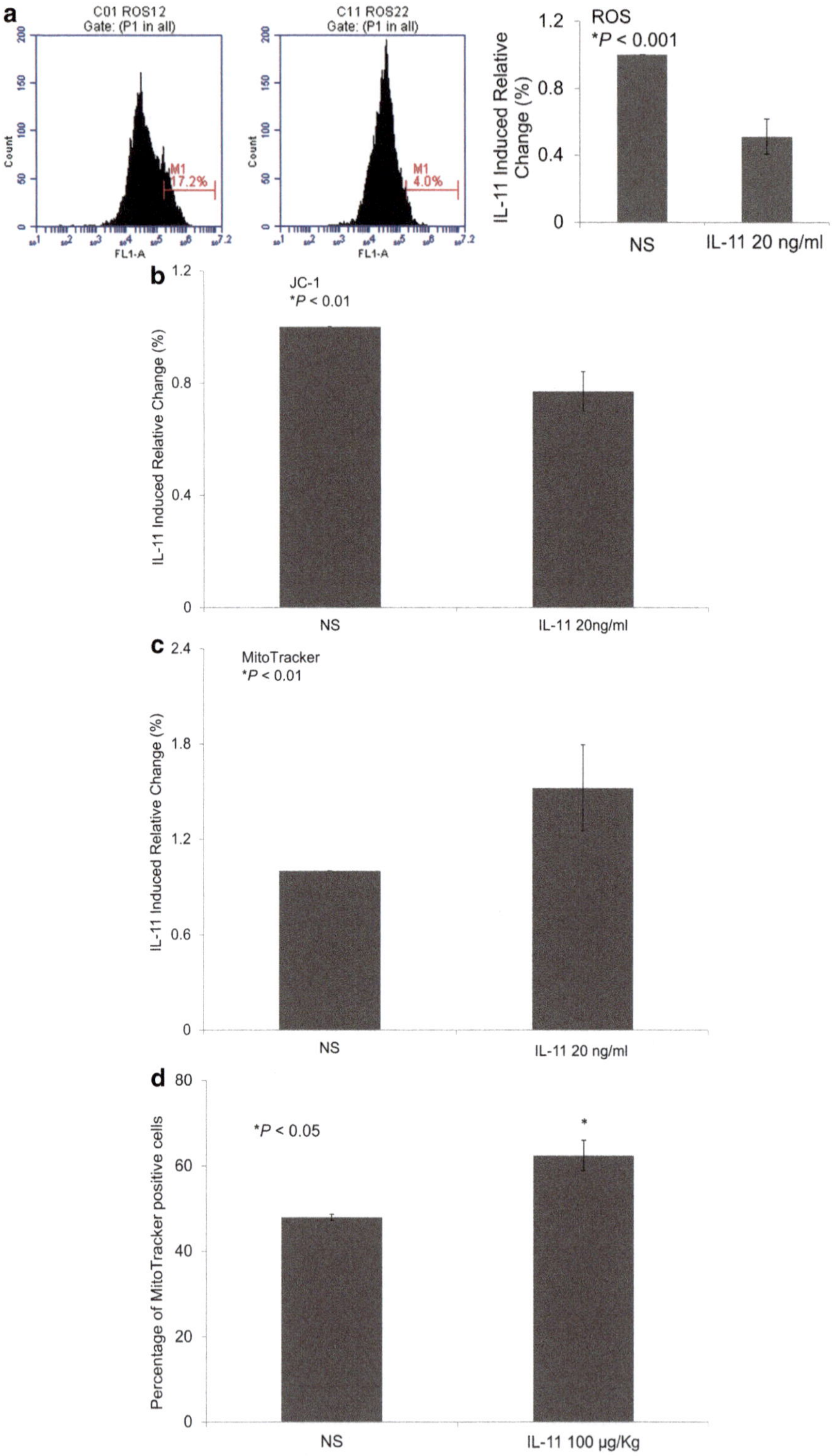

Fig. 35.1 IL-11 protected mitochondria of irradiated BM cells: (**a**) IL-11 (20 ng/ml) reduced ROS production ($P<0.001$); (**b**) IL-11 stabilized MMP, as evidenced by reduced JC-1 transformation

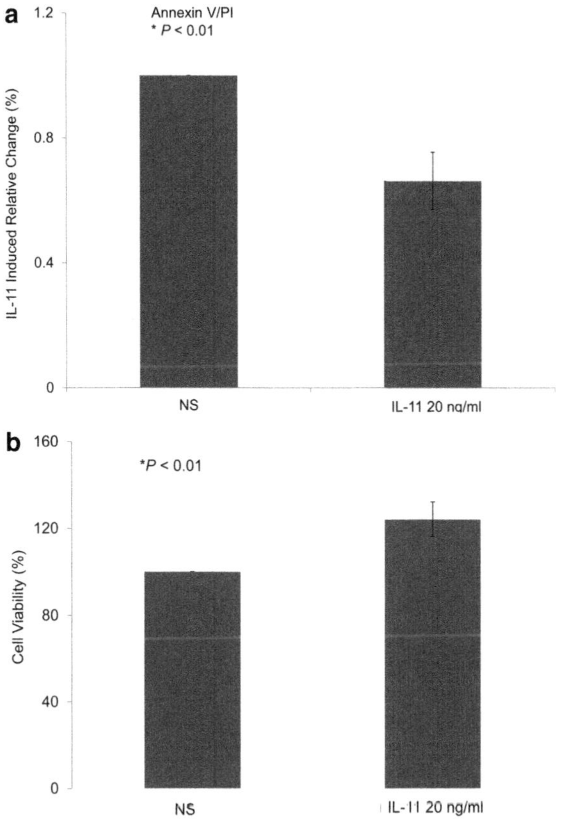

Fig. 35.2 IL-11 reduced apoptosis and enhanced proliferation of irradiated BM cells: (**a**) Annexin V/PI staining indicated that IL-11 (20 ng/ml) reduced apoptosis ($P<0.01$); and (**b**) IL-11 enhanced proliferation of irradiated BM cells ($P<0.01$)

35.4 Discussion

This study explored the radioprotective effect of IL-11. In vitro studies showed that it reduced ROS and maintained MMP and the number and function of irradiated mitochondria. It also reduced apoptosis and enhanced BM cell proliferation. Finally, in vivo studies showed that it reduced immunoregulators (e.g., IL-2 and IL-4) and enhanced BM stimulators (e.g., IL-6 and G-CSF).

Mitochondrial stability and functional integrity provide radiation-damaged cells with the energy to repair DNA, macromolecules, and stem cells in the BM. Our data show that pretreatment with IL-11 prevented radiation-induced loss of mitochondrial stability and integrity, thereby conferring cells with greater radiotolerance.

Fig. 35.1 (continued) ($P<0.01$); (**c**) IL-11 preserved the mitochondrial level, as demonstrated by MitoTracker staining ($P<0.01$); and (**d**) IL-11 (100 μg/kg) protected the BM mitochondrial level from radiation damage in vivo ($P<0.01$)

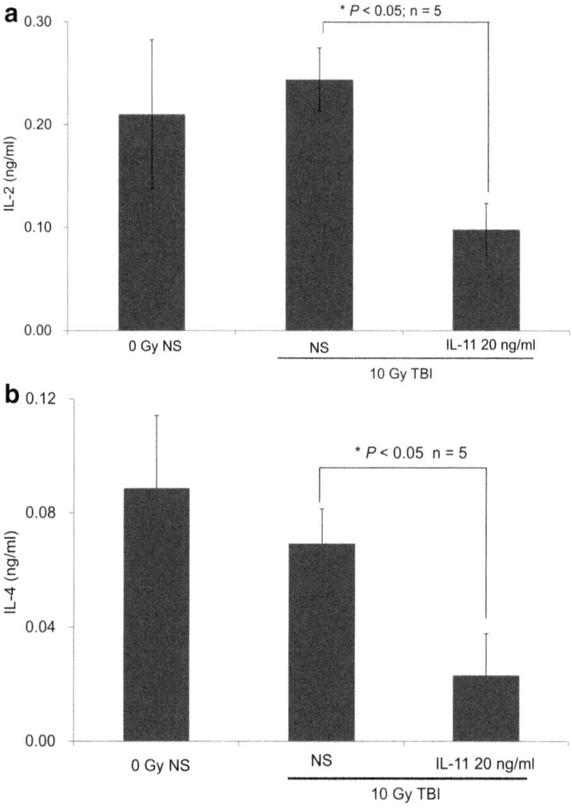

Fig. 35.3 IL-11 suppressed immunoregulators: plasma from mice treated with 100 μg/kg of IL-11 followed by 10-Gy TBI was assayed for different immunoregulators with ELISA. (**a**) IL-11 reduced IL-2 levels ($P<0.05$); and (**b**) IL-11 reduced IL-4 levels ($P<0.05$)

This was evidenced by the reduced apoptosis and enhanced proliferation of irradiated BM cells.

After irradiation, cell death and damage signals can activate the immune system to remove dead cells and to kill the damaged cells to prevent further replication of mutated DNA [8]. To prevent a systemic overreaction, the immune response is tightly controlled by various cytokines and cells; these include IL-2 and IL-4, which promote immune cell maturation and activation [9]. We found that IL-11-treated mice had low levels of IL-2 and IL-4 (Fig. 35.3a and b), indicating that IL-11 is likely a negative regulator for these cytokines.

The radiation-induced depletion of BM stem cells is the major cause of acute radiation syndrome. However, cytokines, such as IL-3, IL-6, IL-11, GM-CSF, CSF, and G-CSF, act synergistically to stimulate BM stem cell regeneration and differentiation. Figure 35.4 shows that IL-11 treatment significantly upregulated IL-6 and G-CSF, the two most important cytokines for the stimulation of BM stem cells, indicating that this treatment may speed up the recovery process and rescue the host.

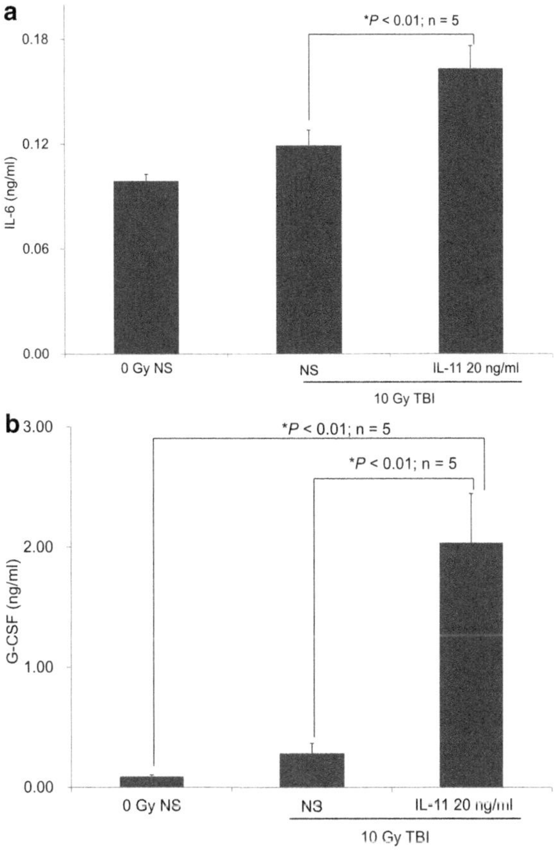

Fig. 35.4 IL-11 enhanced IL-6 and G-CSF production: plasma from mice treated with 100 μg/kg of IL-11 followed by 10-Gy TBI was assayed for different cytokines related to BM regeneration. (**a**) IL-11 enhanced IL-6 production ($P<0.01$); and (**b**) IL-11 enhanced G-CSF production ($P<0.01$)

35.5 Conclusion

This study revealed that IL-11 exerts its radioprotective effect by maintaining mitochondrial functional integrity, suppressing immunoregulators, and increasing cytokines that enhance BM cell regeneration. Due to its possible role in cancer progression, the radioprotective effect of IL-11 has limited utility for the protection of normal tissue for patients undergoing radiotherapy for cancer. However, it could be developed as a potential prophylactic agent for emergency rescuers or mission specialists in the case of a radiation event.

Acknowledgments This project is supported in part by RC1AI0710519, RC2-AI-01075100, RC1-AI0101274 (NIAID/NIH), and Shands Cancer Center startup funds (University of Florida). We thank Dr. Chihray Liu and the medical physics faculty at UF for ensuring dosimetric accuracy in these experiments and Kate Casey-Sawicki for editing this manuscript.

References

1. Klaunig JE, Wang Z, Pu X, Zhou S (2011) Oxidative stress and oxidative damage in chemical carcinogenesis. Toxicol Appl Pharmacol 254(2):86–99
2. Paul SR, Bennett F, Calvetti JA et al (1990) Molecular cloning of a cDNA encoding interleukin 11, a stromal cell-derived lymphopoietic and hematopoietic cytokine. Proc Natl Acad Sci USA 87(19):7512–7516
3. Yin TG, Schendel P, Yang YC (1992) Enhancement of in vitro and in vivo antigen-specific antibody responses by interleukin 11. J Exp Med 175(1):211–216
4. Du XX, Neben T, Goldman S, Williams DA (1993) Effects of recombinant human interleukin-11 on hematopoietic reconstitution in transplant mice: acceleration of recovery of peripheral blood neutrophils and platelets. Blood 81(1):27–34
5. Mace K. Bone marrow chimera protocol. Updated September 2007. Available from http://personalpages.manchester.ac.uk/staff/kimberly.mace/pdfs/MaceBMChimera.pdf. Accessed 4 Sept 2012
6. Leach JK, Van Tuyle G, Lin PS, Schmidt-Ullrich R, Mikkelsen RB (2001) Ionizing radiation-induced, mitochondria-dependent generation of reactive oxygen/nitrogen. Cancer Res 61:3894–3901
7. MitoTracker® Mitochondrion-selective probes. Updated 25 June 2008. Available from http://tools.invitrogen.com/content/sfs/manuals/mp07510.pdf. invitrogen. Accessed 4 Sept 2012
8. Trosko JE (2009) Review paper: cancer stem cells and cancer nonstem cells: from adult stem cells or from reprogramming of differentiated somatic cells. Vet Pathol 46(2):176–193
9. Kondo M, Takeshita T, Ishii N et al (1993) Sharing of the interleukin-2 (IL-2) receptor gamma chain between receptors for IL-2 and IL-4. Science 262(5141):1874–1877

Chapter 36
Tumor Reoxygenation Following Administration of the EGFR Inhibitor, Gefitinib, in Experimental Tumors

Oussama Karroum, Julie Kengen, Vincent Grégoire, Bernard Gallez, and Bénédicte F. Jordan

Abstract It is well recognized that tumor hypoxia is a critical determinant for response to therapy. The effect of an EGFR inhibitor/gefitinib (Iressa®) on tumor oxygenation was monitored daily using in vivo EPR (electron paramagnetic resonance) oximetry on TLT and FSaII tumor models. An increase in pO_2 was shown at a dose of 45 mg/kg i.p. ($n = 4$/group/tumor model). This allowed the identification of a window of reoxygenation in both tumor models (with a maximum between 15 and 20 mmHg after 2 days of treatment). The increase in tumor oxygenation was shown to be the result of a decrease in oxygen consumption. This is the first report on the effect of gefitinib on oxygen consumption by tumor cells and subsequent increase in tumor oxygenation in vivo.

36.1 Introduction

Over the last several years, clinical studies have shown that hypoxia is an independent prognostic indicator of poor patient survival in different tumor types [1]. The oxygenation status of a tumor has been shown to be a pivotal factor in the efficacy of standard radiotherapy [2] because of the so-called oxygen enhancement effect, and the radiation dose required to achieve the same biological effect is about three times higher in the absence of oxygen than in the presence of normal levels of oxygen [3, 4]. The hypoxic microenvironment causes a reduced formation of radiation-induced

O. Karroum • J. Kengen • B. Gallez • B.F. Jordan, Ph.D. (✉)
Biomedical Magnetic Resonance Group, Louvain Drug Research Institute, Université catholique de Louvain, Av.Mounier 73.40, 1200 Brussels, Belgium
e-mail: Benedicte.jordan@uclouvain.be

V. Grégoire
Pole of Molecular Imaging, Radiotherapy and Oncology, Université catholique de Louvain, Brussels, Belgium

S. Van Huffel et al. (eds.), *Oxygen Transport to Tissue XXXV*, Advances in Experimental Medicine and Biology 789, DOI 10.1007/978-1-4614-7411-1_36,

damaging free radicals that fixes damages to DNA, and thus, the overall therapeutic response to radiotherapy is compromised and reduced [5]. A particular area under focus is therefore to combine radiotherapy with co-treatments that are able to transiently increase tumor oxygenation at the time of irradiation. Gefitinib, described as an EGFR inhibitor, has been shown to be able to reduce tumor hypoxia in experimental models [6]. Measurements of hypoxia have been done using in vivo small animal positron emission tomography (PET) imaging with the hypoxia marker [18F] fluoroazomycin arabinoside (FAZA). Gefitinib has also been shown in independent studies to potentiate the effect of radiation therapy [7]. Gefitinib (Iressa or ZD1839) is an orally active, reversible inhibitor of the EGFR tyrosine kinase [8]. In the EU, gefitinib (Iressa) is indicated for the treatment of adult patients with locally advanced or metastatic non-small cell lung cancer (NSCLC). In the current study, we monitored the effect of gefitinib longitudinally on tumor oxygenation, using in vivo electron paramagnetic resonance oximetry, on two distinct experimental tumor models. After identification of a window of reoxygenation by daily measurements of FSaII and TLT tumor pO_2, we investigated the underlying mechanisms responsible for the reoxygenation effect.

36.2 Material and Methods

36.2.1 Animal, Tumor Models, and Treatments

A transplantable liver tumor model (TLT) [9] and a fibrosarcoma (FSaII) [10] were inoculated in the leg of NMRI and C3H/HeOuJIco mice, respectively. For inoculation, approximately 10^6 cells in 0.1 ml of media were injected intramuscularly into the right rear leg. Tumors were allowed to grow up to 8 mm in diameter prior to experimentation. All animal experiments were conducted in accordance with national and university animal care regulations. Gefitinib was purchased from LC Laboratories (MA USA) and was dissolved in DMSO (Invitrogen) (final concentration: 11.25 mg/ml) and delivered daily i.p. at a dose of 45 mg/kg body weight. Control animals were injected with DMSO only.

36.2.2 In Vivo Tumor pO_2 Measurement

In vivo tumor pO_2 was monitored daily using EPR oximetry. The technique relies on the oxygen-dependent broadening of the EPR line width of a paramagnetic oxygen sensor implanted in the tumor [11, 12]. The localized EPR measurements obtained with the 1.2 GHz (or L band) spectrometer correspond to an average of pO_2 values in a volume of 10 mm^3. Data acquisition was performed every day before the injection of gefitinib or vehicle during 4 days, on both TLT and FSaII tumor models (45 mg/kg, $n=4$ for TLT and FSaII); control group received daily DMSO injections ($n=4$ for TLT and FSaII).

36.2.3 Tumor Blood Flow Estimation

Patent blue (Sigma-Aldrich, Belgium) staining was used to obtain a rough estimate of FSaII tumor perfusion [13] at day 2 after treatment with gefitinib ($n=6$) or vehicle (DMSO, $n=14$). This technique involves the injection of 200 μL of a Patent Blue solution (1.25 %) into the tail vein of mice. After 1 min, a uniform distribution of the staining through the body was obtained and mice were sacrificed. Tumors were carefully excised and cut into size-matched halves. Pictures of each tumor cross section were taken with a digital camera. To compare the stained versus unstained area, an in-house program running on Matlab (Mat work, Natick, MA, USA) was developed. The mean of the percentages of the two pictures was then calculated and was used as an indicator of tumor perfusion.

36.2.4 Ex Vivo Oxygen Consumption Rate Evaluation

The method described by James et al. [14] was used. All spectra were recorded on a Bruker EMX EPR spectrometers operating at 9 GHz. FSaII tumor-bearing mice were treated for 2 days with gefitinib as described above ($n=4$ for gefitinib and $n=3$ for vehicle). At day 2 after treatment, the mice were sacrificed and the tumor was excised. FSaII tumors were then dissected in a sterile environment and gently pieced in McCoy's medium. Cells were suspended in 10 % dextran in complete medium. The sealed tubes were placed into quartz EPR tubes, and samples were maintained at 37 °C. As the resulting line width reports on O_2 concentration, oxygen consumption rates were obtained by measuring the O_2 concentration in the closed tube over time and determining the slope of the resulting linear plot.

36.2.5 Statistical Analysis

Results are given as means ± SE values from n animals. Comparisons between groups were made with Student's t test or one-way ANOVA along with post hoc Dunnett's multiple comparison tests when appropriate. P values <0.05 (*), <0.01 (**), or <0.001 (***) were considered significant.

36.3 Results and Discussion

We observed an increase in pO_2 in both TLT and FSaII tumor models from day 1 until day 3 post treatment (this increase was not observed for the control group). The maximal pO_2 was reached after 2 days of treatment. This value was significantly higher than before treatment ($P<0.05$) (Figs. 36.1 and 36.2). We chose day 2 as the "window of reoxygenation" for the rest of the experiments. Since an increase

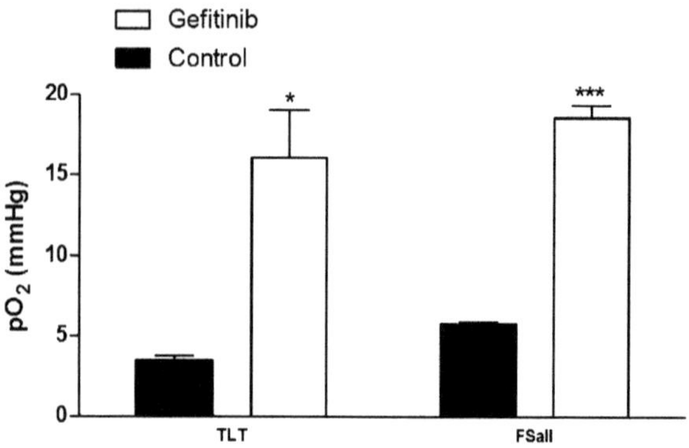

Fig. 36.1 TLT and FSaII tumor oxygenation after 2 days of treatment with gefitinib, assessed by EPR oximetry. Day 2 was selected as the window of reoxygenation for the rest of the experiments

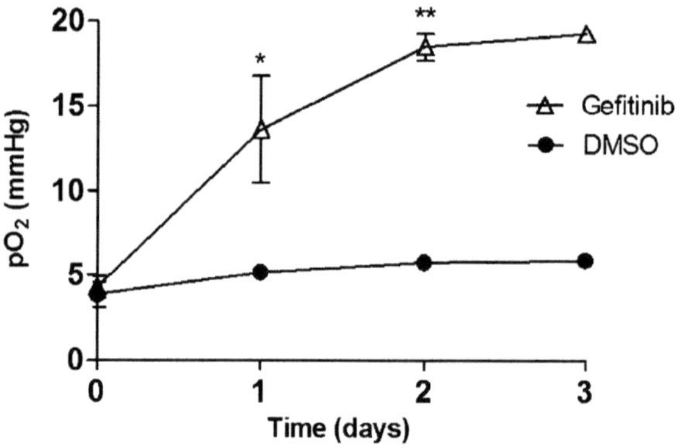

Fig. 36.2 Effect of daily injections of gefitinib (45 mg/kg, $n=4$) on FSaII tumor oxygenation, as monitored using EPR oximetry. Control group received daily DMSO (vehicle) injections ($n=4$). Note the significant increase in tumor oxygenation induced by gefitinib

in tumor oxygenation can be due to both an increase in oxygen supply (blood flow) and/or a decrease in oxygen consumption by tumor cells, both parameters were assessed in the following experiments.

The administration of the drug significantly decreased the oxygen consumption rate of FSaII cells, with a slope of 0.65 μM/min for the control group and 0.26 μM/min for gefitinib ($P<0.01$) (Figs. 36.3 and 36.4).

Here, the main factor that is likely to be responsible for the increase in tumor oxygenation is the decrease in the oxygen consumption rate by tumor cells. This is in accordance with a mathematical model that predicted that modification of oxygen consumption would be much more efficient at alleviating hypoxia than modification of oxygen delivery [15], a theoretical hypothesis that has been further illustrated experimentally in our laboratory using several agents able to target the oxygen consumption by tumor cells [16–18].

In this study, we observed a decrease in oxygen consumption after treatment with gefitinib that might be caused by mitochondrial impairment. Although this remains to be demonstrated in the tumor models used in this study, gefitinib was shown to cause significant mitochondrial membrane depolarization in different breast cancer cell lines [19]. These data suggest a role of the mitochondrial dysfunction as a potential factor at the origin of the decrease in oxygen consumption by tumor cells consecutive to treatment with gefitinib.

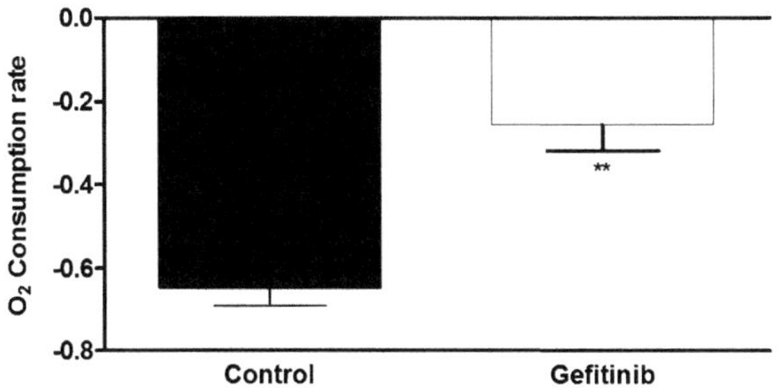

Fig. 36.3 Effect of gefitinib on the rate of oxygen consumption by FSaII tumor cells, as measured ex vivo after 2 days of treatment using EPR oximetry. Note the significant decrease in oxygen consumption rate in gefitinib group ($n=4$) compared to the control group ($n=3$) ($p<0.01$)

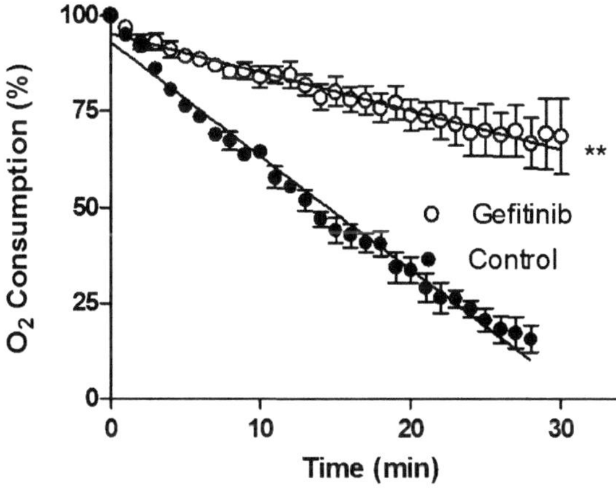

Fig. 36.4 Ex vivo tumor cells' oxygen consumption measurement using EPR oximetry after 2 days of treatment with gefitinib ($n=4$)

The second factor that is able to influence pO_2 is the modification of blood flow (supply) that regulates oxygen delivery to tumors. We measured the ratio (colored area/complete area) observed in tumors after injection of a blue dye. Tumors treated with gefitinib ($n=6$) showed a colored area of 17.4 ± 2.7 % and tumors treated with vehicle ($n=14$) showed a colored area of 11.2 ± 0.9 % (Fig. 36.5). We did not observe any significant modification in blood flow during treatment with gefitinib, contrarily to what we recently observed after administration of the MAPK and VEGFR inhibitor sorafenib [20]. Therefore, the major factor responsible for the increase in tumor oxygenation is the reduction in oxygen consumption by tumor cells.

These results might support the radiosensitizing effect of gefitinib combined with radiation described in other studies [21], where gefitinib was shown to radiosensitize glioma cells by mediated DNA double-strand break repair and enhanced irradiation-induced DNA damage [21]. Increased tumor oxygenation may explain enhancement of irradiation DNA damage [22]; indeed in the presence of oxygen, DNA damage can be stabilized through oxidation of DNA radicals [22].

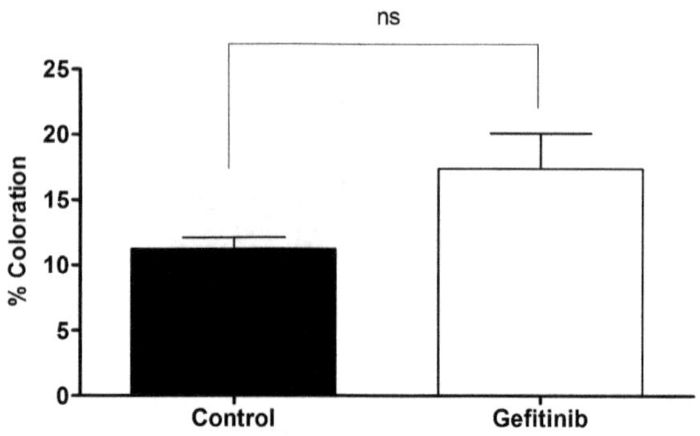

Fig. 36.5 Mean percentage of perfused areas in FSaII tumors assessed by patent blue staining after 2 days of treatment with gefitinib ($n=6$) and control ($n=14$)

To conclude, we observed an increase of pO_2 in TLT and FSaII tumors after 2 days of treatment with gefitinib. The effect is likely to be due to a major decrease in oxygen consumption by tumor cells. This study provides a rationale for using gefitinib as co-treatments for radiation therapy.

Acknowledgments This study was supported by grants from the Belgian National Fund for Scientific Research (FNRS), the Fonds Joseph Maisin, the Saint-Luc Foundation, the "Actions de Recherches Concertées-Communauté Française de Belgique-ARC 09/14-020," and the " Pôle d'Attraction Interuniversitaire PAI VI (P6/38)." OK is "Televie" researcher, and BFJ is Research Associate of the Belgian National Fund for Scientific Research (FNRS).

References

1. Leo C, Giaccia AJ, Denko NC (2004) The hypoxic tumor microenvironment and gene expression. Semin Radiat Oncol 14(3):207–214
2. Vaupel P, Kelleher DK, Thews O (1998) Modulation of tumor oxygenation. Int J Radiat Oncol Biol Phys 42(4):843–848
3. Gray L, Conger A, Ebert M, Hornsey S, Scott OC (1953) The concentration of oxygen dissolved in tissues at time of irradiation as a factor in radiotherapy. Br J Radiol 26(312):638–648
4. Horsman MR, van der Kogel AJ (2009) Therapeutic approaches to tumor hypoxia. In: Joiner M, van der Kogel A (eds) Basic clinical radiobiology, 4th edn. A Hodder Arnold Pub, London, pp 233–245
5. Toustrup K, Sørensen BS, Lassen P et al (2012) Gene expression classifier predicts for hypoxic modification of radiotherapy with nimorazole in squamous cell carcinomas of the head and neck. Radiother Oncol 102(1):122–129
6. Solomon B, Binns D, Roselt P et al (2005) Modulation of intratumoral hypoxia by the epidermal growth factor receptor inhibitor gefitinib detected using small animal PET imaging. Mol Cancer Ther 4(9):1417–1422
7. Bianco C, Tortora G, Bianco R et al (2002) Enhancement of antitumor activity of ionizing radiation by combined treatment with the selective epidermal growth factor receptor-tyrosine kinase inhibitor ZD1839 (Iressa). Clin Cancer Res 8(10):3250–3258
8. Wakeling AE, Guy SP, Woodburn JR et al (2002) ZD1839 (Iressa): an orally active inhibitor of epidermal growth factor signaling with potential for cancer therapy. Cancer Res 62(20): 5749–5754

9. Taper HS, Woolley GW, Teller MN, Lardis MP (1966) A new transplantable mouse liver tumor of spontaneous origin. Cancer Res 26:143–148
10. Volpe JP, Hunter N, Basic I, Milas L (1985) Metastatic properties of murine sarcomas and carcinomas. I. Positive correlation with lung colonization and lack of correlation with s.c. tumor take. Clin Exp Metastasis 3(4):281–294
11. Gallez B, Baudelet C, Jordan BF (2004) Assessment of tumor oxygenation by electron paramagnetic resonance: principles and applications. NMR Biomed 17(5):240–262
12. Gallez B, Jordan BF, Baudelet C, Misson PD (1999) Pharmacological modifications of the partial pressure of oxygen in murine tumors: evaluation using in vivo EPR oximetry. Magn Reson Med 42(4):627–630
13. Sersa G, Cemazar M, Miklavcic D, Chaplin DJ (1999) Tumor blood flow modifying effect of electrochemotherapy with bleomycin. Anticancer Res 19(5B):4017–4022
14. James PE, Jackson SK, Grinberg OY, Schwartz HM (1995) The effects of endotoxin on oxygen consumption of various cell types in vitro: an EPR oximetry study. Free Radic Biol Med 18(4):641–647
15. Secomb TW, Hsu R, Ong ET et al (1995) Analysis of the effects of oxygen supply and demand on hypoxic fraction in tumors. Acta Oncol 34(3):313–316
16. Jordan BF, Gregoire V, Demeure RJ et al (2002) Insulin increases the sensitivity of tumors to irradiation: involvement of an increase in tumor oxygenation mediated by a nitric oxide-dependent decrease of the tumor cells oxygen consumption. Cancer Res 62(12):3555–3561
17. Jordan BF, Christian N, Crokart N, Grégoire V, Feron O, Gallez B (2007) Thyroid status is a key modulator of tumor oxygenation: implication for radiation therapy. Radiat Res 168(4):428–432
18. Jordan BF, Gallez B (2010) Surrogate MR markers of response to chemo- or radiotherapy in association with co-treatments: a retrospective analysis of multi-modal studies. Contrast Media Mol Imaging 5(6):323–332
19. Carloni S, Fabbri F, Brigliadori G et al (2010) Tyrosine kinase inhibitors gefitinib, lapatinib and sorafenib induce rapid functional alterations in breast cancer cells. Curr Cancer Drug Targets 10(4):422–431
20. Karroum O, Kengen J, Danhier P et al (2012) Tumor reoxygenation following administration of Mitogen-Activated Protein Kinase inhibitors: a rationale for combination with radiation therapy. Radiother Oncol 105(1):64–71
21. Kang KB, Zhu C, Wong YL, Gao Q, Ty A, Wong MC (2012) Gefitinib radiosensitizes stem-like glioma cells: inhibition of epidermal growth factor receptor-Akt-DNA-PK signaling, accompanied by inhibition of DNA double-strand break repair. Int J Radiat Oncol Biol Phys 83(1):43–52
22. Jordan BF, Sonveaux P (2012) Targeting tumor perfusion and oxygenation to improve the outcome of anticancer therapy. Front Pharmacol 3:94

Chapter 37
Radiation Affects the Responsiveness of Bone Marrow to G-CSF

Zhenhuan Zhang, Mei Zhang, Wenlong Lv, Luqiang Huang, Liangjie Yin, Shanmin Yang, Jinsheng Hong, Deping Han, Chun Chen, Amy Zhang, Sadasivan Vidyasagar, Paul Okunieff, and Lurong Zhang

Abstract In this study, we investigated the response of irradiated bone marrow cells to granulocyte colony-stimulating factor (G-CSF). Freshly harvested bone marrow cells were treated with either saline (vehicle control) or 20 ng/ml of G-CSF. Thereafter, cells were separated into nonirradiated (no-IR) and irradiated (IR, 0.5 Gy) groups. IR cells exhibited a higher proliferation rate in response to G-CSF, as compared to the no-IR cells. Reduced levels of reactive oxygen species indicated that G-CSF-treated IR cells produced fewer free radicals, as compared to the no-IR cells. The G-CSF-treated IR cells also had a lower apoptotic rate than their no-IR counterparts. Furthermore, G-CSF-treated IR cells exhibited less alteration of mitochondrial membrane potential, as compared to the no-IR cells. Finally, the mitochondrial number increased in the G-CSF-treated IR cells. The radiation-induced increase in plasma IL-6 in vivo could be enhanced by the administration of G-CSF. The data suggest that radiation potentiates the response of bone marrow cells to G-CSF treatment.

Z. Zhang • M. Zhang • L. Huang • L. Yin • S. Yang • A. Zhang • S. Vidyasagar
P. Okunieff • L. Zhang (✉)
Department of Radiation Oncology, UF Shands Cancer Center, University of Florida, Gainesville, FL 32610, USA
e-mail: lurongzhang@ufl.edu

W. Lv • J. Hong • D. Han • C. Chen
Department of Radiation Oncology, First Affiliated Hospital of Fujian Medical University, Fuzhou, Fujian 35001, China

S. Van Huffel et al. (eds.), *Oxygen Transport to Tissue XXXV*, Advances in Experimental Medicine and Biology 789, DOI 10.1007/978-1-4614-7411-1_37,

37.1 Introduction

Granulocyte colony-stimulating factor (G-CSF) is a naturally occurring growth factor that stimulates the bone marrow to produce white blood cells. Therapeutically, recombinant G-CSF is used help patients recover from neutropenia caused by chemotherapy/radiotherapy and nuclear incidents. Because the biological response of cells to drugs is closely related to overall cellular condition, we speculated that normal and irradiated bone marrow (BM) cells would exhibit different levels of responsiveness to G-CSF, which would provide the rationale for its wider clinical use. To test this hypothesis, BM cells from the same mouse were treated with G-CSF and then were subdivided into nonirradiated (no-IR) and irradiated (IR, 0.5 Gy) groups. We analyzed cell proliferation and mitochondrial indices in vitro and cytokine responses in vivo. The IR BM cells exhibited a greater response to G-CSF, as compared to the no-IR cells. Our data suggest that radiation potentiates the response of BM cells to G-CSF treatment.

37.2 Methods

Eight-week-old, male, NIH Swiss mice (National Cancer Institute, Frederic, MD, USA) were used for all experiments. Animal protocols were approved by the Animal Ethics Committee at the University of Florida (Gainesville, FL, USA).

Bone marrow cells harvested from mouse femurs were either seeded in 96-well U-bottom microplates at a density of $5–10 \times 10^3$ cells per well in triplicate (for proliferation) or 5×10^5 /treatment in DMEM containing 10 % heat-inactivated newborn calf serum with 1 % penicillin/streptomycin (for apoptosis and mitochondrial indices). Prepared cells from each mouse were then separated into two treatment groups: (1) saline as vehicle control or (2) 20 ng/ml of mouse recombinant G-CSF. Thereafter, each treatment group was divided into two subgroups: (1) no-IR or (2) IR with 0.5 Gy at a dose rate of 0.98 Gy/min via a 137-cesium source (Best Theratronics Ltd, Ottawa, Canada). After being cultured for 48 h (for proliferation) or overnight (for apoptosis and mitochondrial indices), the BM cells were subjected to:

1. A cell proliferation assay using the standard MTT method. The relative increased proliferation percentage was calculated as: G-CSF IR relative proliferation rate = G-CSF IR MTT/Saline IR MTT, G-CSF no-IR relative proliferation rate = G-CSF no-IR MTT/ Saline no-IR MTT, respectively.
2. A cell apoptosis assay with standard Annexin V/PI staining. The percentage of Annexin V-positive cells was compared between the different groups as: IR relative Annexin V-positive rate = G-CSF IR Annexin V-positivity/Saline IR Annexin V-positivity vs. no-IR relative Annexin V-positive rate = G-CSF no-IR Annexin V-positivity/Saline no-IR Annexin V-positivity.

Mitochondria-specific dyes were used to stain approximately 5×10^5 BM cells from each group prior to analysis with flow cytometry: (1) mitochondrial number with MitoTracker Red CM-H2XRos (Invitrogen Cat # 7513, Grand Island, NY, USA), (2) mitochondrial membrane potential (MMP) with JC-1 (Invitrogen Cat # M34152, 2 μg/100 μl), and (3) reactive oxygen species (ROS) with CM-H2DCFDA (Invitrogen Cat # C6827).

The alteration percentages of the G-CSF groups were presented as: (mean value of G-CSF group/mean value of saline group) × 100 %. The saline groups constituted the baseline and were expressed as 100 %. The alteration percentage induced by G-CSF was compared between the no-IR and IR BM cells from the same mouse manipulated at the same time.

To determine the in vivo effect of G-CSF on an irradiated host, NIH Swiss mice were treated with saline (control) or G-CSF (10 μg/mouse) twice at 12 and 2 h before 10-Gy total body irradiation (TBI) and euthanized 1 h after irradiation. The plasma collected from each group was subjected to an ELISA with a mouse IL-6 kit (R&D Systems, Minneapolis, MN, USA).

An independent Student's *t*-test was used to determine the significance between the control and G-CSF no-IR and IR groups. A P value of <0.05 was regarded as statistically significant.

37.3 Results

Our in vitro results showed that, as compared to the G-CSF no-IR group, the G-CSF IR group exhibited a significant increase in the relative proliferation rate (Fig. 37.1a, b, 135 % vs. 152 %, $P=0.0078$).

Annexin V/PI staining demonstrated that the G-CSF no-IR group and the vehicle control had a similar percentage of apoptotic cells (Fig. 37.2a, b); the G-CSF IR group had a lower percentage of apoptotic cells, as compared to the saline IR group (Fig. 37.2c, d). Using the saline group as a baseline, statistical analysis (Fig. 37.2e) indicated that G-CSF better prevented apoptosis in the IR group, as compared to the no-IR group ($P=0.035$).

MitoTracker staining (Fig. 37.3a, b) showed that the G-CSF IR group had a greater number of mitochondria than did the G-CSF no-IR group ($P<0.01$). JC-1 staining (Fig. 37.4a, b) indicated that G-CSF better stabilized the MMP of the IR group, as compared to the G-CSF no-IR group ($P<0.03$). CM-H2DCFDA staining (Fig. 37.5) demonstrated that G-CSF better reduced the ROS levels in the IR group than in the no-IR group ($P<0.05$).

Our in vivo results indicated that saline-treated IR mice exhibited an increase in IL-6 production, as compared to no-IR mice (Fig. 37.6). G-CSF-treated IR mice exhibited a significant increase in IL-6 production, as compared to the saline-treated IR mice ($P<0.001$). These results suggest that G-CSF might act synergistically with IL-6 to stimulate BM cell regeneration after radiation-induced damage.

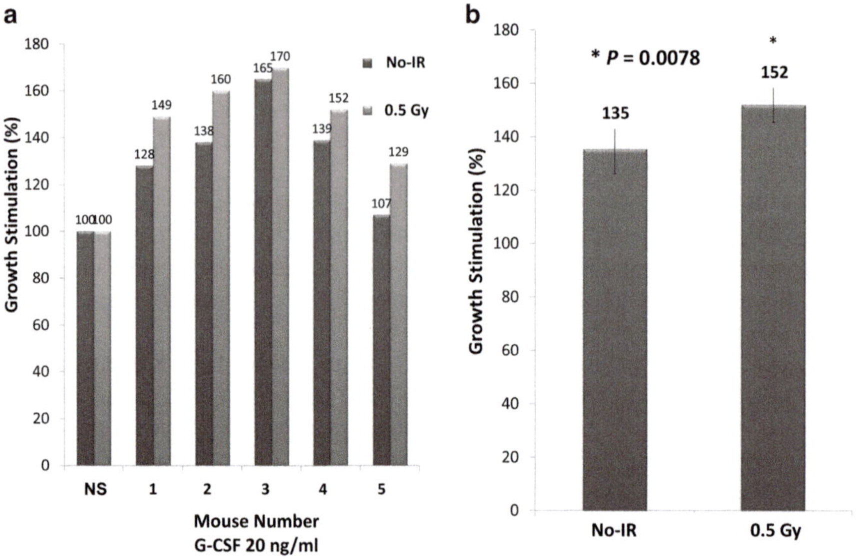

Fig. 37.1 IR BM cells had a high relative proliferation rate in response to G-CSF: (**a**) The individual proliferation rate of IR BM cells increased from 29.0 % to 70.0 %, as compared to no-IR BM cells treated with 20 ng/ml of G-CSF. (**b**) Group statistical analysis indicated that the IR BM cells had a stronger response to G-CSF, as compared to no-IR BM cells (52.0 % vs. 35.0 %, $P=0.0078$)

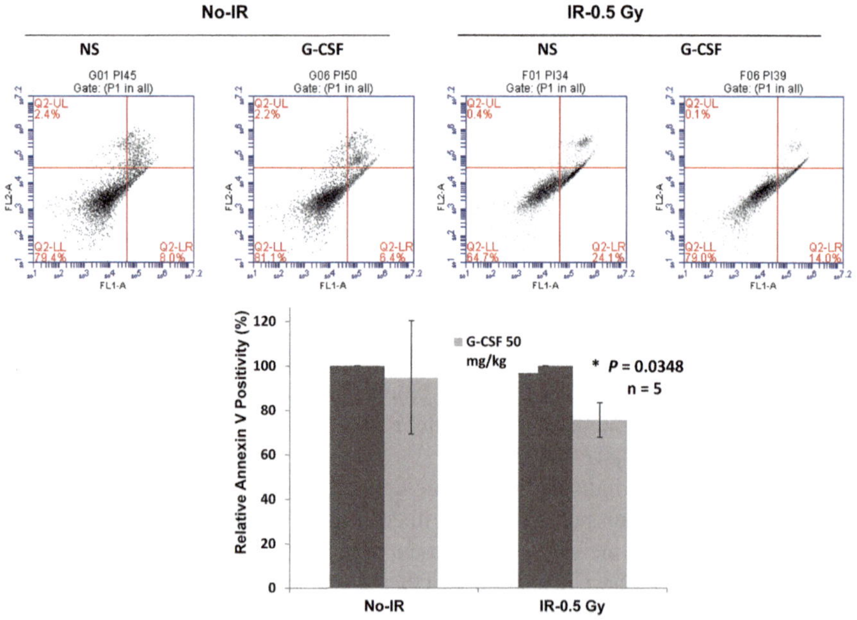

Fig. 37.2 G-CSF better prevented apoptosis in IR BM cells: Group statistical analysis indicated that relatively fewer IR BM cells were apoptotic, as compared to no-IR BM cells ($P=0.0348$, $n=5$)

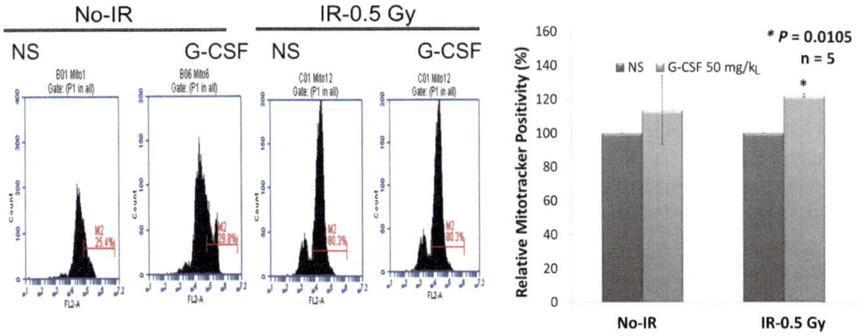

Fig. 37.3 G-CSF better preserved the mitochondria in IR BM cells: Group statistical analysis indicated that IR BM cells exhibited better preservation of mitochondria, as compared to no-IR BM cells ($P = 0.0105$, $n = 5$)

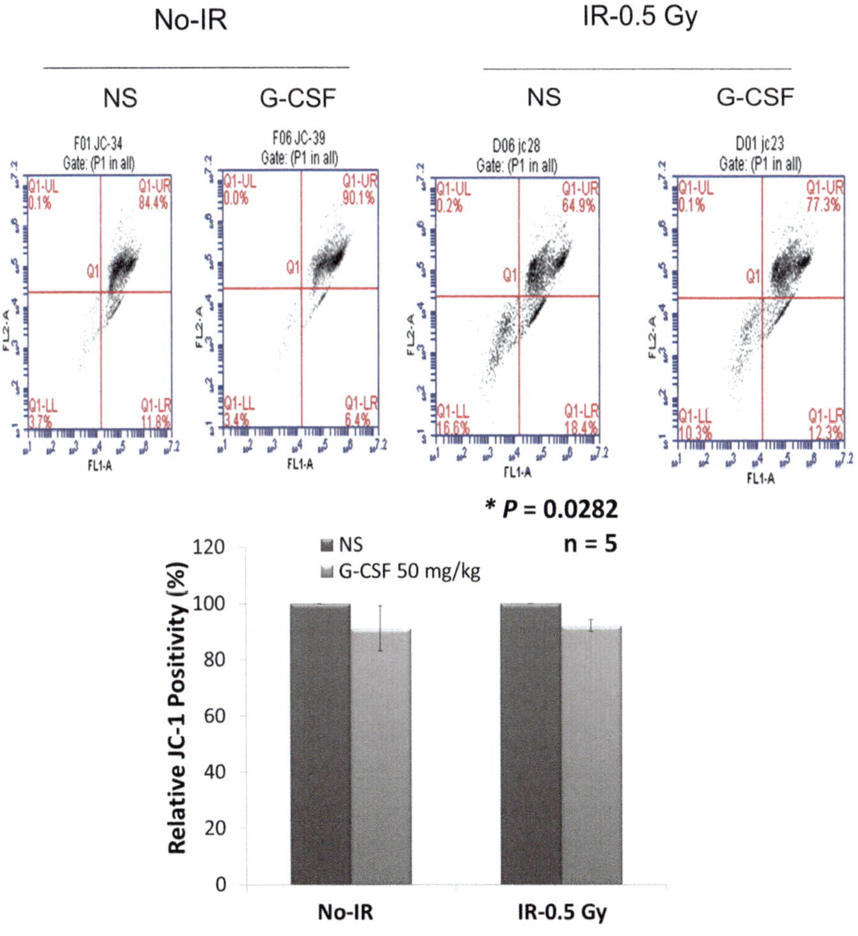

Fig. 37.4 G-CSF better stabilized the MMP of IR BM cells: Group statistical analysis indicated that IR BM cells exhibited better stabilized mitochondria, as compared to no-IR BM cells ($P = 0.028$, $n = 5$)

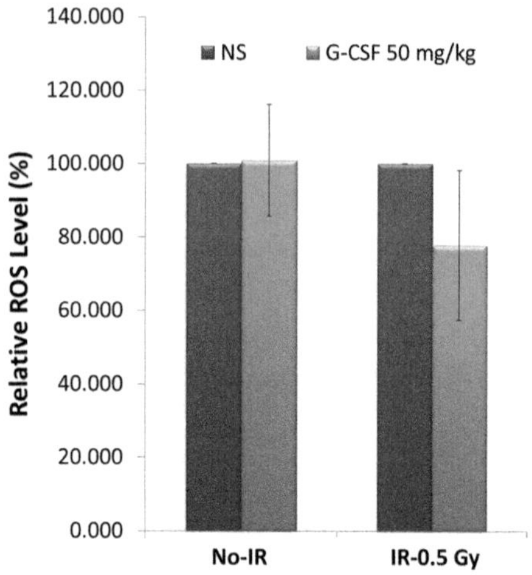

Fig. 37.5 G-CSF better reduced the ROS level in IR BM cells: IR BM cells exhibited relatively less ROS reproduction, as compared to no-IR BM cells ($P<0.05$)

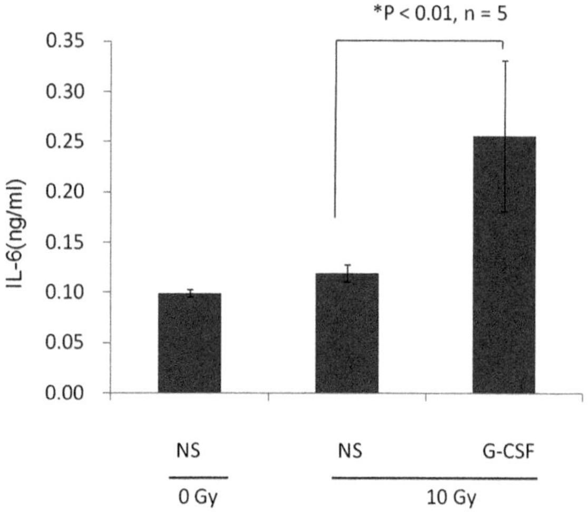

Fig. 37.6 G-CSF significantly increased IL-6 production in IR mice: Mice were treated with G-CSF (10 μg/mouse) 12 and 2 h before 10-Gy TBI. Mouse serum was harvested 1 h after IR. G-CSF further enhanced the IR-induced upregulation of IL-6 (0.10 ± 0.001 ng/ml as baseline; 0.12 ± 0.002 ng/ml with IR stimulation; and 0.26 ± 0.05 ng/ml with G-CSF plus IR, $P<0.01$)

37.4 Discussion

Our in vitro results showed that G-CSF exerted a greater protective effect on IR BM cells than it did on no-IR cells. On the mitochondrial level, G-CSF-treated IR cells experienced reduced ROS levels, less depolarization, and increased preservation. At the cellular level, G-CSF-treated IR cells had a higher relative proliferation rate, and less apoptosis, as compared to saline-treated IR cells. Moreover, our in vivo studies demonstrated that G-CSF-treated IR mice exhibited increased IL-6 levels, as compared to saline-treated IR mice.

To compensate for the radiation-induced loss of cells and their functions, surviving BM cells can become sensitive to "rescue signaling," such as that produced by G-CSF. The molecular mechanism for this phenomenon is not clear. Exposure to G-CSF and radiation might modulate the expression level or binding affinity of the G-CSF receptor [1, 2]. The mechanism might also be related to the synergistic ability of G-CSF and radiation to stimulate the hyperactivation of downstream signaling pathways of G-CSF or its inducible secondary factors (e.g., IL-6), thereby amplifying its protective effects [3–5].

Radiation triggers a clear increase in ROS levels, leading to oxidative stress and mitochondrial dysfunction [6]. A recent study examining H_2O_2-induced oxidative stress in isolated cardiac mitochondria found that G-CSF markedly reduced ROS production and directly prevented mitochondrial swelling and depolarization [6]. Our results are consistent with this finding. Indeed, the action of G-CSF does not seem to be limited to BM cells [6, 7]. The universal effect of G-CSF on various types of critical cells indicates that it may have great potential for wider clinical use.

To protect IR BM cells against apoptosis, G-CSF may (1) suppress p53 functions, which coordinate radiation-induced cell-cycle checkpoint activity and promote apoptosis [8]; (2) inhibit the intragenic caspases pathway [9]; or (3) inhibit extrinsic Fas-triggered apoptosis [10]. In addition, the higher relative proliferation rate of G-CSF-treated IR BM cells may be related to the (1) direct stimulation of BM cells via G-CSF receptor dimerization [11], (2) ability of IR BM cells to escape apoptosis, and (3) indirect stimulation of other factors (e.g., IL-6) triggered by G-CSF.

Cytokines act as a network and compensate for deficient pathways. Our data prove that the host responds to G-CSF by producing IL-6. G-CSF and IL-6 act synergistically to stimulate the production of neutrophils and monocytes [12].

Our in vivo and in vitro models demonstrate that G-CSF protects mitochondrial functions and increases cell regeneration and IL-6 production, especially when a subject is exposed to radiation.

References

1. Sloand EM, Yong AS, Ramkissoon S et al (2006) Granulocyte colony-stimulating factor preferentially stimulates proliferation of monosomy 7 cells bearing the isoform IV receptor. Proc Natl Acad Sci USA 103(39):14483–14488
2. Sultana TA, Harada H, Ito K, Tanaka H, Kyo T, Kimura A (2003) Expression and functional analysis of granulocyte colony-stimulating factor receptors on CD34++ cells in patients with myelodysplastic syndrome (MDS) and MDS-acute myeloid leukaemia. Br J Haematol 121(1): 63–75
3. Marino VJ, Roguin LP (2008) The granulocyte colony stimulating factor (G-CSF) activates Jak/STAT and MAPK pathways in a trophoblastic cell line. J Cell Biochem 103(5): 1512–1523
4. Wang L, Kurosaki T, Corey SJ (2007) Engagement of the B-cell antigen receptor activates STAT through Lyn in a Jak-independent pathway. Oncogene 26(20):2851–2859
5. Gits J, van Leeuwen D, Carroll HP, Touw IP, Ward AC (2006) Multiple pathways contribute to the hyperproliferative responses from truncated granulocyte colony-stimulating factor receptors. Leukemia 20(12):2111–2118

6. Thummasorn S, Kumfu S, Chattipakorn S, Chattipakorn N (2011) Granulocyte-colony stimulating factor attenuates mitochondrial dysfunction induced by oxidative stress in cardiac mitochondria. Mitochondrion 11(3):457–466
7. Harada M, Qin Y, Takano H et al (2005) G-CSF prevents cardiac remodeling after myocardial infarction by activating the Jak-Stat pathway in cardiomyocytes. Nat Med 11(3):305–311
8. Lotem J, Sachs L (1993) Hematopoietic cells from mice deficient in wild-type p53 are more resistant to induction of apoptosis by some agents. Blood 82(4):1092–1096
9. van Raam BJ, Drewniak A, Groenewold V, van den Berg TK, Kuijpers TW (2008) Granulocyte colony-stimulating factor delays neutrophil apoptosis by inhibition of calpains upstream of caspase-3. Blood 112(5):2046–2054
10. Watson RW, O'Neill A, Brannigan AE et al (1999) Regulation of Fas antibody induced neutrophil apoptosis is both caspase and mitochondrial dependent. FEBS Lett 453(1–2):67–71
11. Tsuji K, Ebihara Y (2001) Expression of G-CSF receptor on myeloid progenitors. Leuk Lymphoma 42(6):1351–1357
12. Bhattacharyya M, Ghosh MK (2008) Hemophagoctic lymphohistiocytosis–recent concept. J Assoc Physicians India 56:453–457

Chapter 38
Application of MOBILE (*M*apping of *O*xygen *B*y *I*maging *L*ipids relaxation *E*nhancement) to Study Variations in Tumor Oxygenation

Bénédicte F. Jordan, Julie Magat, Florence Colliez, Elif Ozel, Anne-Catherine Fruytier, Valérie Marchand, Lionel Mignion, and Bernard Gallez

Abstract The aim of the study was to sensitively monitor changes in tumor oxygen using the MOBILE (mapping of oxygen by imaging lipids relaxation enhancement) technique. This method was applied in mammary tumor mouse models on an 11.7T Bruker MRI system. MOBILE was compared with functional imaging R_2*, R_1 of water and with pO_2 measurements (using EPR oximetry and O_2-dependent fluorescence quenching measurements). MOBILE was shown to be capable to monitor changes in oxygenation in tumor tissues.

38.1 Introduction

Tumor hypoxia is acknowledged as a major factor of resistance of solid tumors to treatment [1–3]. Quantitative follow-up of changes in tumor oxygenation can find relevant applications in radiation therapy planning [4, 5] as well as with regard to antiangiogenic and antivascular treatment optimization [6].

Direct quantitative methods, including Eppendorf microelectrodes [7], electron paramagnetic resonance (EPR) oximetry [8], ^{19}F relaxometry [9], or Overhauser-enhanced magnetic resonance imaging (MRI) [10], either are invasive or require the injection of a reporter probe and are currently not clinically applicable. Nowadays, the clinically available armamentarium for this purpose only includes radiolabeled nitroimidazoles detected by positron emission tomography (PET) which prominently accumulate in hypoxic areas, thereby acting as potential markers for tissue hypoxia [11]. Endogenous sources of contrast in MRI include variations of T_1

B.F. Jordan • J. Magat • F. Colliez • E. Ozel • A.-C. Fruytier • V. Marchand
L. Mignion • B. Gallez, Ph.D. (✉)
Biomedical Magnetic Resonance Research Group, Louvain Drug Research Institute, Université Catholique de Louvain, Av.Mounier 73.40, Brussels 1200, Belgium
e-mail: Bernard.gallez@uclouvain.be

S. Van Huffel et al. (eds.), *Oxygen Transport to Tissue XXXV*, Advances in Experimental Medicine and Biology 789, DOI 10.1007/978-1-4614-7411-1_38,

(longitudinal relaxation rate) and T_2* (effective transversal relaxation rate) values in tissue [12]. T_1 discloses sensitivity to dissolved oxygen, which acts as a T_1-shortening paramagnetic contrast agent [13]; T_2* is sensitive to the relative deoxyhemoglobin/oxyhemoglobin (Hb/HbO_2) ratio in vessels [14]. T_2* mapping, also referred to as functional MR imaging or "BOLD" (blood oxygen level dependent) imaging, is sensitive to variations in oxygenation in the vascular compartment and has been successfully applied to monitor changes in tissue oxygenation [15, 16]. However, BOLD-MRI has also demonstrated significant limitations in terms of quantitative relationships between response signal intensity and true changes in tissue pO_2 and is sensitive to changes in total hemoglobin content [17–19]. Recently, changes in tissue oxygen concentrations have been shown to produce changes in relaxation rate R_1 ($=1/T_1$) of water [20–22]. Unfortunately, the technique still suffers from insufficient sensitivity.

In our lab, we have developed an MR method for mapping variations in oxygenation based on changes in relaxation properties of the tissue lipids, which we called "MOBILE," for mapping of oxygen by imaging lipids relaxation enhancement [23]. In vitro, a gain in sensitivity (compared with water) was demonstrated and evaluated to a factor of 7.5 in pure oil and 2.1 and 11.4 in tissue homogenates corresponding to two distinct common lipid peaks with a chemical shift of ~1.3 (fatty acid chains) and ~4.0 ppm (glycerol backbone of triglycerides), respectively [23]. In vivo we also demonstrated the feasibility in several tissues on preclinical and clinical studies. In the present study, our goal was to apply the MOBILE technique to monitor tumor oxygenation during carbogen challenge and to compare this technique with two quantitative measurements of tumor pO_2 (invasive MR compatible fiber-optic fluorescence quenching probes, OxyLite™, and EPR oximetry). We also compared MOBILE to two currently used MR parameters, namely, R_2* and R_1 of water.

38.2 Methods

Syngeneic NT2 and MDA human mammary tumor cells were injected subcutaneously (implanted orthotopically) into the right upper mammary fat pad of 6-week-old FVB/N or nude NMRI female mice (Janvier), respectively ($n=5$/model). Tumors were analyzed when reaching 6 mm in diameter. Animals were anesthetized by inhalation of isoflurane (Forene; Abbot, England; mixed with 21 % oxygen, 1.5 l/h). Respiration rate and body temperature (37.0 °C +/−1.0 °C) were monitored. Studies were undertaken in accordance with the national and local regulations of the ethical committee (agreement number LA 1230467).

Experiments were performed with an 11.7T MRI system (Bruker, BioSpec; Ettlingen, Germany) and with a quadrature volume coil (inner diameter of 40 mm). Three MR measurements of each type (R_1 H_2O, R_1 lipids, R_2*) were acquired sequentially and repeated three times during air breathing. Then, breathing gas was switched to carbogen. Respiratory triggering was employed to avoid motion artifacts. We used the OxyLite™ fiber-optic microprobes for continuously monitoring tumor oxygenation simultaneously with MRI [24]. Independent experiments were also performed to compare MOBILE with EPR oximetry using a similar breathing challenge.

A segmented IR FISP [25] (inversion-recovery fast imaging with steady-state precession) sequence (SSFP FID mode) was used to acquire parametric images of T_1 relaxation time. The acquisition parameters were TR/TE/FA/BW/matrix = 4 ms/1.2 ms/5°/100 kHz/64 × 64, four segments, and a total acquisition time of 1 min 20 s. For the total proton experiment, a series of 100 images were acquired (spaced by a scan repetition time TR = 120 ms) with a slice thickness of 1 mm. For the lipid experiments, we first evaluated the frequency difference in hertz between water and lipid peaks in the ^{1}H spectrum with a single pulse sequence. These offsets were then used as an imaging frequency offset in the same IR FISP protocol. Typically, the lipid peak of interest was approximately 4.0 ppm for tumor as identified experimentally [26]. We added a $\pi/2$ hermite saturation pulse to cancel the water signal. A series of 40 images (spaced by scan repetition time TR = 100 ms) with a slice thickness of 3 mm were acquired. Then, images were fitted using a homemade program in Matlab (the MathWorks, Inc., Natick, MA, USA) to determine the T_1 relaxation in regions of interest (ROIs). Raw data were filtered so as to disregard nonrelevant fits with a T_1 error/T_1 > 30 %. T_1 values were then calculated from the effective T_1* values measured experimentally [27]. Finally, R_1 ($1/T_1$) data were filtered according to the calibration curves obtained in vitro from tissue extracts (R_1 0 % O_2 < measured R_1 < R_1 100 % O_2).

For T_2* measurements, a multi-gradient echo (MGE) sequence was performed with eight echoes (between 3.5 and 31.5 ms) with a total acquisition time of 4 min 48 s. A 256 × 256 matrix was used with TR/flip angle/slice thickness = 1,500 ms/30°/1 mm.

In vivo tumor pO_2 was monitored using electron paramagnetic resonance (EPR) oximetry using charcoal as the oxygen-sensitive probe [8]. EPR spectra were recorded using a 1.1 GHz EPR spectrometer (Magnettech, Berlin, Germany). Calibration curves were made by measuring the EPR line width as a function of the pO_2. Mice were injected in the center of the tumor using the suspension of charcoal (100 mg/ml, 60 μl injected). The tumor under study was placed in the center of the extended loop resonator whose sensitive volume extends 1 cm into the tumor mass. The pO_2 measurements correspond to an average of pO_2 values in this volume of the NT2 tumor models. Charcoal was injected 24 h before experiments. MOBILE and EPR measures were performed the same day, and it was verified that charcoal did not perturb MOBILE or T_1 MRI measurements.

Paired *t*-tests (carbogen versus air breathing), linear fits, and Pearson correlations ($p < 0.05$ (*), $p < 0.01$ (**), or $p < 0.001$ (***) were considered significant) were performed using the GraphPad software.

38.3 Results and Discussion

The basal pO_2 measured with OxyLite in MDA tumors ranged from 2 to 25 mmHg (mean of $pO_2 = 14.15 \pm 5.46$ mmHg), whereas basal pO_2 in NT2 tumors was between 35.5 and 46.7 mmHg (mean of $pO_2 = 37 \pm 0.8$ mmHg). Tumor oxygenation was

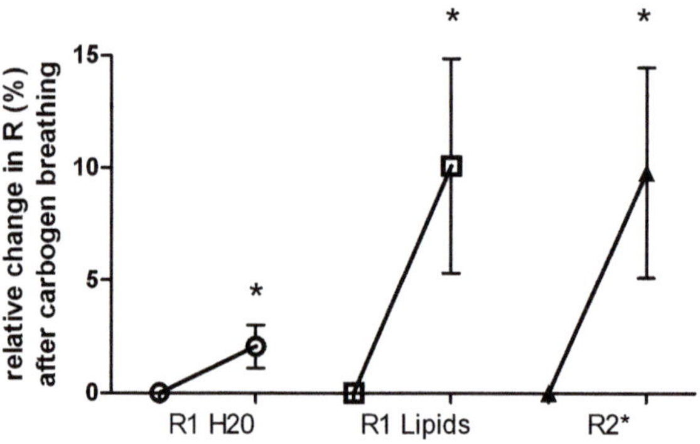

Fig. 38.1 Pooled results of MDA and NT2 tumors ($n = 12$). Relative changes over the whole tumors in relaxation times under air and carbogen breathing conditions for global R1 (water), R1 of lipids, and R2* [R = 1/T]. Note the higher change in R1 of lipids versus global R1 in response to carbogen breathing

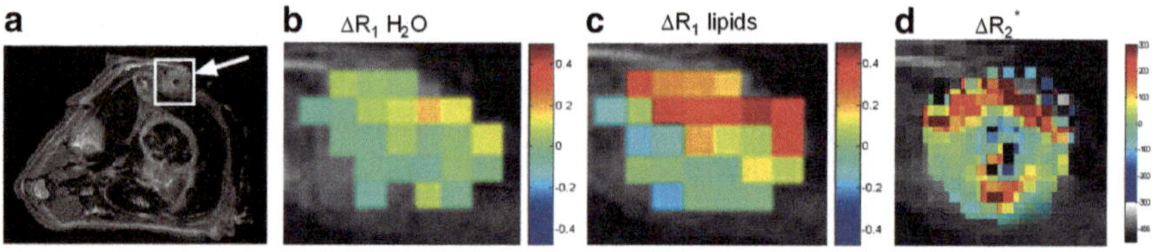

Fig. 38.2 Anatomical transversal MR image of a mouse with an MDA tumor (shown by the *arrow*) with the OxyLite tip (*black spot*) at the center of the tumor (**a**). Typical maps of changes in relaxation times in response to carbogen breathing (**b**, **c**, **d**) (Δ = R carbogen − R air). Color scales highlight the higher sensitivity of the MOBILE technique with a higher proportion of "hot" colors (**c**)

increased in the majority of the tumors (7/8) under carbogen breathing conditions. After carbogen breathing, MDA and NT2 tumors reached a pO_2 around 41.2 ± 2.3 mmHg and 87.7 ± 14.1 mmHg, respectively. Mean relative changes in relaxation rates with respect to the hyperoxic challenge were compared (pooled results for both tumor models) between R_1 H_2O, R_1 lipids and R_2*, with relative changes of 2.1 ± 1.0 % ($p < 0.05$; ΔR_1 H_2O), 10.1 ± 4.8 % ($p < 0.05$; ΔR_1 lipids), and 9.8 ± 4.6 % ($p < 0.05$; ΔR_2*) (Fig. 38.1). Although values are rather dispersed, an increase by a factor of 4.8 was achieved for R_1 lipids compared to R_1 H_2O. The larger dispersion of the data obtained using R_1 lipids compared to R_1 H_2O is explained by the fact that MOBILE is more sensitive to changes in oxygenation than R_1 H_2O and that the tumor is highly heterogeneous (Fig. 38.2): areas with a large change in pO_2 will present a larger change in R_1 lipids compared to R_1 H_2O, while areas with no change in pO_2 will have no change in R_1 (lipids or water). As tumors are heterogeneous (mixture of areas with response and areas without response to hyperoxic challenge), the effect of change in oxygenation leads to a larger distribution in the recorded values when using a method that is highly sensitive. The tip of the OxyLite™ probe could be localized on the MR images (Fig. 38.2a). Typical maps of

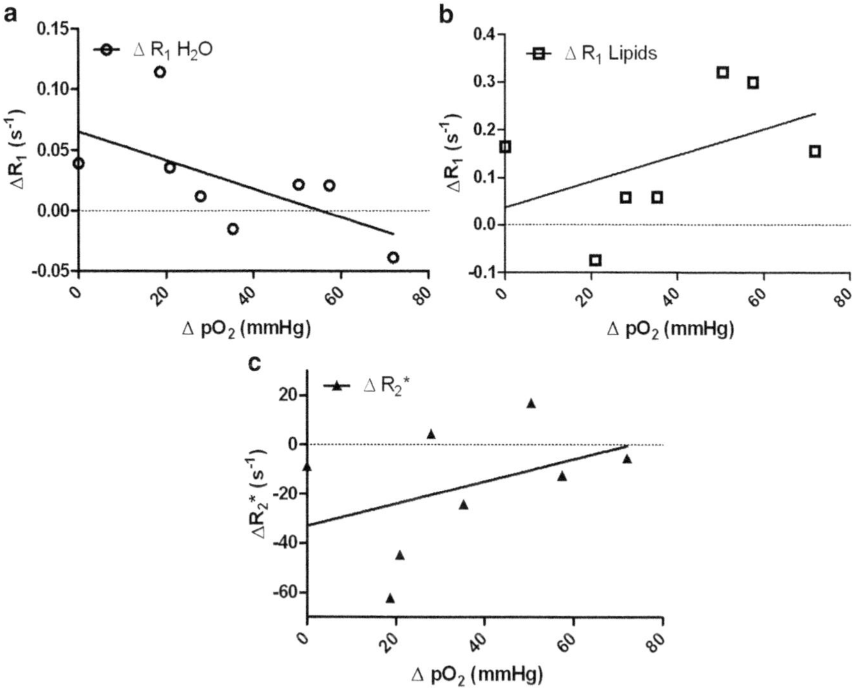

Fig. 38.3 Relation between the quantitative change in tumor pO_2 (assessed by OxyLite probes) and the corresponding changes in relaxation times: global R1 (**a**), R1 of lipids (**b**), and R2* (**c**). Note the positive trend for R1 of lipids

the relative difference in relaxation rates before and during the carbogen challenge were generated in order to compare the sensitivity of the R_1 of lipids (MOBILE), R_1 H_2O, and R_2* methods (Fig. 38.2b–d), indicating that ΔR_1 lipids are more sensitive (more red pixels) than ΔR_1 H_2O on the same tumor. It is possible to compare the quantitative values obtained by fluorescence quenching and the MR relaxation values obtained in the sensitive region of the probe (3*3 pixels with the exclusion of the central pixel, Fig. 38.3a–c), similarly to previously published comparisons between R_2* and pO_2 changes [17]. Some trends were found between ΔR_1 lipids and ΔpO_2, presenting a positive linear fit with a slope of 0.0027 ± 0.0023 ($r^2 = 0.22$, $p = 0.11$). In contrast, ΔR_1 H_2O presented a negative slope versus ΔpO_2 (-0.0012 ± 0.0006; $r^2 = 0.37$, $p = 0.28$), and ΔR_2* presented a positive slope (0.4527 ± 0.4120; $r^2 = 0.17$, $p = 0.31$), while this was expected to be negative (i.e., pO_2 is reported to be linked to $T_2^* = 1/R_2^*$). In a parallel study, we compared EPR oximetry and R_1 lipids and R_1 H_2O in the NT2 tumor model. The basal pO_2 was 8.4 ± 1.1 mmHg and reached 26.1 ± 6.1 mmHg after carbogen breathing. The correlation between R_1 H_2O and pO_2 was not significant ($p = 0.8611$, positive linear fit 0.000409 ± 0.002241; $r^2 = 0.005529$) (Fig. 38.4), while a positive linear significant correlation was found between R_1 lipids and pO_2 (0.01744 ± 0.00656; $r^2 = 0.5407$, $p = 0.0376$) (Fig. 38.4).

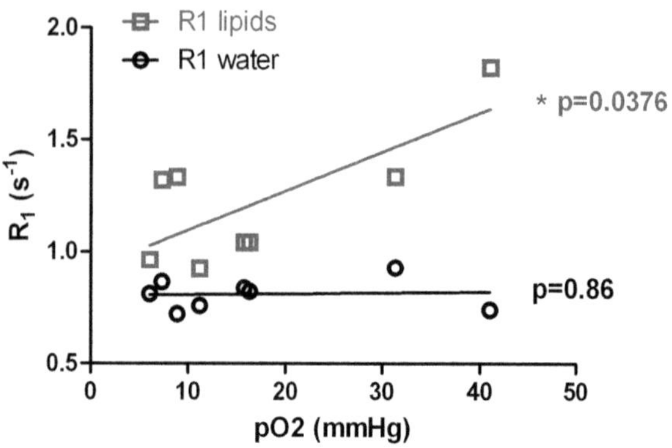

Fig. 38.4 Correlation between global pO_2 measures with global R1 values and lipids R1 values

Overall, these data show that MOBILE is sensitive to changes in tumor oxygenation. Correlations between data sets from MOBILE and direct O_2 measurements suggested that R_1 lipids were significantly correlated with pO_2 measured by EPR. Using OxyLite, we found a positive (but nonsignificant) fit between ΔR_1 lipids and ΔpO_2. Differences observed between EPR and OxyLite measurements can be explained by the difference in the volume sampled, as OxyLite interrogates a very small volume while EPR provides pO_2 measurements over the whole tumor.

Also, ΔR_2*, which was expected to be negatively correlated to pO_2, showed an apparently paradoxical positive fit versus ΔpO_2, thereby emphasizing the multiplicity and complexity of the response of R_2* to changes in oxygenation and perfusion that can be highly dependent on micro/macrovasculature functional parameters [28]. Therefore, information provided by the two relaxation parameters should be complementary. Also, oxygen-enriched gases inhalation was used as a technique to modulate tumor oxygenation. However, actual effects of such gases might vary from one tissue to the other, e.g., according to the presence or absence of CO_2 (i.e., for carbogen). This may explain the dispersion of the results, where tissue tumor heterogeneity is high. MOBILE presents the advantage of allowing longitudinal studies. Interestingly, the short acquisition time (1 min 20 s) of the MOBILE sequence should allow the follow-up of acute hypoxia in tumors in a quantitative and dynamic way, similarly to previous work using ^{19}F-relaxometry or EPR oximetry [29, 30].

38.4 Conclusion

In conclusion, these initial studies suggest that MOBILE has the potential to provide a noninvasive method for obtaining measurements of the variations in tumor oxygenation. Further studies on other tumor models are needed to evaluate if changes in R_1 lipids are quantitatively reflecting changes in pO_2.

Acknowledgments This study was supported by grants from the Belgian National Fund for Scientific Research (FNRS), the Joseph Maisin Foundation, the Saint-Luc Foundation, the "Actions de Recherches Concertées-Communauté Française de Belgique, grant no. ARC 09/14-020," and the "Pôle d'Attraction Interuniversitaire, PAI VI (P6/38)", and the 2011 ISMRM Seed Grant. B.F.J. and P.D.C. are Research Associates of the FNRS and O.F. is a Senior Research Associate of the FNRS.

References

1. Tatum JL, Kelloff GJ, Gillies RJ et al (2006) Hypoxia: importance in tumor biology, noninvasive measurement by imaging, and value of its measurement in the management of cancer therapy. Int J Radiat Biol 82(10):699–757
2. Wilson WR, Hay MP (2011) Targeting hypoxia in cancer therapy. Nat Rev Cancer 11(6): 393–410
3. Begg AC, Stewart FA, Vens C (2011) Strategies to improve radiotherapy with targeted drugs. Nat Rev Cancer 11(4):239–253
4. Rodrigues LM, Howe FA, Griffiths JR, Robinson SP (2004) Tumor R2* is a prognostic indicator of acute radiotherapeutic response in rodent tumors. J Magn Reson Imaging 19(4):482–488
5. Elas M, Bell R, Hleihel D et al (2008) Electron paramagnetic resonance oxygen image hypoxic fraction plus radiation dose strongly correlates with tumor cure in FSa fibrosarcomas. Int J Radiat Oncol Biol Phys 71(2):542–549
6. Robinson SP, Kalber TL, Howe FA et al (2005) Acute tumor response to ZD6126 assessed by intrinsic susceptibility magnetic resonance imaging. Neoplasia 7(5):466–474
7. Milosevic M, Fyles A, Hedley D, Hill R (2004) The human tumor microenvironment: invasive (needle) measurement of oxygen and interstitial fluid pressure. Semin Radiat Oncol 14(3): 249–258
8. Gallez B, Baudelet C, Jordan BF (2004) Assessment of tumor oxygenation by electron paramagnetic resonance: principles and applications. NMR Biomed 17(5):240–262
9. Yu JX, Cui W, Zhao D, Mason RP (2008) Non-invasive physiology and pharmacology using 19F magnetic resonance. In: Tressaud A, Haufe G (eds) Fluorine and health. Elsevier, Amsterdam, pp 197–276
10. Matsumoto S, Yasui H, Batra S et al (2009) Simultaneous imaging of tumor oxygenation and microvascular permeability using Overhauser enhanced MRI. Proc Natl Acad Sci USA 106(42):17898–17903
11. Sharma R (2011) Nitroimidazole radiopharmaceuticals in bioimaging: part I: synthesis and imaging applications. Curr Radiopharm 4(4):361–378
12. Pacheco-Torres J, López-Larrubia P, Ballesteros P, Cerdán S (2011) Imaging tumor hypoxia by magnetic resonance methods. NMR Biomed 24(1):1–16
13. Matsumoto K, Bernardo M, Subramanian S et al (2006) MR assessment of changes of tumor in response to hyperbaric oxygen treatment. Magn Reson Med 56(2):240–246
14. Ogawa S, Lee TM, Kay AR, Tank DW (1990) Brain magnetic resonance imaging with contrast dependent on blood oxygenation. Proc Natl Acad Sci USA 87(24):9868–9872
15. Robinson SP, Howe FA, Griffiths JR (1995) Noninvasive monitoring of carbogen-induced changes in tumor blood flow and oxygenation by functional magnetic resonance imaging. Int J Radiat Oncol Biol Phys 33(4):855–859
16. Baudelet C, Ansiaux R, Jordan BF, Havaux X, Macq B, Gallex B (2004) Physiological noise in murine solid tumours using T2*-weighted gradient-echo imaging: a marker of tumour acute hypoxia? Phys Med Biol 49(15):3389–3411
17. Baudelet C, Gallez B (2002) How does blood oxygen level-dependent (BOLD) contrast correlate with oxygen partial pressure (pO_2) inside tumors? Magn Reson Med 48(6):980–986

18. McPhail LD, Robinson SP (2010) Intrinsic susceptibility MR imaging of chemically induced rat mammary tumors: relationship to histologic assessment of hypoxia and fibrosis. Radiology 254(1):110–118
19. Howe FA, Robinson SP, McIntyre DJ, Stubbs M, Griffiths JR (2001) Issues in flow and oxygenation dependent contrast (FLOOD) imaging of tumours. NMR Biomed 14(7–8):497–506
20. O'Connor JP, Jackson A, Buonaccorsi GA et al (2007) Organ-specific effects of oxygen and carbogen gas inhalation on tissue longitudinal relaxation times. Magn Reson Med 58(3):490–496
21. O'Connor JP, Naish JH, Parker GJ et al (2009) Preliminary study of oxygen-enhanced longitudinal relaxation in MRI: a potential novel biomarker of oxygenation changes in solid tumors. Int J Radiat Oncol Biol Phys 75(4):1209–1215
22. Winter JD, Estrada M, Cheng HL (2011) Normal tissue quantitative T1 and T2* MRI relaxation time responses to hypercapnic and hyperoxic gases. Acad Radiol 18(9):1159–1167
23. Jordan BF, Magat J, Colliez F et al (2012) Mapping of oxygen by imaging lipids relaxation enhancement: a potential sensitive endogenous MRI contrast to map variations in tissue oxygenation. Magn Reson Med. doi:10.1002/mrm.24511 (in press)
24. Jordan BF, Cron GO, Gallez B (2009) Rapid monitoring of oxygenation by 19F magnetic resonance imaging: simultaneous comparison with fluorescence quenching. Magn Reson Med 61(3):634–638
25. Scheffler K, Henning J (2001) T1 quantification with inversion recovery TrueFISP. Magn Reson Med 45(4):720–723
26. Delikatny EJ, Chawla S, Leung DJ, Poptani H (2011) MR-visible lipids and the tumor microenvironment. NMR Biomed 24(6):592–611
27. McGrath DM, Naish JH, O'Connor JP et al (2008) Oxygen-induced changes in longitudinal relaxation times in skeletal muscle. Magn Reson Imaging 26(2):221–227
28. Kim KA, Park MS, Kim IS et al (2012) Quantitative evaluation of liver cirrhosis using T1 relaxation time with 3 tesla MRI before and after oxygen inhalation. J Magn Reson Imaging 36(2):405–410
29. Magat J, Jordan BF, Cron G, Gallez B (2010) Noninvasive mapping of spontaneous fluctuations in tumor oxygenation using 19F MRI. Med Phys 37(10):5434–5441
30. Matsumoto S, Yasui H, Mitchell JB, Krishna MC (2010) Imaging cycling tumor hypoxia. Cancer Res 70(24):10019–10023

Chapter 39
Primo Vascular System and Its Potential Role in Cancer Metastasis

Kyung A. Kang, Claudio Maldonado, Gustavo Perez-Aradia, Ping An, and Kwang-Sup Soh

Abstract The primo vascular system (PVS) is a newly found organ, which is distributed throughout the entire body. The system is composed of nodes storing many small cells and thin vessels branching out from the nodes. Inside the vessel there are multiple subvessels. The PVS is found in and on most organs, including the brain, and interestingly inside some lymph and blood vessels. The PVS is normally difficult to visualize due to its semitransparent optical property and its small size, which may be the main reason why it was not discovered until recently. The diameter of primo vessels (PVs) is in the range of 20–50 μm and the size of a primo node (PN), 100–1,000 μm. The outermost layer of the PVS is more porous than that of blood or lymph capillary vessels, and the nuclei of the PVS endothelial cells are rod shaped. Important PVS properties reported are: in the fluid inside the PVS, there are cells presenting stem cell markers CD133, Oct4, and Nanog, which may imply that this system has a role in regeneration. Another very important finding is its potential relevance to cancer. According to results from an animal study using xenografts of various cancer types (lung, ovarian, skin, gastric cancer, and leukemia), as the tumor grows, the PVS is formed in a high density in the vicinity of the tumor. In addition, it was shown that PVs connect the primary and secondary tumors and that

K.A. Kang (✉)
Department of Chemical Engineering, University of Louisville, Louisville, KY, USA
e-mail: kyung.kang@louisville.edu

C. Maldonado • G. Perez-Aradia
Department of Physiology and Biophysics, University of Louisville, Louisville, KY, USA

P. An • K.-S. Soh
Nano Primo Research Center, Advanced Institute of Convergence Technology, Seoul National University, Suwon, South Korea

S. Van Huffel et al. (eds.), *Oxygen Transport to Tissue XXXV*, Advances in Experimental Medicine and Biology 789, DOI 10.1007/978-1-4614-7411-1_39,

cancer cells were transported via the PVs in an active manner. In this report, we illustrated the formation of the PVS in breast cancer, and using the green fluorescent protein-expressing gastric cancer cell lines, we observed the cancer cell movement from the primary to the secondary sites during the cancer progression.

39.1 Introduction

39.1.1 Primo Vascular System

The primo vascular system (PVS) exists in most, if not all, mammalian organs, forming an extensive network throughout the entire body. The PVS consists of primo nodes (PNs), primo vessels (PVs) that branch out of PNs, transparent fluid inside the PNs and PVs (primo-fluid, P-fluid), and the primo-microcells (P-microcells) stored mainly in PNs. The PVS is very small: the diameter of PVs is in the range of 20–50 μm and the size of a PN is in the range of 100–1,000 μm. The PVS is optically semitransparent (Fig. 39.1). It is, therefore, not easy to observe the system without special training in microsurgery and microscopy, which probably explains why this system was not discovered until the 1960s by Bonghan Kim in North Korea [1, 2]. Shortly after Kim's reports in the 1960s, the PVS research was suddenly discontinued for unknown reasons. In the 2000s, the research was reinitiated by Kwang-Sup Soh in South Korea [3].

The PVS is distinctively different from the blood or lymphatic system. It does not express blood vessel (BV) biomarker CD31 or a lymphatic vessel (LV) biomarker LYVE-1 [4]. The external wall of the PV is porous and, therefore, when a dye (usually trypan blue) is applied to the area suspected with PVS, it gets into the PVs preferentially, allowing the observer to be able to differentiate PVs from BVs and LVs [5]. The cells of the sub-PV possess f-actin along the PV length, indicating the direction of the P-fluid [6]. The average flow speed of the P-fluid was reported

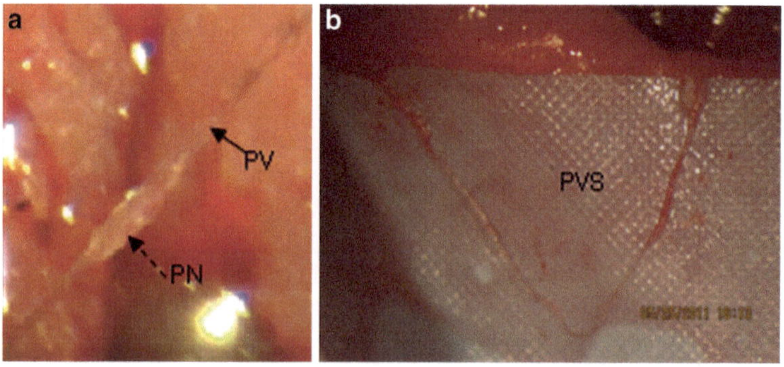

Fig. 39.1 (**a**) A typical set of a PN, and PVs connected to the PN, slightly lifted from the surface of the omentum of a rat, and (**b**) a long, isolated PVS from the rat omentum

to be 0.3 mm/s [6]. The nuclei of the endothelial cell in the sub-PV are rod shaped, and therefore, DAPI (dye specific to DNA or nuclei) is frequently used to confirm the PVS. Typically, the PN is packed with many small cells (P-microcell), and the cells are round and 1.0–2.5 μm in diameter [6]. The cells have small nuclei surrounded by a layer of cytoplasm and a trilaminar plasma membrane. The research team led by Soh [7, 8] confirmed that these small cells possess embryonic stem cell markers, Oct4, Nanog, and CD133. Kwon et al [9]. recently reported that the PVS also contains immune cells, enriched with granulocytes and histiocytes and chromaffin cells. Hong et al. measured the oxygen level of the P-fluid of the PVS formed in the murine melanoma [10] and in the mice embryos [11].

The PVS has been found in all organs in all animals that were studied, including intestine [12, 13], brain [14–16], heart [17], adipose tissue [18], and also within blood [19, 20] and lymphatic vessels [21, 22]. The number of animal species studied for the PVS, up to now, is 7 and the number of organs confirmed for the existence of PVS is 17 [23].

39.1.2 PVS and Cancer

The PVS in/on normal tissues or organs is usually not very densely populated, except for the adipose tissue. However, for four different cancer types (xenograft) in studies reported up to now, the PVS was found densely populated in proximity of the tumor.

The Soh team studied xenografts of human lung cancer NCI-H460 and human ovarian cancer SK-OV-3 [24] and observed dense PVS formed at the sites with the tumor (cancer PVS). The team then designed a unique experiment to elucidate the relationship between the cancer progression and the PVS formation. For the study, they electroporated quantum dots (QDs) into the cancer cells, and the QD-containing live cancer cells were then inoculated to nude mice. Because the QDs were inside these cells, the newly formed cells also possessed QDs and emitted fluorescence, enabling to track cancer cells by fluorescent microscopy. As the cancer progressed, secondary tumors were also formed and both LVs and PVs were generated from the primary tumor. When the fluorescence level was monitored around the area with the tumor, the fluorescence intensity from the cells passing through the PV was approximately twice that from the lymph vessel, indicating that cancer cells were transported through PVs significantly. The result from the fluorescence level emitted from PNs in the ovarian cancer showed the fluorescence at an even greater level than that from the lymph node.

In an attempt to study the origin of the PVS during the cancer progression, Heo and colleagues used cutaneous melanoma B16BL6 (nonfluorescing) into green fluorescence protein (GFP)-expressing host mice [25]. They also found denser PVS around the tumor, and the PVS formed on/in the tumor emitted green fluorescence, implying that it is initiated from the host, although it was not clear whether all PVS formed were from the host because the tumor-initiated PVS would not emit the fluorescence.

The Miller team [26] recently reported that, in a human lymphoma (U937) xenograft, PVS was formed at a higher density. The cells in the cancer PVS expressed CD68, CD45, and lysozyme and possessed the same immune phenotype of the cancer cell U937. The cells upregulated, *by hundred- to thousandfold*, human cancer stem cell-specific transcription factors (particularly CD68, CD45) and an upstream regulator of Nanog that maintains the pluripotent and undifferentiated state of stem cells.

Here, we report the result of PVS formation during the growth of breast cancer xenograft MDA-MB231. We also report the cancer cell tracking from the primary to the secondary tumor, utilizing the GFP-expressing MKN28 human gastric cancer cell line.

39.2 Methods

39.2.1 Observation of PVS in Normal Rat

Sprague Dawley rats were anesthetized with sodium pentobarbital (50 mg/kg i.p.). An incision down the *linea alba* was made to expose the small intestine and omentum. The intestines and omentum were sprayed with a 0.2 % trypan blue solution to identify PVs. Once a PV was identified by the uptake of blue dye, it was carefully dissected with the aid of microforceps, microvessel dilators and a microscope (OP-Mi 7PH Microscope; Carl Zeiss; Germany). Images were taken with a camera (PowerShot SD940 ISl; Canon; USA) mounted on the scope.

39.2.2 PVS of Breast Cancer Xenograft MDA-MB231

Nude mice were inoculated with human MDA-MB231 breasts cancer cells (~10^7 cells) in the right flank, and tumors were allowed to grow for 1–2 months. At the time of PVS observation, mice were euthanized with carbon dioxide and, immediately, a skin incision down the flank was performed to expose the tumor. The tumor was sprayed with a 0.2 % trypan blue solution to identify PVs. The identification and imaging of the PVS were performed using the same method described above.

39.2.3 PVS and GFP-Expressing Gastric Cancer

MKN12 human gastric cells were transfected with GFP-expressing plasmid using Lipofectamine LTX and PLUS reagent (Invitron; Monmouth, UK), following the manufacturer's instruction. The cells transfected with the GFP plasmid were then sorted with FACSAria (BD Biosciences; San Jose, CA, USA). Nude mice were injected with the GFP-expressing cancer cells (~10^7 cells) to the gastric wall, and tumors were allowed to grow for 1–2 months. At the time of the PVS observation, mice were anesthetized with Zoletil/Rompun intraperitoneal injection. Skin incision in the abdominal area was performed to expose the tumor. Fluorescence image was taken using fluorescence microscopy (MVX10; Olympus; Japan) and a monochrome CCD camera (DP30BW; Olympus; Japan) mounted on the scope.

39.3 Results and Discussion

39.3.1 Typical PVS in Normal Rat Model

Figure 39.1a shows a typical set of a PN and PVs connected to the PN, which was isolated from the surface of the omentum of a rat. PNs usually appear along the PV. The PVS on the surface of an organ can be isolated by tracking the PVs, as shown in Fig. 39.1b.

39.3.2 PVS of Breast Cancer Xenograft

Figure 39.2 demonstrates the PVS formed on the human breast cancer type MDA-MB231 xenograft in a nude mouse, lightly stained with trypan blue. As in the cases with the previously studied human lung, ovarian, lymphoma, and melanoma xenografts, the PVS was seen at a higher density than on the tissue without the tumor, which confirms that the PVS is densely formed in the vicinity of the cancerous tumor. This may also imply that some unknown but particular properties of cancer affect the proliferation of cancer PVS, potentially using the PVS for its growth and metastasis, although more extensive studies are required to verify this hypothesis. Assuming that the hypothesis is proven to be true, this property of the cancer PVS may be utilized for early detection of cancer.

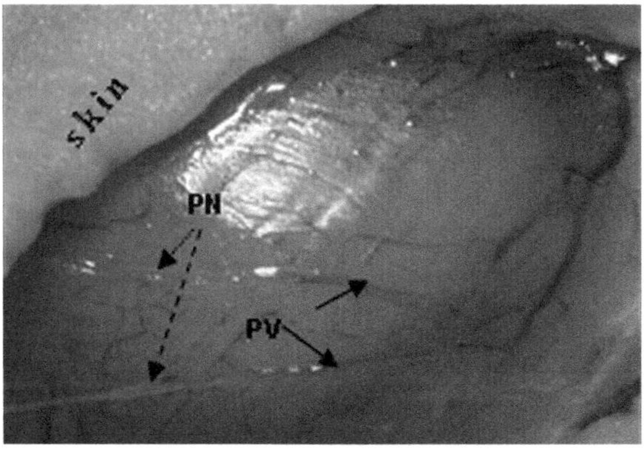

Fig. 39.2 PVs and PNs formed on/in human-to-nude mice xenografts of breast cancer (MDA-MB231), under the skin, using in situ trypan blue staining. *Solid lines*, PVs; *dashed lines*, PNs

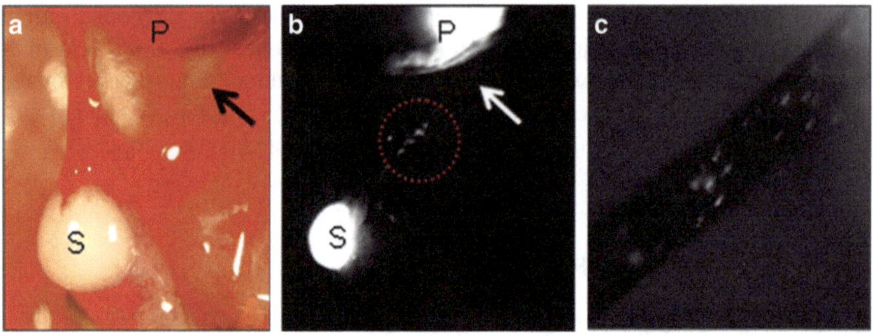

Fig. 39.3 Observation of the relationship between the secondary tumor formation and the PV formed between the primary and the secondary tumors, utilizing GFP-expressing MKN28 human gastric cancer cells. (**a**) Optical microscopy of a PV (indicated by a *black arrow*) connecting the primary (*P*) and the secondary (*S*) tumors. (**b**) Fluorescence microscopy of the same regions is shown in (**a**). The primary and the secondary tumors are fluorescing. In the PV, GFP-expressing cancer cells are present, implying that the cancer cells are being transported via the PV. (**c**) An enlarged image of the circular region of (**b**). GFP-expressing cancer cells inside the PV are more clearly shown

39.3.3 PVS as Cancer Cell Transport Conduit

In the previous studies on the roles of the PVS in cancer cell transport, Soh's group used QD-electroporated cancer cell lines of human lung cancer NCI-H460 and human ovarian cancer SK-OV-3, and the fluorescence generated from the QDs was traced assuming that they represented cancer cells. In this new study, to confirm the cancer cell transport via the cancer PV, GFP-expressing, human gastric cancer cell type MKN28 was xenografted in nude mice (nonfluorescing). As the primary tumor grew large, a secondary tumor was formed, as shown by optical microscopy in Fig. 39.3a. The figure describes the primary and secondary tumors and a PV that connects the two, which is indicated by the black arrow. Figure 39.3b shows an image obtained by fluorescence microscopy for the green fluorescence. Both the primary and secondary tumors were shown to be very bright and there were dots emitting bright green fluorescence from inside the cancer PV connecting the two (indicated by the white arrow), showing that the cancer cells are present inside the PV and are being transported via the PV. This cancer PV itself does not express the fluorescence, potentially indicating that the PV in the tumor is from the host, as in Heo's [24] study result. Figure 39.3c shows the enlarged image of the PV, circled in Fig. 39.3b. The presence of the cancer cells (fluorescence) in the PV is more clearly demonstrated.

39.4 Conclusions

Our study results with the breast cancer MDA-MB231 appeared to confirm that cancer (at least in the case of xenografts) induces dense PVS formation in close proximity to cancerous tumors, which suggests that a cancerous environment triggers cancer PVS formation. If it is a common signature of cancer, then it may be utilized for early detection of cancer.

The study results using GFP-expressing gastric cancer xenograft confirm that many cancer cells are present in and possibly transported via the cancer PVs, as in the previous study results by the Soh team, using QD-containing cancer cell. This result strongly implies that the cancer PVS may have a very important role in cancer metastasis. Future studies will be planned to elucidate the mechanism of the cancer PV formation, to prevent the cancer metastasis via the PVs.

The relationship between the cancer progression and the cancer PVS formation/ cancer cell transportation via the cancer PV is new but very important and deserves immediate multidisciplinary efforts for more complete understanding to develop novel tools for cancer prevention, diagnostic, and treatment.

References

1. Kim BH (1963) On the Kyungrak system. J Acad Med Sci DPR Korea 90:1–35
2. Kim BH (1965) The sanal theory. J Acad Med Sci DPR Korea 108:39–62
3. Soh KS (2012) A brief history of the Bong-Han theory and the primo vascular system. In: Soh KS, Kang KA, Harrison DH (eds) The primo vascular system: its role in cancer and regeneration. Springer, New York, pp 3–5
4. Yoo JS, Ayati MH, Kim HB, Zhang W, Soh KS (2010) Characterization of the primo-vascular system in the abdominal cavity of the lung cancer mouse model and its differences from the lymphatic system. PLoS One 5(4):e9940
5. Ogay V, Bae KH, Kim KW, Soh KS (2009) Comparison of the characteristic features of Bonghan ducts, blood and lymphatic capillaries. J Acupunct Meridian Stud 2(2):107–117
6. Sung B, Kim MS, Lee BC, Yoo JS, Lee SH, Kim YJ, Kim KW, Soh KS (2008) Measurement of flow speed in the channels of novel threadlike structures on the surfaces of mammalian organs. Naturwissenschaften 95(2):117–124
7. Ahn SH, Lee SW, Hwang SY, Kim JH, Sohn IC (2012) Primo-microcell in a primo node as a possible origin of adult stem cells. In: Soh KS, Kang KA, Harrison DH (eds) The primo vascular system: its role in cancer and regeneration. Springer, New York, pp 163–170
8. Ogay V, Soh KS (2012) Identification and characterization of small stem-like cells in the primo vascular system of adult animals. In: Soh KS, Kang KA, Harrison DH (eds) The primo vascular system: its role in cancer and regeneration. Springer, New York, pp 149–156
9. Kwon BS, Ha CM, Yu S, Lee BC, Ro JY, Hwang S (2012) Microscopic nodes and ducts inside lymphatics and on the surface of internal organs are rich in granulocytes and secretory granules. Cytokine 60(2):587–592
10. Hong M, Park SS, Do H, Jhon GJ, Suh M, Lee Y (2011) Primo vascular system of murine melanoma and heterogeneity of tissue oxygenation of the melanoma. J Acupunct Meridian Stud 4(3):e159–e163
11. Hong M, Park SS, Do H, Jhon GJ, Suh M, Lee Y (2012) Study of the primo-vascular system and location-dependent oxygen levels for a mouse embryo. J Nanosci Nanotechnol 12(7): 5168–5172
12. Jiang X, Lee BC, Choi C, Baik KY, Soh KS (2004) Tubular structure of intravascular threadlike structures from rats and rabbits. J Korean Phys Soc 44:1602–1604
13. Shin HS, Johng HM, Lee BC, Cho SI, Soh KS, Baik KY, Yoo JS, Soh KS (2005) Feulgen reaction study of novel threadlike structures (Bonghan ducts) on the surface of mammalian organs. Anat Rec B New Anat 284(1):35–40

14. Dai JX, Lee BC, An P, Su Z, Qu R, Eom KH, Soh KS (2011) In vivo in situ staining of the primo vascular system in ventricles and subarachnoid space of brain by injecting trypan blue into the lateral ventricle. Neural Regen Res 6(28):2171–2175
15. Nam MH, Lim JK, Choi SH, Kim SC, Soh KS (2012) A primo vascular system underneath the superior sagittal sinus in the brain of a rabbit. J Acupunct Meridian Stud 5(5):210–217
16. Lee HS, Park WH, Je A-R, Kweon HS, Lee BC (2012) Evidence for novel structures (primo vessels and primo nodes) floating in the venous sinuses of rat brains. Neurosci Lett 522(2):98–102
17. Lee BC, Kim HB, Sung B, Kim KW, Sohn J, Son B, Chang B-J, Soh KS (2012) Structure of the sinus in the primo vessel inside the bovine cardiac chambers. In: Soh KS, Kang KA, Harrison DH (eds) The primo vascular system: its role in cancer and regeneration. Springer, New York, pp 57–62
18. Lee BC, Kim KW, Soh KS (2009) Visualizing the network of bonghan ducts in the omentum and peritoneum by using trypan blue. J Acupunct Meridian Stud 2(1):66–70
19. Baik KY, Lee J, Lee BC, Johng HM, Nam TJ, Sung B, Cho S, Soh KS (2005) Acupuncture meridian and intravascular bonghan duct. Key Eng Mater 277:125–129
20. Yoo JS, Kim MS, Ogay V, Soh KS (2008) In vivo visualization of bonghan ducts inside blood vessels of mice by using an Alcian blue staining method. Indian J Exp Biol 45(5):336–339
21. Lee BC, Soh KS (2008) Contrast-enhancing optical method to observe a bonghan duct floating inside a lymph vessel of a rabbit. Lymphology 41(4):178–185
22. Johng HM, Yoo JS, Yoon TJ, Shin HS, Lee BC, Lee C, Lee JK, Soh KS (2007) Use of magnetic nanoparticles to visualize threadlike structures inside lymphatic vessels of rats. Evid Based Complement Altern Med 4(1):77–82
23. Soh KS (2012) Current state of research on the primo vascular system. In: Soh KS, Kang KA, Harrison DH (eds) The primo vascular system: its role in cancer and regeneration. Springer, New York, pp 25–40
24. Yoo JS, Kim HB, Won N, Bang J, Kim S, Ahn S, Lee BC, Ahn S, Soh KS (2011) Evidence for an additional metastatic route: in vivo imaging of cancer cells in the primo-vascular system around tumors and organs. Mol Imaging Biol 13(3):471–480
25. Heo C, Hong MY, Jo A, Lee YH, Suh M (2011) Study of the primo vascular system utilizing a melanoma tumor model in a green fluorescence protein expressing mouse. J Acupunct Meridian Stud 4(3):e198–e202
26. Islam MA, Thomas SD, Slone S, Alatassi H, Miller DM (2013) Tumor-associated primo vascular system is derived from xenograft, not host. Exp Mol Pathol 94(1):84–90

Part V
Cell Metabolism

Chapter 40
Pancreaticoduodenectomy Using Perioperative Zymogen Protein C to Help Prevent Blood Clotting: A Trilogy on Increased Patient Safety

Duane F. Bruley, Richard D. Schulick, and Michael B. Streiff

Abstract The blood clotting mechanism is a very important and complex physiologic process. Blood flow must be continuous through the blood vessels to provide essential oxygen and nutrients to the cells of the body. Dr. Melvin H. Knisely (Honorary First President of ISOTT, 1973) named and pioneered research in blood sludging and clotting which led to his nomination for the Nobel Prize by Dr. August Krogh in 1948. Abnormal clotting is a pathological state that can inhibit and prevent normal blood flow, leading to reduced oxygen transport to tissue from the microcirculation. It can result in the death of cells and tissues, including entire organs as well as the patient. Blood clotting and sludging are common occurrences during and after invasive surgery; thus, it is imperative to find safe procedures to reduce or prevent these deadly phenomena. All anticoagulants used today, for clot prevention and dissolution, can cause excessive bleeding that can lead to enormous medical expense to provide control, otherwise causing patient death. Protein C is a natural protein and is the pivotal anticoagulant in the blood. Due to the mechanism of converting the zymogen protein C (ZPC) to active protein C (APC), only when and where it is needed, and their respective half-lives in the body, the natural anticoagulant, antithrombotic characteristics of APC can be utilized without causing bleeds.

D.F. Bruley (✉)
Synthesizer, Inc., 2773 Westminster Road, Ellicott City, MD 21043, USA
e-mail: bruley33@verizon.net

R.D. Schulick
Department of Surgery, University of Colorado, Aurora, CO 80045, USA

M.B. Streiff
Anticoagulation Management Service, Division of Hematology, Department of Medicine, Johns Hopkins Medical Institutions, Baltimore, MD 21205, USA

S. Van Huffel et al. (eds.), *Oxygen Transport to Tissue XXXV*, Advances in Experimental Medicine and Biology 789, DOI 10.1007/978-1-4614-7411-1_40,

40.1 Preamble

This is the final chapter of the trilogy that demonstrates the use of perioperative zymogen protein C to achieve safer patient results in invasive surgery. The same patient has experienced three surgical procedures: (1) total hip replacement, (2) life-threatening internal bleed, and (3) pancreaticoduodenectomy (PD) for pancreatic cancer. These procedures occurred over a period of several years without any complications due to clotting or bleeding. The patient is at very high risk for clotting problems because of age (77 years), previous venous thromboembolism (VTE), protein C deficiency, factor II gene mutation, severe surgery, and pancreatic adenocarcinoma. The percentage chance of VTE occurring increases significantly with the number of risk factors involved. In this case, without appropriate intervention, it was predicted that the patient had a 100 % risk of having a VTE.

The result is especially meaningful since there was no clotting or out-of-the-ordinary bleeding, even though the PD procedure is one of the most invasive of abdominal surgeries. The management of congenital and acquired thrombophilia with multimodality antithrombotic therapy is recommended for most surgeries. It is strongly suggested that ZPC be considered, along with standard practice anticoagulation, for any patient with protein C deficiency. It remains to be seen if all invasive surgeries would be safer with the use of ZPC.

40.2 Introduction

This chapter is a continuation of the application of the hypothesis for increased patient safety using ZPC for invasive medical procedures. Although we have proposed this hypothesis since the early 1980s, the first formal presentation of this strategy was delivered at the 2007 ISOTT meeting held in Uppsala, Sweden. The resulting paper first appeared in the Springer Publishing Company Experimental Medicine and Biology series, Oxygen Transport to Tissue [1]. This hypothesis was based on the attributes of the blood anticoagulant ZPC. Two important characteristics result in its action locally rather than globally in the body. First, the mechanism for ZPC to activated protein C (APC) occurs only where and when the APC is needed. Second, the half-life of ZPC in the body is approximately 8–10 h, and the half-life of APC is on the order of minutes. These two attributes allow local anticoagulant activity without exacerbating bleeding. Therefore, we defined ZPC as the "silver bullet" of anticoagulation [2].

This study represents the third chapter in a trio of clinical testing of the hypothesis. The same patient experienced all three of the major surgeries using perioperative ZPC. The first surgery was a total hip replacement (12 Nov 2007). The second was for the repair of a life-threatening internal bleed at the site where a flat polyp was removed from a location near the cecum (15 Sept 2008). The bleed was partially due to the administration of heparin and warfarin and began as the patient, who was on lifetime warfarin, was progressing toward an INR of 2.0 [3, 4]. This chapter involving the PD surgery procedure for pancreatic cancer utilizing perioperative ZPC (1 Oct 2010) is the final chapter of the trilogy. These cases illustrate the importance of controlling abnormal functions (pathology) leading to tissue deprivation of oxygen, which is of particular interest to the objectives of the International Society on Oxygen Transport to Tissue (ISOTT).

ISOTT research focuses on all processes of normal and pathological oxygen transport within the blood and tissue [5, 6]. Foremost of interest is blood hemostasis to ensure optimal tissue oxygenation. This includes the coagulation cascade, particularly the significance of natural anticoagulants and substitutes for them in cases of pathology. The phenomenon of "blood sludging" (agglutination) was first observed by Dr. Melvin H. Knisely via his quartz rod crystal optical technology [7]. Our initial studies included mathematical modelling of oxygen transport in the microcirculation along with polarographic microelectrode oxygen sensing [8–10]. This work was complemented with studies on anti-adhesive drugs to enhance tissue oxygenation [11].

Historically, protein C (PC) was not so named until Stenflo isolated the protein from bovine plasma and labeled it protein C because it was the third peak in the chromatograph elution [12]. Originally, activated protein C (APC) was coined autoprothrombin IIA (APIIA) because it was believed to be derived from prothrombin [13]. Twelve years later, it was shown that APIIA's precursor was not only distinct from prothrombin, but it inhibited the factor Xa-mediated activation of prothrombin as well as facilitating fibrinolysis [14]. Shortly after Stenflo isolated the bovine PC, Seegers verified it to be identical to APIIA [15]. Three years later, Kisiel was able to purify human PC [16].

Zymogen protein C (ZPC) is the pivotal anticoagulant and antithrombotic in the human blood coagulation cascade [17]. PC is a glycoprotein with a molecular weight of 62,000 Da. Human ZPC is synthesized in the liver as a single-chain precursor and circulates in the blood primarily as a two-chain inactive zymogen until it is activated by proteolytic cleavage. The protein is a serine protease that requires vitamin K for normal biosynthesis [13]. It is a member of the vitamin K-dependent (VKD) family also consisting of coagulation proteins such as factors VII, IX, proteins S and Z, and prothrombin.

ZPC is a trace protein in human blood with a concentration of 4 μg/mL [12, 18, 19]. Serious problems can occur when the PC level in the blood is lowered. For example, patients deficient in PC are at risk of deep vein thrombosis (DVT) [20] and other clotting complications, resulting in tissue oxygen deprivation, some of which can be life threatening. When these blood clots break away from the surface of the vein and enter the bloodstream, they will induce strokes, heart attacks, and pulmonary embolisms. Venous thrombosis is one of the most frequent complications in medical patients [21]. It is the most common cause of death in patients undergoing major orthopaedic operations. In the United States, it has been estimated that 300,000 hospitalizations and 50,000 deaths occur [21], and this amounts to millions and possibly billions of dollars in medical expenses annually.

Although very complex, three main pathways are involved in regulating coagulation [22]. One pathway that utilizes heparin-like molecules and antithrombin III results in the inhibition of coagulation proteases. A second pathway, referenced as lipoprotein-associated coagulation inhibitor, or extrinsic pathway inhibitor, blocks the activity of factor VIIa-tissue factor complex [23]. The third and most important pathway involves APC which neutralizes factors Va and VIIIa [17]. These major pathways function together to inhibit both the proteases and regulatory proteins (cofactors) of the clotting system. It has been shown through clinical studies that antithrombin III, protein C, and factor S deficiencies all exhibit thrombotic complications [20, 24].

Although inefficient, PC can be activated by thrombin alone. This activation process can be enhanced by a factor of at least 1000 via a complex between thrombin and a membrane protein called thrombomodulin on the surface of endothelial cells [17, 25]. The activated PC is a potent serine protease that regulates blood coagulation by forming a complex with protein S (PS) on both endothelial and platelet surfaces. This deactivates factors Va and VIIIa, thus preventing generation of the enzymes factors Xa and thrombin [26, 27].

Cofactor PS circulates in the blood as a free agent and in complex with C4bBP which is a regulatory protein of the complement system. The PS-C4bBP complex is not functional as a cofactor for activated PC in factor Va inactivation and therefore downregulates the effectiveness of PC in the inactivation of Va and VIIIa. Va is required to produce thrombin. Once thrombin is produced, it activates factor I (fibrinogen) to form fibrin, which synthesizes a soft clot. VIIIa is then required to solidify the clot. So by inactivating Va and VIIIa, you downregulate the clotting process.

At present time, heparin and warfarin are used to treat ZPC deficiency and other hypercoagulable conditions. The disadvantage of these drugs is that both have dangerous side effects. Excessive internal bleeding is a major problem that can result from their use, possibly causing a stroke or major organ failure. Additionally, warfarin-induced skin necrosis and heparin-induced skin necrosis can lead to amputation of extremities and death. Also, pregnant women cannot use warfarin.

Previous animal testing [28–31] and clinical trials indicate that PC is an effective anticoagulant/antithrombotic for many medical indications without harmful side effects. This unique feature is a function of the mechanistic behavior of PC in the body. Differing from all other anticoagulants, PC circulates the blood in an inactive form and is activated only at the site where it is needed and at the time when it is needed. Also, ZPC can be used at greater than normal blood concentrations without bleeding complications.

PC has many clinical applications. Not only can it be used to treat genetically deficient patients, but it can also be used to treat septic shock [32], hip and knee replacement patients, warfarin-induced skin necrosis patients, heparin-induced thrombocytopenia, patients doing fibrinolytic therapy, and patients undergoing angioplasty or suffering from unstable angina, etc. [33]. Additionally, research has shown that the use of safe anticoagulants could lower the rate of strokes in the USA from 80,000 to 40,000 per year, reduce patient complications, and save the medical industry an estimated $600 million per year [34].

ZPC concentrate has been shown to be successful for the prevention and treatment of thrombosis in individuals with inherited and acquired PC deficiency and to avoid the problems associated with fresh and frozen plasma administration [35–38]. When one considers that PC is the only known anticoagulant/antithrombotic without bleeding side effects and 1 in 300 people are hereditary PC deficient [21], it is easy to see the enormous benefit of having inexpensive PC available to medical patients. Additionally, there are innumerable patients with acquired PC deficiency.

40.3 Experimental Medical Procedure

The patient experienced painless obstructive jaundice and was found to have a mass in the periampullary area. The tumor biopsy revealed adenocarcinoma. As a result, it was proposed that the mass be surgically removed. The procedure recommended was pancreaticoduodenectomy, most often referred to as the Whipple surgery, named after Dr. Allen O. Whipple, who was the first American to perform it and reported it in 1935 [39]. This operation was actually first performed by Dr. Walther Kausch of Germany in 1909 and reported in 1912 [40]. This is a very extensive operation that involves more than just removing parts of the pancreas. Because the pancreas is anatomically connected to other organs and ducts, multiple organs are involved in the operation. The surgery typically removes the head of the pancreas, the duodenum, the common bile duct, the gall bladder, and often, part of the stomach.

The surgical procedure was complicated by the fact that the patient had multiple risk factors for a VTE, including the patient's age (77 years), pancreatic cancer, major surgery, previous VTE, protein C deficiency, and factor II mutation. This combination of risk factors results in a markedly increased risk of VTE formation during and after surgery. Evidence shows that patients with five or more risk factors approach a 100 % chance of developing VTE.

To prevent blood clotting for the management of congenital and acquired thrombophilia, a multimodality, antithrombotic therapy was used [41]. The patient was on lifetime warfarin, which was stopped 5 days before surgery and replaced with Lovenox until 24 h before surgery. Six hours prior to surgery, PC concentrate at 50 μ/kg IV was administered every 6 h until after surgery, then every 8 h for 24 h, and then every 12 h until post-op day 8. Two days after surgery, warfarin was administered orally, and heparin was administered 5,000 units SQ every 8 h until post-op day 5, then Lovenox 10 units/kg/h IV to aPTT 50–65 s. The Lovenox was discontinued when the INR reached a value of 2.0. Thus, the VTE prophylaxis consisted of pre- and postoperative administration of ZPC, warfarin, and heparin. However, only ZPC was administered in the 2+ days encompassing the surgery.

40.4 Outcome

According to medical statistical data, this patient had a probability of experiencing a VTE that approached 100 %, in particular because of the radical invasiveness of this surgery and the multiple risk factors present. However, the patient had no thrombotic events and no bleeding complications throughout the entire procedure and recovery. This demonstrates the value of multimodality VTE prophylaxis with ZPC as the bridge. The patient moved on to warfarin therapy and continues to function normally without incidence.

40.5 Production Research Note

As presented in this trilogy of chapters, it would be beneficial to future surgical patients as well as the many hereditary and acquired protein C-deficient patients to develop innovative upstream and downstream bioprocessing strategies for the low-cost, high-volume production of ZPC. This would make ZPC available to a wider patient population.

The two existing protein C products (Ceprotin, Baxter International; Xigris, Eli Lilly) are so expensive that they are rarely used even in cases where patient survival is in question. If cost were not an issue, the zymogen could be used for prophylactic treatment of PC deficiency, as well as other disease states or in standard medical and surgical procedures. The significant examples of perioperative procedures utilizing ZPC to prevent clotting without exacerbating bleeding as presented in this trilogy support the need to produce inexpensive ZPC.

There is ongoing research toward achieving this goal. The three sources of PC that are available from upstream bioprocessing include rDNA cell culture technology [42], blood plasma [43–46], and transgenic animals [47, 48]. Our research continues to optimize the downstream processing for these raw materials via immobilized metal affinity chromatography (IMAC) [49–58]. The correct combination of ion exchange, IMAC, and absorption and elution buffers are being investigated for the optimal bio-downstream processing.

40.6 Conclusion

The three chapters of this trilogy establish the need for multimodality VTE prophylaxis utilizing ZPC as a bridge during invasive medical procedures. This is an important indicator of the need to produce a low-cost ZPC for many invasive medical procedures and utilization by heterozygote protein C-deficient patients. The ultimate goal would be either to produce this low-cost ZPC product or to design a substitute that mimics ZPC's "silver bullet" ability to act as an anticoagulant without increasing bleeding [2].

Acknowledgments The authors would like to express their appreciation to E. Eileen Thiessen for the preparation of the PowerPoint slide presentation and the production of this chapter.

References

1. Bruley DF (2009) Zymogen protein C concentrate for safer heterozygote surgery, "I am a guinea pig!". Adv Exp Med Biol 645:115–121
2. Bruley DF, Streiff MB (2013) Nature's "silver bullet" for anticoagulation: mechanism of zymogen protein C to activated protein C. Adv Exp Med Biol 765:15–21
3. Bruley DF, Mears SC, Streiff MB (2010) Safer surgery using zymogen protein C concentrate. Adv Exp Med Biol 662:439–445
4. Bruley DF, Jagannath SB, Streiff MB (2011) Zymogen protein C to prevent clotting without bleeding during invasive medical procedures. Adv Exp Med Biol 701:91–97
5. Bruley DF (2012) The history of ISOTT. Adv Exp Med Biol 737:1–9
6. Bruley DF (2008) ISOTT: roots, founding and beyond. Adv Exp Med Biol 614:1–8
7. Goro FW (1948) Blood sludge. Life Mag 24(22):49–59
8. Thews G (1960) Oxygen diffusion in the brain. A contribution to the question of the oxygen supply of the organs. Pflugers Arch 271:197–226
9. Reneau DD, Bruley DF, Knisely MH (1967) A mathematical simulation of oxygen release, diffusion and consumption in the capillaries and tissue of the human brain. In: Chemical engineering in medicine and biology. Plenum Press, New York, pp 135–241
10. Artigue RS, Bruley DF (1984) The transport of oxygen, glucose, carbon dioxide and lactic acid in the human brain: mathematical models. Adv Exp Med Biol 159:1–4
11. Bicher HI, Bruley DF, Knisely MH (1973) Anti-adhesive drugs and tissue oxygenation. Adv Exp Med Biol 37:657–667
12. Stenflo J (1976) A new vitamin K-dependent protein. J Biol Chem 251(2):355–363
13. Mammen EF, Thomas WR, Seegers WH (1960) Activation of purified prothrombin to autoprothrombin II (platelet cofactors II or autoprothrombin IIA). Thromb Diath Haemorrh 5:218–249
14. Marcianik E (1972) Inhibitor of human blood coagulation elicited by thrombin. J Lab Clin Med 79(6):924–934
15. Seegers WH, Novoa E, Henry RL, Hassouna HI (1976) Relationship of "new" vitamin K-dependent protein C and "old" autoprothrombin II-a. Thromb Res 8(5):543–552
16. Kisiel W (1979) Human plasma protein C: isolation, characterization and mechanism of activation by a thrombin. J Clin Invest 64(3):761–769
17. Esmon CT (1989) The roles of protein C and thrombomodulin in the regulation of blood coagulation. J Biol Chem 264(9):4743–4746
18. Fernlund P, Stenflo J (1982) Amino acid sequence of the light chain of bovine protein C. J Biol Chem 257(20):12170–12179
19. Kisiel W, Carnfield WM, Ericsson LH, Davie EW (1977) Anticoagulant properties of bovine plasma protein C following activation by thrombin. Biochemistry 16(26):5824–5831
20. Clouse LH, Comp PC (1986) The regulation of hemostasis: the protein C system. N Engl J Med 314(20):1298–1304
21. Bertina RM (1988) Protein C and related proteins: biochemical and clinical aspects. (Contemporary issues in haemostasis and thrombosis). Churchill Livingstone, New York
22. Esmon CT (1990) Regulation of coagulation: the nature of the problem. In: Bruley DF, Drohan WN (eds) Protein C and related coagulants. Gulf, Houston, p 3
23. Rapaport SI (1989) Inhibition of factor VIIa/tissue factor-induced blood coagulation: with particular emphasis upon a factor Xa-dependent inhibitory mechanism. Blood 73(2):359–365
24. Rosenberg R, Marcum JA (1985) Heparin-like molecules and thrombotic disease. ASAIO J 8:215

25. Zushi M, Gomi K, Yamamoto S, Maruyama I, Hayashi T, Suzuki K (1989) The last three consecutive epidermal growth factor-like structures of human thrombomodulin comprise the minimum functional domain for protein C-activating cofactor activity and anticoagulant activity. J Biol Chem 264(18):10351–10353
26. Walker FJ, Sexton PW, Esmon CT (1979) The inhibition of blood coagulation by activated protein C through the selective inactivation of activated factor V. Biochim Biophys Acta 571(2):333–342
27. Fulcher CA, Gardiner JE, Griffin JH, Zimmerman TS (1984) Proteolytic inactivation of human factor VIII procoagulant protein by activated human protein C and its analogy with factor V. Blood 63(2):486–489
28. Colucci M, Stassen JM, Collen D (1984) Influence of protein C activation on blood coagulation and fibrinolysis in squirrel monkey. J Clin Invest 74(1):200–204
29. Comp P, Esmon CT (1981) Generation of fibrinolytic activity by infusion of activated protein C into dogs. J Clin Invest 68(5):1221–1228
30. Gruber A, Griffin JH, Harker LA, Hanson SR (1989) Inhibition of platelet-dependent thrombosis formation by human activated protein C in a primate model. Blood 73(3):639–642
31. Okijima K, Koga S, Kaji M et al (1990) Effect of protein C and activated protein C on coagulation and fibrinolysis on normal human subjects. Thromb Haemost 63(1):48–53
32. Sharma GR, Gerlitz B, Berg DT et al (2008) Activated protein C modulates chemokine response and tissue injury in experimental sepsis. Adv Exp Med Biol 614:83–91
33. Bruley DF, Drohan WN (1990) Protein C and related anticoagulants, Advances in applied biotechnology series. Gulf Publishing Company (Portfolio Publishing Company, The Woodlands
34. Winslow R (1995) Increased use of blood-thinning drugs could prevent 40,000 strokes a year. Wall St J
35. Vukovich T, Auerger K, Weil J, Engelmann H, Knöbl P, Hadorn HB (1988) Replacement therapy for a homozygous protein C deficiency-state using a concentrate of human protein C and S. Br J Haematol 70(4):435–440
36. Dreyfus M, Magny JF, Bridey F et al (1991) Treatment of homologous protein C deficiency and neonatal purpura fulminans with a purified protein C concentrate. N Engl J Med 325(22):1565–1568
37. Manco-Johnson M, Nuss R (1992) Protein C concentrate prevents peripartum thrombosis. Am J Hematol 40(1):69–70
38. Marlar RA, Montgomery RR, Broakman AW (1989) Diagnosis and treatment of homozygous protein C deficiency. Report of the working party on homozygous protein C deficiency of the subcommittee on protein C and protein S, international committee on thrombosis and haemostasis. J Pediatr 114(4 Pt 1):528–534
39. Harvard Health Letter (2009) The Whipple procedure: better outcomes for pancreatic cancer surgery. www.health.harvard.edu Apr 2009
40. Whiple AO, Parsons WB, Mullins CR (1935) Treatment of carcinoma of the ampulla of vater. Ann Surg 102(4):763–779
41. Streiff MB (2011) "In the Eye of a Thrombotic Storm": management of congenital and acquired thrombophilia with multimodality anti-thrombotic therapy. ISOTT Meeting 2011, Georgetown University, Washington, DC
42. Ganesh RS,Grinnell BW et al (2008) Activated protein C modulates chemokine response and tissue injury in experimental sepsis. Adv Exp Med Biol 614:83–92
43. Wu H, Bruley DF, Kang KA (1998) Protein C separation from human blood plasma Cohn fraction IV-1 using immobilized metal affinity chromatography. Adv Exp Med Biol 454:697–704
44. Wu H, Bruley DF (1999) Homologous blood protein separation using immobilized metal affinity chromatography: protein C separation from prothrombin with application to the separation of factor IX and prothrombin. Biotechnol Prog 15(5):928–931
45. Wu H, Bruley DF (2002) Chelator, metal ion and buffer studies for protein C separation. Comp Biochem Physiol A Mol Integr Physiol 132(1):213–220
46. Tadepalli SS, Bruley DF, Kang KA et al (1997) Immobilized metal affinity chromatography process identification and scale-up for protein C production. Adv Exp Med Biol 428:31–43

47. Velander WH, Johnson JL, Page RL et al (1992) High-level expression of a heterologous protein in the milk of transgenic swine using the cDNA encoding human protein C. Proc Natl Acad Sci USA 89(24):12003–12007
48. Drohan WH, Wilkins TD, Lattime E et al (1994) A scalable method for the purification of recombinant human protein C from the milk of transgenic swine. In: Advance bioprocess engineering. Kluwer Academic Publishers, Dordrecht, pp 501–507
49. Porath J (1992) Immobilized metal ion affinity chromatography. Protein Expr Purif 3(4):263–281
50. Porath J, Carlsson J, Olsson I, Belfrage G (1975) Metal chelate affinity chromatography, a new approach to protein fractionation. Nature 258(5536):598–599
51. Andersson L (1991) Recognition of phosphate groups by immobilized aluminum (III) ions. J Chromatogr A 539(2):327–334
52. Andersson L, Porath J (1986) Isolation of phosphoproteins by immobilized metal (Fe^{3+}) affinity chromatography. Anal Biochem 154(1):250–254
53. Ramadan N, Porath J (1985) Fe3+– hydroxamate as immobilized metal affinity-adsorbent for protein chromatography. J Chromatogr 321(1):93–104
54. Sulkowski E (1988) Immobilized metal ion affinity of proteins on IDA-Fe3+. Makromol Chem Macromol Symp 17(1):335–348
55. Wong JW, Albright RL, Wang NH (1991) Immobilized metal ion affinity chromatography (IMAC) chemistry and bioseparations applications. Sep Purif Rev 20(1):49–106
56. Winzerling JJ, Patrick B, Porath J (1992) How to use immobilized metal ion affinity chromatograph. Methods 4(1):4–13
57. Thiessen EE, Bruley DF (2003) Theoretical studies of IMAC interfacial phenomena for the purification of protein C. Adv Exp Med Biol 540:183
58. Lee J, Thiessen EE, Bruley DF (2005) Anticoagulant/anti-inflammatory/antithrombotic protein C production-metal ion/protein interfacial interaction in immobilized metal affinity chromatography. Adv Exp Med Biol 566:381–387

Chapter 41
Inhibition of Mammalian Target of Rapamycin Induces Renal Mitochondrial Uncoupling in Rats

Ebba Sivertsson and Malou Friederich-Persson

Abstract The mechanisms underlying diabetic nephropathy are not fully understood. However, recent research indicates mitochondria dysfunction as a contributing factor. Mammalian target of rapamycin (mTOR) is a known regulator of mitochondria function and could therefore also be involved in the development of diabetic nephropathy. The present study investigates the role of mTOR for controlling the function of mitochondria isolated from normal and diabetic rat kidneys. Control and streptozotocin-induced diabetic rats were treated with the mTOR inhibitor rapamycin (0.2 mg/day) by oral gavage for 14 days, after which mitochondria function was investigated using high-resolution respirometry. Mitochondrial uncoupling was defined as increased oxygen usage unrelated to ATP production. mTOR inhibition induced mitochondria uncoupling in control rats, but did not affect the already occurring uncoupling in kidney mitochondria from diabetic animals. Inhibition of mTOR using rapamycin induces mitochondria uncoupling in control rats, suggesting a role of mTOR as a moderator of mitochondria efficiency. No effect of mTOR inhibition was observed in mitochondria from diabetic animals, suggesting that there are other pathways in addition to the mTOR pathway regulating mitochondria function in diabetes. The functional significance of the mTOR pathway in regulating mitochondria efficiency warrants further attention.

E. Sivertsson (✉) • M. Friederich-Persson
Department of Medical Cell Biology, Division of Integrative Physiology,
Uppsala University, Uppsala, Box 571 751 23, Sweden
e-mail: Ebba.sivertsson@mcb.uu.se

S. Van Huffel et al. (eds.), *Oxygen Transport to Tissue XXXV*, Advances in Experimental Medicine and Biology 789, DOI 10.1007/978-1-4614-7411-1_41,

41.1 Introduction

Diabetic nephropathy is a common complication of diabetes mellitus and the leading cause of end-stage renal disease [1], resulting in dialysis and/or kidney transplantation. Despite considerable research, the molecular mechanisms behind the development of diabetic nephropathy are not yet fully understood.

Common clinical features in diabetic nephropathy are mesangial cell proliferation, glomerular hypertrophy, and thickening of the glomerular basement membrane [1–3], ultimately resulting in fibrosis and chronic renal failure [2]. However, it is known that early metabolic changes such as deficient oxygen handling [4], diminished microcirculation [5], and altered mitochondria function [6] precede these structural changes. Furthermore, diabetic nephropathy is associated with increased oxidative stress causing reduced renal oxygen tension [4] and bioavailability of nitric oxide [7]. As the mitochondria are a major source of oxidative stress in the cell, changes in mitochondrial function may affect renal oxygenation and contribute to kidney tissue hypoxia, mechanisms that are strongly connected with development of diabetic nephropathy [8, 9].

A known regulator of mitochondria function is the mammalian target of rapamycin (mTOR) [10, 11], which is a highly conserved serine/threonine kinase controlling cell growth and proliferation. mTOR exists in two forms: mTOR complex 1 and 2 (mTORC1, 2) when associated with the protein raptor or rictor, respectively. Activation of mTORC2 stimulates cytoskeletal reorganization and inhibits apoptosis [12]. mTORC1 interacts with the transcription factor yin-yang one through the peroxisome proliferator-activated receptor coactivator 1α (PGC-1α) and stimulates transcription of genes involved in mitochondria oxidative phosphorylation [11]. Furthermore, mTOR signaling is stimulated by nutrients, such as glucose and amino acids, leading to increased oxidative phosphorylation and increased mitochondrial oxygen consumption [13], and is therefore believed to play a role in the development of diabetic nephropathy. Rapamycin, a specific mTOR inhibitor, is clinically used as an immunosuppressant to inhibit rejection following organ transplantation [1–3]. Its immunosuppressant action is mediated through inhibition of mTORC1, subsequently preventing the activation of T- and B-cells [14] but also preventing coactivation of PGC-1α and transcription of mitochondria oxidative genes [11], possibly affecting mitochondria function.

The present study investigates the role of mTOR for controlling the function of mitochondria isolated from normal and diabetic rat kidneys.

41.2 Methods

All animal procedures were carried out according to National Institutes of Health guidelines and approved by the local animal ethics committee. All chemicals were from Sigma-Aldrich, St Louis, MO, unless otherwise stated. Statistical comparisons were performed using Student's unpaired *t*-test. $p < 0.05$ was considered significant and all values are presented as mean ± SEM.

41.2.1 Diabetes Induction and Treatment with Rapamycin

Diabetes was induced in age-matched male Sprague–Dawley rats (Charles River, Sulzfeld, Germany) with streptozotocin (55 mg/kg bw, iv, tail vein). Blood glucose levels were determined 24 h after injection using a test regent strip (Abbott, Alameda, CA, USA) from blood samples obtained from a cut of the tip of the tail and considered diabetic if blood glucose levels increased above 15 mmol/L and remained elevated. Rapamycin (0.2 mg/day) was administered by oral gavage for 2 weeks to the control and diabetic animals with corresponding control groups left untreated (n=3–6/group).

41.2.2 Mitochondrial Isolation and In Vitro Measurements of Oxygen Consumption

The rats were decapitated and the kidneys rapidly excised and placed in ice-cold buffer A (in mM: 250 sucrose, 10 HEPES, pH 7.4, 300 mOsm/kg H_2O). The cortex was dissected on ice and homogenized in ice-cold buffer A with a Potter-Elvehjem homogenizer (approximately 600 rpm). The homogenate was centrifuged at 800×g (4 °C, 10 min) and the supernatant centrifuged at 8000×g (4 °C, 10 min). The resulting pellet was washed and dissolved in 200 μl buffer A. Oxygen consumption was measured using an Oxygraph-2 k (Oroboros Instruments, Innsbruck, Austria). Mitochondria were added to the chamber containing 2 ml of air-equilibrated buffer B (in mM: 70 sucrose, 220 mannitol, 5 $MgCl_2$, 1 EGTA, 1 g/L BSA, 5KPO^-_4, 10 HEPES, pH 7.4, 300 mOsm/kg H_2O) and 48 μM sodium palmitate, and the basal oxygen consumption recorded after which 10 mM glutamate was added to achieve state 4 respiration. Subsequently, 300 μM ADP was added to achieve the ADP-stimulated O_2 consumption (state 3 respiration) and the respiratory control ratio (RCR, an indication of mitochondria viability) calculated as state 3 over state 4 respiration. Oxygen consumption unrelated to ATP synthesis was recorded after incubation with oligomycin (ATP synthase inhibitor, 12 μg/mg protein). The recorded oxygen consumption was corrected for protein concentration and expressed as pmol/min/mg protein.

41.3 Results

Rapamycin administration resulted in elevated blood glucose levels and reduced body weight in control animals, but not in diabetic animals (Table 41.1). mTOR inhibition did not affect RCR in either control or diabetic animals (Fig. 41.1).

Isolated mitochondria from treated controls had increased oxygen consumption after incubation with oligomycin compared to untreated controls (Fig. 41.2). However, oxygen consumption in treated diabetic animals remained unaffected (Fig. 41.2).

Table 41.1 Blood glucose levels and body weight of control and diabetic animals with or without treatment with rapamycin

Group	Treatment	Blood glucose (mmol/L)	Body weight (g)
Control	–	5.0±0.2	483±6
	Rapamycin	14.0±1.0*	422±20*
DM	–	25.5±1.3	346±20
	Rapamycin	27.4±0.2	319±17

*Denotes $p<0.05$ compared to untreated controls

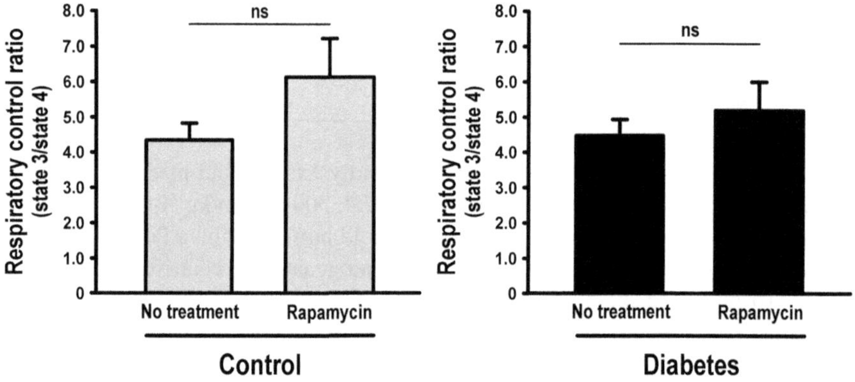

Fig. 41.1 Respiratory control ratio in mitochondria isolated from kidneys of control (*left panel*) and diabetic (*right panel*) animals with or without chronic mTOR inhibition for 14 days

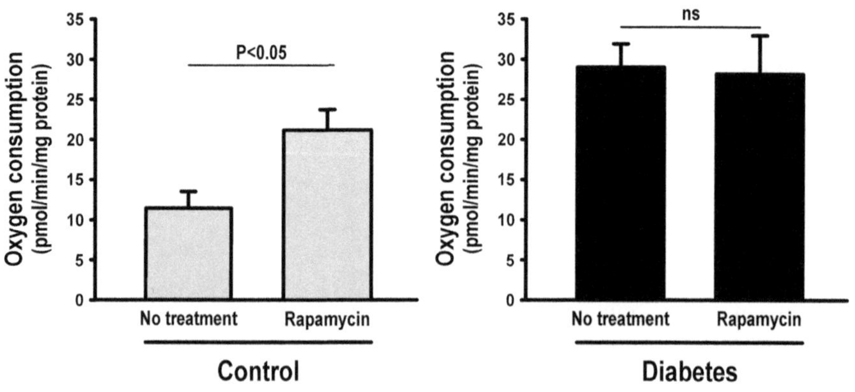

Fig. 41.2 Oxygen consumption after incubation with oligomycin in mitochondria isolated from kidneys of control (*left panel*) and diabetic (*right panel*) animals with or without chronic mTOR inhibition for 14 days

41.4 Discussion

In the present study, mTOR inhibition in control animals resulted in increased mitochondria oxygen consumption during oligomycin incubation, demonstrating mitochondria uncoupling, i.e., decreased efficiency of the mitochondria. Mitochondria uncoupling is the process of protons being transported across the mitochondria inner membrane without resulting in the production of ATP. However, mitochondrial uncoupling results in decreased mitochondria membrane potential, which may have beneficial effects by reducing the production of reactive oxygen species. This is accomplished at the expense of increased mitochondrial oxygen consumption to sustain sufficient cellular ATP levels. There are several different proteins that can cause mitochondrial uncoupling, including uncoupling protein (UCP)-1–5 and the adenine nucleotide translocator (ANT) [6, 15, 16]. Interestingly, UCP-2 protein expression is elevated and activity increased in mitochondria from diabetic kidneys, which has been shown to result in increased mitochondrial oxygen usage [6, 17].

One limitation of the protocol used in this study is that rapamycin administration to the control animals caused hyperglycemia. Hyperglycemia is a registered side effect of rapamycin and has been reported in patients as dose-dependent [18, 19]. It cannot be ruled out that this contributed to the reported findings.

Interestingly, no effect of mTOR inhibition was observed in mitochondria from diabetic rats, suggesting that there are other pathways in addition to the mTOR pathway regulating mitochondria function in diabetic kidneys. One mechanism known in the diabetic kidney is the regulation by oxidative stress of UCP-2, resulting in mitochondria uncoupling [6, 17].

In conclusion, inhibition of mTOR results in uncoupling of mitochondria isolated from control kidneys but not in mitochondria from diabetic kidneys. The different effects of mTOR inhibition in control and diabetic animals are interesting and will be investigated in further studies. Additionally, to solidify a role of the mTOR pathway in regulating mitochondrial uncoupling and the development of diabetic nephropathy, a lower dose of rapamycin, which does not cause hyperglycemia in control animals, is needed.

References

1. Mori H, Inoki K, Masutani K et al (2009) The mTOR pathway is highly activated in diabetic nephropathy and rapamycin has a strong therapeutic potential. Biochem Biophys Res Commun 384(4):471–475
2. Lloberas N, Cruzado JM, Franquesa M et al (2006) Mammalian target of rapamycin pathway blockade slows progression of diabetic kidney disease in rats. J Am Soc Nephrol 17(5): 1395–1404
3. Yang Y, Wang J, Qin L et al (2007) Rapamycin prevents early steps of the development of diabetic nephropathy in rats. Am J Nephrol 27(5):495–502
4. Palm F, Cederberg J, Hansell P, Liss P, Carlsson PO (2003) Reactive oxygen species cause diabetes-induced decrease in renal oxygen tension. Diabetologia 46(8):1153–1160

5. Nordquist L, Palm F (2007) Diabetes-induced alterations in renal medullary microcirculation and metabolism. Curr Diabetes Rev 3(1):53–65
6. Friederich M, Fasching A, Hansell P, Nordquist L, Palm F (2008) Diabetes-induced upregulation of uncoupling protein-2 results in increased mitochondrial uncoupling in kidney proximal tubular cells. Biochim Biophys Acta 1777(7–8):935–940
7. Palm F, Friederich M, Carlsson PO, Hansell P, Teerlink T, Liss P (2008) Reduced nitric oxide in diabetic kidneys due to increased hepatic arginine metabolism: implications for renomedullary oxygen availability. Am J Physiol Renal Physiol 294(1):F30–F37
8. Friederich M, Hansell P, Palm F (2009) Diabetes, oxidative stress, nitric oxide and mitochondria function. Curr Diabetes Rev 5(2):120–144
9. Palm F (2006) Intrarenal oxygen in diabetes and a possible link to diabetic nephropathy. Clin Exp Pharmacol Physiol 33(10):997–1001
10. Schieke SM, Phillips D, McCoy JP et al (2006) The mammalian target of rapamycin (mTOR) pathway regulates mitochondrial oxygen consumption and oxidative capacity. J Biol Chem 281(37):27643–27652
11. Cunningham JT, Rodgers JT, Arlow DH, Vazquez F, Mootha VK, Puigserver P (2007) mTOR controls mitochondrial oxidative function through a YY1-PGC-1alpha transcriptional complex. Nature 450(7170):736–740
12. Belibi F, Ravichandran K, Zafar I, He Z, Edelstein CL (2011) mTORC1/2 and rapamycin in female Han: SPRD rats with polycystic kidney disease. Am J Physiol Renal Physiol 300(1): F236–F244
13. Schieke SM, Finkel T (2006) Mitochondrial signaling, TOR, and life span. Biol Chem 387(10–11):1357–1361
14. Morath C, Arns W, Schwenger V et al (2007) Sirolimus in renal transplantation. Nephrol Dial Transplant 22(8):Viii61–Viii65
15. Echtay KS (2007) Mitochondrial uncoupling proteins–what are their physiological role? Free Radic Biol Med 43(10):1351–1371
16. Garlid KD, Jaburek M, Jezek P (2001) Mechanism of uncoupling protein action. Biochem Soc Trans 29(Pt 6):803–806
17. Persson MF, Franzén S, Catrina SB et al (2012) Coenzyme Q10 prevents GDP-sensitive mitochondrial uncoupling, glomerular hyperfiltration and proteinuria in kidneys from db/db mice as a model of type 2 diabetes. Diabetologia 55(5):1535–1543
18. Groth CG, Bäckman L, Morales JM et al (1999) Sirolimus (rapamycin)-based therapy in human renal transplantation: similar efficacy and different toxicity compared with cyclosporine. Sirolimus European Renal Transplant Study Group. Transplantation 67(7):1036–1042
19. Teutonico A, Schena PF, Di Paolo S (2005) Glucose metabolism in renal transplant recipients: effect of calcineurin inhibitor withdrawal and conversion to sirolimus. J Am Soc Nephrol 16(10):3128–3135

Chapter 42
Molecular Hydrogen Consumption in the Human Body During the Inhalation of Hydrogen Gas

Akito Shimouchi, Kazutoshi Nose, Tomoe Mizukami, Dock-Chil Che, and Mikiyasu Shirai

Abstract Inhaling or ingesting hydrogen (H_2) gas improves oxidative stress-induced damage in animal models and humans. We previously reported that H_2 was consumed throughout the human body after the ingestion of H_2-rich water and that the H_2 consumption rate (V_{H_2}) was 1.0 μmol/min/m^2 body surface area. To confirm this result, we evaluated V_{H_2} during the inhalation of low levels of H_2 gas. After measuring the baseline levels of exhaled H_2 during room air breathing via a one-way valve and a mouthpiece, the subject breathed low levels (160 ppm) of H_2 gas mixed with purified artificial air. The H_2 levels of their inspired and expired breath were measured by gas chromatography using a semiconductor sensor. V_{H_2} was calculated using a ventilation equation derived from the inspired and expired concentrations of $O_2/CO_2/H_2$, and the expired minute ventilation volume, which was measured with a respiromonitor. As a result, V_{H_2} was found to be approximately 0.7 μmol/min/m^{2}BSA, which was compatible with the findings we obtained using H_2-rich water. V_{H_2} varied markedly when pretreatment fasting to reduce colonic fermentation was not employed, i.e., when the subject's baseline breath hydrogen level was 10 ppm or greater. Our H_2 inhalation method might be useful for the noninvasive monitoring of hydroxyl radical production in the human body.

A. Shimouchi (✉) • K. Nose • T. Mizukami • M. Shirai
National Cerebral and Cardiovascular Research Center, Osaka, Japan
e-mail: ashimouc@res.ncvc.go.jp

D.-C. Che
Department of Chemistry, Graduate School of Science, Osaka University, Toyonaka, Osaka, Japan

S. Van Huffel et al. (eds.), *Oxygen Transport to Tissue XXXV*, Advances in Experimental Medicine and Biology 789, DOI 10.1007/978-1-4614-7411-1_42,

42.1 Introduction

Hydroxyl radicals (OH) among reactive oxygen species (ROS) are highly reactive and deleterious on biological systems. Therefore, the estimation of oxygen radical production is clinically important because oxygen radicals are closely associated with numerous diseases such as reperfusion injury, metabolic syndrome, and ischemic heart disease. Because OH radicals are short-lived in the body, numerous studies have focused on the development of analytical methods for assaying their reaction end products. However, very few reports have estimated the in vivo production rate of ROS in the human body [1].

On the other hand, it has been reported that inhaling or ingesting hydrogen (H_2) gas improves oxidative stress-induced damage in the brain [2], heart [3], liver [4], and other organs [5]. Several lines of evidence suggest that among ROS, exogenous H_2 is selectively trapped by OH, although the reaction mechanism remains unclear. These reports suggest that exogenous H_2 binds to oxygen radicals. In fact, Ohsawa et al. [1] demonstrated an arterial-venous H_2 concentration difference during the inhalation of H_2 in rats and suggested that the H_2 had been incorporated into tissues.

However, very few studies have reported direct evidence as to whether exogenous H_2 is consumed in the human body. We previously hypothesized that the H_2 consumption observed after the ingestion of H_2-rich water might be associated with oxygen radical production in the human body. Therefore, we examined H_2 consumption after the ingestion of H_2-rich water and reported that 59 % of the H_2 molecules were exhaled and 38 % or less were consumed in the human body; thus, the H_2 consumption rate was determined to be 1.0 μmol/min/m^2 body surface area [6].

To confirm our previous result and reduce the time required to assess subjects' H_2 consumption rates, the present study evaluated H_2 consumption during the inhalation of low levels of H_2 gas.

42.2 Methods

42.2.1 *Subject and Experimental Setup*

A 55-year-old adult volunteer participated in this study. For 7 of the 11 experiments, the subject refrained from consuming food, supplements, and drugs for at least 15 h before the experiments in order to decrease colonic fermentation. The subject was allowed to drink water during this period. The remaining four experiments were performed after lunch.

After measuring their baseline levels of exhaled H_2 (baseline F_{EH_2}) during room air breathing via a one-way valve and mouthpiece, the subject breathed low levels (140–180 ppm) of H_2 gas mixed with purified artificial air.

On the experimental day, the subject rested in a sitting position, wore a nose clip and mouthpiece, and breathed room air for 6 min in order to allow us to measure the

Table 42.1 Parameter abbreviations

Baseline F_{EH_2}	Fractional concentration of breath H_2 during room air breathing
$F_{IH_2}, F_{IO_2}, F_{ICO_2}$	Fractional concentrations of inhaled H_2, O_2, and CO_2 during H_2 inhalation
$F_{EH_2}, F_{EO_2}, F_{ECO_2}$	Fractional concentrations of exhaled H_2, O_2, and CO_2 during H_2 inhalation
V_I	Inspired minute volume during H_2 inhalation
V_E	Expired minute volume during H_2 inhalation
V_{H_2}	H_2 consumption during H_2 inhalation

baseline F_{EH_2} for 6 min and then breathed low levels of H_2 gas mixed with artificial air. The expiratory minute volume was continuously measured using a respiromonitor (RS330, Minato Medical Science Co., Ltd., Osaka, Japan). Every 2 min, exhaled breath was collected for 30 s in a Douglas bag, and a breath sample was immediately transferred to a gas-tight glass syringe so that H_2 analysis could be performed using a biogas analyzer (TRIlyzer 3000, Taiyo Ltd, Osaka, Japan). Prior to the present experiment, we confirmed that there was no significant loss of H_2 from the subject's respiratory circuit.

42.2.2 Calculations

When colonic fermentation is reduced by overnight starvation, it can be assumed that the baseline F_{EH_2} remains constant during H_2 inhalation. Therefore, the H_2 consumption rate (V_{H_2}) is determined as follows:

$$V_{H_2} = V_I\, F_{IH_2} - V_E\left(F_{EH_2} - \text{Baseline } F_{EH_2}\right) \tag{42.1}$$

where V_I and V_E denote the inspired and expired minute ventilation volumes, respectively. All volumes were measured under standard pressure, temperature, and dry conditions. The fractional H_2 concentrations of inspired and expired breath are expressed as F_{IH_2} and F_{EH_2} , respectively. The abbreviations used for the other parameters are listed in Table 42.1. The ventilation equation for inert gases was as follows:

$$V_I\left(1 - F_{IO_2} - F_{ICO_2}\right) = V_E\left(1 - F_{EO_2} - F_{ECO_2}\right) \tag{42.2}$$

Substituting V_I from Eq. 42.1 into Eq. 42.2 gives Eq. 42.3 for V_{H_2} as follows:

$$V_{H2} = \left\{\frac{1 - F_{EO_2} - F_{ECO_2}}{1 - F_{IO_2} - F_{ICO_2}} \cdot F_{IH_2} - (F_{EH_2} - \text{Baseline}F_{EH_2})\right\} \cdot V_E \tag{42.3}$$

Therefore, V_{H_2} can be easily calculated by measuring baseline F_{EH_2} , F_{IH_2} , F_{IO_2} , F_{ICO_2} , F_{EH_2} , F_{EO_2} , F_{ECO_2} , V_I and V_E .

42.3 Results

When colonic fermentation was reduced by overnight starvation, i.e., when the subject's baseline F_{EH_2} was <10 ppm, the H_2 concentration of their inspired breath (F_{IH_2}) was 164.4 ± 13.7 ppm, and after the inhalation of H_2 gas, the concentration of exhaled H_2 (F_{EH_2}) increased to a similar level (164.8 ± 12.2 ppm) within 4–6 min. However, as shown in Table 42.2, the subject's H_2 consumption rate (V_{H_2}) was calculated to be 0.71 ± 0.47 μmol/min/m^2 body surface area or 16.3 ± 10.8 nmol/min/kg body weight, which agreed well with the results we previously obtained using hydrogen-rich water.

In the other four trials, which were not preceded by starvation, when the subject did not maintain their baseline F_{EH_2} below 10 ppm, V_{H_2} varied more than when the baseline F_{EH_2} level was < 10 ppm, as shown in Fig. 42.1.

42.4 Discussion

The present study confirmed that H_2 was consumed in the human body during the inhalation of low levels of H_2 gas. V_{H_2} varied markedly when the baseline F_{EH_2} level was 10 ppm or greater; however, when the baseline F_{EH_2} level was reduced to <10 ppm by overnight starvation, the mean V_{H_2} rate was compatible with previous results obtained with H_2- rich water.

H_2 molecules are not involved in metabolic pathways in the human body, except those that occur in the bacterial flora in the colon. We previously reported that pre-treatment with antibiotics did not affect V_{H_2} [6]. Furthermore, we observed that H_2 loss from the skin surface was negligible and that the administration of vitamin C decreased V_{H_2} after the ingestion of H_2-rich water in a dose-dependent manner [6]. Thus, we confirmed that H_2 was consumed after the ingestion of H_2 water. The present study detected a similar V_{H_2} during H_2 inhalation. It has been reported that H_2 is a weak but selective scavenger of hydroxyl radicals. Therefore, these pieces of evidence lead us to speculate that exogenous H_2 binds to OH radical and that V_{H_2} reflects the OH production rate in the human body, at least to some extent.

Physicochemical studies have reported the temperature dependence of rate constants in the reaction of $H_2 + OH \rightarrow H + H_2O$ from 250 to 1,050 K [7, 8]. The experimental activation energy is 4.0 kcal/mol for this reaction [9], suggesting that the reaction proceeds even at room temperature (298 K), although the reaction rate is slow. It is likely that the $H_2 + OH \rightarrow H + H_2O$ reaction could take place in the mitochondrial temperature and be accelerated due to its special biological properties. However, the detailed reaction mechanism in living cells remains unclear. Thus, further basic studies are needed to elucidate these reactions.

In Eq. 42.1, it is assumed that the baseline F_{EH_2} was constant at lower levels, i.e., after colonic fermentation had been decreased by food intake restriction prior to the experiment. As shown in Fig. 42.1, when the baseline F_{EH_2} was 10 ppm or higher, V_{H_2} varied markedly. This variability seemed to depend upon the change in breath

Table 42.2 Hydrogen consumption rate obtained by repeated experiments and related parameters. See Table 42.1 for an explanation of the parameter abbreviations

Parameter	Baseline	F_{IH_2}	F_{EH_2}	F_{IO_2}	F_{ICO_2}	F_{EO_2}	F_{ECO_2}	V_E	V_{H_2}	V_{H_2}	V_{H_2}
Units	F_{EH_2} ppm	ppm	ppm	%	%	%	%	L/min	μmol/min	μmol/min/m²BSA	nmol/min/kgBW
1st	0.9	183.2	183.9	22.00	0.00	17.6	3.60	8.43	0.66	0.31	7.1
2nd	0.5	175.5	174.5	22.00	0.00	17.9	3.57	9.82	1.23	0.58	13.4
3rd	0.6	168.8	164.6	22.00	0.00	17.5	3.34	9.51	2.29	1.08	24.8
4th	2.2	166.1	165.7	22.00	0.00	17.0	3.46	10.26	1.51	0.71	16.4
5th	5.9	158.9	159.8	22.00	0.00	18.9	3.29	10.63	2.39	1.13	26.0
6th	8.0	157.7	159.7	22.00	0.00	17.2	3.40	9.56	2.57	1.21	28.0
7th	2.1	140.8	145.1	22.00	0.00	17.2	3.37	8.98	−0.12	−0.06	−1.3
Mean	2.9	164.4	164.8	22.00	0.00	17.81	3.43	9.60	1.50	0.71	16.3
SD	2.9	13.7	12.2	0.00	0.00	0.23	0.12	0.74	1.00	0.47	10.8

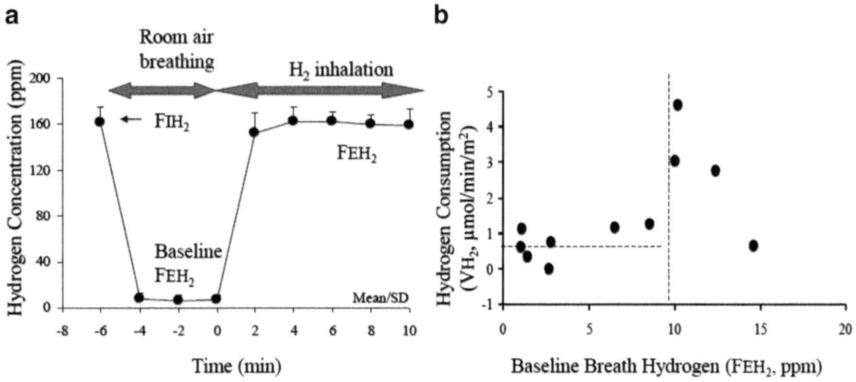

Fig. 42.1 *Left*: Changes in the exhaled hydrogen concentration during the breathing of room air and low levels of hydrogen gas (see Table 42.1 for an explanation of the abbreviations). *Right*: Relationship between hydrogen consumption (V_{H_2}) and baseline breath hydrogen (F_{EH_2}) levels

H_2 induced by abdominal fermentation during the measurement period. It is well recognized that a higher concentration of breath H_2 is caused by the acceleration of colonic fermentation accompanied by increased contraction of the colon, the presence of undigested food in the colon, and the resultant changes in internal pressure [10]. Therefore, we consider that the present method should be used in reduced colonic fermentation conditions, i.e., in clinical settings involving surgery, intensive care, or health screening. Further refinement is needed to ameliorate the inconvenience of this method, as is clinical evidence that indicates that V_{H_2} reflects ROS production throughout the whole body.

The present method took 10–20 min to complete, which is significantly shorter than the V_{H_2} measurement time in our H_2-rich water method (more than 60 min). As there is no current method for directly measuring whole body OH production, the present method could be used for real-time monitoring as an indirect index that reflects the OH production rate in the whole human body. Further studies are needed to clarify the clinical significance of our V_{H_2} measurement method.

42.5 Conclusion

We have developed a new method for estimating H_2 consumption in the whole human body involving the inhalation of low levels of H_2 gas. Repeated measurements indicated that the H_2 consumption rate was approximately 0.7 μmol/min/m^{2}BSA, which was compatible with that obtained using H_2-rich water. Hydrogen consumption might be closely associated with oxygen radical production in the human body.

Acknowledgments This study was supported by the Japan Society for the Promotion of Science (Grant-in-Aid 21240057, 21659211, 24659288) and the Intramural Research Fund of the National Cerebral and Cardiovascular Center (22-4-5, 22-1-5). The authors have no conflicts of interest to report. We thank the volunteers who participated in this study.

References

1. Halliwell B, Gutterridge JMC (2007) Free radicals in biology and medicine, 4th edn. Oxford University Press, Oxford
2. Ohsawa I, Ishikawa M, Takahashi K et al (2007) Hydrogen acts as a therapeutic antioxidant by selectively reducing cytotoxic oxygen radicals. Nat Med 13(6):688–694
3. Hayashida K, Sano M, Ohsawa I et al (2008) Inhalation of hydrogen gas reduces infarct size in the rat model of myocardial ischemia-reperfusion injury. Biochem Biophys Res Commun 373(1):30–35
4. Fukuda K, Asoh S, Ishikawa M, Yamamoto Y, Ohsawa I, Ohta S (2007) Inhalation of hydrogen gas suppresses hepatic injury caused by ischemia/reperfusion through reducing oxidative stress. Biochem Biophys Res Commun 361(3):670–674
5. Ohta S (2011) Recent progress toward hydrogen medicine: potential of molecular hydrogen for preventive and therapeutic applications. Curr Pharm Des 17(22):2241–2252
6. Shimouchi A, Nose K, Shirai M, Kondo T (2012) Estimation of molecular hydrogen consumption in the human whole body after the ingestion of hydrogen-rich water. Adv Exp Med Biol 737:245–250
7. Ravishankara AR, Nicovich JM, Thompson RL, Tuliyt FP (1981) Kinetic study of the reaction of OH with H, and D, from 250 to 1050 K. J Phys Chem 85(17):2498–2503
8. Smith IWM, Zelmer R (1974) Rate measurements of reactions of OH by resonance absorption. Part 3.-Reactions of OH with H2, D2, and hydrogen and deuterium halides. J Chem Soc Faraday Trans 270:1045–1056
9. Zhang DH, Light JC (1966) A six dimensional quantum study for atom–triatom reactions: the $H+H_2O \rightarrow H_2+OH$ reaction. J Chem Phys 104:4544–4553
10. Levitt MD, Bond JH, Levitt DG (1981) Gastrointestinal gas. In: Johnson LR (ed) Physiology of the gastrointestinal tracts. Raven, New York, pp 1301–1315

Chapter 43
Oxidative Metabolism: Glucose Versus Ketones

Allison Prince, Yifan Zhang, Colleen Croniger, and Michelle Puchowicz

Abstract The coupling of upstream oxidative processes (glycolysis, beta-oxidation, CAC turnover) to mitochondrial oxidative phosphorylation (OXPHOS) under the driving conditions of energy demand by the cell results in the liberation of free energy as ATP. Perturbations in glycolytic CAC or OXPHOS can result in pathology or cell death. To better understand whole body energy expenditure during chronic ketosis, we used a diet-induced rat model of ketosis to determine if high-fat-carbohydrate-restricted "ketogenic" diet results in changes in total energy expenditure (TEE). Consistent with previous reports of increased energy expenditure in mice, we hypothesized that rats fed ketogenic diet for 3 weeks would result in increased resting energy expenditure due to alterations in metabolism associated with a "switch" in energy substrate from glucose to ketone bodies. The rationale is ketone bodies are a more efficient fuel than glucose. Indirect calorimetric analysis revealed a moderate increase in VO_2 and decreased VCO_2 and heat with ketosis. These results suggest ketosis induces a moderate uncoupling state and less oxidative efficiency compared to glucose oxidation.

A. Prince • C. Croniger • M. Puchowicz (✉)
Departments of Nutrition, Schools of Medicine and Engineering, Case Western Reserve University, 10900 Euclid Ave., W-G48, Cleveland, OH 44106-4954, USA
e-mail: prince.allison@gmail.com; colleen.croniger@case.edu; map10@case.edu

Y. Zhang
Departments of Biomedical Engineering, Schools of Medicine and Engineering, Case Western Reserve University, 10900 Euclid Ave., W-G48, Cleveland, OH 44106-4954, USA
e-mail: yifan.zhang@case.edu

S. Van Huffel et al. (eds.), *Oxygen Transport to Tissue XXXV*, Advances in Experimental Medicine and Biology 789, DOI 10.1007/978-1-4614-7411-1_43,

43.1 Introduction

Our complete understanding of the regulation and coupling of glucose consumption to oxidative metabolism is essential to discerning the pathophysiology of many mitochondrial-related diseases. Normal metabolism involves the ability of the organism to adapt to variations in fed and fasting conditions, as well as changes in hormonal control. Abnormal metabolism often involves nutritional and enzymatic deficiencies resulting in "metabolic-related" diseases. All mammalian cells consume glucose as an oxidative substrate through glycolysis by producing pyruvate (or lactate). Under aerobic conditions, the entry of pyruvate into the citric acid cycle (CAC) results in its complete oxidation to CO_2 and H_2O.

Although glucose is the major metabolic fuel, alternate energy substrates such as ketone bodies, can serve as efficient metabolic fuel sources for most organs (except liver). Ketone bodies are produced as a result of hepatic fatty acid oxidation (via beta-oxidation) during conditions of glucose sparing. The ability of the liver to generate ketone bodies as a supplemental or alternative fuel for other organs is especially important under certain nutritional conditions, such as with fasting, starvation, feeding of a ketogenic diet (high fat, carbohydrate restricted) and development. The use of ketogenic diets in the human population continues to receive attention due to its neuroprotective benefits [1], especially in children and adolescents with epilepsy [2]. Evidence for the mechanisms of the ketogenic diet continues to be explored in animal models. While it is documented in humans and rodents that the ketogenic diet results in loss of overall body weight, what remains unclear is its effect on energy expenditure in relation to total body mass and lean body mass. Studies in mice have reported that following the ketogenic diet, mice had higher energy expenditure compared to counterparts fed the standard chow control diet [3, 4]. Conversely, when the ketogenic diet was fed to obese humans, the result was a decrease in energy expenditure following a significant decrease in weight [5].

We sought to determine if there is an increase in energy expenditure in a rat model of ketosis as previously observed in mice. VO_2, VCO_2, and heat (energy expenditure) were measured over 22 h (light-dark cycle) in rats fed standard (STD) or ketogenic (KG) diets following a 3-week dietary regime.

43.2 Methods

The experimental protocol employed by this study was approved by the Institutional Animal Care and Use Committee (IACUC) at Case Western Reserve University.

43.2.1 Animals and Diet Regime

Male Wistar rats (~ 28 days old; 150 ± 10 g) were purchased from Charles River Laboratories International, Inc. After 1 week of acclimation to the CWRU animal facility climate, rats were fasted overnight to initiate ketosis and then placed on either STD ($n = 4$) or KG ($n = 4$) diets (STD Teklab 8664; KG no. D12369b, Research Diets, New Brunswick, NJ, USA). Ketosis was confirmed via ketone body analysis (BHB) with tail blood sample and ketone meter. The macronutrient composition of the diets (% kcal), STD versus KG, are as follows: 27.5 versus 89.5 fat, 52.6 versus 0.1 carbohydrate, and 20.0 versus 10.4 protein.

43.2.2 Energy Expenditure and Body Composition Measurements

Indirect calorimetry was performed using the Oxymax system (Columbus Instruments' Comprehensive Lab Animal Monitoring System (CLAMS), Columbus, OH). VO_2, VCO_2, heat (energy expenditure), and the respiratory quotient (RQ) were determined. All measurements were taken without fasting; animals had free access to food and water. The experiments ran over 22 h, from approximately 12 p.m. to 10 a.m. (12-h dark cycle, 6 p.m. to 6 a.m.). Fat-free mass was estimated in each animal using dilution analysis; deuterium-labeled water was injected into the peritoneal cavity, and the dilution was determined by mass spectrometry analysis of the total body water (^2H_2O) assayed from a small blood sample [6].

43.2.3 Statistical Analysis

Data are expressed as mean ± SD. Results were assessed using two-tailed unpaired Student's *t*-test. Significance was defined as $P < 0.05$.

43.3 Results

43.3.1 Physiological Variables

The number of days rats were on the diet prior to the experiment and corresponding changes in body weight and composition are shown in Table 43.1. The average weight of the KG-fed rats trended lower than the STD diet group.

Table 43.1 Physiological variables

	Standard diet (STD)	Ketogenic diet (KG)
Weight (g)	365.9 ± 26	331.0 ± 30
Δ weight (g)	165.2 ± 12	145. 0 ± 13
Fat-free mass (%)[a]	61.4 ± 2.0	60.3 ± 2.7
Lean body mass (g)[a]	224.5 ± 1.2	199.6 ± 0.9
Age (days)	74.0	72.3
Length of time on diet (days)	23.5	21.3

Δ Change in weights after 3 weeks of feeding diets compared to day 1
[a]Fat-free and lean body mass was estimated using the deuterated water-dilution method [6]

43.3.2 Energy Expenditure Using Indirect Calorimetry

Rats on the KG diet had significantly higher levels of oxygen consumption (VO_2) and lower levels of carbon dioxide production (VCO_2) compared to the STD diet when normalized to kg body weight, Fig. 43.1a. A similar response was observed when data were normalized per animal (data not shown). As expected, the respiratory quotient during the light-dark cycles were lower in the KG diet group (KG, RQ = 0.71 vs. STD, RQ = 0.94; $P < 0.05$) which was consistent with increased fat oxidation compared to carbohydrate oxidation, Fig. 43.1b.

43.4 Discussion

In this study, 3 weeks of feeding the ketogenic diet resulted in decreased total body energy expenditure with increased fat oxidation. The change in energy expenditure is most likely not reflective of decrease in body mass. Kennedy et al. [4] described comparable body weights in both ketogenic and calorie-restricted mice despite differences in calories consumed. However, evaluation of 24-h heat production using the CLAMS apparatus confirmed that ketogenic-diet-fed animals had 11 % higher energy expenditure than the STD mice (kcal/h). Another study in human adolescents found no significant changes in resting energy expenditure [2].

Our findings of increased VO_2 and decreased VCO_2 suggest a slight uncoupling of oxygen consumption to ATP production (as measured by decreased energy expenditure). This is consistent with increased uncoupling protein activity and KG diet regime [7, 8], through the activation of mitochondrial uncoupling proteins. In summary, our preliminary results of a 3-week feeding of a KG diet in rats suggest ketosis induces an overall decrease in energy expenditure which we postulate may be linked to increased uncoupling protein activity. This response could be advantageous towards reducing reactive oxygen species during oxidative stress conditions, such as with stroke. However, the effects of diet-induced ketosis on individual organ systems, such as brain and liver, continue to be explored [9, 10].

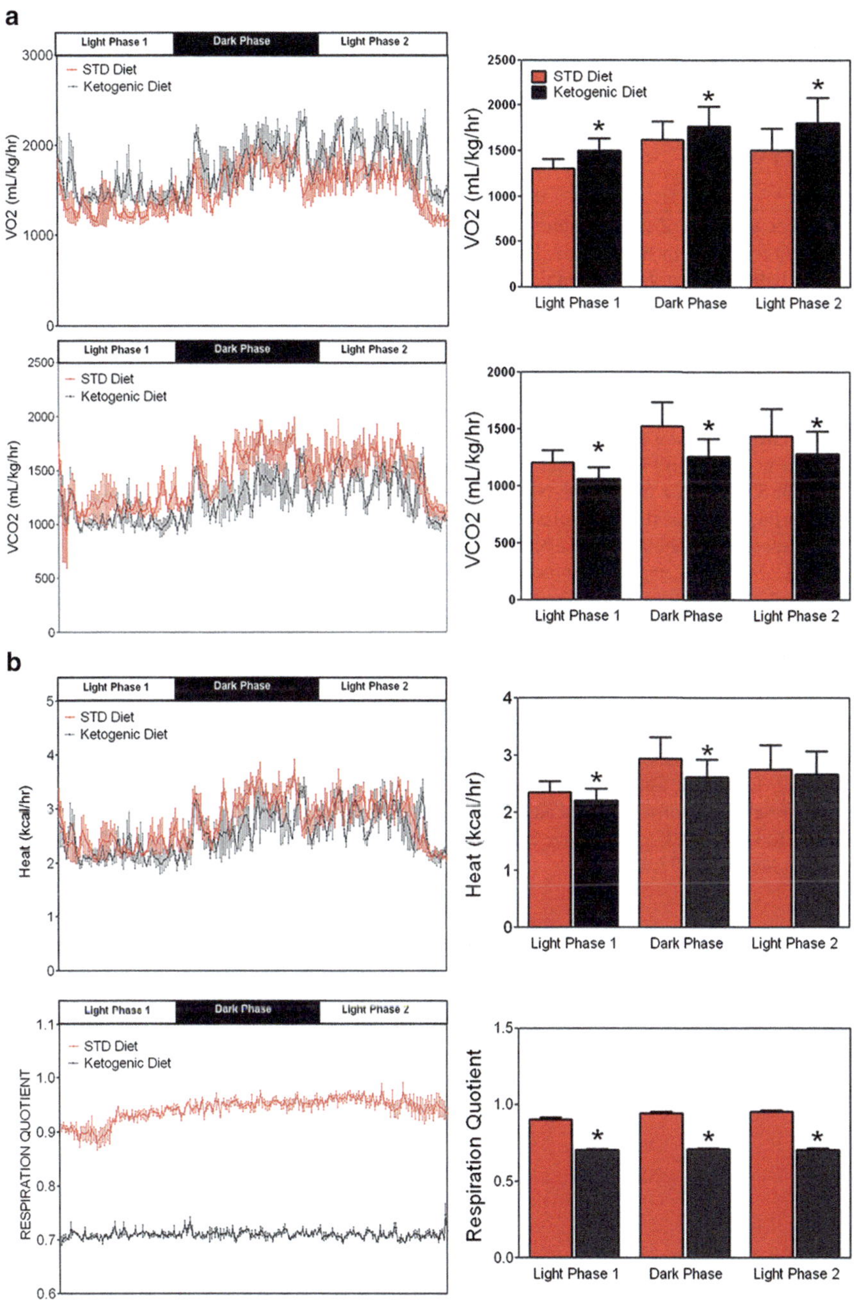

Fig. 43.1 Indirect calorimetry was assessed by measurements of VO_2, VCO_2 (ml/kg/h) (panel **a**), heat (kcal/kg/h), and respiratory quotient (RQ) (panel **b**). Values are mean ± SE; * indicates significant difference ($P<0.05$) compared to STD diet group

Acknowledgments This research has been supported by the National Institutes of Health, Mouse Metabolic Phenotyping Center, MMPC U24 DK76174.

References

1. Puchowicz MA, Zechel J, Valerio J et al (2008) Neuroprotection in diet induced ketotic rat brain following focal ischemia. J Cereb Blood Flow Metab 28(12):1907–1916
2. Tagliabue A, Bertoli S, Trentani C, Borrelli P, Veggiotti P (2012) Effects of the ketogenic diet on nutritional status, resting energy expenditure, and substrate oxidation in patients with medically refractory epilepsy: a 6-month prospective observational study. Clin Nutr 31(2):246–249
3. Jornayvaz F, Jurczak M, Lee H et al (2010) A high-fat, ketogenic diet causes hepatic insulin resistance in mice, despite increasing energy expenditure and preventing weight gain. Am J Physiol Endocrinol Metab 299(5):E808–E815
4. Kennedy A, Pissios P, Otu H et al (2007) A high-fat, ketogenic diet induces a unique metabolic state in mice. Am J Physiol Endocrinol Metab 292(6):E1724–E1739
5. Willi S, Oexmann M, Wright N, Collop N, Key LL Jr (1998) The effects of a high-protein, low-fat, ketogenic diet on adolescents with morbid obesity: body composition, blood chemistries, and sleep abnormalities. Pediatrics 101(1 Pt 1):61–67
6. McCabe B, Bederman I, Croniger C, Millward C, Norment C, Previs S (2006) Reproducibility of gas chromatography-mass spectrometry measurements of 2H labeling of water: application for measuring body composition in mice. Anal Biochem 350(2):171–176
7. Veech R (2004) The therapeutic implications of ketone bodies: the effects of ketone bodies in pathological conditions: ketosis, ketogenic diet, redox states, insulin resistance, and mitochondrial metabolism. Prostaglandins Leukot Essent Fatty Acids 70(3):309–319
8. Sullivan P, Rippy N, Dorenbos K, Concepcion R, Agarwal A, Rho J (2004) The ketogenic diet increases mitochondrial uncoupling protein levels and activity. Ann Neurol 55(4):576–580
9. Li Q, Tomcik K, Zhang S, Puchowicz MA, Zhang GF (2012) Dietary regulation of catabolic disposal of 4-hydroxynonenal analogs in rat liver. Free Radic Biol Med 52(6):1043–1053
10. Xu K, LaManna JC, Puchowicz MA (2012) Neuroprotective properties of ketone bodies. Adv Exp Med Biol 737:97–102

Part VI
System Modelling

Chapter 44
Modelling Blood Flow and Metabolism in the Piglet Brain During Hypoxia-Ischaemia: Simulating pH Changes

Tharindi Hapuarachchi, Tracy Moroz, Alan Bainbridge, David Price, Ernest Cady, Esther Baer, Kevin Broad, Mojgan Ezzati, David Thomas, Xavier Golay, Nicola J. Robertson, and Ilias Tachtsidis

Abstract We describe the extension of a computational model of blood flow and metabolism in the piglet brain to investigate changes in neonatal intracellular brain pH during hypoxia-ischemia (HI). The model is able to simulate near-infrared spectroscopy (NIRS) and magnetic resonance spectroscopy (MRS) measurements obtained from HI experiments conducted in piglets. We adopt a method of using ^{31}P-MRS data to estimate of intracellular pH and compare measured pH and oxygenation with their modelled counterparts. We show that both NIRS and MRS measurements are predicted well in the new version of the model.

44.1 Introduction

Experimental studies have shown a shift in brain pH following hypoxia-ischemia (HI) – the deprivation of oxygen supply [1]. HI is a major cause of perinatal brain injury [2]. Modest changes in pH can result in alterations to protein structures and therefore affect the function of membrane channels and enzymes crucial to many vital cellular functions.

T. Hapuarachchi (✉) • T. Moroz
CoMPLEX, University College London, London, UK
e-mail: t.hapuarachchi@ucl.ac.uk

A. Bainbridge • D. Price • E. Cady
Medical Physics and Bioengineering, University College London, London, UK

E. Baer • I. Tachtsidis
Department of Medical Physics and Bioengineering, University College London, London, UK

K. Broad • M. Ezzati • N.J. Robertson
Institute for Women's Health, University College London, London, UK

D. Thomas • X. Golay
Institute of Neurology, University College London, London, UK

S. Van Huffel et al. (eds.), *Oxygen Transport to Tissue XXXV*, Advances in Experimental Medicine and Biology 789, DOI 10.1007/978-1-4614-7411-1_44,

Piglets are often used as models of human neonates in experimental studies involving anoxic and/or hypoxic and ischaemic insults. In order to investigate HI and to better understand the results from these experiments, we have built a computational model of blood flow and metabolism in the neonatal piglet brain (BrainPiglet) [2]. This model is an adaptation and extension of an earlier model of the adult human brain [3]. The model is used to simulate near-infrared spectroscopy (NIRS) and magnetic resonance spectroscopy (MRS) data – two non-invasive methods used to monitor brain tissue oxygenation, haemodynamics and metabolism during HI experiments. We have recently extended the model further, by simulating carotid artery occlusion [4] and intracellular brain pH (BrainPiglet v2). In this chapter, we (i) describe the main dynamics of intracellular H^+ ions incorporated in order to model pH, (ii) explain the methods used to obtain an estimate of brain pH from ^{31}P-MRS measurements and (iii) validate the model by comparing pH data from HI experiments in piglets with model-simulated pH.

44.2 Experimental Methods and Protocol

All experiments were done under UK Home Office Guidelines (Animals [Scientific Procedures] Act, 1986) and were approved by the Institute of Neurology, University College London. In this study, 1-day-old piglets were ventilated and anaesthetised. Inflatable occluders were surgically placed around the carotid arteries, and arterial partial pressures of oxygen and carbon dioxide, blood glucose and heart rate were maintained at a normal level. Baseline MRS and NIRS were first acquired before transient HI was induced for ~1 h, by inflating the occluders and reducing fractional inspired oxygen (FiO_2) to 12 % (normal value 21 %). The occluders were released 10–20 min after β-NTP (a correlate of ATP) had reduced by ~70 %, and FiO_2 was subsequently increased to normalise blood saturation. ^{31}P-MRS and broadband NIRS were acquired every 1 min during the baseline period, during HI and for a further ~2 h to monitor recovery from HI [5]. This study is ongoing. As of August 2012, data were obtained from 22 piglets.

NIRS measures changes in the concentrations of oxy- and deoxy-haemoglobin in blood (HbO_2, HHb). Variations in cerebral blood volume are marked by changes in total (oxy- and deoxy-) haemoglobin concentration. We use MRS (either proton (^{1}H) or phosphorus (^{31}P)) to observe variations in by-products of cellular metabolism such as inorganic phosphate (Pi), phosphocreatine (PCr), adenosine triphosphate (ATP) and lactate (a marker of anaerobic metabolism). More specifically, we employed ^{31}P-MRS to estimate intracellular pH via the chemical shifts of Pi, phosphoethanolamine (PEt) and ATP (pH_{Pi}, pH_{PEt} and pH_{ATP} respectively). We have used the following titration curves for pH_{Pi} and pH_{PEt} [5]:

$$pH_{pi} = 6.77 + \log_{10}\left(\frac{\delta_{P_i} - 3.29}{5.68 - \delta_{P_i}}\right),\ pH_{PEt} = 5.625 + \log_{10}\left(\frac{\delta_{PEt} - 3.190}{6.946 - \delta_{PEt}}\right) \quad (44.1)$$

where δ_{Pi} is the chemical shift difference between PCr resonance and an amplitude weighted mean of the Pi resonances [5]. PEt has been observed to be a major component of the phosphomonoester (PME) peak, and hence, δ_{PEt} is calculated as the chemical shift in PME relative to PCr [5]. Consequently, the mean of pH_{Pi} and pH_{PEt} was adopted as the overall pH_{Pi-PEt} measurement. We used the MAGPAC programme (magnesium and pH from ATP calculation [5]) to calculate pH_{ATP} from the chemical shifts of α-NTP, β-NTP and γ-NTP. In addition, we continuously record systemic variables such as arterial blood pressure (P_a), arterial oxygen saturation (SaO_2), breathing rate and heart rate.

44.3 Model

We have developed a mathematical model of blood flow and metabolism, placing emphasis on the physiology of the brain. It consists of a set of algebraic relations and differential equations, describing cerebral blood flow and oxygenation and oxygen and energy metabolism on a cellular level. This system of equations incorporates ~100 parameters and ~25 variables. P_a, SaO_2 and arterial carbon dioxide (P_aCO_2) are, where available, used as inputs. The model is then able to simulate changes in NIRS-measured HbO_2 and HHb and MRS-measured Pi, PCr and ATP. It also models changes in the cerebral metabolic rate of oxygen consumption ($CMRO_2$) and lactate. Blood flow is modelled as three compartments – arteries and arterioles, capillaries and veins – with varying conductances and radii. We recently added an extra compartment to represent the supply of blood into the arteries [4]. By varying the radius of this new compartment, we can simulate the carotid artery occlusion that results in ischemia. In order to simulate pH in our model, we altered eight reactions as detailed in Table 44.1, to represent the main dynamics of H^+ions. Changes are shown in bold. Mitochondrial and cytoplasmic protons are modelled separately as H_m and H_{cyt}. Similarly, we also differentiated between cytoplasmic and mitochondrial NAD and NADH concentrations. p1, p2 and p3 represent the number of protons pumped in each reaction, a3r the concentration of reduced cytochrome a3 and $Cu_{A,o}$ the concentration of oxidised cytochrome-c-oxidase. Cr represents creatine, gluc glucose, Lac lactate, and Py Pyruvate.

For simplicity, we have kept the total concentrations of cytochrome and NAD at either oxidised or reduced state constant. Therefore, only one oxidisation state is included in the model equations above. We have not modelled the conversion of flavin adenine dinucleotide (FAD) in the TCA cycle, which utilises mitochondrial protons. To compensate, we incorporated 5/3 H_m to the left hand side of the

Table 44.1 Mitochondrial and cytoplasmic reactions modified to simulate intracellular pH

Mitochondria	Cytoplasm
(i) Oxidative phosphorylation	(iv) Glycolysis
$2Cu_{A'o} + \left(p1 + \frac{5}{3}\right)H_m \rightarrow 2NAD + (p1 + 4)H_{cyt}$	$2ADP + 2P_i + gluc + 2NAD_{cyt} \rightarrow 2ATP + 2Py + 4H_{cyt}$
$P_2H_m \rightarrow 4Cu_{A'o} + 4a3r + p_2H_{cyt}$	
$o_2 + 4Cu_{A'o} + p_2H_m \rightarrow$	
(ii) Tricarboxylic acid cycle	(v) PCr to ATP conversion:
$py + 5NAD + H_{cyt} \rightarrow 4H_m$	$PCr + ADP + H_{cyt} \rightarrow ATP + Cr$
(iii) Protons reenter mitochondria (via leak and complex V): $H_{cyt} \rightarrow H_m$	(vi) Pyruvate to lactate conversion $py + H_{cyt} \rightarrow Lac + NAD_{cyt}$

Table 44.2 New parameters and their values

Parameter	Description	Value	Source
Km_glucNN	Km for NAD in the caricature of glycolysis	1.0	[3]
Keq_MAshut	Equilibrium constant for the malate-aspartate shuttle	10.0	[3]
NADcytn	Normal concentration of NAD in the cytoplasm	359	[6]
NADHcytn	Normal concentration of NADH in the cytoplasm	50	[6]

oxidative phosphorylation equation (Table 44.1 (i)). On the right hand side of the equation, an additional four protons are pumped into the cytoplasm by complex II. The rate of reaction for glycolysis (Table 44.1 (iv)) was also altered to account for the new reactants. Similar to the mitochondrial proton buffer in the previous model [2], we have added a simple proton buffer in the cytoplasm. Furthermore, we introduced the malate-aspartate shuttle, which transports electrons produced by glycolysis in the cytoplasm across the NADH-impermeable mitochondrial membrane to be used in oxidative phosphorylation. During this enzyme-driven process, NADH in the cytoplasm is oxidised to NAD, while NAD in the mitochondrial matrix is reduced to NADH. We simplified this system and modelled it as a mass action reaction (44.2). The rates for the forward and backward reactions are k_MAshut and k_nMAshut, respectively (Eqs. 44.4 and 44.3). Table 44.2 lists the new parameters that have been added to the model, as necessitated by the changes above.

$$H_{cyt} + NAD \rightarrow NAD_{cyt} + H_m \tag{44.2}$$

$$k\_nMAshut = \frac{k\_MAshutNADH}{Keq\_MAshutNADH_{cyt}} \tag{44.3}$$

$$k\_MAshut = \frac{\frac{2}{3}CMRO_{2r}NADH_{cyt}}{NADH_{cytr}NAD_rH_{cyt} - Keq\_MAshut^{-2}\ NAD_{cyt}NADH_rH_r} \quad (44.4)$$

44.4 Results

The steady-state output of the model for cerebral blood flow (CBF) and cytoplasmic and mitochondrial pH with increasing SaO_2 are illustrated in Fig. 44.1. Normal average brain pH is ~7 [1]. P_a and SaO_2 data from the piglet experiments were used as inputs into the model. $PaCO_2$ was not recorded in this instance; however, as the piglets were ventilated with controlled CO_2 concentrations, we have assumed $PaCO_2$ remains constant at 40 mmHg. Due to space constraints, we present in Fig. 44.2 results from only one piglet (LWP180). We used the Morris method to determine the most influential parameters and the SciPy Powell method to detect the optimum values of these parameters to achieve a good fit [7]. Consequently, we increased the values of three parameters in our model; the normal total haemoglobin concentration (Xtot_n) was increased from 5.40 to 6.298 mM, the concentration of cytochrome c oxidase (CCO) in tissue (cytox_tot_tis) from 0.0022 to 0.004257 and the normal oxidised fraction of Cu_A (a_frac_n) from 0.67 to 0.75. Other piglets show similar results.

44.5 Discussion

The BrainPiglet model has been extended to simulate the biochemistry affecting intracellular brain pH and shown to successfully predict ^{31}P-MRS pH measurements in addition to other metabolic changes. The steady-state simulations (Fig. 44.1) may

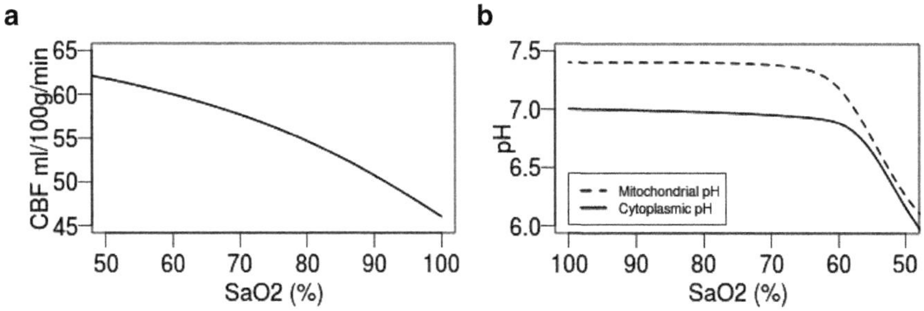

Fig. 44.1 Steady-state model simulations of (**a**) cerebral blood flow (*CBF*) and (**b**) pH against arterial oxygen saturation (SaO_2)

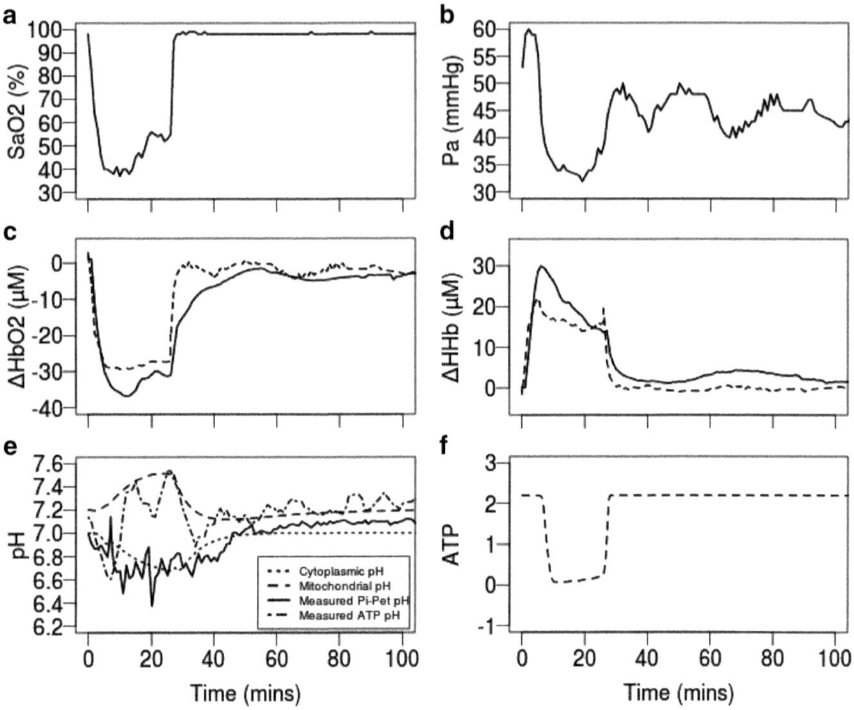

Fig. 44.2 Measured arterial oxygen saturation (SaO_2, **a**) and blood pressure (P_a, **b**) used as inputs to the model; NIRS and MRS measurements (*solid line*) from one piglet (LWP180) compared with modelled results (*dashed* and *dotted lines*) (**c–e**); simulated ATP (**f**). pH_{ATP} in (**e**) calculated from MRS measurements of α-NTP, β-NTP and γ-NTP

be used to validate the behaviour of the model. In Fig. 44.1a, the model replicates a common relationship between CBF and SaO_2, similar to results published earlier [2]. The drop in pH_{Pi-PEt} seen in Fig. 44.1b is indicative of acidosis brought about by a deprivation of oxygen. We have shown successful simulations of metabolic and pH changes in the neonatal brain of one piglet during HI (Fig. 44.2). For a good fit of modelled to measured HbO_2 and HHb three parameters were optimised – the total concentration of haemoglobin in the blood, CCO and the fraction of blood flowing through the carotid arteries were slightly increased. This implies there is a higher concentration of oxygen supplied to the cell than previously modelled. Such biological parameters may also vary from one individual to another and so can be altered to suit each individual patient. We must note that the comparison of measured and simulated pH is not as straightforward. Although we specifically model intracellular cytoplasmic and mitochondrial pH, ^{31}P-MRS provides an average estimate of brain pH in a select area comprised of blood, tissue and various cells. It is not yet possible to clearly distinguish between the different components with this technique of measurement. It may have been anticipated that the modelled cytoplasmic pH fits well with the measured pH_{Pi-PEt} data, as Pi is said to concentrate in the cytoplasm. We also observe in our pH_{ATP} measurement, albeit noisy, an alkaline rise which is concurrent with the modelled mitochondrial pH during HI. These changes occur in tandem with a drop in proton motive force across the mitochondrial membrane – less cytoplasmic protons flow back into the mitochondria, rendering the cytoplasm more acidic.

There are a number of limitations to our model. The oxygen-haemoglobin dissociation curve determines the binding affinity of haemoglobin to oxygen, acting as a biological buffer. However, this rate varies with changes in blood pH; we hope to model this concept in future. In addition, there is possibly a greater variation in pH throughout the brain than that observed in the measurements.

We have effectively modified our model of neonatal brain metabolism and circulation to simulate brain pH. By investigating further, we hope to gain a better understanding of physiological processes during oxygen deprivation. In due course, we aim to adapt the model to the human neonatal brain.

Acknowledgments We would like to thank the Wellcome Trust (088429/Z/09/Z) for financial support of this work. The first author is supported by her doctoral centre CoMPLEX, UCL.

References

1. Robertson NJ, Cowan FM, Cox IJ, Edwards AD (2002) Brain alkaline intracellular pH after neonatal encephalopathy. Ann Neurol 52(6):732–742
2. Moroz T, Banaji M, Robertson NJ, Cooper CE, Tachtsidis I (2012) Computational modelling of the piglet brain to simulate near-infrared spectroscopy and magnetic resonance spectroscopy data collected during oxygen deprivation. J R Soc Interface 9(72):1499–1509
3. Banaji M, Tachtsidis I, Delpy D, Baigent S (2005) A physiological model of cerebral blood flow control. Math Biosci 194(2):125–173
4. Moroz T, Hapuarachchi T, Banaji M et al (2012) Modelling blood flow and metabolism in the piglet brain during hypoxic-ischaemia: simulating brain energetics (this volume)
5. Cady EB, Iwata O, Bainbridge A, Wyatt JS, Robertson NJ (2008) Phosphorus magnetic resonance spectroscopy 2 h after perinatal cerebral hypoxia-ischemia prognosticates outcome in the newborn piglet. J Neurochem 107(4):1027–1035
6. Williamson DH, Lund P, Krebs HA (1967) The redox state of free nicotinamide-adenine dinucleotide in the cytoplasm and mitochondria of rat liver. Biochem J 103(2):514–527
7. Jones E, Oliphant E, Peterson P et al (2001) SciPy: open source scientific tools for Python. http://www.scipy.org. Accessed 20 July 2012

Chapter 45
Modelling Blood Flow and Metabolism in the Piglet Brain During Hypoxia-Ischaemia: Simulating Brain Energetics

Tracy Moroz, Tharindi Hapuarachchi, Alan Bainbridge, David Price, Ernest Cady, Ether Baer, Kevin Broad, Mojgan Ezzati, David Thomas, Xavier Golay, Nicola J. Robertson, Chris E. Cooper, and Ilias Tachtsidis

Abstract We have developed a computational model to simulate hypoxia-ischaemia (HI) in the neonatal piglet brain. It has been extended from a previous model by adding the simulation of carotid artery occlusion and including pH changes in the cytoplasm. Here, simulations from the model are compared with near-infrared spectroscopy (NIRS) and phosphorus magnetic resonance spectroscopy (MRS) measurements from two piglets during HI and short-term recovery. One of these piglets showed incomplete recovery after HI, and this is modelled by considering some of the cells to be dead. This is consistent with the results from MRS and the redox state of cytochrome-c-oxidase as measured by NIRS. However, the simulations do not match the NIRS haemoglobin measurements. The model therefore predicts that further physiological changes must also be taking place if the hypothesis of dead cells is correct.

T. Moroz (✉) • T. Hapuarachchi
CoMPLEX, University College London, London, UK
e-mail: t.moroz@ucl.ac.uk

A. Bainbridge • D. Price • E. Cady
Medical Physics and Bioengineering, University College London Hospitals, London, UK

E. Baer • I. Tachtsidis
Department of Medical Physics and Bioengineering, University College London, London, UK

K. Broad • M. Ezzati • N.J. Robertson
Institute for Women's Health, University College London, London, UK

D. Thomas • X. Golay
Institute of Neurology, University College London, London, UK

C.E. Cooper
School of Biological Sciences, University of Essex, Colchester, UK

S. Van Huffel et al. (eds.), *Oxygen Transport to Tissue XXXV*, Advances in Experimental Medicine and Biology 789, DOI 10.1007/978-1-4614-7411-1_45,

45.1 Introduction

Hypoxia-ischaemia (HI) is a major cause of brain damage in neonates. Piglets are often used as models to investigate the processes occurring during HI and to test treatments. We have previously developed a computational model to simulate oxygen deprivation in the neonatal piglet brain [1]. This model has been extended to allow simulations of HI induced by carotid artery occlusion. We are able to use the model to compare with data from near-infrared spectroscopy (NIRS) and magnetic resonance spectroscopy (MRS). These two non-invasive modalities have been used simultaneously to monitor newborn piglets subjected to HI. The model allows the measurements to be analysed together and the relationships between them to be explored.

45.2 The Model

The model simulates circulation and metabolism in the neonatal brain. It is an extension of a model which has previously been used to investigate anoxia in piglets [1]. A schematic diagram of the model is shown in Fig. 45.1. The metabolic part of the model simulates metabolites both in the cytoplasm and the mitochondria. The mitochondrial part of the model focuses on the redox state of the electron transport chain, in particular cytochrome-c-oxidase (CCO). The cytoplasmic part of the model focuses on energy metabolism and includes simplified descriptions of glycolysis and ATP use. The model is able to simulate the variables which are

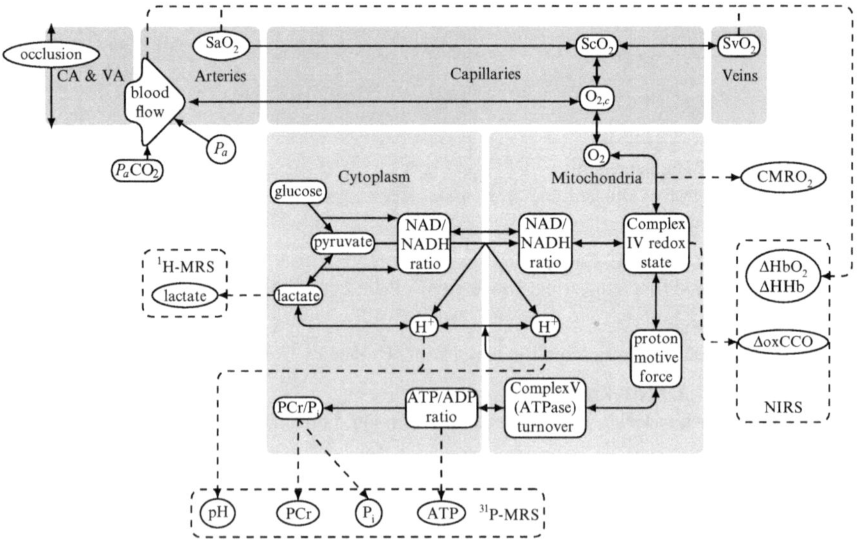

Fig. 45.1 Schematic diagram of the model. CA and VA refer to the carotid and vertebral arteries

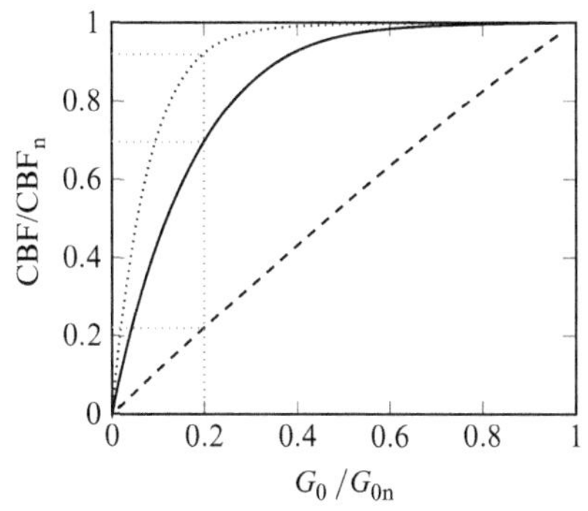

Fig. 45.2 CBF versus conductance of the supplying arterial compartment for G_f= 10.0 (*dotted*), G_f=4.0 (*solid*) and G_f=0.1 (*dashed*). Both variables are shown as fractions of their normal value. Complete occlusion of the carotid arteries is equivalent to reducing the conductance to 0.2

measured by MRS including ATP, phosphocreatine (PCr), inorganic phosphate (P_i) and lactate concentrations. It has also been extended to simulate pH changes in the cytoplasm.

The circulatory part of the model allows simulation of the NIRS haemogolobin measurements. It has been extended to allow simulation of carotid artery occlusion. This was done by adding an extra compartment to represent all the arteries supplying the brain. The main arteries responsible for this are the carotid arteries: in adult humans it is estimated that 80 % of the cerebral blood supply flows through them [2]. In the model, this fraction (c_f) determines the conductance of the supplying arterial compartment (G_0) during carotid artery occlusion by

$$G_0 = G_{0,n}(1 - c_f) \tag{45.1}$$

where $G_{0,n}$ is the conductance when there is no occlusion which is set by

$$G_0 = G_f G_n \tag{45.2}$$

where G_n is the normal conductance of the cerebral arterial compartment. The ratio G_f is difficult to obtain from the literature, so was set by examining the results of the simulations. The change in modelled cerebral blood flow (CBF) as a function of $G_0/G_{0,n}$ is shown in Fig. 45.2. Three different values for the fraction G_f are shown. When the G_f is large, the CBF remains high until the conductance is only a small fraction of its normal value. When G_f is small, the relationship becomes more linear. From examining this curve, a value of 4 was chosen for G_0.

45.3 Methods

The simulations were compared with modelled data from experiments involving piglets less than 24-h-old. The piglets were anaesthetised and mechanically ventilated. Their arterial oxygen saturation (SaO_2) and mean arterial blood pressure (MABP) were continuously monitored. The piglets were also monitored with NIRS to measure the change in concentration of oxyhaemoglobin (ΔHbO_2), deoxyhaemoglobin (ΔHHb) and oxidised cytochrome-c-oxidase (ΔoxCCO). In addition, measurements of nucleotide triphosphate (NTP) which is mainly ATP, PCr and P_i were recorded as a fraction of the exchangeable phosphate pool (EPP) by ^{31}P-MRS. After 10 min of baseline measurements, vascular occluders surrounding both carotid arteries were inflated and the inspired oxygen fraction (FiO_2) was reduced to 12 %. When the β-NTP peak had fallen to 50 % of its baseline value, FiO_2 was titrated to maintain the β-NTP peak between 30 % and 50 % of its baseline height for 12.5 min. Following this, the occluders were deflated and FiO_2 was returned to normal. Measurements were continued for approximately another 2 h.

The measured SaO_2 and MABP were used as inputs to the model, and its outputs were compared with the NIRS and MRS measured variables. A Morris sensitivity analysis was used to identify which parameters had the most important effect on fitting the modelled signals to the measured signals. The results showed that the most important parameters were those representing the concentration of the measured quantities, i.e., the blood haemoglobin concentration, the tissue concentration of cytochrome-c-oxidase and the normal concentrations of ATP, PCr and P_i. These parameters were adjusted to best match the modelled and measured signals for the individual piglets.

Not all piglets showed recovery of the ΔoxCCO signal and the ^{31}P-MRS signals following the insult. One hypothesis to explain this is that some of the cells have died. In order to simulate this, the model was altered so that a fraction of the cells d were treated as dead following the insult. In these cells, CCO was assumed to be completely reduced and all exchangeable phosphate was assumed to be in the form of P_i. It was also assumed that no oxygen was consumed in the dead cells, so that the modelled rate of oxygen transfer from the capillaries to the mitochondria was reduced to $1 - d$ of its normal rate. Several of the model outputs were also changed:

$$\begin{aligned}
&\text{output NTP / EPP} = \frac{(1-d)[\text{ATP}]}{[\text{EPP}]} \\
&\text{output PCr / EPP} = \frac{(1-d)[\text{PCr}]}{[\text{EPP}]} \\
&\text{output P}_\text{i}\text{ / EPP} = \frac{(1-d)[\text{Pi}]}{[\text{EPP}]} + d \\
&\text{output }\Delta\text{oxCCO} = (1-d)\Delta\text{oxCCO} - d\text{oxCCO} \\
&\text{output CMRO}_2 = (1-d)\text{CMRO}_2.
\end{aligned} \tag{45.3}$$

45.4 Results

Figure 45.3 shows the simulated and measured signals for a piglet (LWP180) which showed recovery following HI. The fraction of dead cells d was set to 0. Figure 45.4 shows the same signals but for a piglet (LWP188) which did not recover. For these simulations, d was set to 0.4.

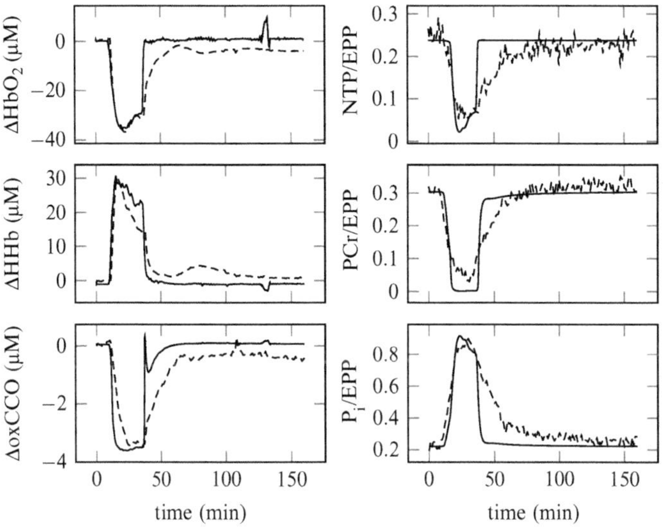

Fig. 45.3 A comparison between modelled (*solid*) and measured (*dashed*) signals from NIRS (*left*) and MRS (*right*) from a single piglet (LWP180)

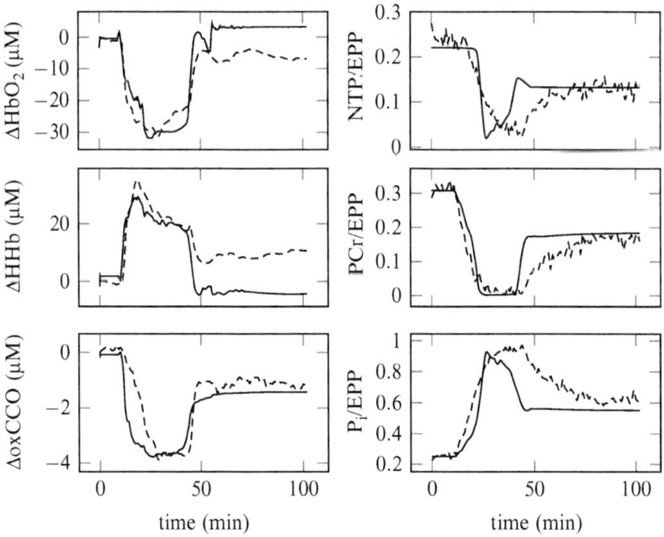

Fig. 45.4 Modelled signals (*solid*) compared with measured signals (*dashed*) from NIRS (*left*) and MRS (*right*) from a piglet (LWP188) which did not recover following HI. The simulations use a value of $d = 0.4$ after the insult

45.5 Discussion

The model has been used to simulate NIRS and MRS measurements during HI. The model is able to simulate carotid artery occlusion. It is known that with only one carotid artery occluded, there is no change in CBF in piglets. Measurements of CBF when both arteries are occluded (and there is no change in oxygen saturation) include 75 % [3] and 45 % [4] of the baseline value. However, these experiments also involved changes in blood pressure. The modelled value lies between these two values, but more data are necessary to validate this part of the model.

The model is well able to simulate the magnitude of changes during HI. The time course of all the metabolic signals show that the model is predicting the recovery of these signals to baseline faster than is seen in the measured signals. A possible reason for this is that there are physiological changes occurring during HI which are not modelled.

The difference in recovery time is even more pronounced in the piglet which did not fully recover. However, the final values of the modelled ΔoxCCO, NTP/EPP, PCr/EPP and P_i/EPP are similar to their measured equivalents. This is consistent with a fraction of the cells being dead. The model allows the consequences of this assumption on other signals to be investigated. It predicts that the overall rate of oxygen metabolism ($CMRO_2$) would drop compared to baseline, which would cause the oxygen extraction fraction to fall and hence ΔHbO_2 to rise and ΔHHb to fall as seen in Fig. 45.4. However, this is not what is seen in the measurements, which suggests that there are other physiological changes occurring after HI if the assumption of cell death is correct. Possibilities for this include a large increase in $CMRO_2$ in the functioning cells, perhaps caused by mitochondrial uncoupling, or that blood may no longer be perfusing the whole brain. Alternatively, the cells may not be dead but functioning at a reduced capacity, or spatial differences between the measurements and pattern of cell death may give misleading results. Finally, the experimental results may have been affected by changes in the haematocrit of the piglet. Further investigation with the model and analysis of data from more piglets will help to answer these questions.

Acknowledgments The authors would like to thank the Wellcome Trust (088429/Z/09/Z) for the financial support of this work.

References

1. Moroz T, Banaji M, Robertson NJ, Cooper CE, Tachtsidis I (2012) Computational modelling of the piglet brain to simulate near-infrared spectroscopy and magnetic resonance spectroscopy data collected during oxygen deprivation. J R Soc Interface 9(72):1499–1509
2. Edvinsson L, Mackenzie E, Mcculloch J (1992) Cerebral blood flow and metabolism. Raven, New York
3. Kurth C, Levy W, McCann J (2002) Near-infrared spectroscopy cerebral oxygen saturation thresholds for hypoxia-ischemia in piglets. J Cereb Blood Flow Metab 22(3):335–341
4. Oriot D, Beharry K, Gordon JB, Aranda JV (1995) Ascorbic acid during cerebral ischemia in newborn piglets. Acta Paediatr 84(6):621–626

Chapter 46
Mathematical Modelling of Near-Infrared Spectroscopy Signals and Intracranial Pressure in Brain-Injured Patients

David Highton, Jasmina Panovska-Griffiths, Martin Smith, and Clare E. Elwell

Abstract Raised intracranial pressure (ICP) is a key concern following acute brain injury as it may be associated with cerebral hypoperfusion and poor outcome. In this research we describe a mathematical physiological model designed to interpret cerebral physiology from neuromonitoring: ICP, near-infrared spectroscopy and transcranial Doppler flow velocity. This aims to characterise the complex dynamics of cerebral compliance, cerebral blood volume, cerebral blood flow and their regulation in individual patients. Analysis of data from six brain-injured patients produces cohesive predictions of cerebral biomechanics suggesting reduced cerebral compliance, reduced volume compensation and impaired blood flow autoregulation. Patient-specific physiological modelling has the potential to predict the key biomechanical and haemodynamic changes following brain injury in individual patients, and might be used to inform individualised treatment strategies.

46.1 Introduction

Primary brain injury causes changes in cerebral biomechanics, haemodynamics and metabolism which threaten cerebral perfusion and potentiate secondary hypoxic-ischaemic injury. Raised intracranial pressure (ICP) is a key component of cerebral pathophysiology, and ICP monitoring is the most prevalent neuromonitoring modality used to guide therapy, because intracranial hypertension is frequently associated

D. Highton (✉) • M. Smith
Neurocritical Care Unit, University College London Hospitals, London, UK
e-mail: d.highton@ucl.ac.uk

J. Panovska-Griffiths • C.E. Elwell
Department of Medical Physics and Bioengineering, University College London, London, UK

S. Van Huffel et al. (eds.), *Oxygen Transport to Tissue XXXV*, Advances in Experimental Medicine and Biology 789, DOI 10.1007/978-1-4614-7411-1_46,

with cerebral hypoperfusion and poor outcome. ICP is not a discrete physiological process but a combination of complex underlying pathophysiology involving cerebral compliance, blood flow regulation, cerebral blood volume regulation and cerebrospinal fluid dynamics [1]. Because of the rigid confines of the skull, an increase in any volume (brain, blood volume, cerebrospinal fluid) must be compensated by reduction in another. However, these compensatory mechanisms are often exhausted following brain injury, leading to low compliance, exponential increases in ICP and brain hypoperfusion. Understanding these processes and their effect on cerebral oxygen delivery holds the key to providing optimal therapy after brain injury.

Increasingly, mathematical physiological modelling has been used to understand these complex processes in individual patients; this can synthesise physiological predictions from complex datasets that signal processing-based techniques cannot. Neuromonitoring has the capacity to report different aspects of these processes: (1) ICP is commonly measured invasively with a sensor placed into brain parenchyma, (2) transcranial Doppler ultrasound of the middle cerebral artery (Vmca) is a non-invasive surrogate of cerebral blood flow, and (3) near-infrared spectroscopy (NIRS) is a non-invasive optical technique that measures haemoglobin species that reflect cerebral haemodynamic and metabolic function.

Ursino et al. [2] extensively investigated a simplified biomechanical model describing ICP and Vmca dynamics for clinical data, but a major limitation of this model is the absence of any neuromonitoring surrogate of blood volume. We have previously described the BrainSignals model [3] of cerebral haemodynamics, oxygenation and metabolism used for the interpretation of Vmca and NIRS data. This does not simulate conditions of reduced compliance or ICP because it is predominantly used for interpreting data from healthy volunteers. The aim of this research is to describe a model which can be used to interpret clinical recordings of ICP, Vmca and NIRS in brain-injured patients simulating and reporting measures of the reduced compliance which may inform therapy.

46.2 Methods

We have extended the BrainSignals model, described in [3] adding a dynamic representation of ICP and cerebral blood volume using equations described in [2, 4]. The principal modifications are:

- ICP is simulated from the interaction between cerebral blood volume, the non-linear intracranial pressure-volume relationship and the dynamics of cerebrospinal fluid production/absorption.
- Cerebral arterial volume is described by a single arterial resistance and volume which responds to systemic blood pressure and carbon dioxide and oxygen delivery, and represents a balance between intra-arterial pressure, vessel elastance and ICP.
- Venous volume is described as a Starling resistor in two compartments representing the intracerebral veins and venous sinuses. Volume changes are predominantly mediated through distal resistance induced by ICP changes.

The key model parameters of physiological significance with regard to the biomechanical function are:

- Cerebral elastance (*k_E*): this describes the exponential nature of the cerebral pressure-volume relationship. Increasing elastance results in dramatic increases in the exponential rise in ICP with volume changes.
- Cerebrospinal fluid conductance (*G_o*): this describes the resistance to fluid absorption into the venous circulation and represents the ability of cerebrospinal fluid absorption to compensate for volume and ICP changes. A low *G_o* impairs the effectiveness of compensation for volume changes and elevates ICP.
- Cerebral autoregulation: controls the ability to maintain stable cerebral blood flow. As cerebral perfusion pressure falls, arterial vasodilation maintains flow. The parameter *k_aut* describes the strength of arterial dilation, while *P_an* modifies the pressure at which autoregulation becomes effective. This activity is a key determinant of cerebral blood volume and flow, thus is essential to simulate this scenario.

46.2.1 Model Testing

Six sedated, ventilated, brain-injured patients demonstrating spontaneous blood pressure variability were identified from a larger multimodal monitoring study. The patients were selected on the basis of providing clear changes in monitored Vmca, NIRS and ICP for model interpretation. NIRS monitoring was performed using the NIRO 100 (Hamamatsu Photonics KK, Japan) ipsilateral to Vmca (DWL Doppler Box, Compumedics, Germany) and intraparenchymal ICP monitoring. NIRS-measured normalised total haemoglobin index (nTHI) was derived using spatially resolved spectroscopy and represents an absolute scaled estimate of total haemoglobin concentration ([HbT]), a surrogate of cerebral blood volume. Invasive arterial blood pressure from a radial artery catheter, end tidal CO_2 ($ETCO_2$) and pulse oximetry data were collected using an IntelliVue monitor (Philips, N.V., Netherlands). Signals were filtered with a low-pass 0.1-Hz fifth-order Butterworth filter to remove respiratory variation and high-frequency noise, synchronised and downsampled to 1 Hz. Arterial blood pressure, pulse oximetry and $ETCO_2$ (approximating arterial CO_2) were used as model inputs. These produced simulated outputs for ICP, Vmca and [HbT] which were compared with their measured counterparts ICP, Vmca and nTHI.

Optimisation was performed by a simulated annealing algorithm comparing the difference between measured signals and simulated outputs for ICP, Vmca and [HbT]. Optimal values for model parameters (*k_E*, *G_o*, *k_aut*, *P_an*) resulting from this process summarise the physiological dynamics that best characterise the measured signals. The difference between measured signals and simulated outputs is described as the mean absolute difference between the two. Improvement in the model fit after optimisation is expressed as the percentage difference between simulated outputs and measured signals at basal parameter settings and optimised parameter settings, divided by their basal value. Signal processing-derived indices are also reported for comparison: (1) RAP is an index of cerebral volume compensation calculated from the correlation of the first harmonic of the ICP waveform with the mean ICP and reflects pathophysiology encoded by *k_E* and *G_o*, and (2) the mean velocity index (Mx) is an index of cerebral autoregulation calculated from the linear correlation coefficient between blood pressure and Vmca and reports a similar process to *k_aut*.

46.3 Results

Our modified model reproduced ICP, Vmca and nTHI signals with acceptable accuracy through modifying a limited set of parameters that characterise cerebral biomechanics (*k_E*, *G_o*, *k_aut*, *P_an*). The mean absolute difference for Vmca (2.62+/−1.52 cm/s), ICP (1.77+/−1.28 mmHg) and nTHI (0.013+/−0.015 au) is small and within the error of the measurement techniques where this has been defined [5], indicating acceptable model performance. An example dataset is shown in Fig. 46.1 demonstrating the difference in model output at basal parameter values and post-optimisation. The individual parameter values suggest impaired autoregulation (*k_aut*, [75 % basal]), increased elastance (*k_E*, [355 % basal]) and reduced cerebrospinal fluid compensation (*G_o*, [22 % basal]) are required to explain the measured signals.

Individual dataset results are shown in Table 46.1. Variation in all parameters was required to characterise the physiological processes that explain the measured ICP, Vmca and nTHI. In datasets 2–6, this achieved a high level of accuracy. Dataset 1 was more problematic, and there was a much larger difference between measured and modelled nTHI (0.044 au) and Vmca (5.55 cm/s). The mean parameter values demonstrated increased elastance (*k_E*, [673 % basal]) and impaired cerebrospinal fluid conductance (*G_o* [21 % basal]) which is consistent with the mean RAP (0.79+/−0.17), indicating impaired volume compensatory reserve. Impaired autoregulation is also predicted (*k_aut* [61 % basal]) and (*P_an* [95 % basal]). This agrees with the mean Mx (0.40+/−0.30) suggesting impaired autoregulation.

46.4 Discussion

We have demonstrated a simplified model of cerebral biomechanics, haemodynamics and metabolism which can be applied to clinical data from brain-injured patients and predicts key biomechanical features of cerebral pathophysiology. Parameter values for elastance (*k_E*), cerebrospinal fluid conductance (*G_o*) and cerebral autoregulation (*k_aut*, *P_an*) constitute a fingerprint of cerebral dynamics from individual brain-injured patients which summarise the function of the system and are in agreement with signal processing-based measures of cerebral volume

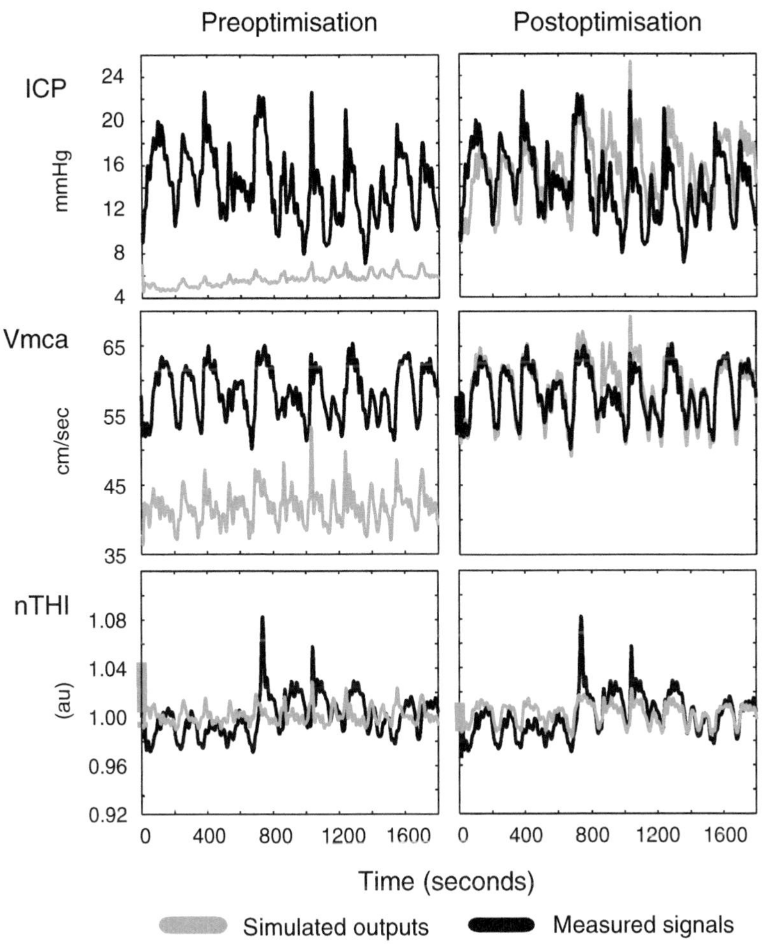

Fig. 46.1 Measured signals and simulated outputs are shown demonstrating the difference between measured ICP, Vmca and nTHI and simulated outputs at basal parameter values (pre-optimisation) and at optimal values (post-optimisation) for *k_aut*, *P_an*, *k_E*, *G_o* (patient 2 Table 46.1). These parameter values suggest impaired cerebral autoregulation (*k_aut*<1.0), increased elastance (*k_E*>0.11) and reduced cerebrospinal fluid conductance (*G_o*<0.0019) are required to explain the measured signals, a finding consistent with brain injury

compensatory reserve (RAP), autoregulation (Mx) and the expected pathophysiology in this group. Model-enhanced interpretation progresses beyond the capacity of derived indices, predicting true physiological processes, and might be used clinically in two separate ways. Firstly, these parameters constitute biomarkers of cerebral compliance, volume compensation and autoregulation which indicate pathology and might deliver prognostic information or guide treatment intervention [6]. They also define the function of a complex system: using this fingerprint, interventions

Table 46.1 Summary of model performance predicting ICP, Vmca and nTHI and parameter values required to do this. Individual parameter values representing autoregulation (*k_aut*, *P_an*) and cerebral compliance (*k_E*, *G_o*) are shown for each patient (basal values in brackets). Mean parameter values suggest impaired autoregulation (reduced *k_aut*) and reduced cerebral compliance (elevated *k_E*, reduced *G_o*). This process leads to a small mean absolute difference between measured and modelled signals (ΔVmca, ΔICP, ΔnTHI) and improvement in the difference compared with basal parameter values

Patient	*k_aut* (1.0)	*P_an* (100)	*k_E* (0.11)	*G_o* (0.0019)	ΔVmca (cm/s)	ΔICP (mmHg)	ΔnTHI (au)
1	0.63	87.49	0.17	7.06×10^{-4}	5.55	0.58	0.044
2	0.75	80.06	0.39	4.26×10^{-4}	1.38	2.72	0.010
3	1.00	99.99	0.15	7.13×10^{-4}	2.93	0.81	0.007
4	0.01	100.00	0.17	2.00×10^{-4}	1.96	0.65	0.010
5	0.99	99.99	0.26	1.99×10^{-4}	2.06	3.88	0.004
6	0.25	100.00	3.31	2.14×10^{-4}	1.87	0.98	0.003
Mean (sd)	0.61 (0.40)	94.59 (8.70)	0.74 (1.26)	4.10×10^{-4} (2.48×10^{-4})	2.62 (1.52)	1.77 (1.28)	0.013 (0.015)
Improvement (%)					80 % (14 %)	79 % (12 %)	31 % (29 %)

such as changes to blood pressure, carbon dioxide and oxygenation might be investigated in silico to assess the effect on cerebral oxygen delivery to guide appropriate clinical interventions [5, 7].

Metabolic aspects involve additional complexity and will require additional work to devise a computationally effective analytic approach which is able to explain the metabolic components in NIRS signals (oxyhaemoglobin, deoxyhaemoglobin, cytochrome c oxidase). One dataset in this investigation was more difficult to simulate; this could be related to our model (oversimplification, model inaccuracies, ineffective optimisation) or monitoring factors (optical complexity in injured brain affecting NIRS signals, regional variation in physiology). It is unrealistic to expect a simplified model to be applicable to all situations, and this has been a recognised limitation of other models. Our study examined only six patients, but this allowed extensive assessment of model performance, which would be challenging in greater numbers. Future research is required to investigate this model in larger numbers, particularly validating in a number of physiological contexts.

Model-enhanced interpretation of cerebral compliance and autoregulation delivers cohesive physiological predictions based on ICP, Vmca and NIRS that summarises biomechanics of brain injury in individual patients. Clinical intervention based on fully informed physiological prediction may facilitate individualised therapy, maximising the benefit of available treatments.

Acknowledgments This work was undertaken at University College London Hospitals and partially funded by the Department of Health's National Institute for Health Research Centres funding scheme. Support was also been provided by the Medical Research Council and Wellcome Trust. The authors are indebted to the medical and nursing staff of the Neurocritical Care Unit at the National Hospital for Neurology and Neurosurgery and to the study patients and their families.

References

1. Czosnyka M, Smielewski P, Timofeev I et al (2007) Intracranial pressure: more than a number. Neurosurg Focus 22(5):E10
2. Ursino M, Lodi C (1997) A simple mathematical model of the interaction between intracranial pressure and cerebral hemodynamics. J Appl Physiol 82(4):1256–1269
3. Banaji M, Mallet A, Elwell CE, Nicholis P, Cooper CE (2008) A model of brain circulation and metabolism: NIRS signal changes during physiological challenges. PLoS Comp Biol 4(11):e1000212
4. Banaji M, Tachtsidis I, Delpy D, Baigent S (2005) A physiological model of cerebral blood flow control. Math Biosci 194(2):125–173
5. Ursino M, Ter Minassian A, Lodi CA, Beydon L (2000) Cerebral hemodynamics during arterial and CO2 pressure changes: in vivo prediction by a mathematical model. Am J Physiol Heart Circ Physiol 279(5):H2439–H2455
6. Highton D, Panovska-Griffiths J, Ghosh A et al (2013) Modelling cerebrovascular reactivity: a novel near-infrared biomarker of cerebral autoregulation? Adv Exp Med Biol 765:87–93
7. Wakeland W, Agbeko R, Vinecore K, Peters M, Goldstein B (2009) Assessing the prediction potential of an in silico computer model of intracranial pressure dynamics. Crit Care Med 37(3):1079–1089

Chapter 47
Dependence on NIRS Source-Detector Spacing of Cytochrome *c* Oxidase Response to Hypoxia and Hypercapnia in the Adult Brain

Christina Kolyva, Arnab Ghosh, Ilias Tachtsidis, David Highton, Martin Smith, and Clare E. Elwell

Abstract Transcranial near-infrared spectroscopy (NIRS) provides an assessment of cerebral oxygen metabolism by monitoring concentration changes in oxidised cytochrome *c* oxidase Δ[oxCCO]. We investigated the response of Δ[oxCCO] to global changes in cerebral oxygen delivery at different source-detector separations in 16 healthy adults. Hypoxaemia was induced by delivery of a hypoxic inspired gas mix and hypercapnia by addition of 6 % CO_2 to the inspired gases. A hybrid optical spectrometer was used to measure frontal cortex light absorption and scattering at discrete wavelengths and broadband light attenuation at 20, 25, 30 and 35 mm. Without optical scattering changes, a decrease in cerebral oxygen delivery, resulting from the reduction in arterial oxygen saturation during hypoxia, led to a decrease in Δ[oxCCO]. In contrast, Δ[oxCCO] increased when cerebral oxygen delivery increased due to increased cerebral blood flow during hypercapnia. In both cases the magnitude of the Δ[oxCCO] response increased from the detectors proximal (measuring superficial tissue layers) to the detectors distal (measuring deep tissue layers) to the broadband light source. We conclude that the Δ[oxCCO] response to hypoxia and hypercapnia appears to be dependent on penetration depth, possibly reflecting differences between the intra- and extracerebral tissue concentration of cytochrome *c* oxidase.

C. Kolyva (✉) • I. Tachtsidis • C.E. Elwell
Department of Medical Physics and Bioengineering, University College London, London, UK
e-mail: ckolyva@medphys.ucl.ac.uk

A. Ghosh • D. Highton • M. Smith
Neurocritical Care Unit, University College Hospitals, London, UK

S. Van Huffel et al. (eds.), *Oxygen Transport to Tissue XXXV*, Advances in Experimental Medicine and Biology 789, DOI 10.1007/978-1-4614-7411-1_47,

47.1 Introduction

Transcranial near-infrared spectroscopy (NIRS) provides a measure of cerebral oxygen delivery by monitoring concentration changes in oxygenated ($\Delta[HbO_2]$) and deoxygenated haemoglobin (Δ[HHb]), non-invasively. Concentration changes in oxidised cytochrome *c* oxidase (Δ[oxCCO]) can also be derived with NIRS. Cytochrome *c* oxidase (CCO) is the terminal electron acceptor in the mitochondrial respiratory chain, and being responsible for over 95 % of oxygen metabolism, it is instrumental in aerobic ATP synthesis and in maintaining mitochondrial function [1]. Since in the short term the total concentration of CCO does not change, changes in the Δ[oxCCO] signal track changes in the CCO redox state, which essentially reflects the balance between cerebral energy supply and demand [2]. Thus, Δ[oxCCO] is an appealing target for bedside monitoring, for the assessment of regional cerebral metabolic status and oxygen utilisation.

Technical complexities are associated with the measurement of Δ[oxCCO] in the adult brain, in the presence of significantly higher concentrations of haemoglobin, most notably the possible interference of changes in optical scattering with the NIRS measurements [3, 4] and the insufficient chromophore separation by the algorithm used to convert optical density into concentration changes [3, 4]. A hybrid optical spectrometer (pHOS) with the capacity for measurements at multiple interoptode distances (and thus at multiple depths) and accompanying algorithm designed to address the above issues have recently been developed by our group [5]. The capacity for multi-distance Δ[oxCCO] recordings would contribute considerably to the interpretation of this measurement, by determining if there is a distance/depth-dependent response of Δ[oxCCO] in the adult head [6–8].

The aim of the present study was to investigate the multi-depth response of Δ[oxCCO] to global changes in cerebral oxygen delivery driven by systemic hypoxia and hypercapnia. We hypothesised that Δ[oxCCO] would show an incremental response with increasing source-detector separation, mirroring potential differences in the extra- and intracranial distribution of this chromophore.

47.2 Methods

47.2.1 Study Population

A total of 16 adult healthy volunteers were studied (Table 47.1). The studies were approved by the local ethics committee and all subjects provided written informed consent.

Table 47.1 Patient demographics and systemic variables

	Hypoxia		Hypercapnia	
n	15		12	
Age	30 (22–35)		30 (25–34)	
Gender	10 male		8 male	
	Baseline	**End challenge**	**Baseline**	**End challenge**
SpO_2 *(%)*	97 (93–100)	80 (70–86)*	98 (93–100)	97 (93–100)
$EtCO_2$ *(kPa)*	5.5 (3.9–6.4)	5.2 (4.2–6.4)*	5.5 (4.5–6.0)	7.8 (6.3–9.2)*
Vmca *(Δ% from rest)*	0	14.3 (−5.4–53.2)*	0	57.4 (21.8–87.0)*

Table entries are mean (range)
*$P<0.05$

47.2.2 *Protocol*

Hypoxaemia was induced by delivery of a hypoxic gas mix, using a sequential gas delivery circuit. Following 5 min of air inhalation (‘*baseline*’), nitrogen was added to the inspired gas and titrated to produce a progressive reduction in arterial oxygen saturation (SpO_2) to 80 %, whilst maintaining constant end-tidal partial pressure of carbon dioxide ($EtCO_2$). SpO_2 was sustained at 80 % for 5 min, before returning the inspired gas to room air. Upon reaching normoxia, 5 min of baseline completed the sequence. Hypercapnia was induced by addition of 6 % CO_2 to the inspired gas mix, after 5 min of initial baseline. This mix was inhaled for another 5 min, before the inspired CO_2 fraction was returned to 0. When normocapnia was restored, 5 min of baseline concluded the protocol.

47.2.3 *Instrumentation*

The pHOS, described in more detail elsewhere, was used for frontal near-infrared measurements during changes in the composition of the inspired gases [5]. The pHOS combines frequency domain (FD) and broadband (BB) components and can measure light absorption and scattering at discrete wavelengths (690, 750, 790 and 850 nm), together with BB light attenuation in the range 504–1,068 nm. Each pHOS optode incorporates an FD channel (source-detector spacing 30 and 35 mm) and BB channel (source-detector spacing 20, 25, 30 and 35 mm). One optode positioned ipsilateral to other cerebral monitoring was used. Each sampling cycle of the pHOS lasts 3.2 s, and BB and FD measurements are sequential. Systemic recordings included beat-to-beat SpO_2, continuous measurements of inspired/expired oxygen and CO_2 partial pressure and middle cerebral artery flow velocity, measured with transcranial Doppler ultrasonography.

47.2.4 Data Analysis

Data analysis was performed in Matlab (version R2010b, Mathworks). Δ[HbO_2], Δ[HHb] and Δ[oxCCO] were determined from the 780–900 nm portion of the BB attenuation change data using the UCLn algorithm, which is based on the modified Beer-Lambert law. Differential pathlength factor (DPF) was assumed to be 6.26 [9] and its wavelength dependence was accounted for [10]. Changes in total haemoglobin concentration were defined as Δ[HbT] = Δ[HbO_2] + Δ[HHb] and in haemoglobin difference as Δ[Hbdiff] = Δ[HbO_2] − Δ[HHb]. The concentrations were linearly detrended for removal of baseline drift and low-pass filtered with a fifth-order Butterworth filter (cut-off frequency 0.08 Hz). The absorption (μ_a) and reduced scattering (μ_s') coefficients were quantified from the FD measurements.

Inspired and end-tidal gas partial pressures were derived from the positive and negative envelopes of the partial pressure waveforms. Separately for each volunteer, the beginning of the induction to hypoxia/hypercapnia and end of hypoxia/hypercapnia were identified from the O_2/CO_2 envelopes, with the period between the two points denoted as 'challenge'. To enable data averaging across subjects despite potential variation in the timing of their response, the challenges were the split into eight phases; hypoxia was split into four equal phases corresponding to the gradual fall in SpO_2 and another four corresponding to the plateau of 80 % SpO_2, whilst hypercapnia was split in eight equal phases. These eight phases represented time intervals '2–9' of each challenge. Time interval '1' corresponded to the baseline immediately prior to the induction, whilst time intervals '10–17' covered the period after the end of the challenge; all these intervals mirrored in duration interval '2'. Representative optical and systemic data for each of the 17 time intervals were derived by averaging the last 9*3.2 s worth of data of the corresponding 17 data segments.

47.2.5 Statistical Analysis

SPSS was used (version 18.0, IBM). Normality was assessed with Q-Q plots. Repeated measures ANOVA tests with Greenhouse-Geisser correction determined whether the group means overall changed significantly between time points 1 and 17. Post hoc tests with Bonferroni corrections established the time points with a statistically significant change compared to point 1. Average data are expressed as mean ± SD and statistical significance was assumed at $P < 0.05$.

47.3 Results

Summary demographic data, separately for the two challenges, are given in Table 47.1. The table also includes group data at baseline and at the end of the challenge (time point 9) for a number of systemic parameters.

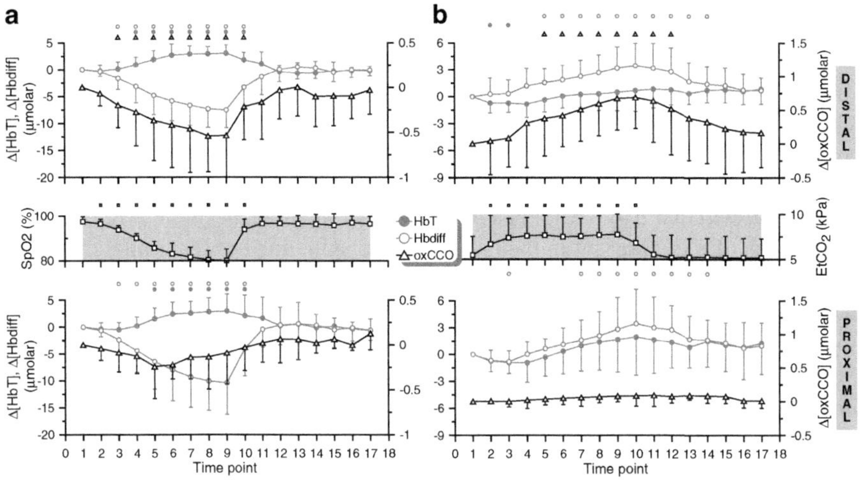

Fig. 47.1 (**a**) Grand averages of the time courses of Δ[HbT], Δ[Hbdiff] and Δ[oxCCO] measured by the detectors distal (*top*) and proximal (*bottom*) to the light source, over the 15 volunteers that participated in the hypoxia challenge. The corresponding arterial oxygen saturation trace (SpO_2) is also provided for reference. The small symbols on top of each plot indicate statistical significance with respect to point 1 ($P<0.05$) for the parameters plotted in matching symbols. (**b**) Grand averages of the time courses of Δ[HbT], Δ[Hbdiff] and Δ[oxCCO] measured over the 12 volunteers that participated in the hypercapnia challenge. The corresponding end-tidal CO_2 trace ($EtCO_2$) is also provided for reference. The small symbols on top of each plot indicate statistical significance with respect to point 1 ($P<0.05$)

Group grand averages of the time courses of Δ[oxCCO], Δ[HbT] and Δ[Hbdiff] as measured during hypoxia by the two extreme detectors are shown in Fig. 47.1a. During hypoxia, the detector distal to the light source recorded a decrease in Δ[oxCCO] ($P<0.001$), in agreement with previous studies [6], which was accompanied by an increase in Δ[HbT] ($P<0.001$) and decrease in Δ[Hbdiff] ($P<0.001$). In terms of directional changes, the findings were qualitatively similar for all detectors, but the magnitude of the Δ[oxCCO] response to hypoxia gradually increased from the proximal to the distal detectors.

Figure 47.1b displays the group time courses of Δ[oxCCO], Δ[HbT] and Δ[Hbdiff] during hypercapnia. The distal detector registered an increase in Δ[oxCCO] ($P<0.001$), in agreement with previous studies [11], accompanied by an increase in both Δ[HbT] ($P<0.05$) and Δ[Hbdiff] ($P<0.001$). These trends were similar for all detectors, but the magnitude of the Δ[oxCCO] response to hypercapnia gradually increased from the proximal to the distal detectors.

No changes in μ_s' (P=NS for all wavelengths) were measured during hypoxia or hypercapnia.

47.4 Discussion and Conclusions

These first multi-depth Δ[oxCCO] measurements during global changes in cerebral oxygen delivery have revealed an increase in the amplitude of the Δ[oxCCO] response with increasing penetration depth. This dependence is most likely mirroring differences in the concentration distribution of CCO in the adult head. It has been suggested that Δ[oxCCO] is a brain-specific signal on account of higher concentrations of CCO present in the brain than the skin, due to higher mitochondrial density [6–8], but evidence for the existence of such a distribution in the adult head has not been provided before, partly because the necessary data were not technologically possible to obtain in humans in vivo.

With measurements from the distal detector therefore being representative of cerebral events, our results support the findings of previously published studies that were conducted with a single source-detector pair and could thus not confirm with certainty that the changes in the head Δ[oxCCO] signal they reported during manipulation of cerebral oxygen delivery were indeed of cerebral origin [6, 11]. In healthy adults, a decrease in cerebral oxygen delivery, induced by a reduction in arterial oxygen saturation during moderate hypoxia, was followed by a 0.24 μmolar decrease (median) in Δ[oxCCO], indicating reduced cellular oxygen availability [6]. We measured a decrease of 0.55 μmolar (median), with the same interoptode spacing as [6] (3.5 cm). Analogously, an increase in oxygen delivery, via increased cerebral blood flow during hypercapnia, was accompanied by a 0.25 ± 0.17 μmolar increase in Δ[oxCCO], indicating that at normoxic normocapnia CCO is not fully oxidised [11]. In agreement with [11], our data show an increase of 0.69 ± 0.46 μmolar at the same interoptode spacing (3.5 cm). Some animal data suggest that CCO is fully oxidised in normoxia [12]. However, our studies were carried out in healthy awake humans rather than anaesthetised animals, which may have a significant impact on the relationship between oxygen supply and demand and CCO oxidation.

The pHOS system has been specifically optimised for monitoring Δ[oxCCO], by combining measurements of light absorption and scattering at discrete wavelengths with multi-distance measurements of BB light attenuation. No change in optical scattering (a potential confounding influence on Δ[oxCCO] measurements) was measured. Moreover, the use of BB light for resolving chromophores over hundreds of wavelengths gave confidence that the algorithm used to convert optical density to concentration changes would provide sufficient chromophore separation.

We conclude that the Δ[oxCCO] response to hypoxia and hypercapnia appears to be dependent upon penetration depth, possibly reflecting differences between the intra- and extracerebral tissue concentration of CCO.

Acknowledgments The authors would like to thank the volunteers who participated in this study and acknowledge funding from the UK Medical Research Council and Central London Research Network. This work was undertaken at University College London Hospitals and partially funded by the Department of Health's National Institute for Health Research Centres funding scheme.

References

1. Richter OM, Ludwig B (2003) Cytochrome c oxidase–structure, function, and physiology of a redox-driven molecular machine. Rev Physiol Biochem Pharmacol 147:47–74
2. Smith M (2011) Shedding light on the adult brain: a review of the clinical applications of near-infrared spectroscopy. Philos Transact A Math Phys Eng Sci 369(1955):4452–4469
3. Cooper CE, Cope M, Springett R et al (1999) Use of mitochondrial inhibitors to demonstrate that cytochrome oxidase near-infrared spectroscopy can measure mitochondrial dysfunction noninvasively in the brain. J Cereb Blood Flow Metab 19(1):27–38
4. Cooper CE, Springett R et al (1997) Measurement of cytochrome oxidase and mitochondrial energetics by near-infrared spectroscopy. Philos Trans R Soc Lond B Biol Sci 352:669–676
5. Kolyva C, Tachtsidis I, Ghosh A et al (2012) Systematic investigation of changes in oxidized cerebral cytochrome c oxidase concentration during frontal lobe activation in healthy adults. Biomed Opt Express 3(10):2550–2566
6. Tisdall MM, Tachtsidis I, Leung TS, Elwell CE, Smith M (2007) Near-infrared spectroscopic quantification of changes in the concentration of oxidized cytochrome c oxidase in the healthy human brain during hypoxemia. J Biomed Opt 12(2):024002
7. Smith M, Elwell C (2009) Near-infrared spectroscopy: shedding light on the injured brain. Anesth Analg 108(4):1055–1057
8. Kakihana Y, Matsunaga A, Yasuda T, Imabayashi T, Kanmura Y, Tamura M (2008) Brain oxymetry in the operating room: current status and future directions with particular regard to cytochrome oxidase. J Biomed Opt 13(3):033001
9. Duncan A, Meek JH, Clemence M et al (1995) Optical pathlength measurements on adult head, calf and forearm and the head of the newborn infant using phase resolved optical spectroscopy. Phys Med Biol 40(2):295–304
10. Essenpreis M, Elwell CE, Cope M, van der Zee P, Arridge SR, Delpy DT (1993) Spectral dependence of temporal point spread functions in human tissues. Appl Opt 32(4):418–425
11. Tachtsidis I, Tisdall MM, Leung TS et al (2009) Relationship between brain tissue haemodynamics, oxygenation and metabolism in the healthy human adult brain during hyperoxia and hypercapnea. Adv Exp Med Biol 645:315–320
12. Hoshi Y, Hazeki O, Kakihana Y, Tamura M (1997) Redox behavior of cytochrome oxidase in the rat brain measured by near-infrared spectroscopy. J Appl Physiol 83(6):1842–1848

Chapter 48
Modeling Hemoglobin Nitrite Reductase Activity as a Mechanism of Hypoxic Vasodilation?

Zimei Rong and Chris E. Cooper

Abstract The brain's response to hypoxia is to increase cerebral blood flow (CBF). However, the molecular mechanism underpinning this phenomenon is controversial. We have developed a model to simulate brain blood flow and oxygen metabolism called BRAINSIGNALS. This model is primarily designed to assist in the interpretation of multimodal noninvasive clinical measurements. However, we have recently used this model to test the feasibility of a range of molecular mechanisms proposed to explain hypoxic vasodilation. An increase in the concentration of the vasodilator nitric oxide (NO) at low pO_2 is a feature of many such mechanisms. One model suggests that mitochondrial cytochrome *c* oxidase (CCO) catalyzes the metabolism of NO. This metabolism declines at low pO_2, resulting in an increase in the steady-state levels of NO and a consequent increase in CBF. Using BRAINSIGNALS we were able to model this effect. However, the increases in NO and CBF occurred at far lower pO_2 values than predicted from physiological data (Rong et al. 2013 Adv Exp Med Biol. 765, 231–238). The aim of the present study was to test an alternative mechanism, one that actively generates NO as pO_2 drops, namely, the reduction of nitrite to NO by deoxyhemoglobin. In this mechanism, NO synthesis has a maximum of NO production near the hemoglobin p50. The addition of this mechanism resulted in a significantly better fit to the experimental data of the $CBF(PaO_2)$ curve.

Z. Rong (✉) • C.E. Cooper
School of Biological Sciences, University of Essex, Colchester CO4 3SQ, UK
e-mail: zimeirong@hotmail.com

S. Van Huffel et al. (eds.), *Oxygen Transport to Tissue XXXV*, Advances in Experimental Medicine and Biology 789, DOI 10.1007/978-1-4614-7411-1_48,

48.1 Introduction

Hypoxia is a pathological situation in which the whole body or an isolated organ suffers from insufficient oxygen supply. In most tissues of the body, the response to hypoxia is vasodilation. This relationship has been studied in many systems from human [1] to animals, such as dogs [2] and ducks [3]. The precise shape of the CBF(PaO_2) curve in the brain *in vivo* is subject to considerable experimental variation – and in the case of humans for obvious reasons, the lower limits are not readily explored. For the purposes of model simulation, the dog system provides some of the most reliable and accurate data [2] and was therefore used in this study.

Many hypotheses have been suggested to explain the mechanism of hypoxic vasodilation. However, none of them have gained universal acceptance. We have developed a BRAINSIGNALS model [4] to simulate hypoxia of the brain, cerebral blood flow increasing as pO_2 decreases [5]. We feel this model can be a useful test bed to compare different molecular mechanisms. In our previous work we used this model to test the idea that nitric oxide (NO) was the key mediator of hypoxic vasodilation. The BRAINSIGNALS model incorporates an increase in the radius of the blood vessel with decreasing pO_2. Therefore, we incorporated an NO dependence on the size of the vessel radius [5], resulting in an NO concentration dependence on CBF.

It is not clear why NO levels should increase at low pO_2, given that NO production by the enzyme nitric oxide synthase (NOS) requires oxygen. One model suggests that mitochondrial cytochrome *c* oxidase catalyzes the metabolism of NO [6]. This metabolism declines at low pO_2, resulting in an increase in the overall NO levels and consequently an increase in CBF. Recently we used BRAINSIGNALS to test the feasibility of this mechanism [5]. However, using the literature value for the oxygen K_m for NOS, we were unable to model the data adequately. The decrease in NO production at low pO_2 dominated the decreased mitochondrial metabolism of NO resulting in a fall, rather than a rise in the steady-state NO concentration. We were able to induce a hypoxic rise in NO and hence vasodilation if we lowered the NOS oxygen K_m. However, even then the threshold where CBF rose significantly occurred at a very low pO_2, inconsistent with the experimental data. We therefore decided to test an alternative NO generation mechanism whereby deoxyhemoglobin actively produces NO via the reduction of nitrite in an allosteric manner, resulting in maximal NO production near the hemoglobin-oxygen p50 [7]. The aim of this study is to incorporate this mechanism into BRAINSIGNALS model to test how the simulated CBF(PaO_2) curve fits to the experimental data.

48.2 Mathematical Model

We wish to reproduce this experimental hypoxic vasodilation with a mathematical model. First, we digitized the experimental data from McDowall reporting hypoxic vasodilation in the dog [2]. We then used a BRAINSIGNALS model [4] which is available online [8] to simulate hypoxic vasodilation. BRAINSIGNALS is a two-compartment model. The first compartment is the cerebral blood vessel, which is approximated as a tube with single radius *r*, which is determined from the equilibrium of forces acting on the vessel walls. The cerebral blood flow *CBF* is linked to *r* as

$$CBF = \left(P_a - P_v\right) K_G r^4 \tag{48.1}$$

The second compartment is the mitochondria, where oxygen is reduced. The full description of mitochondrial activity is complex. First, NADH transfers electrons to Cu_{Ao} with a reaction rate f_1, forming Cu_{Ar} and NAD^+. Cu_{Ar} then transfers electrons to cytochrome a_{3o} with a rate of f_2, forming a_{3r} and Cu_{Ao}. In the final step, a_{3r} transfers electrons to O_2 with a rate of f_3. During the reaction, a_{3r} is oxidized to a_{3o} and oxygen is reduced to water. In the BRAINSIGNALS model, the whole metabolism process was simplified to four differential equations for four components, CuA_o, $Cyta_{3r}$, O_2, and H (proton), as

$$\mathrm{d}\left[\mathrm{CuA_o}\right] / \mathrm{d}t = 4\left(f_2 - f_1\right) \tag{48.2}$$

$$\mathrm{d}\left[\mathrm{Cytoa_3}\right] / \mathrm{d}t = 4\left(f_2 - f_3\right) \tag{48.3}$$

$$\mathrm{d}\left[\mathrm{O_2}\right] / \mathrm{d}t = J_{\mathrm{O2}} / Vol_{mit} - f_3 \tag{48.4}$$

$$\mathrm{d}\left[\mathrm{H^+}\right] / \mathrm{d}t = \left(L - p_1 f_1 - p_2 f_2 - p_3 f_3\right) / Vol_{Hi} \tag{48.5}$$

where the notations were defined previously [4] and also in the following equations.

There are two links between the two compartments: cerebral blood vessel and mitochondria. The first one is oxygen delivery from blood to mitochondria as

$$J_{\mathrm{O2}} = CBF\left(\left[\mathrm{HbO_{2a}}\right] - \left[\mathrm{HbO_{2v}}\right]\right) = D_{\mathrm{O2}}\left(\left[\mathrm{O_{2a}}\right] - \left[\mathrm{O_{2v}}\right]\right) \tag{48.6}$$

The second link is stimuli η linked to radius of blood vessel *r*, which is simplified from several equations from the literature [4] as

$$\tanh\left(\frac{\eta}{2}\right) = \frac{\dfrac{\left((P_a + P_v)/2 - P_{ic}\right) r}{\sqrt{r^2 + 2r_o h_o + h_o^2} - r} - \sigma_{eo}\left(\exp\left(\dfrac{K_\sigma (r - r_o)}{r_o}\right) - 1\right) + \sigma_{coll}}{\dfrac{T_{\max o}}{\sqrt{r^2 + 2r_o h_o + h_o^2} - r} \exp\left(-\left|\dfrac{r - r_m}{r_t - r_m}\right|^{n_m}\right)} - 1 \tag{48.7}$$

The relationship between η and *r* is monotonic, i.e., with *r* increasing, η decreases monotonically. η is linearly proportional to the O_2 or NO concentration, which is filtered to take account of time delays from the concentration change to the blood vessel widening. This delay is approximated with a first-order filter as

$$\mathrm{d}v_x / \mathrm{d}t = \left(x - v_x\right) / \tau_x \quad \text{where} \quad x = \mathrm{O_2} \text{ or } \mathrm{NO} \tag{48.8}$$

In our previous work on modeling of NO-mediated hypoxic vasodilation, we assumed that NO is generated by nitric oxide synthase and removed by CCO. Here we retain that NO is removed by CCO and assume that NO is generated by deoxy-hemoglobin shown in Fig. 48.1a as

$$\mathrm{HbFe^{2+}}\left(\mathrm{deoxyHb}\right)+\mathrm{NO_2}^{-}+\mathrm{H}^{+}\rightarrow\mathrm{NO}+\mathrm{HbFe^{3+}}+\mathrm{OH}^{-} \tag{48.9}$$

The rate of NO production is $d[NO]/dt=k_{NO}[Hb][NO_2^-]$ and k_{NO} absorbs oxygen dependence of $[Hb^{2+}]$ into this coefficient. This reaction rate is oxygen concentration dependent for a combination of reasons. The concentration of the substrate deoxyhemoglobin increases as pO_2 falls. However, the intrinsic reaction rate of nitrite with the deoxyheme group is lower in the form of the hemoglobin tetramer (T state) that predominates at low pO_2. This results in a complicated "inverted U" dependence of NO production on pO_2. In order to describe this phenomenon mathematically, we need to use the Monod-Wyman-Changeux (MWC) model of the allosteric transition in hemoglobin. Three parameters in the MWC model [9] are: a dimensionless oxygen concentration $\alpha=pO_2/K_R$, a conformation equilibrium constant $L=T_0/R_0$, and a dissociation constant ratio $c=K_R/K_T$, where K_R and K_T are defined as the microscopic dissociation constants of oxygen from these states of hemoglobin, respectively. k_R and k_T are the microreaction rate constants of the unliganded R-state and T-state heme sites with nitrite, respectively:

$$k_{NO}=\frac{(1+\alpha)^3 k_R + L(1+c\alpha)^3 k_T}{(1+\alpha)^4 + L(1+c\alpha)^4} \tag{48.10}$$

48.3 Results and Discussions

The normal cerebral blood flow of dog was reported as 47 ml/min/100 g [10]. Conventionally hemoglobin-oxygen affinity is characterized experimentally by the Hill parameters. For example, oxygen affinity of dog hemoglobin was represented by Hill coefficients $n=2.63$ and $p50=31$ mmHg [11]. However, the full MWC model is necessary to describe the hemoglobin nitrite reductase activity. However, it is not trivial to obtain the MWC parameters L, c, and K_R from the Hill parameters. The MWC model parameters $L=100{,}605$, $c=0.0138179$, and $K_R=1.61253$ mmHg were obtained from fitting the maximum gradient of the Hill plot ($n=\max[d\log(Y/(1-Y))/d\log(\alpha)]=2.63$) and the oxygenation $Y(p50=31\text{ mmHg})=[\alpha(1+\alpha)^3+Lc\alpha(1+c\alpha)^3]/[(1+\alpha)^4+L(1+c\alpha)^4]=0.5$.

There is no consensus value for the intrinsic rate constants for NO production by R-state and T-state hemoglobin [12]. However, we were able to obtain reasonable

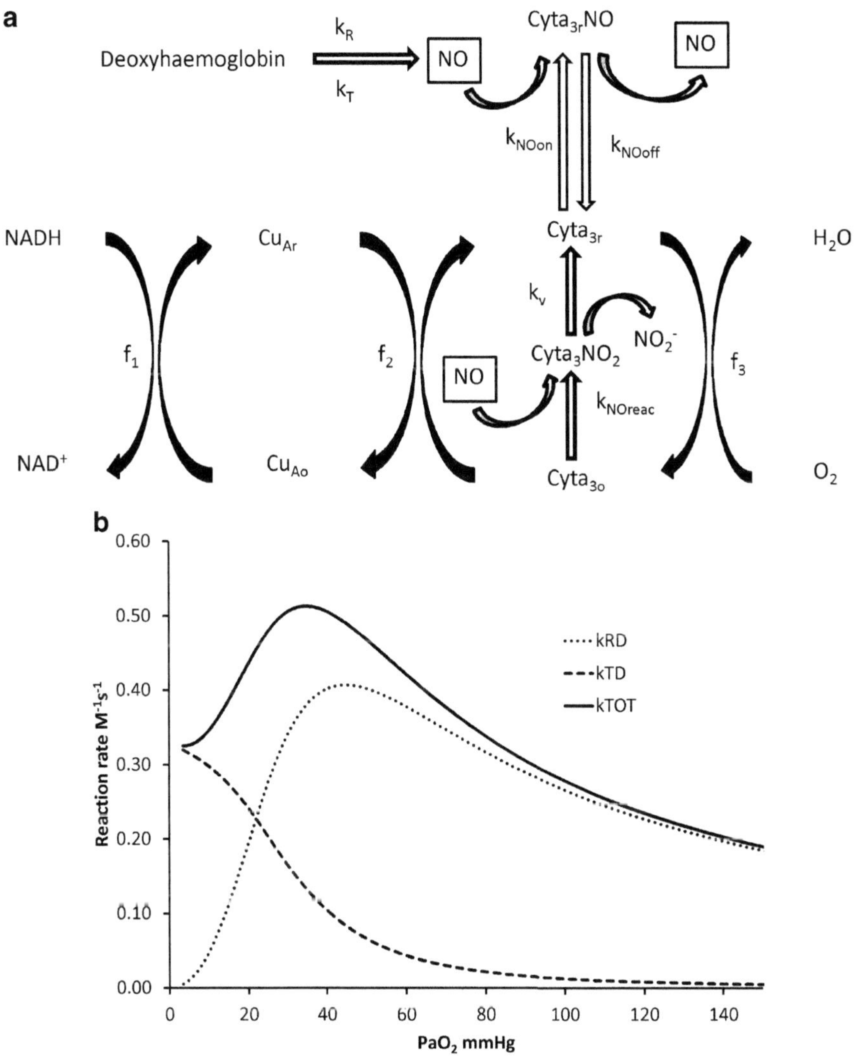

Fig. 48.1 (**a**) Schematic diagram of the changes to the BRAINSIGNALS model. The chemical reaction rates for NO generation by deoxyhemoglobin are $k_R = 18$ M^{-1} s^{-1} and $k_T = 0.33$ M^{-1} s^{-1}. The chemical reaction rates for competitive inhibition of cytochrome oxidase by NO are $k_{NOon} = 0.04$ nM^{-1} s^{-1} and $k_{NOoff} = 0.16$ s^{-1}. The chemical reaction rates for uncompetitive inhibition are $k_{NOreac} = 2.0 \times 10^{-4}$ nM^{-1} s^{-1} and $k_v = 0.75$ s^{-1}. (**b**) The macro bimolecular nitrite reductase activity of hemoglobin as a function of oxygen partial pressure. The *solid line* represents the overall rate constant and is the sum of the R-state rate constant (*dotted line*) and T-state rate constant (*dashed line*). For calculation of the nitrite reduction rate constant, the MWC model parameters used were $L = 100{,}605$ and $c = 0.0138179$ and $K_R = 1.61253$ mmHg

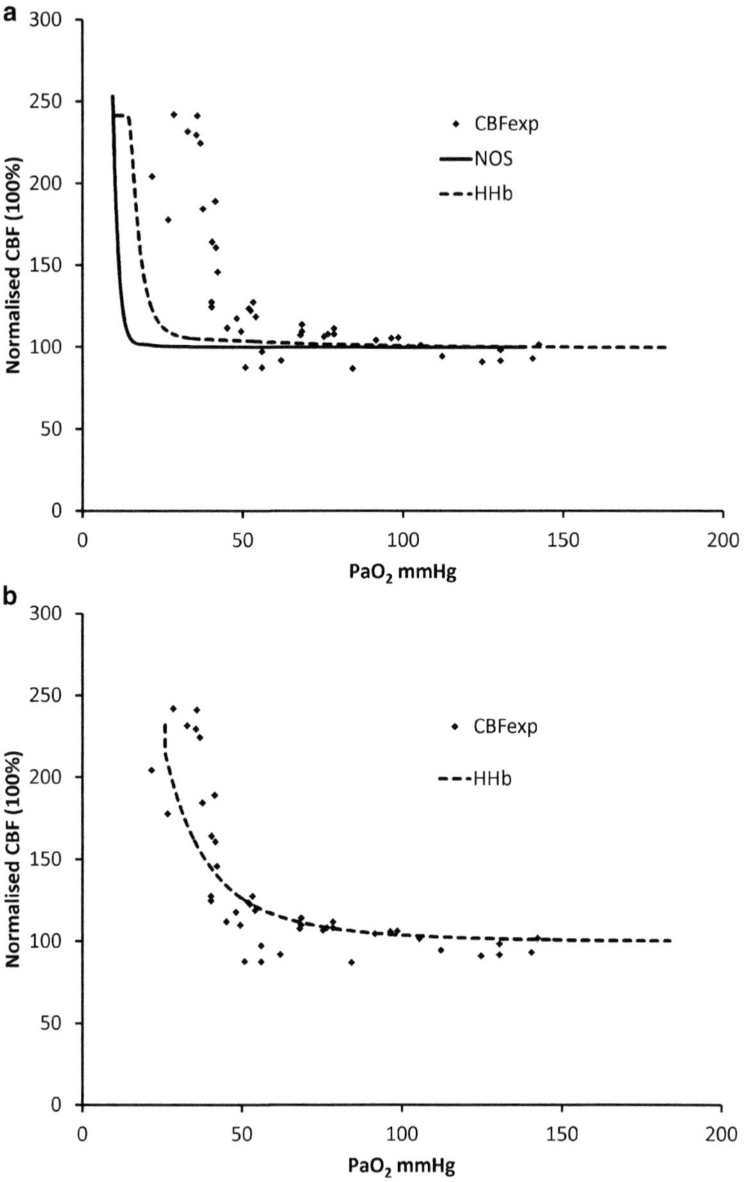

Fig. 48.2 (**a**) Experimental CBF(PaO_2) in dogs was reported by McDowall [2]. The *solid line* is the simulated CBF(PaO_2) using the BRAINSIGNALS model with NO generated by NOS (K_m=0.00003 mM) [5] and the *dashed line* with NO generated from nitrite by deoxyhemoglobin as Eq. 48.10 (k_R=18 $M^{-1}s^{-1}$ and k_T=0.33 $M^{-1}s^{-1}$). (**b**) As 2A but for the fit k_T (only) was changed to 6.6 M^{-1} s^{-1}

fits to literature data using k_R=18 M^{-1} s^{-1} and k_T=0.33 M^{-1} s^{-1}. Combining these values with the MWC parameters previously calculated, the nitrite reductase activity of deoxyhemoglobin can be expressed as a function of pO_2 (Fig. 48.1b). As expected, the contribution of the overall nitrite reductase activity from deoxyhemoglobin molecules in T-state hemoglobin increases monotonically as pO_2 falls, whereas that from the deoxyhemes in the R state at first increases and then decreases as the concentration of this form of the tetramer falls off at low pO_2.

An experimental CBF(PaO_2) in dogs (dotted curve) was constructed from McDowall [2]. We then incorporated the nitrite reductase mechanism of NO production into the BRAINSIGNALS model and simulated hypoxic vasodilation (Fig. 48.2). NO removal was modeled in a cytochrome *c* oxidase-dependent manner as described previously [5]. The solid line in Fig. 48.2a is the simulated CBF(PaO_2) curve with NO levels generated by NOS and removed by cytochrome *c* oxidase [5]; the dashed line incorporates NO production from nitrite by deoxyhemoglobin rather than from NOS. We can see that there is now an improvement in the fit, but the threshold where CBF rises significantly is still at too low a pO_2.

As there are no current data reporting on dog hemoglobin nitrite reductase activity, the values of k_R and k_T (Fig. 48.2a) were simply set from measurements of the human protein. Therefore, we felt justified in heuristically varying k_R and k_T to see if we could obtain an improved fit. A simple change of k_T (alone) to 6.6 $M^{-1}s^{-1}$ was sufficient to result in a close fit of the model to the data (Fig. 48.2b). We are currently investigating whether such a large (20-fold) interspecies difference in k_T is biochemically reasonable.

48.4 Conclusions

During this study we modified our two compartment BRAINSIGNALS model to derive a single equation linking external stimuli to the radius of the blood vessel. We then derived a compact equation to express the hemoglobin nitrite reductase activity. Using these equations we determined that hypoxic vasodilation in the brain can be modeled assuming nitric oxide is produced by hemoglobin and consumed by mitochondrial cytochrome *c* oxidase and that the resulting steady-state concentration controls the blood vessel diameter. However, it is as yet not clear that this represents a unique solution to the problem nor that the precise rate constants required to fit the data are biochemically reasonable.

Acknowledgments This work is financially supported by the Leverhulme Trust (F/00 213/T).

References

1. Shimojyo S, Scheinberg P, Kogure K, Reinmuth OM (1968) The effect of graded hypoxia upon transient cerebral blood flow and oxygen consumption. Neurology 18(2):127–133
2. McDowall DG (1966) Interrelationship between blood oxygen tensions and cerebral blood flow. In: Payne JP, Hill DW (eds) Oxygen measurements in blood and tissue and their significance. Churchill, London, pp 205–219
3. Grubb B, Colacino JM, Schmidt-Nielsen K (1978) Cerebral blood flow in birds: effect of hypoxia. Am J Physiol 234(3):H230–H234
4. Banaji M, Mallet A, Elwell CE, Nicholls P, Cooper CE (2008) A model of brain circulation and metabolism: NIRS signal changes during physiological challenges. PLoS Comput Biol 4(11):e1000212
5. Rong Z, Banaji M, Moroz T, Cooper CE (2013) Can mitochondrial cytochrome oxidase mediate hypoxic vasodilation via nitric oxide metabolism? Adv Exp Med Biol 765:231–238
6. Palacios-Callender M, Hollis V, Mitchison M et al (2007) Cytochrome c oxidase regulates endogenous nitric oxide availability in respiring cells: a possible explanation for hypoxic vasodilation. Proc Natl Acad Sci USA 104(47):18508–18513
7. Gladwin MT, Kim-Shapiro DB (2008) The functional nitrite reductase activity of the hemeglobins. Blood 112(7):2636–2647
8. BRAINCIRC: an open source modelling environment http://braincirc.sourceforge.net. Brain signal model: http://www.medphys.ucl.ac.uk/braincirc/download/repos/NIRSmodel.html (2011)
9. Monod J, Wyman J, Changeux JP (1965) On the nature of allosteric transitions: a plausible model. J Mol Biol 12:88–118
10. Heistad DD, Marcus ML, Abboud FM (1978) Role of large arteries in regulation of cerebral blood flow in dogs. J Clin Invest 62(4):761–768
11. Pagel PS, Hettrick DA, Montgomery MW et al (1998) RSR13, a synthetic modifier of haemoglobin-oxygen affinity, enhances the recovery of stunned myocardium in anesthetised dogs. J Pharmacol Exp Ther 285:1–8
12. Patel RP, Hogg N, Kim-Shapiro DB (2011) The potential role of the red blood cell in nitrite-dependent regulation of blood flow. Cardiovasc Res 89:507–515

Part VII
Measurement Technologies

Chapter 49
Development of a Hybrid Microwave-Optical Tissue Oxygenation Probe to Measure Thermal Response in the Deep Tissue

Allann Al-Armaghany, Kenneth Tong, and Terence S. Leung

Abstract The design of a new non-invasive hybrid microwave-optical tissue oxygenation probe is presented, which consists of a microwave biocompatible antenna and an optical probe. The microwave antenna is capable of inducing localised heat in the deep tissue, causing tissue blood flow and therefore tissue oxygenation to change. These changes or thermal responses are measured by the optical probe using near-infrared spectroscopy. Thermal responses provide important information on thermoregulation in human tissue. The first prototype of the biocompatible antenna was developed and placed on the human calf for in vivo experiments. The measured results include oxy-, deoxy- and total haemoglobin concentration changes (ΔHbO_2/ΔHHb/ΔHbT), tissue oxygenation index and the normalised tissue haemoglobin index for two human subjects. Both ΔHbO_2 and ΔHbT show an increase during 5 min of microwave exposure. The thermal response, defined as the ratio of the increase in ΔHbT to the time duration, is 7.7 μM/s for subject 1 (fat thickness = 6.8 mm) and 18.9 μM/s for subject 2 (fat thickness = 5.0 mm), which may be influenced by the fat thicknesses. In both subjects, ΔHbO_2 and ΔHbT continued to increase for approximately another 70 s after the microwave antenna was switched off.

A. Al-Armaghany (✉)
Department of Electronic and Electrical Engineering,
University College London, London, UK

137 Kiln Place, London NW5 4AP, UK
e-mail: A.armaghany@gmail.com; a.al-armaghany@ucl.ac.uk

K. Tong
Department of Electronic and Electrical Engineering,
University College London, London, UK

T.S. Leung
Department of Medical Physics and Bioengineering,
University College London, London, UK

S. Van Huffel et al. (eds.), *Oxygen Transport to Tissue XXXV*, Advances in Experimental Medicine and Biology 789, DOI 10.1007/978-1-4614-7411-1_49,

49.1 Introduction

The body temperature of living organism is maintained at a constant level via metabolic processes. Excessive heat can be carried away by the increased blood flow, as a result of dilated local blood vessels. The aim of this study was to investigate the change in blood volume and oxygenation caused by the controlled temperature elevation in the deep tissue including skin, fat and muscle, which can be achieved by the newly developed hybrid microwave-optical probe. The non-invasive hybrid probe exploits a new type of biocompatible antenna to induce a local temperature rise in muscle. The heating of tissues causes vasodilation and changes the local tissue oxygenation, which can be measured by an optical probe using near-infrared spectroscopy. Current non-invasive thermoregulation investigations into vascular diseases, such as Raynaud's phenomenon and systemic sclerosis, involve only external heating on the skin [1]. The capability of heating deep tissues such as muscle from within will provide new information about the pathophysiology of these vascular diseases. Another potential application is to aid the differentiation of certain types of tumour from normal tissues, e.g. upon heating, certain tumour blood flow (and therefore tissue oxygenation) has less capacity to increase compared to normal tissue because tumour blood vessels are already extremely dilated even without heating [2]. The hybrid probe can also improve the understanding of microwave diathermy, a therapy technique commonly used to treat muscle suffering from sports injuries, to help optimise its therapeutic efficacy and avoid undesirable side effects [3]. In our application, the peak temperature required for local deep tissue heating is 41 °C (for comparison, a hot bath is approximately 40–41 °C). It has been shown that a temperature of 41 °C is enough to initiate an increased muscle blood flow [3]. As the proposed applicator is not temperature controlled and relies on theoretical and simulation predictions, the induced heating is 4 °C below the margins of 45 °C for safety reasons.

49.2 Material and Methods

49.2.1 Microwave Antenna Design

Unlike the conventional microwave diathermy applicator, the hybrid microwave antenna is compact, conformal and low power [4]. The hybrid probe is to be placed on the skin surface of either the forearm or the calf. The design of the microwave antenna requires careful consideration of the dielectric properties of the skin, fat and muscle layers. Apart from the microwave frequency, the shape and dimension of the various components (the metallic patch, dielectric substrate and superstrate) in the microwave antenna can all affect the distribution of heat in the skin, fat and muscle. In particular, a type of superstrate, which acts as a coupling layer to the skin and is an electrical and thermal insulator, has been chosen to minimise any microwave

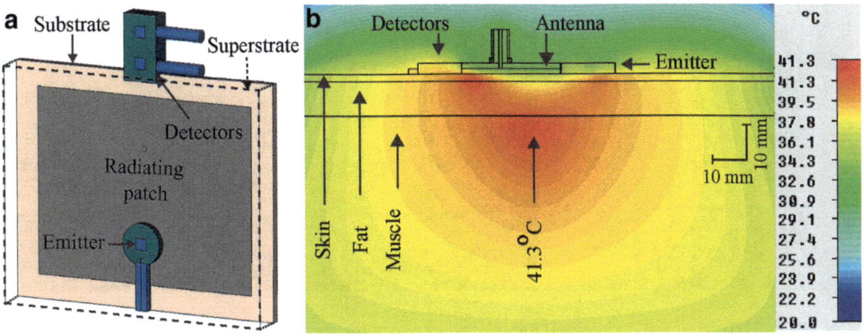

Fig. 49.1 Simulation model: (**a**) antenna design with the integrated detectors and emitter and (**b**) transient thermal distribution at the peak point inside the muscle

reflections at the skin surface to avoid skin overheating. Through computer simulations and lab experiments, the design of the hybrid probe, which consists of the microwave antenna and the optical source and detectors, has been optimised as shown in Fig. 49.1a.

49.2.2 The Biological Thermal Model

Human tissues generate heat by metabolism, and blood flow can carry heat away. These factors need to be considered in our application. A series of simulations were performed using the transient and thermal solver of CST Microwave Studio 2102, based on the finite integration technique. To obtain an accurate approximation of the thermal distribution, it is essential to include the electrical and thermal properties of the tissues. The conversion of electromagnetic energy into heat in biological tissue was modelled by the Pennes bioheat equation, which takes many factors into consideration, including the specific heat capacity, density, thermal conductivity, body temperature, blood temperature metabolic heat generation and capillary blood perfusion coefficient [5].

The simulation model in Fig. 49.1b shows the temperature distribution in the biological tissue model. The model consists of three layers, i.e. the skin, fat and muscle layers, with the corresponding thicknesses of 2, 10 and 50 mm. The simulation has open thermal boundaries at a background temperature of 20 °C and the simulation corresponds to 20 min after microwave exposure. The simulated two-dimensional thermal distribution in the model has taken metabolism and perfusion into consideration. The peak temperature is confined to a local area in the muscle at a maximum temperature of 41.3 °C, which is sufficient to cause dilation in the blood vessels. The peak temperature was recorded at 17 mm under the skin layer. It is worth mentioning that thermal distribution in the biological tissue varies as a function of fat thickness and operating wavelength, and there are particular cases that hot spots can be formed in the fat layer [6].

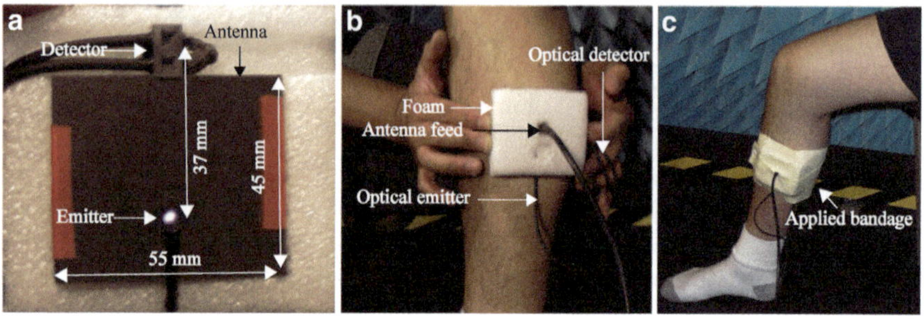

Fig. 49.2 Hybrid probe design and the in vivo experiment: (**a**) the biocompatible antenna with the integrated optical emitter and detector, (**b**) the hybrid probe housed inside a foam casing being placed on the calf and (**c**) the hybrid probe secured on the calf with bandage during the experiment

49.2.3 The Optical Probe

The changes in tissue oxygenation caused by tissue heating are measured using an optical monitor based on near-infrared spectroscopy. This kind of optical monitor has been widely used to measure tissue oxygenation in the muscle and brain. In this preliminary study, a commercially available NIRO-100 tissue oxygenation monitor (Hamamatsu Photonics KK, Japan) was used which can measure $\Delta HbO_2/\Delta HHb$, tissue oxygenation index (TOI) and normalised tissue haemoglobin index (nTHI) by exploiting near-infrared light at three wavelengths (775, 813 and 853 nm). The version of the optical probe used here employed optical fibres to transmit and collect light at the measurement site and contained no metallic parts. The source-detector spacing was set to 37 mm to ensure sufficient light penetration. The differential pathlength factor (DPF) of 4.94 was used in the modified Beer-Lambert law for the conversion of measured data [7].

49.2.4 In Vivo Experiment Protocol

The hybrid probe was housed in rectangular foam for minimal interference with the antenna as shown in Fig. 49.2. The rectangular foam was not included in the simulation, so the actual skin/foam interface temperature may be higher than the simulation result because of the lower thermal dissipation in air, but it would not significantly alter the deep muscle heating induced by the microwaves. The hybrid probe was placed and wrapped with bandage around the calf muscle as shown in Fig. 49.2c. Two adult subjects participated in the experiments. The fat thickness at the measurement site was measured with an ultrasound scanner for both subjects and shown in Table 49.1. The experiment procedure consists of (i) an initial rest, duration of 1 min, (ii) followed by 5 min of microwave exposure and (iii) then 5 min of rest with the microwave switched off. The study was approved by the UCL Ethics Committee.

49.3 Results and Discussion

The results of one subject are depicted in Fig. 49.3 which show both ΔHbO_2 and ΔHbT increased during the 5 min of microwave exposure. The thermal response has been defined as the ratio of the total increase of ΔHbT to the time duration (5 min) and can be calculated by taking the gradient of a regressed straight line which best fitted the portion of ΔHbT when microwave was on. After the microwave was switched off, both ΔHbO_2 and ΔHbT continued to rise for approximately another 70 s which has been defined as 'latency'. During this time, blood vessels remained dilated until blood perfusion takes sufficient heat away that tissue temperature is back to normal, at which point, blood vessels began to constrict and ΔHbT to drop. The latency may therefore be related to perfusion and thermoregulation. Both the

Table 49.1 Measured results for each subject

Subject	Fat layer thickness (mm)	Thermal response ΔHbT (μM/s)	Latency (s)
1	6.8	7.7×10^{-3}	70.0
2	5.0	18.9×10^{-3}	69.5

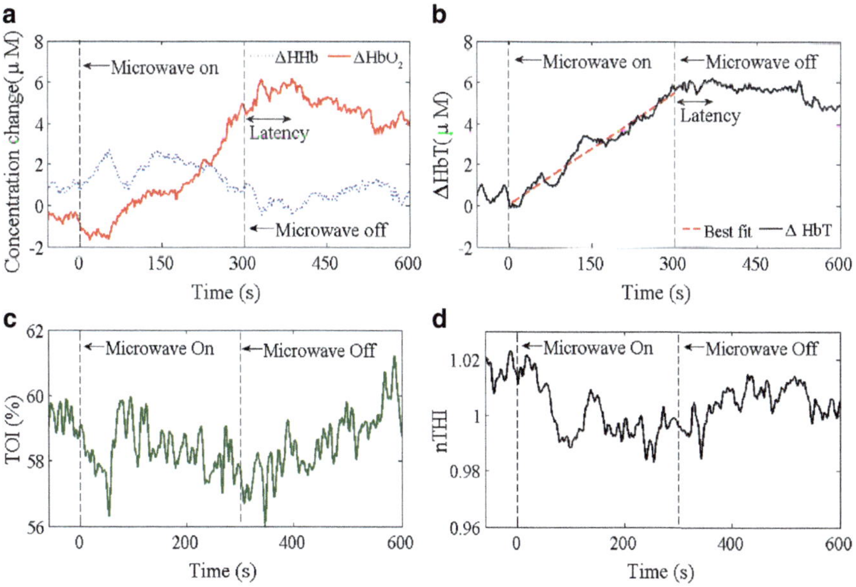

Fig. 49.3 Measured time series of a subject: (**a**) concentration change in oxy-/deoxyhaemoglobin, (**b**) total haemoglobin concentration, (**c**) tissue oxygenation index and (**d**) normalised tissue haemoglobin index

thermal response and latency were listed in Table 49.1. The thermal response of subject 1 is lower than that of subject 2. Apart from a genuine difference in physiology, another explanation is the partial volume effect in the ΔHbT signal caused by the two subjects' different fat thicknesses. The fact that ΔHHb shows no significant change in Fig. 49.3a indicates that vasodilation occurs primarily in the arterial blood vessels. The TOI measurement in Fig. 49.3c shows numerous oscillations with no particular patterns with microwave heating. The nTHI as depicted in Fig. 49.3d shows a decrease during microwave exposure; this seemingly suggests that there was a reduction in blood volume which contradicts the earlier results. However, this decreasing nTHI could be caused by a localised blood volume increase which led to the (first) optical detector nearer to the emitter recording a higher optical attenuation than the other (second) optical detector did (recall nTHI is calculated based on attenuation slope recorded by two closely placed detectors [8]). This would result in a smaller attenuation slope and therefore a seemingly lower nTHI. This effect is more prominent when the local blood volume increase is closer to the first detector which in our case means a more superficial heating.

49.4 Conclusions

The aim of this study was to develop a hybrid microwave-optical probe to investigate the thermal response in deep tissue including skin, fat and muscle. The design of the hybrid probe was based on extensive computer simulations that have taken metabolism and perfusion into consideration. Our simulation results have shown that the current design can successfully induce a local temperature rise to around 41.3 °C in the muscle. This temperature is sufficient to cause a thermal response, i.e. increased blood flow, and therefore, tissue oxygenation changes, which can be measured by the optical probe. The first prototype of the hybrid probe has been developed and used for in vivo measurements on two subjects. Upon the application of microwaves, a thermal response in terms of an increasing ΔHbT has been successfully detected. In future study, the microwave antenna will be cooled to minimise any heating in the skin and fat layers which will allow the thermal response of muscle to be investigated exclusively.

Acknowledgments This work was partly funded by EPSRC (Grant Code EP/G005036/1). The authors would also like to thank Hamamatsu Photonics KK for loaning the NIRO-100 monitor.

References

1. Herrick AL, Clark S (1998) Quantifying digital vascular disease in patients with primary Raynaud's phenomenon and systemic sclerosis. Ann Rheum Dis 57(2):70–78
2. Song CW (1984) Effect of local hyperthermia on blood flow and microenvironment: a review. Cancer Res 44(10):4721s–4730s

3. Giombini A, Giovannini V, Di Cesare A et al (2007) Hyperthermia induced by microwave diathermy in the management of muscle and tendon injuries. Br Med Bull 83:379–396
4. Goats GC (1990) Physiotherapy treatment modalities microwave diathermy. Microwaves 24(4):212–218
5. Pennes HH (1998) Analysis of tissue and arterial blood temperatures in the resting human forearm. J Appl Physiol 85(1):5–34
6. Guy A, Lehmann J (1974) Therapeutic applications of electromagnetic power. Proc IEEE 62(1):55–75
7. Duncan A, Meek JH, Clemence M et al (1995) Optical pathlength measurements on adult head, calf and forearm and the head of the newborn infant using phase resolved optical spectroscopy. Phys Med Biol 40(2):295–304
8. Suzuki S, Takasaki S, Ozaki T, Kobayashi Y (1999) A tissue oxygenation using NIR spatially resolved spectroscopy. Proc SPIE 3597:582–592

Chapter 50
Oxygen-Sensitive Quantum Dots for Possible Nanoscale Oxygen Imaging in Cultured Cells

Kenji Higashi, Takashi Jin, and Eiji Takahashi

Abstract Quantum dots (QDs) are the semiconductor crystal with a nanometer particle size that emit fluorescence of a size-dependent wavelength. In this study, we examined whether L-cysteine-capped CdTe quantum dots (QD580, diameter ~4 nm) might be used as an optical probe for intracellular oxygen (O_2) in cultured cells. QD580 was successfully introduced in cultured COS-7 cells by incubating cells with 10 nM QD580 for 5–60 min at 37 °C. Cells were exposed to 20 % O_2 (0.5 h), then 0.5 % O_2 or 20 % O_2 (1 h), and finally 20 % O_2 (0.5 h) gases. We found significant increases in the fluorescence intensity at 0.5 % O_2. However, when compared with QD580 in buffer solution, QD580 fluorescence in cells was considerably weak and vulnerable to repeated excitation light exposures. The present study demonstrated the potential of L-cysteine-capped CdTe QDs as a nanoscale probe for intracellular O_2 in cultured cells. Further improvement of the QD is necessary for quantitative assessment of O_2 in the cell.

50.1 Introduction

Quantum dots (QDs) are the semiconductor crystal with a nanometer particle size. Quantum dots have been introduced in bioscience as a novel fluorescent probe [1]. Recently, Xia et al. [2] demonstrated that CdTe QDs covered with L-cysteine exhibit reversible changes in both the magnitude and wavelength of the fluorescence

K. Higashi • E. Takahashi (✉)
Advanced Technology Fusion, Graduate School of Science and Engineering, Saga University, Honjo 1, Saga 840-8502, Japan
e-mail: eiji@cc.saga-u.ac.jp

T. Jin
Laboratory for Nano-Bio Probes, RIKEN Quantitative Biology Center, Osaka, Japan

S. Van Huffel et al. (eds.), *Oxygen Transport to Tissue XXXV*, Advances in Experimental Medicine and Biology 789, DOI 10.1007/978-1-4614-7411-1_50,

according to the concentration of O_2 in the buffer solution. Then, we questioned whether these QDs might be a novel probe for intracellular O_2 in cultured cells. Thus, the aim of the present study is to demonstrate that the fluorescence of L-cysteine-capped CdTe QDs loaded into cultured cells changes according to the cellular O_2 level.

50.2 Methods and Results

50.2.1 General

L-cysteine-capped CdTe QDs were synthesized by the bottom-up approach and kept in phosphate buffer solution (5 μM, pH = 10 ~ 11). We used the L-cysteine-capped CdTe QDs with a diameter of ~4 nm (QD580). In the buffer solution, the emission wavelength was ~578 nm with 73 nm FWHM.

A fibroblast cell line COS-7 was cultured on 13-mm diameter, 0.15-mm thickness cover glasses (CultureCoverGlass®, Matsunami) in DMEM (Sigma) supplemented with 10 % fetal calf serum and antibiotics. The QD580-loaded cells on the CultureCoverGlass® were covered with a 10-μl hepes-Tyrode buffer solution (pH = 7.4 at 37 °C) and placed in the airtight measuring cuvette in which humidified gas containing 0.1 ~ 20 % O_2 was supplied at 2 ml/min. The measuring cuvette was placed on the stage of an inverted microscope (IX-71, Olympus). All the experiments were conducted at 37 °C.

The QD was excited at 440 nm and the fluorescence image at 595 nm was captured using a 16-bit cooled CCD camera (SV512, PixelVision) attached to the microscope. Due to the relatively weak QD fluorescence in the cell, exposure duration was set to 5 s. All the image processing was carried out off-line using IPLab (Scanalytics) software.

50.2.2 QD Loadings

To optimize QD580 loading in COS-7 cells, we first assessed the cytotoxicity of the QD by examining morphological changes in the cell (using the phase contrast optics in the microscope) while changing QD580 concentration in the culture medium (10 nM or 50 nM). Then, the magnitude of QD580 fluorescence was determined for 5-, 15-, 30-, or 60-min incubation with 10 nM QD580. Prolonged incubation with QD580 damaged the cell. Changes in the cell shape were evident in > 1-h incubation with 50 nM QD580 or > 3-h incubation with 10 nM QD580. For 10 nM QD580, the magnitude of the fluorescence in the cell was comparable irrespective of the loading duration between 5 and 60 min, indicating very rapid uptake of QD580 in COS-7 cells (Fig. 50.1). From these results, the cells were incubated with 10 nM QD580 for 30 min in the experiments described below.

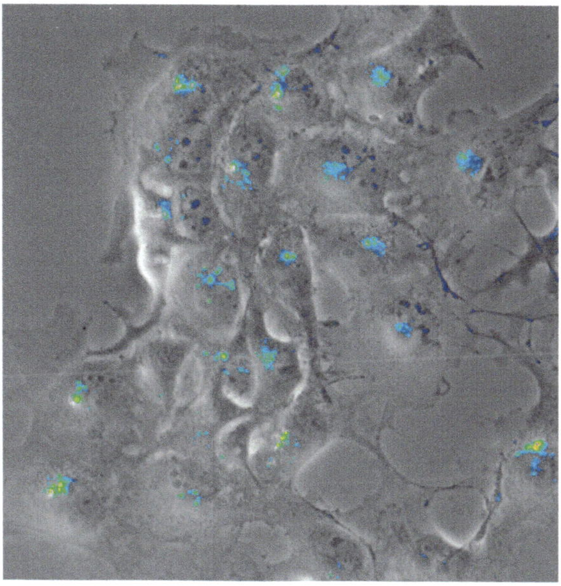

Fig. 50.1 Intracellular localization of QD580 in COS-7 cells. The QD580 fluorescent image was superimposed on the phase contrast image. Because the particle size is extremely small (~4 nm), the present phase contrast image only visualizes QDs accumulated to a sufficient amount in intracellular structures such as lysosomes

50.2.3 Photo Bleaching

In most inorganic fluorescent dyes, the fluorescence gradually diminishes after repeated exposure to excitation light (photo bleaching). Photo bleaching can be a considerable artifact in the fluorescence measurement. Thus, we determined the magnitude of photo bleaching in L-cysteine-capped CdTe QDs with ~3-nm diameter (QD545). In COS-7 cells loaded with QD545, a 5-s exposure to the excitation light (440 nm) was repeated every 1 min, 5 times. Oxygen concentration in the measuring cuvette was either 0.5 % or 20 %. The fluorescence was considerably attenuated after repeated excitations in such a way that the magnitude of the fluorescence levelled off to <50 % of the initial intensity after five exposures. Interestingly, attenuated QD545 fluorescence appeared to recover when consecutive excitations were separated by longer time intervals: at 0.5 % O_2, the fluorescence dropped to 78 ± 8 % of the initial level for a 1-min time interval between excitations, while for a 10-min interval, the fluorescence significantly increased to 91 ± 12 % of the initial level (t-test, $p < 0.05$). The results were similar in QD580.

50.2.4 O_2 Dependence of QD Fluorescence

Next, we examined whether QD580 fluorescence changes according to the O_2 level in the cell. The cells were superfused with 20 % O_2 gas for 30 min and the first QD580 fluorescence measurement was conducted. Then, the O_2 concentration in the superfusion gas was reduced to 0.5 % and sustained for 1 h so that the cellular

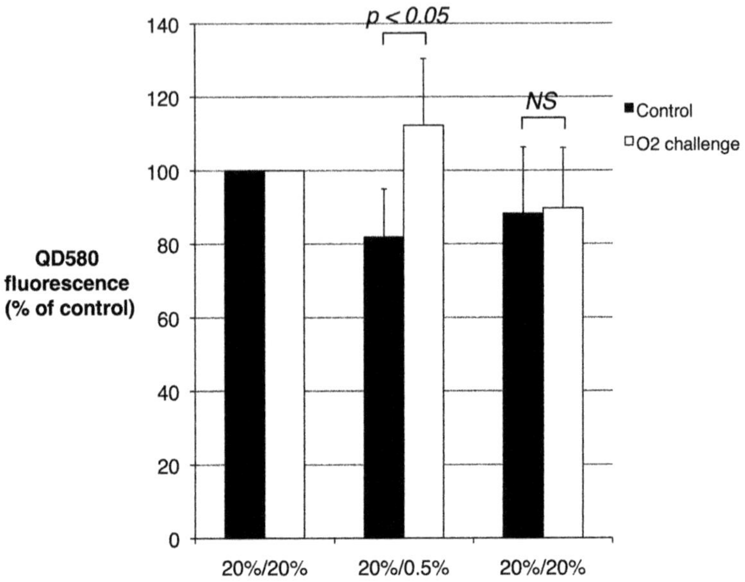

Fig. 50.2 Elevation of QD580 fluorescence in O_2 challenge. Data are represented in mean ± SD. Differences in the fluorescence level were tested using unpaired *t*-test where $p<0.05$ is considered significant. *NS* not significant

O_2 was equilibrated with the superfusion gas, and the second measurement was conducted. Lastly, the cell was superfused with 20 % O_2 for 30 min and the fluorescence measurement was conducted ("O_2 challenge"). As the control to the "O_2 challenge," similar fluorescence measurements were carried out with fixing the O_2 concentration at 20 %.

Figure 50.2 illustrates changes in the magnitude of QD580 fluorescence. In the control experiment, QD580 fluorescence decreased to 82 ± 13 % of the first measurement ($n=25$), whereas in the "O_2 challenge," (0.5 % O_2) QD580 fluorescence showed a significant increase (112 ± 18 % of the first measurement, $n=31$) compared to the control. Additionally, in the "O_2 challenge" experiment, the fluorescence intensity at 605 nm was significantly increased compared to the fluorescence at 585 nm, indicating a rightward shift of the fluorescence peak during hypoxia.

50.3 Discussions

In the present study, we demonstrated that L-cysteine-capped CdTe QDs can be easily introduced in COS-7 cells, and the magnitude of QD580 fluorescence in COS-7 cells changes according to O_2 transitions between 0.5 % and 20 %. These results indicate a potential of L-cysteine-capped CdTe QDs as an optical probe for intracellular O_2. However, for ultimate nanoscale intracellular O_2 imaging, optical characteristics of QD580, particularly when residing inside cells, await thorough investigation. For example, we noticed that the fluorescence intensity of QD580 in cells is considerably low compared with that in the buffer solution. Comparing at a similar local concentration, the magnitude of the QD580 fluorescence in COS-7 cells was only 3 % of QD580 in the buffer solution. Furthermore, while photo bleaching of QD580 fluorescence was minimal in buffer solution, the fluorescence of QD580 in cells appeared quite sensitive to repeated exposure to excitation light. We presume that the intracellular milieu has a considerable effect on the optical properties of QDs, particularly those covered with organic molecules such as amino acids as in the present study.

Another technical challenge is the quantification of the O_2 levels using intracellular QD580. In the present study, we compared QD580 fluorescence between 20 % O_2 and 0.5 % O_2 where QD580 fluorescence at 20 % O_2 was regarded as a reference. In this case, because more than 30 min was required for cellular O_2 to reach the equilibrium to the test gas, these two separate fluorescence measurements at two different cellular O_2 levels are relatively time consuming and not suitable for practical uses. This problem may be resolved by the use of two different QDs with different O_2 sensitivity and fluorescent wavelength. For example, we simultaneously loaded L-cysteine-capped CdTe QDs (O_2 sensitive, fluorescence at 580 nm) and glutathione-capped CdSe/ZnSe QDs (O_2 insensitive, fluorescence at 540 nm). Fluorescence of each QDs was successfully separated spatially in the fluorescent image of the cell. Thus, O_2-dependent changes in QD580 fluorescence may be quantitatively assessed by a single fluorescence measurement where QD540 fluorescence is used as a reference.

In conclusion, the present study certainly demonstrated a potential of L-cysteine-capped CdTe QDs as a nanoscale probe for intracellular O_2 in cultured cells. Further improvement of the QD is necessary for quantitative assessment of O_2 in the cell.

Acknowledgments This work was supported by KAKENHI 23500520 to ET.

References

1. Chan WC, Nie S (1998) Quantum dot bioconjugates for ultrasensitive nonisotopic detection. Science 281(5385):2016–2018
2. Xia Y, Zhang T, Diao X, Zhu C (2007) Measurable emission color change: size-dependent reversible fluorescence quenching of CdTe quantum dots by molecular oxygen. Chem Lett 36(2):242–243

Chapter 51
Boron Tracedrug Design for Neutron Dynamic Therapeutics for LDL

Hitoshi Hori, Yoshijiro Nazumi, and Yoshihiro Uto

Abstract We describe our solution for removal of the low-density lipoprotein (LDL) depot contained in proteins and lipids as a 'druggable' target for atherosclerotic cardiovascular diseases by neutron dynamic therapy (NDT), which we developed using boron tracedrugs for NDT against bovine serum albumin as a model protein. Thus, we examined, among our developed boron tracedrugs, a boron-containing curcuminoid derivative UTX-51, to destroy freshly isolated human LDL dynamically under irradiated thermal neutron to obtain a decreased intensity of band of LDL treated with UTX-51 and thermal neutron irradiation in their SDS-PAGE and electrophoresis analysis. These results suggest that UTX-51 might be a novel candidate of 'beyond chemical' therapeutic agents for atherosclerotic cardiovascular disease.

51.1 Introduction

We are currently studying the medicinal chemistry of boron tracedrugs we have developed, which are examples of next-generation 'on-demand traceable' drugs [1–3]. In the early stages of drug discovery and development, medicinal chemistry researchers made a great effort to evaluate the pharmacokinetics (PK) of drug candidates. Recently, increased demand for PK studies has encouraged researchers to develop drugs with superior traceability. Traditionally, radiolabeled compounds have been studied for these purposes. These, however, have two inherent problems: their half-life and the specific regulations for use by experimental facilities. These problems have increased the need for medicinal chemists to develop traceable drugs that do not require radioisotope labeling themselves. We had the idea of developing

H. Hori (✉) • Y. Nazumi • Y. Uto
Department of Life System, Institute of Technology and Science, Graduate School, The University of Tokushima, Minamijosanjimacho-2, Tokushima 770-8506, Japan
e-mail: Hori@bio.tokushima-u.ac.jp

S. Van Huffel et al. (eds.), *Oxygen Transport to Tissue XXXV*, Advances in Experimental Medicine and Biology 789, DOI 10.1007/978-1-4614-7411-1_51,

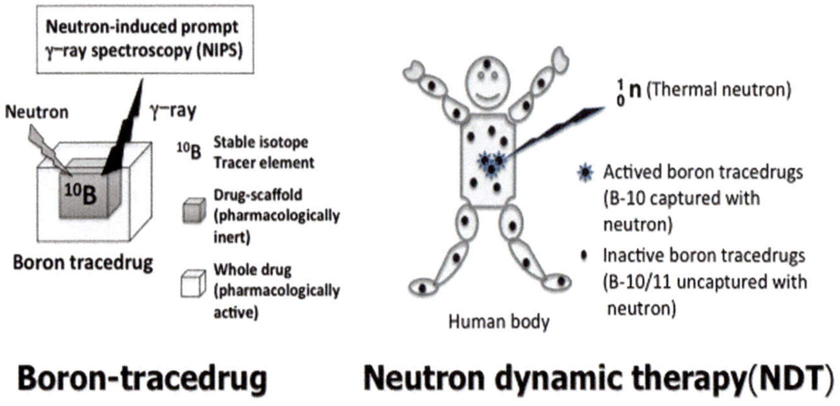

Fig. 51.1 Schematic images of boron tracedrugs and neutron dynamic therapy (*NDT*)

tracedrugs with boron atoms embedded in their scaffold. This idea was based on our previous drug design studies for the development of hypoxia-targeting boron-10 (^{10}B) carrier compounds for boron neutron capture therapy (BNCT) [4]. ^{10}B, a naturally occurring and stable isotope (19.9 %), possesses neutron capture activity that produces a prompt γ-ray when irradiated by a thermal neutron. The ^{10}B concentration can be measured by neutron-induced prompt γ-ray spectroscopy (NIPS) to detect the actual localization of boron tracedrugs.

Here, we discuss our boron tracedrug approach to the development of dynamic drugs for neutron dynamic therapy (NDT); we termed these drugs 'neutron dynamic therapeutics'. We present an overview of our concept of a boron tracedrug for NDT (Fig. 51.1). Neutrons attack ^{10}B in boron tracedrugs bound to macromolecules such as proteins, DNA/RNAs, sugars, and lipids and generate a vast amount of energy that subsequently decomposes the boron tracedrugs and their adjacent molecules.

We previously designed the boron tracedrugs UTX-42, UTX-43, and UTX-44, which possess antioxidant potency [5]. In order to explore their additional potential and the destructive physical power exerted by weak thermal neutrons, thermal neutron irradiation of bovine serum albumin (BSA) treated with UTX-42 and UTX-44 and the newly designed boron tracedrugs UTX-47, UTX-50, and UTX-51 was performed to cause destructive dynamic damage of BSA during thermal neutron irradiation [5, 6].

Atherosclerosis causes lethal disease due to cerebral and myocardial infarction and its prevention is a very important issue in industrial countries. Its key player, low-density lipoprotein (LDL), accumulates inside the artery walls with increased endogenous cholesterol that can impede blood flow and thereby compromise transportation of oxygen and nutritional elements to organs. This depot should be removed for the prevention and treatment of atherosclerosis. We describe our solution for removal of the LDL depot contained in proteins and lipids by neutron dynamic therapy using boron tracedrugs for NDT against BSA as a model protein. Thus, we examined, among our developed boron tracedrugs, a boron-containing curcuminoid derivative UTX-51, to destroy a freshly isolated human LDL dynamically under irradiated thermal neutron.

51.2 Materials and Methods

51.2.1 UTX-51 and Materials

UTX-51 (Fig. 51.2) was synthesized in our laboratory and all other chemicals were purchased from Tokyo Chemical Industry Co. Ltd, Wako Pure Chemical Industries Ltd., and Sigma-Aldrich.

51.2.2 Molecular Orbital Calculations

The *ab initio* MO calculation was performed with B3LYP hybrid density functional in conjugation with the 3-21G (d) basis set using the Gaussian 03 suite programs [7]. Visualization of the molecular geometries calculated by Gaussian 03 was carried out with GaussView 3.0 [8].

51.2.3 Neutron Irradiation

UTX-51 was used for the LDL irradiation study. The final concentration of the boron tracedrugs was 100 μM (diluted with pH 7.0, wide-range buffer from 2.0 mM boron tracedrug stock solution in DMSO), and 10 mg/ml BSA stock solution was made up in phosphate-buffered saline (PBS). Thermal neutron irradiation was performed. The neutron fluence was measured from the radioactivation of gold foils at the front

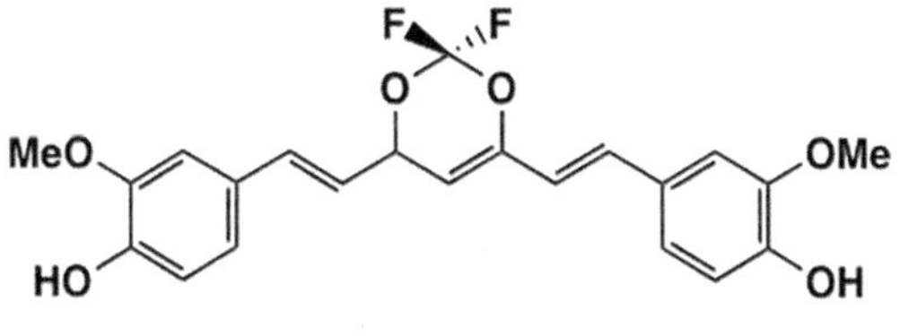

UTX-51

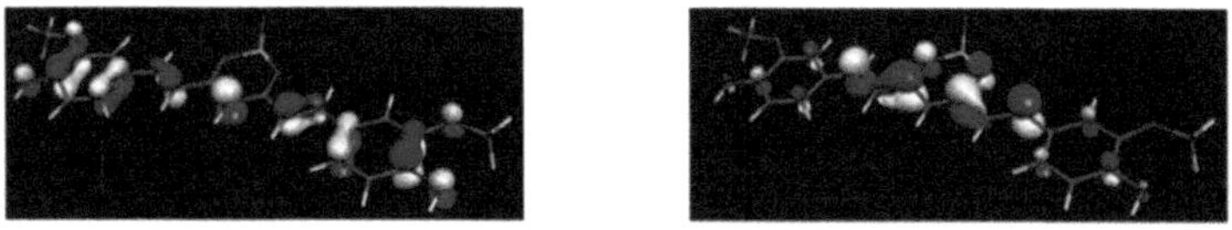

Fig. 51.2 Boron tracedrugs, a boron-containing curcuminoid derivative, UTX-51, and its molecular orbitals, HOMO and LUMO, and their eigenvalues (eV)

of the sample tubes, and the average neutron fluence determined from the measured values was used. Contaminating γ-ray doses, including secondary γ-rays, were measured with thermoluminescence dosimeter powder at the front of the sample tubes. The absorbed dose was calculated using the flux-to-dose conversion factor [9].

51.3 Results and Discussion

We designed boron-containing curcuminoid UTX-51 as a model compound of boron tracedrug for NDT to destroy LDL and LDL-related lipoproteins. We thought that the boron-containing curcuminoid compound, UTX-51, could be an effective candidate as a model compound of boron tracedrug for NDT because of curcumin, which binds a variety of proteins [10, 11]. Since curcumin may bind its targets in a non-competitive manner, its affinity does not need to be in the low molar range, such as nanomolar concentration, to cause a destructive effect by neutron dynamic therapy. First, we designed boron-containing curcuminoid UTX-51, compared to those of its parent natural product curcumin, using molecular orbital (MO)-based molecular modelling method using Gaussian 03 with B3LYP/6-31G (d). The lowest unoccupied MO (LUMO) and the highest occupied MO (HOMO) of UTX-51, as shown in Fig. 51.2, were not localized at the boron connection site like curcumin. Its energy levels of the LUMOs (E_{LUMO}) and HOMO (E_{HOMO}) had general values. Thermal neutron irradiation was conducted at the Kyoto University Research Reactor Institute (KURRI). SDS-PAGE was performed to detect the decomposition of LDL (46 μg) treated with UTX-51and thermal neutron irradiation. Electrophoresis-based analysis indicated a decrease in the intensity of the bands for LDL treated with UTX-51 (at doses of 2 and 20 nmol) and thermal neutron irradiation as shown in Fig. 51.3.

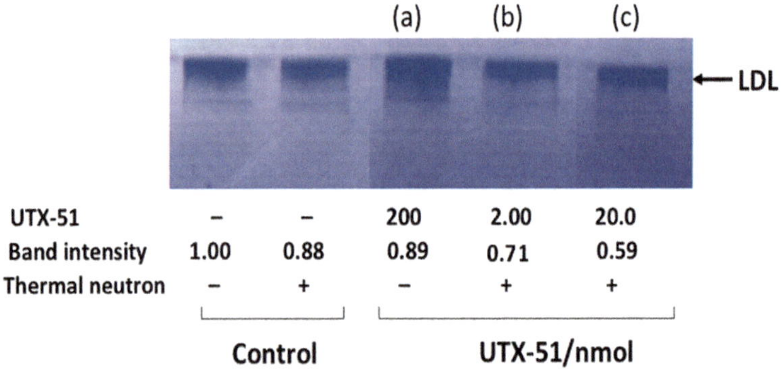

Fig. 51.3 Dynamic effects of boron tracedrug UTX-51 to human LDL under thermal neutron irradiation. Irradiation condition: UTX-51: 2.00, 20.0 nmol; LDL, 46 μg; Thermal neutron (absorbed dose 1.9 Gy; time 45 min). Human LDL treated with boron tracedrug UTX-51 on SDS-PAGE gel stained with Coomassie Brilliant Blue (*CCB*) and band intensity. Band intensities were measured with SWEDAY JustTLC software (control = 1.00). (**a**) ^{10}B/LDL = 2,500:1; (**b**) ^{10}B/LDL = 25:1; (**c**) ^{10}B/LDL = 250:1

51.4 Conclusions

To the best of our knowledge, we present data showing for the first time that the boron tracedrug UTX-51 causes destructive dynamic damage of LDL during thermal neutron irradiation. We also suggest that boron tracedrugs can be used as neutron dynamic therapeutics for LDL-related pathological associated disease and other protein-protein interaction diseases [12, 13].

Acknowledgments We thank Dr. Koji Ono and his colleagues (Research Reactor Institute, Kyoto University, Japan) for their support during this thermal neutron irradiation study.

References

1. Hori H, Uto Y, Nakata E (2012) Boron tracedrugs challenge for neutron dynamic therapy. Anticancer Res 32(6):2235–2239
2. Hori H, Uto Y, Nakata E (2010) Medicinal electronomics bricolage design of hypoxia-targeting antineoplastic drugs and invention of boron tracedrugs as innovative future-architectural drugs. Anticancer Res 30(9):3233–3242
3. Hori H, Uto Y, Nakata E (2011) Medicinal bricolage design of hypoxia-targeting drugs and invention of boron trace-drugs as next-generation universal drugs. Shikoku Acta Med 67(9): 7–14
4. Nagasawa H, Uto Y, Kirk KL, Hori H (2006) Design of hypoxia-targeting drugs as new cancer chemotherapeutics. Biol Pharm Bull 29(12):2335–2342
5. Nakata E, Koizumi M, Yamashita Y, Uto Y, Hori H (2012) Boron tracedrug: design, synthesis and pharmacological activity of phenolic BODIPY-containing antioxidants as traceable next generation drug model. Adv Exp Med Biol 737:301–306
6. Nakata E, Koizumi M, Yamashita Y et al (2011) Design, synthesis and destructive dynamic effects of BODIPY-containing and curcuminoid boron tracedrugs for neutron dynamic therapy. Anticancer Res 31(7):2477–2481
7. Frisch MJ, Trucks GW, Schlegel HB et al (2004) Gaussian Inc. Gaussian 03, Revision B.04, Wallingford
8. Flükiger HP, Lüthi S, Portmann J (2000) Weber. Swiss National Supercomputing Centre CSCS, Manno
9. Caswell RS, Coyne JJ, Randolph ML (1980) Kerma factors for neutron energies below 30 MeV. Radiat Res 83(2):217–254
10. Goel A, Kunnumakkara AB, Aggarwal BB (2008) Curcumin as "Curecumin": from kitchen to clinic. Biochem Pharmacol 75(4):787–809
11. Esatbeyoglu T, Huebbe P, Ernst IMA, Chin D, Wagner AE, Rimbach G (2012) Curcumin–from molecule to biological function. Angew Chem Int Ed Engl 51(22):5308–5332
12. Jeon YH, Lee JY, Kim S (2012) Chemical modulators working at pharmacological interface of target proteins. Bioorg Med Chem 20(6):1893–1901
13. Thiel P, Kaiser M, Ottmann C (2012) Small-molecule stabilization of protein-protein interactions: an underestimated concept in drug discovery? Angew Chem Int Ed Engl 51(9): 2012–2018

Chapter 52
New Method of Analyzing NIRS Data from Prefrontal Cortex at Rest

Wakana Ishikawa, Masakaze Sato, Yukikatsu Fukuda, Takashi Matsumoto, Naohiro Takemura, Takeo Tsujii, and Kaoru Sakatani

Abstract The aim of this study was to develop a simple technique for objective assessment of mental stress levels by measuring hemoglobin concentration changes in the bilateral prefrontal cortex (PFC) at rest, employing two-channel near-infrared spectroscopy (NIRS). Each subject was instructed to think about nothing in particular for 3 min and then to complete the State-Trait Anxiety Inventory (STAI) test. Next, NIRS measurements were taken and the left/right asymmetry of PFC activity at rest was evaluated by calculating the proposed Laterality Index at Rest (LIR). There was a significant positive correlation between the LIR and STAI score in 39 subjects. The present method allowed evaluation of mental stress level from NIRS data in the PFC at rest.

52.1 Introduction

The incidence of stress-induced psychological and somatic diseases has been increasing rapidly in industrialized societies, and it is important to clarify the neurophysiological mechanisms of stress response in order to establish effective stress management methods. A simple, noninvasive method for assessment of stress response and for evaluation of the efficacy of relaxation methods is required for this purpose. We have previously used near-infrared spectroscopy (NIRS) for investigation of neurophysiological mechanisms of mental stress and to evaluate relaxation methods. We found that the prefrontal cortex (PFC) plays an important role in stress

W. Ishikawa (✉) • M. Sato • Y. Fukuda • T. Matsumoto
Department of Electrical Engineering and Bioscience, Waseda University, Tokyo, Japan
e-mail: ishikawa10@matsumoto.eb.waseda.ac.jp

N. Takemura • T. Tsujii • K. Sakatani
Department of Neurological Surgery, Division of Optical Brain Engineering, Nihon University School of Medicine, Tokyo, Japan

S. Van Huffel et al. (eds.), *Oxygen Transport to Tissue XXXV*, Advances in Experimental Medicine and Biology 789, DOI 10.1007/978-1-4614-7411-1_52,

response; asymmetry of PFC activity measured by NIRS correlated with behavioral and somatic responses to mental stress [1–4]. In this study, we employed NIRS to evaluate the asymmetry of PFC activation during activity at rest without any task, using a newly developed parameter, Laterality Index at Rest (LIR). We also investigated the correlation between LIR and anxiety level evaluated with the State-Trait Anxiety Inventory (STAI) test.

52.2 Materials and Methods

52.2.1 Experimental Settings

The study population comprised 39 subjects (29 women; 10 men); 19 were 20–24 years old and 20 were 60–79 years old. Written informed consent was obtained from each subject on forms approved by the ethical committee of Nihon University School of Medicine. Each subject was seated in a comfortable chair in a dimmed room, and we measured oxy- and deoxy-Hb concentration changes in the bilateral PFC with a two-channel NIRS (PNIRS-10, Hamamatsu Photonics K.K., Japan). The NIRS probes were set symmetrically on the forehead; the positioning is similar to the midpoint between electrode positions Fp1/Fp3 (left) and Fp2/Fp4 (right) of the international 10–20 system. One trial consisted of the following steps. First, each subject completed the STAI questionnaire before NIRS measurements. Second, calibration of the equipment was performed. Third, the subject was told that the preparation period (1 min) would begin. Fourth, the subject was instructed to rest quietly for 3 min: rest period. This corresponds to the analysis period. Figure 52.1 schematically depicts the experimental protocol.

52.2.2 Data Analysis

In order to analyze left/right asymmetry of PFC activity at rest, we calculated the laterality scores. Consider

$$\Delta oxyR_{\min} = \min_{t \in analysis\ interval} \Delta oxyR_t \tag{52.1}$$

$$\Delta oxyL_{\min} = \min_{t \in analysis\ interval} \Delta oxyL_t \tag{52.2}$$

where $\Delta oxyR_t$ and $\Delta oxyL_t$ denote oxy-Hb concentration changes of the right and the left PFC. The quantities defined by Eqs. 52.1 and 52.2 are the variations with respect to their minimum values, so that they are always nonnegative. Based on these quantities, we defined the *Laterality Index at Rest* (*LIR*) as follows:

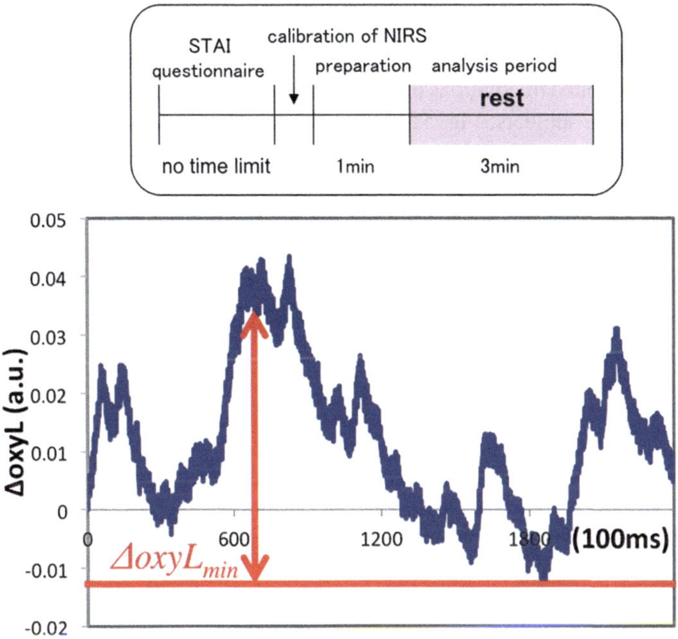

Fig. 52.1 The experimental protocol is illustrated in the *left panel*. Calibration refers to the fact that the hemoglobin concentration values are set to reference value so that measurement can start. The *panel* on the *right* shows typical traces of $\Delta oxyL_t$ and the minimum value of $\Delta oxyL_{min}$. As exemplified by the *red arrow*, the quantities defined by Eq. 52.2 are the variations with respect to the minimum value. The concept is the same for $\Delta oxyR_t$

$$LIR = \frac{\sum_{t \in analysis\ interval} \left((\Delta oxyR_t - \Delta oxyR_{\min}) - (\Delta oxyL_t - \Delta oxyL_{\min}) \right)}{\sum_{t \in analysis\ interval} \left((\Delta oxyR_t - \Delta oxyR_{\min}) + (\Delta oxyL_t - \Delta oxyL_{\min}) \right)} \quad (52.3)$$

The numerator of Eq. 52.3 consists of the difference between oxy-Hb concentration changes of the right and the left PFC summed over the analysis period (3 min). It is convenient if such index is normalized in such a way that the resulting numerical values are in $[-1,+1]$. The normalization constant is defined by the sum, instead of the difference, of oxy-Hb concentration changes of the right and the left PFC. It should be noted that if we had used $\Delta oxyR_t$ and $\Delta oxyL_t$ per se, instead of the variations from their minimum values, then the denominator could be zero or near zero where the target quantity diverges. It should also be noted that if the normalization constant becomes negative, then the index does not make much sense and makes it difficult to interpret the index. The quantities defined are schematically illustrated in Fig. 52.1. A positive LIR indicates that the right PFC is more active at rest than the left PFC, on average, while a negative LIR indicates that the left PFC is more active at rest than the right PFC, on average. We then analyzed the relation between LIR and STAI state scores.

52.3 Results

All subjects exhibited fluctuations of oxy-Hb in the bilateral PFC at rest. The amplitude varied among the subjects; Fig. 52.2 shows typical examples of $\Delta oxyR_t$ and $\Delta oxyL_t$ in subjects with high (60) and low (28) STAI scores. There was a significant positive correlation between exp (LIR) and STAI scores ($r=0.513$, $p=0.0008$). One reason for considering exp (LIR) instead of LIR is that we found that the correlation coefficient of STAI with exp (LIR) was larger than that with LIR. Generally, properties of a random variable change under coordinate changes. Fig. 52.3a shows the scatter

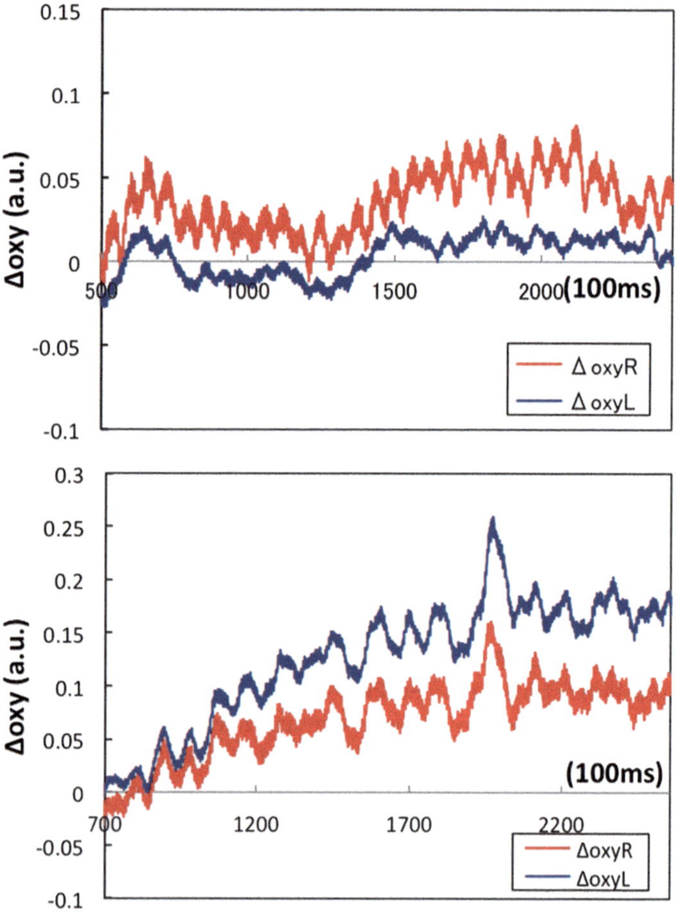

Fig. 52.2 The *left panel* shows typical examples of $\Delta oxyR_t$ and $\Delta oxyL_t$ in a subject with a high STAI score (60). The *right panel* shows typical examples of $\Delta oxyR_t$ and $\Delta oxyL_t$ in a subject with a low STAI score (28)

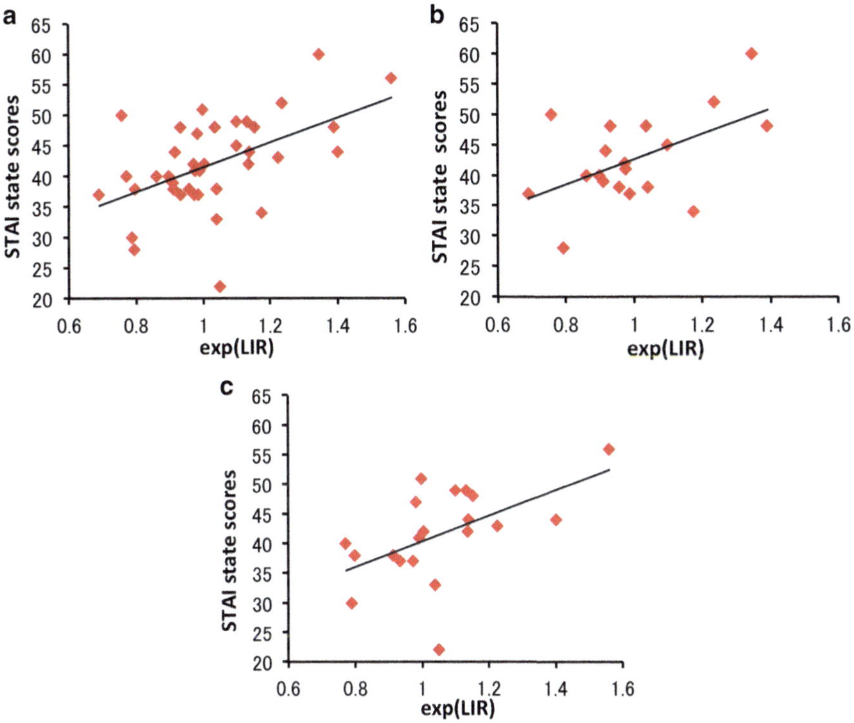

Fig. 52.3 (**a**) All STAI versus exp (LIR). Scatter plot of values of exp (LIR) against STAI test scores of all 39 subjects. Larger values of exp (LIR) indicate that the right PFC was more active at rest than the left PFC, corresponding to a higher anxiety level. (**b**) Scatter plot of exp (LIR) against STAI score for 19 young subjects. (**c**) Scatter plot of exp (LIR) against STAI score for 20 older subjects. Correlation lines are shown in *black*. Correlation coefficient of STAI with exp (LIR) was larger than that with LIR. Generally, the properties of a random variable change under coordinate changes

plot of exp (LIR) against STAI score for all 39 subjects. This indicates that the right PFC was more active at rest than the left PFC, corresponding to a higher anxiety level. In order to examine a possible effect of aging on the correlation between LIR and STAI, we analyzed the correlation in the young group ($n=19$, 20–24 years) and the older group ($n=20$, 60–79 years) separately. In the young group there was a significant positive correlation between the exp (LIR) and STAI state score ($r=0.525$, $p=0.021$) (Fig. 52.3b). In the older group we found a similar positive correlation between the exp (LIR) and STAI state scores ($r=0.536$, $p=0.015$) (Fig. 52.3c). This suggests that aging has no significant effect on the correlation. In the present project, at least, we have not found an index of interest for deoxy-Hb.

52.4 Discussion

Subjects with right-dominant oxy-Hb changes at rest, evaluated in terms of LIR, showed higher STAI scores, while those with left-dominant oxy-Hb changes at rest showed lower STAI scores. In NIRS activation studies, changes of oxy-Hb during activation imply evoked changes of regional cerebral blood flow (rCBF) in response to neuronal activation, since changes in oxy-Hb are correlated with changes in rCBF [5]. NIRS measurements during the resting condition have shown that a change in oxy-Hb occurred without a change in total Hb [6, 7]. In addition, simultaneous measurement of NIRS and electroencephalography (EEG) showed an increase of oxy-Hb when the EEG showed alpha 2 wave (10–13 Hz) and a decrease of oxy-Hb when the EEG showed alpha 1 wave (7–9 Hz) [7]. These observations indicate that changes in oxy-Hb at rest reflect changes of neuronal activity at rest. The left/right asymmetry of PFC activity is correlated with specific emotional responses to mental stress and personality traits [8–10]. It has been reported that positive and negative affective stimuli shift the asymmetry in PFC activity. For example, film-induced fear increases relative right-sided PFC activation [10], whereas induced positive affective stimuli elicit an opposite pattern of asymmetric activation [11]. In addition, asymmetry in PFC activity at rest is correlated with the emotional state [12]. Kemp et al. demonstrated that patients with major depressive disorder exhibited reduced left frontal activity in the resting state compared with normal controls [12]. These results suggest that asymmetry in PFC activity at the resting state can predict the emotional state. The PFC plays an important role in mediating somatic responses to stress via projections to the neuroendocrine and autonomic centers in the medial hypothalamus [13]. Interestingly, regulation of stress response is differentially mediated by the right and left PFC, which is similar to regulation of emotional responses. Here, NIRS data demonstrated that right-dominant PFC activity during stress tasks was associated with hyperactivity of the stress response system, while left-dominant PFC activity was associated with small stress responses [1–4]. However, further studies are needed to clarify the relation between the right/left asymmetry of PFC activity at rest and the stress response system. Finally, it should be noted that concentration changes in oxy-Hb measured by NIRS reflect blood flow changes in both intra- and extracerebral tissues including the skull and skin. Recent studies demonstrated that the scalp-related hemodynamic changes are locked into the functional activation tasks [14]. In addition, some of the oxy-Hb oscillation could be systemically driven. For example, systemic blood pressure fluctuations significantly altered NIRS recordings through expression in extracranial tissues and within the brain itself [15]. Further studies are necessary to clarify these effects on the present NIRS data.

References

1. Tanida M, Sakatani K, Takano R, Tagai K (2004) Relation between asymmetry of prefrontal cortex activities and the autonomic nervous system during a mental arithmetic task: near infrared spectroscopy study. Neurosci Lett 369(1):69–74
2. Tanida M, Katsuyama M, Sakatani K (2007) Relation between mental stress-induced prefrontal cortex activity and skin conditions: a near infrared spectroscopy study. Brain Res 1184: 210–216
3. Tanida M, Katsuyama M, Sakatani K (2008) Effects of fragrance administration on stress-induced prefrontal cortex activity and sebum secretion in the facial skin. Neurosci Lett 432(2):157–161
4. Sakatani K, Tanida M, Katsuyama M (2010) Effects of aging on activity of the prefrontal cortex and autonomic nervous system during mental stress task. Adv Exp Med Biol 662: 473–478
5. Hoshi Y, Kobayashi N, Tamura M (2001) Interpretation of near-infrared spectroscopy signals: a study with a newly developed perfused rat brain model. J Appl Physiol 90(5):1657–1662
6. Hoshi Y, Tamura M (1997) Fluctuations in the cerebral oxygenation state during the resting period in functional mapping studies of the human brain. Med Biol Eng Comput 35(4): 328–330
7. Hoshi Y, Kosaka S, Xie Y, Kohri S, Tamura M (1998) Relationship between fluctuations in the cerebral hemoglobin oxygenation state and neuronal activity under resting conditions in man. Neurosci Lett 245(3):147–150
8. Davidson RJ, Jackson DC, Kalin NH (2000) Emotion, plasticity, cortex, and regulation: perspectives from affective neuroscience. Psychol Bull 126(6):890–909
9. Canli T, Zhao Z, Desmond JE, Kang E, Gross J, Gabrieli JD (2001) An fMRI study of personality influences on brain reactivity to emotional stimuli. Behav Neurosci 115(1):33–42
10. Fischer H, Andersson JL, Furmark T, Wik G, Fredrikson M (2002) Right-sided human prefrontal brain activation during acquisition of conditioned fear. Emotion 2(3):233–241
11. Ahern GL, Schwartz GE (1985) Differential lateralization for positive and negative emotion in the human brain: EEG spectral analysis. Neuropsychologia 23(6):745–755
12. Kemp AH, Griffiths K, Felmingham KL et al (2010) Disorder specificity despite comorbidity: resting EEG alpha asymmetry in major depressive disorder and post-traumatic stress disorder. Biol Psychol 85:350–354
13. Tsigos C, Chrousos CP (2002) Hypothalamic-pituitary-adrenal axis, neuroendocrine factors and stress. J Psychosom Res 53(4):865–871
14. Kirilina E, Jelzow A, Heine A et al (2012) The physiological origin of task-evoked systemic artifacts in functional near infrared spectroscopy. Neuroimage 61(1):70–81
15. Minati L, Kress IU, Visani E, Medford N, Critchley HD (2011) Intra- and extra-cranial effects of transient blood pressure changes on brain near-infrared spectroscopy (NIRS) measurements. J Neurosci Methods 197(2):283–288

Chapter 53
Radiation Oxygen Biology with Pulse Electron Paramagnetic Resonance Imaging in Animal Tumors

Gage Redler, Martyna Elas, Boris Epel, Eugene D. Barth, and Howard J. Halpern

Abstract The reduced oxygen in tumors (hypoxia) generates radiation resistance and limits tumor control probability (TCP) at radiation doses without significant normal tissue complication. Modern radiation therapy delivery with intensity-modulated radiation therapy (IMRT) enables complex, high-dose gradient patterns, which avoid sensitive human tissues and organs. EPR oxygen images may allow selection of more resistant parts of a tumor to which to deliver more radiation dose to enhance TCP. EPR O_2 images are obtained using injected narrow-line, low relaxation rate trityl spin probes that enable pulse radiofrequency EPR O_2 images of tumors in the legs of mice, rats, and rabbits, the latter exceeding 4 cm in size. Low relaxation rates of trityls have enabled novel T_1-, rather than T_2-, based oximetry, which provides near absolute pO_2 imaging. Tomographic image formation and filtered back projection reconstruction are used to generate these images with fixed, linear stepped gradients. Images obtained both with T_2 and T_1 oximetric images have demonstrated the complex in vivo mechanism explaining the unexpected efficacy of TNFerade, a radiation-inducible adenoviral construct to locally produce TNF-induced vascular as well as radiation damage [1, 2]. The unexpected efficacy of large-dose radiation fractions is seen to be due to an interaction between host microvasculature and tumor cells producing a prompt (15 min) postradiation hypoxia, paralyzing tumor cell repair, and sensitizing tumors. Finally, cure of tumors treated to a single 50 % control dose shows a significant dependence on EPR

G. Redler • B. Epel • E.D. Barth • H.J. Halpern (✉)
Center for EPR Imaging In Vivo Physiology, Chicago, IL, USA

Department of Radiation Oncology, University of Chicago, Chicago, IL, USA
e-mail: h-halpern@uchicago.edu

M. Elas
Center for EPR Imaging In Vivo Physiology, Chicago, IL, USA

Department of Radiation Oncology, University of Chicago, Chicago, IL, USA

Department of Molecular Biology, Jagiellonian University, Krakow, Poland

S. Van Huffel et al. (eds.), *Oxygen Transport to Tissue XXXV*, Advances in Experimental Medicine and Biology 789, DOI 10.1007/978-1-4614-7411-1_53,

O_2 image hypoxic fractions, best shown with the fraction of voxels less than 10 Torr (HF10). We show that these O_2 images provide a quantitative basis for measuring tumor and normal tissue response to abnormally low O_2 levels. Measurements of vascular endothelial growth factor (VEGF) production in a specific syngeneic mouse fibrosarcoma, FSa versus fraction of tissue voxels with pO_2 less than 10 Torr, produced a slope of 0.14 pg VEGF protein/mg total protein/% HF10. We argue that this quantification may be diagnostic of tumor versus normal tissue, and it may be etiologic in the development of malignancy.

53.1 Introduction

There are at least two reasons why images of local pO_2 in living animal tissues might be of general interest. The first is the importance of lack of oxygen in a large number of disease processes, including myocardial infarction, stroke, and cancer. The second reason for interest in local pO_2 oxygen images lies in the insight, in vivo, that such images may provide between an environmental stimulus or condition and the genetic response to it at the tissue and organismal level.

Hypoxia in cancer has been associated with resistance to genotoxic cancer treatments. Hypoxic resistance to radiation has been known for over a century, starting with the historic observations of Schwarz in 1909 [3]. The systematics of this evolved in the 1920s and 1930s with the work of Holthusen [4], as well as Gray, Conger, and others in the 1940s and 1950s [5], the quantification of the effect in cellular systems [6], and the appreciation of its applicability to human cancers, in particular the more recent work by Vaupel et al. [7–10].

The molecular biology of genetically induced cellular signalling has been the focus of most biologists for the past five decades [11]. Over the past two decades, the homeostatic centrality of the maintenance of adequate molecular oxygen has been recognized in the role of hypoxia-inducible factor, HIF, a master regulator of organismal signalling response to oxygen inadequacy [12]. The information leading to these conclusions has been obtained on either the cellular level or on the whole animal level. Heterogeneity of oxygen levels has not been addressed, particularly in tumors. The heterogeneity of tumor pO_2, shown in EPR O_2 images [13, 14] as well as needle electrode measurements of tumors, e.g., [15], requires the use of images to fully elucidate the relationship between the stimulating condition, hypoxia, and the genetic response.

We argue that in vivo images provide a robust means by which quantitative assessment of the unperturbed local oxygen environment can be made and then registered with the molecular signals that respond to the environment. This will form the basis of a quantitative stimulus–response model of the molecular response to the environment. Local abnormalities in this stimulus–response relationship may be etiologic in disease process. Such abnormalities may be based in abnormal transcription, as is assumed by most molecular biologists. This ultimately assumes that the only active agent in disease process is the genome. However, we argue that this is a simplified model, and it ignores a number of aspects of tissue and cell environment that can be disrupted and contribute to disease process. For example, vessel dysfunction, particularly microvessel dysfunction, can interfere with intercellular communication as part of paracrine and endocrine signalling processes. Intracellular protein glycation markedly affects endothelial function [16] and can have a similar effect on autocrine signalling and carcinogenesis [17–19]. These mechanisms may interfere with diffusion or translocation of signals. They will then change the relationship between the environmental trigger and the molecular response. They will change, for example, the dose–response relationship between local concentrations of oxygen and the HIF1-associated signal response elements. By concentrating solely on the relationship between hypoxia and HIF1 response and disregarding other possible environmental factors, we can approximate such a response as a simple linear function. In this simplified model, HIF1-associated signal (e.g., VEGF) would increase linearly as the pO_2 diminishes. There may be a normal slope or intercept in this relationship. Disease process may be reflected in alterations of the slope or intercept seen in the correlation between pO_2 and signal.

53.2 Methods

Electron paramagnetic resonance (EPR) O_2 imaging (**EPROI**) gives quantitative localized pO_2 images of various tumors in syngeneic mice, rats, and rabbits [20–22]. The oxygen broadening of narrow EPR spectral lines, or, equivalently, the increase in relaxation rates of electron magnetization, reports the pO_2 with 1–2 Torr resolution in image voxels as small as 1 mm^3 [13, 23]. We have found remarkable freedom from toxicity in 283 mice injected with the pO_2 reporter molecule OX063 (GE Healthcare) used to acquire the data reported here. OX063 is extracellular in its distribution and is rapidly cleared from the body through renal excretion with a half-life of ~5 min. It appears to be selectively retained in tumors with a half-life of ~30 min. This bodes well for eventual application for human subjects. OX063 and the partially deuterated O_2 reporter $OX063_{d24}$, also known as OX071, have very limited dependence on viscosity. We have found that, by using pulse sequences that image the longitudinal relaxation rate, R_1 ($R_1 = 1/T_1$, where T_1 is the longitudinal relaxation time), self-relaxation of the trityl spin probe is reduced to well within the 1 Torr uncertainties of our image voxel pO_2 values. EPROI:

1. Correlates point by point with Oxylite measurements [13]
2. Significantly sharpens the tumor cure prediction (along with dose in bivariate analysis) [20]
3. Independently distinguishes sensitive from resistant animal tumors treated with a single-dose magnitude, the 50 % tumor control dose, TCD_{50}, in two tumor types
4. Provides in vivo measurement of the rapid postradiation hypoxia induced via vascular apoptosis in response to large fraction radiation [24]

Other in vivo oxygen or hypoxia imaging modalities include ^{19}F MRI and ^{18}F-misonidazole and ^{64}Cu-ATSM positron emission tomography (PET). PET can be used for human studies. ^{19}F MRI is presently used only in animal studies. Neither of these other imaging modalities or other available in vivo animal O_2 images provide such a combination of accuracy in the voxel pO_2, lack of confounding biologic variability, and low level of invasiveness and toxicity.

Hypoxia in EPROI has been locally correlated with concentrations of the hypoxia protein vascular endothelial growth factor (VEGF) in biopsies stereotactically registered with the pO_2 images. These methods of registration and VEGF quantification have been described previously [21].

53.3 Results

In Fig. 53.1, obtained with natural isotopic abundance OX063 as the oxygen reporter, we show the correlation between the fraction of voxels with pO_2 less than 10 Torr (HF10) and the concentration of VEGF in biopsies obtained from 12 fibrosarcomas grown intramuscularly in the legs of C3H mice to volumes of 0.5 ml (10 mm equivalent diameters), as previously described [21]. For each biopsy, the absolute concentration of VEGF was determined from the specimen. There were ~75 pO_2 image voxels to obtain statistics from for each biopsy sample. This allowed us to compute the HF10, which appeared to be the most reliable statistic among mean pO_2, median pO_2, HF10, HF5, and HF2.5. Uncertainties in the HF10 and in

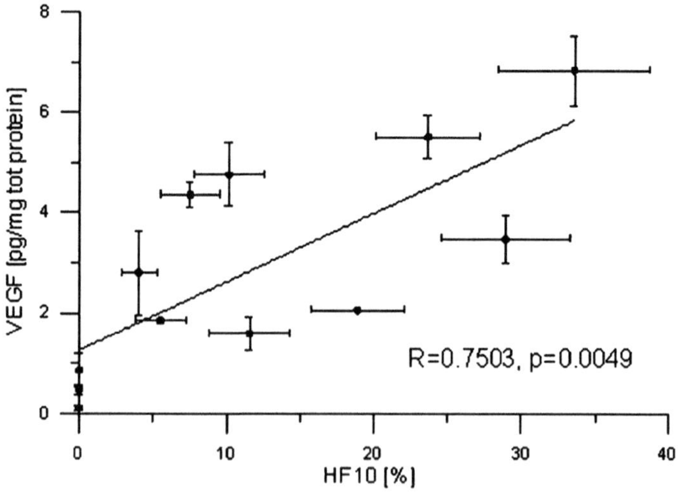

Fig. 53.1 Correlation between hypoxia, defined by the percentage of voxels in the biopsy with pO_2 less than 10 Torr as determined from the EPR oxygen image, and the VEGF concentration in pictograms per microgram of total protein

the VEGF concentration from each sample are shown in Fig. 53.1. The Pearson product–moment correlation coefficient, R = 0.75, provides a basis for establishing the fraction of variation in the samples that is due to true interdependence. The regression slope of the tumor tissue response is 0.14 pg VEGF/mg tumor tissue/%HF10. This is a measure, although undoubtedly a simplified one, of the signal response by tumor cells to produce more vascular endothelial cells in response to local hypoxia, as measured by percentage HF10. The heterogeneity of tumor pO_2, as shown in EPROI [13, 14] as well as needle electrode measurements of tumors [15], requires the use of images to define local tumor pO_2 and HF10. Because of the large variation of pO_2 and HF10 between the 20 μl biopsy volumes obtained in the study (typically 2 per tumor, one high and one low pO_2), a relatively small number of samples achieved significance, as seen in Fig. 53.1.

53.4 Discussion and Conclusions

This correlation between stereotactic biopsy-derived VEGF concentrations and biopsy sample HF10 values from registered EPROI provided statistically significant correlation between the two values. We argue that this slope, 0.14 pg VEGF/mg tumor tissue/%HF10, is a measure of tissue and tumor response to hypoxia. Although normal tissue measurements remain to be generated, we argue that this may be not only different, and a diagnostic of the difference between tumor and normal tissue, but that it may be etiologic in the development of malignancy. The dysfunctional chaos of tumor growth may stimulate either a reduced or an enhanced tissue response. EPROI may provide the basis for quantification of this response and introduce a new, quantitative aspect for evaluating tumor versus normal tissue response to the hypoxic environment.

Acknowledgments Supported by NIH grants P41 EB002034 and R01 CA98575.

References

1. Hallahan DE, Mauceri HJ, Seung LP et al (1995) Spatial and temporal control of gene therapy using ionizing radiation. Nat Med 1(8):786–791
2. Haney CR, Parasca AD, Fan X et al (2009) Characterization of response to radiation mediated gene therapy by means of multimodality imaging. Magn Reson Med 62(2):348–356
3. Schwarz G (1909) Über Desensibilisierung gegen Röntgen- und Radiumstrahlen. Munchner Medizinische Wochenschrift 56:1217–1218
4. Holthusen H (1921) Beitrage zur biologie der strahlenwirkung. Pflugers Arch 187:1–24
5. Gray LH, Conger AD, Ebert M, Hornsey S, Scott OC (1953) The concentration of oxygen dissolved in tissues at the time of irradiation as a factor in radiotherapy. Br J Radiol 26(312): 638–648
6. Howard-Flanders P (1958) Physical and chemical mechanisms in the injury of cells by ionizing radiation. In: Lawrence JH, Tobias CA (eds) Advances in biology and medical physics, vol 13. Academic, New York, pp 553–603

7. Thomlinson RH, Gray LH (1955) The histological structure of some human lung cancers and the possible implications for radiotherapy. Br J Radiol 9(4):539–563
8. Gatenby RA, Kessler HB, Rosenblum JS et al (1988) Oxygen distribution in squamous cell carcinoma metastases and its relationship to outcome of radiation therapy. Int J Radiat Oncol Biol Phys 14(5):831–838
9. Hockel M, Schlenger K, Mitze M, Schäffer U, Vaupel P (1996) Hypoxia and radiation response in human tumors. Semin Radiat Oncol 6(1):3–9
10. Brizel DM, Scully SP, Harrelson JM et al (1996) Tumor oxygenation predicts for the likelihood of distant metastases in human soft tissue sarcoma. Cancer Res 56(5):941–943
11. Alberts B, Johnson A, Lewis J et al (2008) Molecular biology of the cell, 5th edn. Garland Science, New York
12. Semenza GL (1988) Hypoxia-inducible factor 1: master regulator of O_2 homeostasis. Curr Opin Genet Dev 8(5):588–594
13. Elas M, Ahn KH, Parasca A et al (2006) Electron paramagnetic resonance oxygen images correlate spatially and quantitatively with oxylite oxygen measurements. Clin Cancer Res 12(14 Pt 1):4209–4217
14. Elas M, Williams BB, Parasca A et al (2003) Quantitative tumor oxymetric images from 4D electron paramagnetic resonance imaging (EPRI): methodology and comparison with blood oxygen level-dependent (BOLD) MRI. Magn Reson Med 49(4):682–691
15. Kallinowski F, Zander R, Hoeckel M, Vaupel P (1990) Tumor tissue oxygenation as evaluated by computerized-pO_2-histography. Int J Radiat Oncol Biol Phys 19(4):953–961
16. Wautier JL, Schmidt AM (2004) Protein glycation: a firm link to endothelial cell dysfunction. Circ Res 95(3):233–238
17. Tafani M, Schito L, Pellegrini L et al (2011) Hypoxia-increased RAGE and P2X7R expression regulates tumor cell invasion through phosphorylation of Erk1/2 and Akt and nuclear translocation of NF-kappa B. Carcinogenesis 32(8):1167–1175
18. Kang R, Loux T, Tang D et al (2012) The expression of the receptor for advanced glycation endproducts (RAGE) is permissive for early pancreatic neoplasia. Proc Natl Acad Sci USA 109(18):7031–7036
19. Rojas A, Figueroa H, Morales E (2010) Fueling inflammation at tumor microenvironment: the role of multiligand/rage axis. Carcinogenesis 31(3):334–341
20. Elas M, Bell R, Hleihel D et al (2008) Electron paramagnetic resonance oxygen image hypoxic fraction plus radiation dose strongly correlates with tumor cure in FSa fibrosarcomas. Int J Radiat Oncol Biol Phys 71(2):542–549
21. Elas M, Hleihel D, Barth ED et al (2011) Where it's at really matters: in situ in vivo vascular endothelial growth factor spatially correlates with electron paramagnetic resonance pO_2 images in tumors of living mice. Mol Imaging Biol 13(6):1107–1113
22. Epel B, Haney CR, Hleihel D, Wardrip C, Barth ED, Halpern HJ (2010) Electron paramagnetic resonance oxygen imaging of a rabbit tumor using localized spin probe delivery. Med Phys 37(6):2553–2559
23. Epel B, Sundramoorthy SV, Mailer C, Halpern HJ (2008) A versatile high speed 250-MHz pulse imager for biomedical applications. Concept Magn Reson Part B Magn Reson Eng 33B(3):163–176
24. Thin TH, García-Barros M, Fuller J, et al (in preparation) Epigenetic regulation of homologous recombination by microvascular dysfunction controls tumor cure by single dose radiation. Cell

Chapter 54
Wavelength Selection for the Improvement of the Signal-to-Noise Ratio for Imaging of Haemoglobin Oxygenation with RGB Reflectometry

André Steimers, Sarina Steinke, and Matthias Kohl-Bareis

Abstract We demonstrate the optimisation of wavelengths for the imaging of cortical haemoglobin oxygenation with broadband RGB reflectometry. Wavelengths were chosen in order to minimise the likely crosstalk and optimise the signal-to-noise ratio by simulating effects of different combinations of wavelengths on the condition number of the resulting extinction coefficient matrices. The results obtained were evaluated experimentally for four combinations of commercially available LED combinations and compared with data from the literature.

54.1 Introduction

When using RGB (red, green, blue) reflectometry to record small changes in haemoglobin oxygenation associated with neurometabolic and neurovascular coupling [1], the quality of the calculated concentrations of oxygenated and deoxygenated haemoglobin (oxyHb, deoxyHb) in terms of signal-to-noise ratio (SNR) and crosstalk depends mainly on the wavelength-dependent experimental parameters like the sensitivity spectra of the detector and light source as well as the tissue under interrogation. While the extinction spectra of haemoglobin and the optical path length are given by the optical properties of the tissue and the sensitivity spectra of the detector is given by available CCDs, the light source can be adapted to improve the performance of the setup. Therefore, the choice of the illumination system and its emission spectrum is a crucial factor for RGB reflectometry.

To optimise the wavelengths in the visible range, we first simulated the effects of combinations of wavelengths with different bandwidths on the condition number as a measure of SNR and crosstalk for a three-wavelength system

A. Steimers (✉) • S. Steinke • M. Kohl-Bareis
RheinAhr Campus Remagen, University of Applied Sciences Koblenz, Remagen, Germany
e-mail: steimers@rheinahrcampus.de

S. Van Huffel et al. (eds.), *Oxygen Transport to Tissue XXXV*, Advances in Experimental Medicine and Biology 789, DOI 10.1007/978-1-4614-7411-1_54,

and for two different RGB cameras. After this theoretical examination, a condition number analysis was done for commercial high-power colour LEDs. Furthermore, the theoretical findings were evaluated and compared for different combinations of high-power LEDs to confirm the connection of condition number with crosstalk and noise by measurements on a tissue-simulating phantom and biological tissue.

54.2 Methods

Crosstalk is mainly caused by using the wrong sensitivity spectra for the light source, detector and optical path length, and this has the same effect as calculating the haemoglobin concentrations with the wrong extinction spectra. On the other hand, noise is mainly caused by physical effects like changes in the intensity of the illumination due to thermal fluctuations or ambient light. Again, the influence of these effects can be reduced by an appropriate choice of wavelengths.

The calculation of changes in haemoglobin concentrations depends in large parts on the extinction coefficient matrix **E**. For RGB reflectometry, **E** is determined by the haemoglobin extinction spectra and the integral of the product of the optical characteristics of the imaging system and the tissue over the broad wavelength bands for the RGB detectors [2]. Haemoglobin concentration changes are then obtained by matrix inversion from experimental attenuation changes. This illustrates the importance of **E** as it includes all basic parameters and the description of the physiological and physical conditions of the measurement setup. From the parameters that define **E**, mainly the light source spectra can be varied for optimisation. To find the best combination of wavelengths for a three-wavelength spectroscopy system, the condition number of the matrix **E** is analysed as an objective parameter for its quality. For any matrix, the condition number is defined as the ratio of its largest singular value to the smallest singular value [3]. Here we calculated C as the inverse of the condition number (cond) of **E**, ($C = 1/\text{cond}(\mathbf{E})$). A value of C close to 1 indicates a well-conditioned matrix, while a value close to 0 signifies that higher errors and crosstalk are likely. Therefore, the aim is to find a combination of three wavelengths that gives a high value of C for the corresponding **E** matrix.

54.3 Simulation of a Three-Wavelength System

For a three-wavelength system, C was calculated for all permutations of the wavelengths λ_1, λ_2 and λ_3, with each wavelength varied between 400 and 700 nm. The spectra of the simulated light sources were assumed to be of Gaussian shape with full widths at half maximum (FWHM) varying from 3 to 49 nm. The spectral detector characteristic of two CCD cameras (A311fc, Basler AG, Germany and AD-080GE, Jai A/S, Denmark) was included.

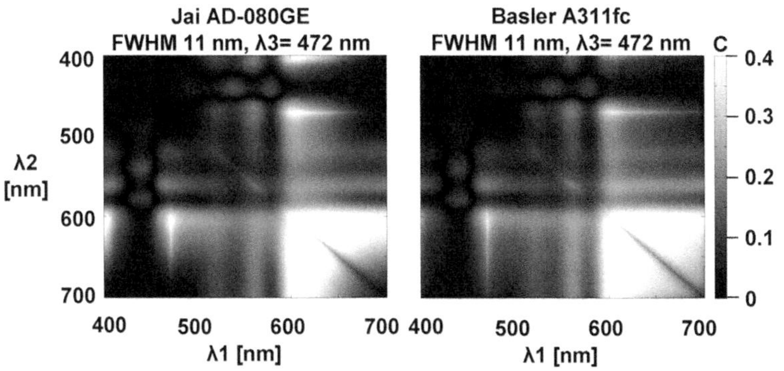

Fig. 54.1 Inverse condition numbers C as function of λ_1 and λ_2 and a fixed $\lambda_3=472$ nm with a FWHM of 11 nm for sensitivity spectra of the cameras Jai AD-080GE and Basler A311fc

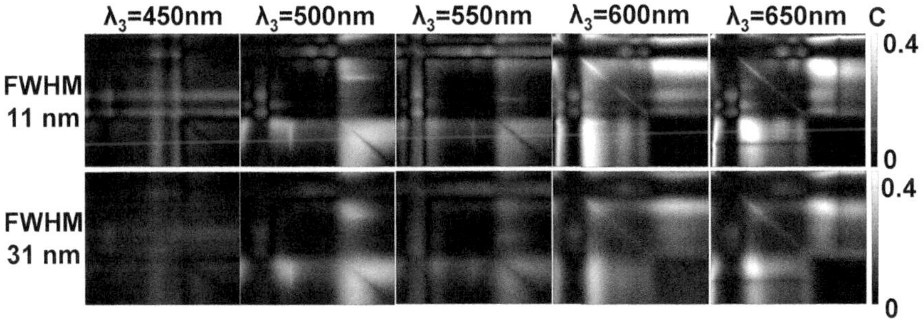

Fig. 54.2 Inverse condition number C as a function of λ_1 (x-axis, 400–700 nm) and λ_2 (y-axis, 700–400 nm) for fixed λ_3 between 450 and 650 nm, with FWHM of 11 and 31 nm. The sensitivity spectra of the camera Jai AD-080GE were used

In Fig. 54.1, C is shown for both cameras as a function of λ_1 and λ_2 and a FWHM of 11 nm. As it gives the highest value of C, the third wavelength was fixed at $\lambda_3=472$ nm. From the figure, it can be read that the highest C, and therefore the most promising combination, is for λ_1 around 600 nm and λ_2 between 600 and 700 nm. The differences in C for the two CCDs are caused by the different sensitivity spectra of the cameras. Therefore, the spectral characteristic of the illumination system has to be adapted to the CCD.

One more choice of inverse condition numbers as greyscale images for two permutated and one static λ3 for the Jai CCD is given in Fig. 54.2. A set of images for the FWHMs of 11 and 31 nm is given for the static λ3 of 450, 500, 550, 600 and 650 nm. When considering the FWHM of the wavelengths, it is noticeable that with growing bandwidths the inverse condition numbers decrease. Another effect is that the areas of good condition numbers seem to smear out due to the increasing effects of overlapping spectra.

54.4 Selection of Commercial LEDs Based on Matrix Condition

The condition number analysis served as a first guess for the selection of commercial high-power colour LEDs with an electrical power of at least 3 W and wavelength bands separated by at least 30 nm for using the Jai CCD as detector. For this, published emission spectra were used as well as spectra measured with a lab spectrometer (USB 2000, Ocean Optics Inc., Dunedin, USA). Four LED modules M1–M4 were looked at in detail: M1 is a complete LED module (LZ4, LED Engin) with an integral electrical power of 10 W. Module M2 is a combination of three LEDs with emission peaks separated by more than 50 nm. In the spectral band between 500 and 630 nm, the extinction coefficients of oxyHb and deoxyHb have a very specific shape; therefore, M3 was chosen as the best combination of wavelengths within this band. M4 was chosen as the best combination with C higher than for M3. Here the best combination is defined as the best compromise between a high C and luminous flux. The measurement showed that the real characteristics of the LEDs differ from those stated in the data sheets. For this reason, the calculations for the four modules were performed again with the measured values for their peak wavelength and FWHM.

In Table 54.1a, a comparison is given for different LEDs and for the measured values (lab) and those published in the data sheets (pub) as well as the inverse condition numbers. The C values are markedly different for the modules M1–M4 and for the published and measured LED emission spectra. This indicates that a measurement of the LED spectra is recommended for a best RGB-reflectometry system.

Table 54.1 Commercial LEDs of the assembled modules M1–M4. The peak wavelengths and the FWHM of their emission spectra are given based on own measurements (lab) and the data sheets (pub). The values of C are given when including the JAI CCD detector spectra

		λ (nm)		FWHM (nm)		C	
M	LED	lab	pub	lab	pub	lab	pub
M1	LED Engin LZ4 blue	460	460	34	20	0.093	0.194
	LED Engin LZ4 amber	594	590	15	20		
	LED Engin LZ4 red	630	625	17	20		
M2	Philips LUXEON Rebel b	480	470	32	20	0.118	0.127
	Avago Tech. ASMT-Ax3x g	523	525	46	20		
	Avago Tech. ASMT-Ax3x a	608	590	18	15		
M3	Avago Tech. ASMT-Ax3x g	523	525	46	20	0.182	0.149
	Avago Tech. ASMT-Ax3x a	608	590	18	15		
	Osram Platinum Dragon r	650	625	22	20		
M4	Luxeon LUXEON Rebel b	483	470	33	20	0.326	0.303
	Avago Tech. ASMT-Ax3x a	608	590	18	15		
	LED Engin LZ1 r	647	623	23	15		

54.5 Experimental Evaluation of Choice of Wavelengths

To back up the findings of the condition number analysis, we experimentally tested the magnitude of errors induced by ambient light. To this end, all LEDs of the four modules M1–M4 were adjusted to the same radiant flux, a white balance of the CCD camera was done and the gain adjusted for equal count rates for the R-, G- and B-detector channels. In the first test, the surface of a silicone, tissue-simulating phantom of known optical properties was imaged, followed by imaging of the skin of a subject's arm during arterial cuff occlusion, i.e. during large changes in haemoglobin concentrations. Ambient light was induced by an additional LED (PL6N-3LFx 3 W RGB, ProLightOpto Tech. Corp., Taiwan) by either simultaneously ('white disturbance') or sequentially ('RGB disturbance') switching on its RGB emitters. This disturbance was adjusted to change the CCD count rates by 10 %. The deviations of the haemoglobin values by this additional illumination were averaged for oxyHb and deoxyHb and are summarised in Table 54.2 as errors in haemoglobin in arbitrary units.

The errors in haemoglobin due to the ambient light (Table 54.2) show large variations for the modules M1–M4. When averaging the data for each module, there is a strong correlation with the condition number, i.e. 1/C (correlation coefficient 0.98 for lab-based C). When the condition numbers are based on the published spectra (see Table 54.1), the correlation is much weaker (correlation coefficient = 0.30).

54.6 Implications for Imaging Systems

The described condition number analysis gives indications for the quality of an RGB system for the assessment of haemoglobin, and the experimental data support this claim. In the literature, the common approach for the imaging of cortical haemoglobin is a combination of a monochromatic CCD and a filter wheel setup that allows a sequential illumination with a number of narrowband wavelengths. Therefore, the condition number assessment has been extended to some of the systems described in the literature when the emission wavelength characteristics are given.

Table 54.2 Errors in haemoglobin (arbitrary units; *a.u.*) induced by ambient light of the evaluation measurements for modules M1–M4 for phantom data and in vivo skin data during arterial cuff occlusion. For comparison, the C values of Table 54.1 are given

		Errors in haemoglobin (a.u.)			
		Phantom		In vivo	
M	C lab	White	R+G+B	White	R+G+B
M1	0.093	1.825	4.278	2.277	4.003
M2	0.118	1.247	3.495	1.665	3.110
M3	0.182	1.013	3.030	1.124	2.943
M4	0.326	0.778	2.192	1.034	2.428

Table 54.3 Comparison of the inverse condition numbers C for different filter wheel setups used in the literature, a previous RGB system [2] and four LED modules M1–M4 (in conjunction with CCD AD-080GE, Jai) for evaluation measurements. The systems are ordered for ascending C, where a high C value indicates a high data quality

Setup	Peak wavelengths (nm)	C
M1	460, 594, 630	0.093
Sakagushi et al. [4]	510, 540, 560, 580	0.096
M2	480, 607, 523	0.118
Prakash et al. [5]	560, 570, 577, 610	0.142
Steimers et al. [2]	470, 530, 625	0.149
M3	517, 607, 647	0.182
Dunn et al. [1]	560, 570, 580, 590, 600, 610	0.185
Hillman et al. [6]	472, 532, 570, 610	0.203
M4	483, 607, 647	0.326

The results are summarised in Table 54.3, where the different systems are sorted for ascending C. According to this table, our previous RGB-reflectometry setup [2] is comparable to some and weaker than other systems of the literature. When optimising the illumination spectra and selecting module M4, the results indicate that the RGB approach is comparable to or better than what has been used in the literature.

For this reason, it seems reasonable to make use of the simplicity and reliability of a RGB-reflectometry setup for imaging of cortical haemoglobin.

References

1. Dunn AK, Devor A, Bolay H et al (2003) Simultaneous imaging of total cerebral hemoglobin concentration, oxygenation, and blood flow during functional activation. Opt Lett 28(1):28–30
2. Steimers A, Gramer M, Ebert B, Füchtemeier M, Royl G, Leithner C, Dreier J P, Lindauer U, Kohl-Bareis M (2009) Imaging of cortical haemoglobin concentration with RGB reflectometry. Proc SPIE 7368, Clinical and Biomedical Spectroscopy, 736813
3. Cheney W, Kincaid D (2008) Numerical mathematics and computing. Brooks/Cole, Pacific Grove, pp 321–322
4. Sakagushi K, Tachibana T, Furukawa S et al (2007) Experimental prediction of the wavelength-dependent path-length factor for optical intrinsic signal analysis. Appl Opt 46(14):2769–2777
5. Prakash N, Biag JD, Sameer SA et al (2007) Temporal profiles and 2-dimensional oxy-, deoxy-, and total-hemoglobin somatosensory maps in rat versus mouse cortex. Neuroimage 37(1):S27–S36
6. Hillman EMC (2007) Optical brain imaging in vivo: techniques and applications from animal to man. J Biomed Opt 12(5):051402

Chapter 55
Improving Pulse Oximetry Accuracy by Removing Motion Artifacts from Photoplethysmograms Using Relative Sensor Motion: A Preliminary Study

R.W.C.G.R. Wijshoff, M. Mischi, P.H. Woerlee, and R.M. Aarts

Abstract To expand applicability of pulse oximetry in low-acuity ambulatory settings, the impact of motion on extracted parameters as saturation (SpO_2) and pulse rate (PR) needs to be reduced. We hypothesized that sensor motion relative to the skin can be used as an artifact reference in a correlation canceller to reduce motion artifacts in photoplethysmograms (PPGs), in order to improve SpO_2 and PR measurements. This has been proven true in in vivo measurements, where forehead PPGs have been obtained while subjects are walking on a treadmill and relative sensor motion has been measured via self-mixing interferometry. By using relative motion in a normalized least mean square algorithm, the standard deviation of SpO_2 and PR errors is on average reduced by 31 % and 13 %, respectively.

55.1 Introduction

Pulse oximetry is widely applied in medical care to measure peripheral arterial oxygen saturation (SpO_2) and pulse rate (PR). SpO_2 is derived from red (RD) and infrared (IR) photoplethysmograms (PPGs), and PR from either.

At present, use of pulse oximeters is spreading in low-acuity ambulatory settings, e.g., to obtain temporal SpO_2 patterns of chronic lung disease (CLD) patients during activities of daily living. Temporal SpO_2 patterns can be used to advance

R.W.C.G.R. Wijshoff (✉) • M. Mischi • R.M. Aarts
Department of Electrical Engineering, Signal Processing Systems group, Eindhoven University of Technology, Den Dolech 2, Eindhoven 5612 AZ, The Netherlands
e-mail: R.W.C.G.R.Wijshoff@tue.nl; M.Mischi@tue.nl; R.M.Aarts@tue.nl

P.H. Woerlee
Patient Care Solutions, Philips Research, High Tech Campus 34, Eindhoven 5656 AE, The Netherlands
e-mail: Pierre.Woerlee@philips.com

S. Van Huffel et al. (eds.), *Oxygen Transport to Tissue XXXV*, Advances in Experimental Medicine and Biology 789, DOI 10.1007/978-1-4614-7411-1_55,

understanding of CLD and to improve oxygen therapy [1–3]. In addition, heart rate variability (HRV) can be derived from a PPG, e.g., to assess dynamic characteristics of the patient's autonomic nervous system [4].

As PPGs are highly susceptible to motion, ambulatory oximetry and PPG derived HRV can be severely affected by motion artifacts [1, 4]. We hypothesized that motion artifacts in PPGs correlate with sensor motion relative to the skin. Therefore, relative sensor motion can be used as a reference for motion artifacts in a correlation canceller, in order to reduce motion artifacts in PPGs. This enables more reliable beat-to-beat SpO_2 and PR measurements. Relative sensor motion is measured via self-mixing interferometry (SMI) in a laser diode, which has been added to a commercially available forehead pulse oximetry sensor.

This chapter focuses on reflective pulse oximetry sensors, because these sensors are most practical to measure relative sensor motion. Commercially available forehead sensors have been employed to obtain PPGs during walking from healthy volunteers. Forehead measurements have the advantage of being less affected by vasoconstriction, which can cause weakly pulsatile PPGs to be measured peripherally, e.g., in the fingers [5]. Additionally, forehead sensors are more stable during motion as compared to finger sensors [6].

55.2 Methods

Forehead PPGs were measured in five healthy male volunteers (35.6 ± 11.3 years) while walking on a treadmill at speeds between 4 and 8 km/h to generate repetitive motion artifacts that can be expected in ambulatory settings. Each speed was maintained for 2 min and was preceded and followed by a 1-min period during which the subject stood still. Figure 55.1 gives an overview of the measured signals. Forehead

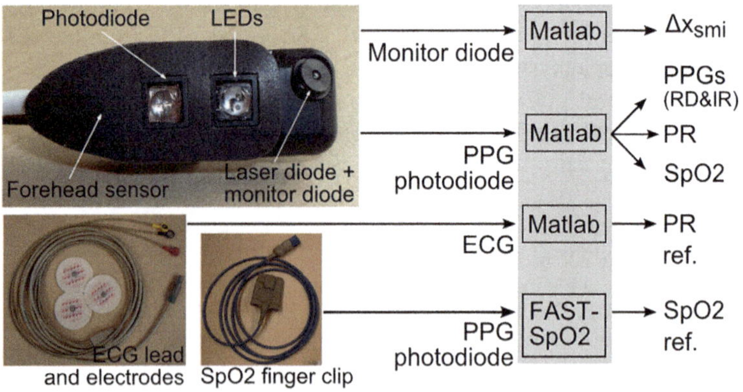

Fig. 55.1 Overview of the measured signals: forehead PPGs (RD and IR) are obtained with a commercial sensor to which a laser diode has been added to measure relative sensor motion; PR and SpO_2 are derived from the forehead PPGs; HR is determined from a single-lead ECG as PR reference; an SpO_2 reference is obtained from the index finger using the FAST-SpO_2 algorithm

PPGs were obtained with a Nellcor Oxisensor II RS-10, which was attached using an adhesive and a headband. SpO_2 and PR were derived on a beat-to-beat basis: SpO_2 via the ratio of the DC-normalized RD and IR pulse amplitudes [7] and PR from the IR PPG by defining a pulse at the average of the time instances of the systolic slope and the diastolic and systolic portions. References for SpO_2 and PR were obtained via an SpO_2 clip on the index finger and a single-lead ECG, respectively. Motion of the reference SpO_2 clip was minimized during walking. The Philips FAST-SpO_2 algorithm was used to obtain the reference SpO_2. Heart rate (HR) was derived from the R-peaks in the ECG.

To measure motion of the forehead sensor with respect to the skin via SMI, an 850 nm vertical-cavity surface-emitting laser diode (VCSEL) with a monitor diode was added to the commercial forehead sensor (Fig. 55.1). The algorithm to measure motion via SMI is explained in detail in [8] and will only be described shortly. During relative motion, the laser light that is backscattered by the skin is Doppler shifted with Doppler phase φ_d. Furthermore, the laser injection current is amplitude modulated. Therefore, when the Doppler-shifted light re-enters the laser cavity, the monitor diode measures a signal proportional to $\sin(\varphi_d)$ around the modulation frequency and a signal proportional to $\cos(\varphi_d)$ around the second harmonic of the modulation frequency. After demodulating these signals to baseband and normalizing their amplitudes, the sensor motion Δx_{smi} can be determined from the Doppler phase by counting the number of Doppler cycles:

$$\Delta x_{smi}[k] = \frac{1}{2\pi}\,\text{unwrap}\left(\arctan 2\left(\frac{\sin(\varphi_d[k])}{\cos(\varphi_d[k])}\right)\right) \approx \sum_{l=0}^{k} f_d[l]\Delta\tau, \qquad (55.1)$$

with discrete time index k, Doppler frequency $f_d[k]$ (Hz), and sampling time $\Delta\tau$ (s). Further, arctan2(.) refers to an arctangent implementation that takes into account in which quadrant the Doppler signals are located. Each phase change of 2π rad in $\varphi_d[k]$ corresponds to a full Doppler cycle, and the sign of the phase change depends on the direction of motion. Motion cannot be measured in absolute units, because the time-varying angle between laser beam and skin is not exactly known.

Relative motion is subsequently used in a normalized least mean square (NLMS) algorithm [9] to reduce the motion artifacts in the PPGs (Fig. 55.2). Here, all signals are down-sampled to 100 Hz. First, the PPG photodiode signal $v_{PD}[k]$ is passed through a 0.3-Hz high-pass filter (HPF) and delayed. Second, the motion artifact estimate is subtracted, giving $e_o[k]$. The artifact estimate is obtained from $\Delta x_{smi\text{-}s}[k] = 10^{-5}\Delta x_{smi}[k]$ (of comparable magnitude as $v_{PD}[k]$) via FIR filter $\mathbf{h}[k]$ of $N_f = 101$ coefficients. The low frequent signal is added to $e_o[k]$ again to determine SpO_2 from the cleaned PPGs. The optimum FIR filter is obtained iteratively via:

$$\mathbf{h}[k+1] = \mathbf{h}[k] + \frac{\mu}{a + \left\|\Delta\mathbf{x}_{smi-s}[k]\right\|^2} e_o[k]\Delta\mathbf{x}_{smi-s}[k], \qquad (55.2)$$

with step-rate parameter $\mu = 5.10^{-3}$ and $a = 10^{-6}$. The filtered PPG has been delayed by $(N_f - 1)/2$ samples to allow for symmetrical FIR filters.

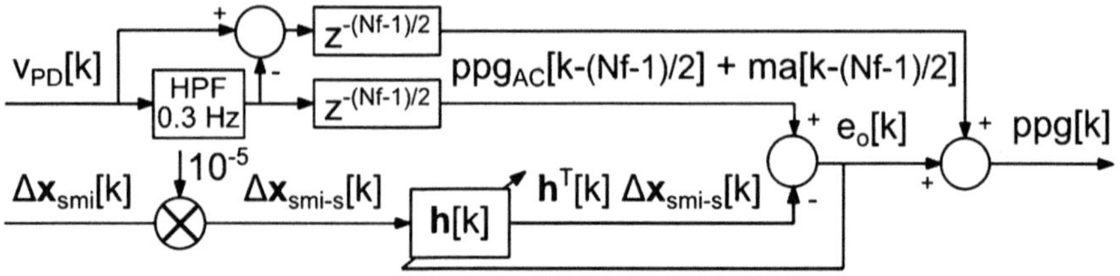

Fig. 55.2 NLMS structure: motion artifacts ma[k] are reduced in the photodiode signal $v_{PD}[k]$ by subtracting the artifacts which are estimated from relative sensor motion $\Delta x_{smi}[k]$ filtered by an FIR filter of N_f coefficients

55.3 Results

As shown in the red PPG's spectrogram in Fig. 55.3e, walking causes additive motion components at the full swing (±1 Hz) and step frequencies (±2 Hz), as well as their harmonics. These motion components result in an amplitude modulation of the PPG as shown by the dashed curve in Fig. 55.3a. Consequently, PR and SpO_2 oscillate around the references, as shown by their differences ΔPR (circles, Fig. 55.3b) and ΔSpO_2 (circles, Fig. 55.3c), SpO_2 showing the strongest oscillation. Furthermore, relative sensor motion (Fig. 55.3d) indicates that the sensor moves at the step rate.

By using relative sensor motion as an artifact reference in an NLMS algorithm (Fig. 55.2), motion artifacts can be largely removed from the PPGs, as shown by the spectrogram of the cleaned red PPG in Fig. 55.3f. Here, a strong reduction of the step-rate component at ±2 Hz has been achieved, and a small reduction of the motion components at ±1 Hz and ±3 Hz. As a result, the cleaned PPG's amplitude is more stable (solid curve in Fig. 55.3a), and the oscillatory pattern in PR and SpO_2 determined from the cleaned PPGs is reduced (stars in Fig. 55.3b and c, respectively). The magnitude frequency response of the converged FIR filter that achieves this artifact reduction shows resonances at ±1 Hz, ±2 Hz, and ±3 Hz (Fig. 55.3g).

The effect of motion on PR and SpO_2 and the performance of the NLMS algorithm vary over the subjects (Fig. 55.4). The difference with the reference (mean ± SD) for PR (ΔPR) and SpO_2 (ΔSpO_2) is determined in baseline when the subject is standing still before and after walking (triangles), during walking (circles), and after the NLMS algorithm has been applied (stars). Only measurement results are shown (12 out of 25) in which PR did not coincide with step rate and in which the RMS amplitude of Δx_{smi} during motion was at least three times larger compared to baseline. Motion increases the PR SD most in subjects 1 and 5 and by 264 % on average; it does not introduce a bias. Motion increases the SpO_2 SD in all subjects and by 42 % on average, except for subject 3 in which the SpO_2 SD decreases during motion. In subject 5 also the baseline SpO_2 bias varies over the measurements. Removal of motion artifacts reduces the SpO_2 SD in all cases and by 31 % on average, but the PR SD only in seven out of 12 cases and by 13 % on average.

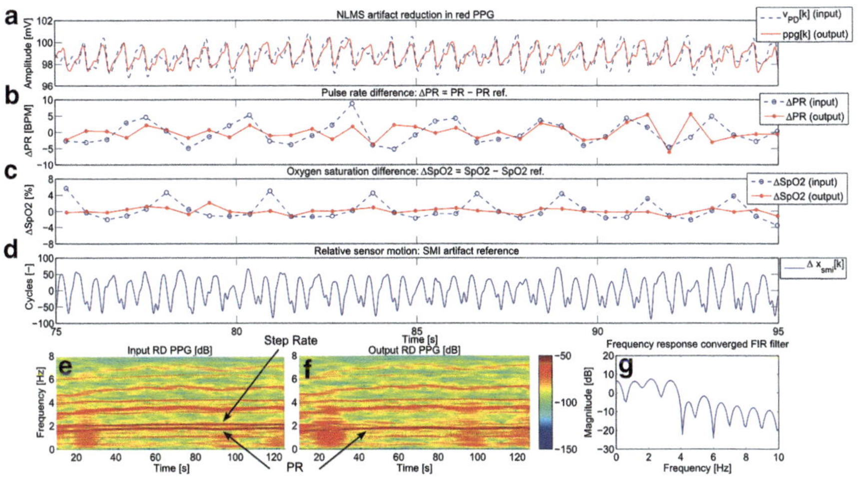

Fig. 55.3 PPG motion artifacts in subject 5 caused by walking at 6 km/h (**a**, *dashed curve*) can be reduced strongly (**a**, *solid curve*) via NLMS by using relative sensor motion (**d**) as an artifact reference; comparing the difference between PR and PR ref. (ΔPR) (**b**) and SpO_2 and SpO_2 ref. (ΔSpO_2) (**c**) resulting from the corrupted (*circles*) and cleaned (*stars*) PPGs shows a larger improvement for SpO_2 than for PR; spectrograms of the corrupted (**e**) and cleaned (**f**) RD PPG show a reduction of motion components at ±1 Hz, ±2 Hz, and ±3 Hz, and the magnitude frequency response of the converged FIR filter shows resonances at these frequencies (**g**)

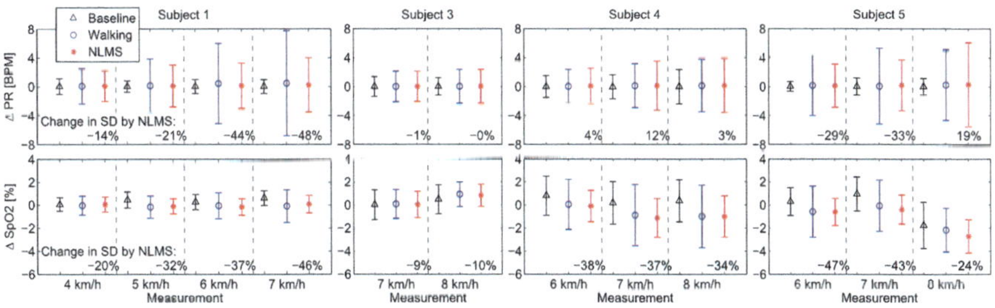

Fig. 55.4 Difference (mean ± SD) between PR and PR ref. (ΔPR), and SpO_2 and SpO_2 ref. (ΔSpO_2), in baseline when the subject is standing still (*triangles*), during walking (*circles*), and after NLMS artifact reduction (*stars*); change in SD as a result of NLMS artifact reduction is shown as a percentage

55.4 Discussion

The NLMS algorithm has only been applied to measurements in which PR and step rate did not coincide and in which the relative motion signal was sufficiently strong. When PR and step rate coincide, the algorithm removes both motion artifact and PPG. This drawback is inherent to correlation cancelation. A relative motion signal of poor quality may result from a small sensor to forehead motion. Differences in relative motion may result from intersubject differences in walking and variations in sensor contact force. As the same type of headband is used for all subjects, the head's circumference determines the force exerted on the sensor.

Figure 55.3 shows that the steady motion component in the PPG can be strongly reduced via the NLMS algorithm. However, this implementation of the NLMS algorithm is unable to correct for fast and relatively large sensor displacements, as happens at 20 s in Fig. 55.3e and f. Furthermore, in some subjects, PR is observed via SMI, in which case the NLMS algorithm can affect the PR component in the PPG.

The forehead and finger SpO_2 time traces are not identical in shape, even during baseline, which complicates their synchronization. Therefore, ΔSpO_2 is partly caused by synchronization issues, e.g., causing ΔSpO_2 in subject 3 to be smaller during motion compared to baseline. Even so, Fig. 55.4 shows that walking has a moderate effect on PR and SpO_2 derived from forehead PPGs, which can be a result of step rate being close to PR and the stable measurement site [6]. Further, Fig. 55.4 shows that the improvement in PR is smaller compared to the improvement in SpO_2. This indicates that the algorithm can improve the amplitude of the PPG waveform, but cannot perfectly recover its shape. Finally, intersubject differences in the effect of walking on PPGs, PR, and SpO_2 can be a result of differences in sensor positioning and fixation, skin dynamics, anatomy, and the way of walking.

55.5 Conclusions

If PR and the motion frequency do not coincide, relative sensor motion measured via SMI can be used as an artifact reference in an NLMS algorithm to reduce motion artifacts in a PPG, which can improve SpO_2 and PR measurements. Furthermore, the forehead is a stable position to measure PPGs during walking, as walking has a moderate effect on forehead PPGs and derived SpO_2 and PR.

Acknowledgments This work was supported by NL Agency, IOP Photonic Devices, IPD083359 HIP – Hemodynamics by Interferometric Photonics.

References

1. Barratt CW, Vyas H, Hayes-Gill BR, Crowe JA, Flatman D (2007) Detection of previously unrecognized daytime desaturation in children with chronic lung disease. J Med Eng Technol 31(2):101–108
2. Cutaia M (2002) Ambulatory monitoring of oxygen saturation in chronic lung disease: optimizing long-term oxygen therapy. Clin Pulm Med 9(6):297–305
3. Zhu Z, Barnette RK, Fussell KM, Michael Rodriguez R, Canonico A, Light RW (2005) Continuous oxygen monitoring–a better way to prescribe long-term oxygen therapy. Respir Med 99(11):1386–1392

4. Lu S, Zhao H, Ju K et al (2008) Can photoplethysmography variability serve as an alternative approach to obtain heart rate variability information? J Clin Monit Comput 22(1):23–29
5. Branson RD, Mannheimer PD (2004) Forehead oximetry in critically ill patients: the case for a new monitoring site. Respir Care Clin N Am 10(3):359–367
6. Yamaya Y, Bogaard HJ, Wagner PD, Niizeki K, Hopkins SR (2002) Validity of pulse oximetry during maximal exercise in normoxia, hypoxia and hyperoxia. J Appl Physiol 92(1):162–168
7. Palreddy S (1997) Signal processing algorithms. In: Webster JG (ed) Design of pulse oximeters. Taylor & Francis Group, New York
8. Wijshoff RWCGR, Mischi M, Veen J, Van der Lee AM, Aarts RM (2012) Reducing motion artifacts in photoplethysmograms by using relative sensor motion: phantom study. J Biomed Opt 17(11):117007–1–15
9. Haykin S (1996) Least-mean-square algorithm. In: Adaptive filter theory, 3rd edn. Prentice-Hall, Upper Saddle River

Chapter 56
Measuring the Vascular Diameter of Brain Surface and Parenchymal Arteries in Awake Mouse

Yuta Sekiguchi, Kazuto Masamoto, Hiroyuki Takuwa, Hiroshi Kawaguchi, Iwao Kanno, Hiroshi Ito, Yutaka Tomita, Yoshiaki Itoh, Norihiro Suzuki, Ryo Sudo, and Kazuo Tanishita

Abstract The present study reports a semiautomatic image analysis method for measuring the spatiotemporal dynamics of the vessel dilation that was fluorescently imaged with either confocal or two-photon microscope. With this method, arterial dilation induced by whisker stimulation was compared between cortical surface and parenchymal tissue in the vibrissae area of somatosensory cortex in awake Tie2-GFP mice in which the vascular endothelium had genetically expressed green fluorescent protein. We observed that a mean arterial diameter during a pre-stimulus baseline state was 39 ± 7, 19 ± 1, 16 ± 4, 17 ± 4, and 14 ± 3 μm at depths of 0, 100, 200, 300, and 400 μm, respectively. The stimulation-evoked dilation induced by mechanical whisker deflection (10 Hz for 5 s) was 3.4 ± 0.8, 1.8 ± 0.8, 1.8 ± 0.9, 1.6 ± 0.9, and 1.5 ± 0.6 μm at each depth, respectively. Consequently, no significant differences were observed for the vessel dilation rate between the cortical surface and parenchymal arteries: 8.8 %, 9.9 %, 10.9 %, 9.2 %, and 10.3 % relative to their

Y. Sekiguchi
Department of Integrated Design Engineering, Keio University, Yokohama, Japan

Molecular Imaging Center, National Institute of Radiological Sciences, Chiba, Japan

K. Masamoto, Ph.D. (✉)
Molecular Imaging Center, National Institute of Radiological Sciences, Chiba, Japan

Center for Frontier Science and Engineering, University of Electro-Communications, 1-5-1 Chofugaoka, Chofu, Tokyo 182-8585, Japan
e-mail: masamoto@mce.uec.ac.jp

H. Takuwa • H. Kawaguchi • I. Kanno • H. Ito
Molecular Imaging Center, National Institute of Radiological Sciences, Chiba, Japan

Y. Tomita • Y. Itoh • N. Suzuki
Department of Neurology, School of Medicine, Keio University, Tokyo, Japan

R. Sudo • K. Tanishita
Department of Integrated Design Engineering, Keio University, Yokohama, Japan

Department of System Design Engineering, Keio University, Yokohama, Japan

S. Van Huffel et al. (eds.), *Oxygen Transport to Tissue XXXV*, Advances in Experimental Medicine and Biology 789, DOI 10.1007/978-1-4614-7411-1_56,

baseline diameters, respectively. These preliminary results demonstrate that the present method is useful to further investigate the quantitative relationships between the spatiotemporally varying arterial tone and the associated blood flow changes in the parenchymal microcirculation to reveal the regulatory mechanism of the cerebral blood flow.

56.1 Introduction

Cerebral arteries of the cortical surface and parenchymal tissue are known to have anatomically different features. Firstly, a surface artery consists of threefold thicker layers of smooth muscle cells as compared to the parenchymal arteries [1]. Secondly, a surface artery is situated apart from the parenchymal tissue [2], whereas a parenchymal artery is closely surrounded with brain cells (i.e., astrocyte and microglia). Thirdly, a surface artery has the responsibilities to prevent blood-stealing effects on the adjacent vascular territories when one particular region is activated (i.e., functional hyperemia) [3]. On the other hand, a parenchymal artery is dedicated to control blood supply predominantly for a specific vascular territory [4]. These differences suggest the coordinated control mechanism with the surface and parenchymal arteries in balancing blood supply to meet locally varying energy demand in the brain tissue.

In our previous study, we proposed a hypothesis that cortical surface vascular tone and parenchymal blood flow are individually coordinated [5]. However, no direct measurements of the vessel dynamics in the parenchymal tissue where the blood flow was regulated were performed. In the present study, we developed a method for measuring the spatiotemporally varying vessel diameter that was fluorescently imaged with either confocal or two-photon microscope. And then, we compared the arterial dilation between the cortical surface and parenchymal tissue in the vibrissae area of mouse somatosensory cortex during an induction of whisker stimulation [6].

56.2 Methods

56.2.1 Animal Experiment

Animal use and experimental protocols were approved by the Institutional Animal Care and Use Committee in National Institute of Radiological Sciences. A total of nine male STOCK Tg[Tie2GFP] 287Sato/J mice (20–32 g, 7–11 weeks) [7], in which the vascular endothelium had genetically been expressed with green fluorescent protein (GFP), were used for the experiments with a chronic cranial window over the somatosensory cortex [8]. The animal was placed on a custom-made fixation device with maintaining awake conditions during microscopic imaging experiments [6]. Sulforhodamine 101 was injected intraperitoneally to mice to fluorescently label blood plasma [9]. Cortical surface vasculature was imaged with a confocal mode (488-nm excitation), whereas parenchymal vasculature was imaged with a two-photon excitation mode (900 nm) using laser scanning fluorescent microscope system (TCS-SP5 MP, Leica). Spatiotemporal dynamics of the arterial response to mechanically induced single whisker stimulation (10 Hz with 1.2-mm deflection repeated for 5 s) was measured at depths of 0, 100, 200, 300, and 400 μm from the cortical surface in the activated somatosensory barrel cortex with a pixel resolution of 0.5 μm per pixel. In each measurement, a total of 20-s continuous acquisition (0.15 s per frame) was performed (5 s for pre-stimulus baseline, 5 s during stimulation, and 10 s for poststimulus). A total of either 8 or 16 trials were repeatedly performed with an intertrial interval of 30 s, and the image was averaged over the trials to improve signal-to-noise ratio. A baseline diameter was measured by averaging the diameter over 5-s pre-stimulus periods, and stimulus-evoked vasodilation was the diameter averaged for 1–6 s from the onset of stimulation minus the baseline diameter. After completion of the vascular imaging experiments, cerebral blood flow (CBF) responses to whisker stimulation (the same paradigm as the vascular imaging experiments) were also measured with laser-Doppler flowmetry (LDF; FLO-C1, OMEGA FLO) over the site where the vascular response was measured.

56.2.2 *Image Analysis*

The offline semiautomatic image analysis was conducted with a custom-written MATLAB code as follows:

(i) Image binarization and morphological filtering
A fixed rectangular region of interest (ROI) was manually set for the raw images (Fig. 56.1). Median filter over 3 by 3 pixels was applied, and the images were binarized by adjusting a threshold intensity every 1 % from 1 % to 100 % of the maximum intensity in the image. To remove small dot noises, a binary closure (dilation followed by erosion) was also applied.

(ii) Fitting with a parallelogram or ellipse for diameter measurements
For the surface arteries, an edge of the vessel was fitted with a parallelogram, and the vessel diameter (i.e., an external side of the vascular endothelium) was measured by dividing an area of the fitted parallelogram by a length of the center line of the longitudinal vessel trace (Fig. 56.1a). For the parenchymal arteries, an area of the vessel was approximated with an ellipse, and the diameter (i.e., an inner side of the vessel) was measured with a minor axis of the ellipse (Fig. 56.1b).

(iii) Criteria for thresholding
A threshold intensity applied for the image binarization was further tested, and only the data that meet the following criteria were used.

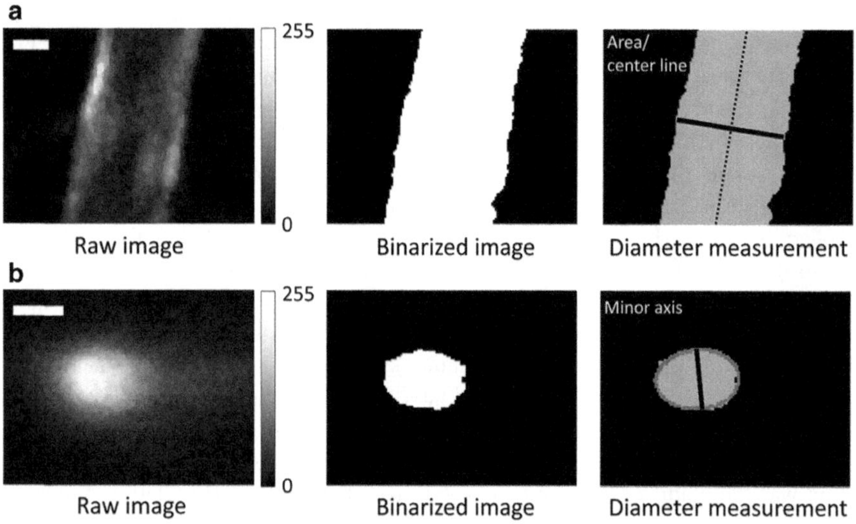

Fig. 56.1 An image analysis for measuring the vessel diameter. A representative raw image (*left*) of the target surface and parenchymal arteries and its binarized one (*center*) are shown in panels **a** and **b**, respectively. A scale bar indicates 10 μm. On the cortical surface (**a**), a vessel was located in parallel with the image plane, and thus the diameter was calculated by dividing the vessel area (*gray area*) by a length of the approximated center line (*a black dotted line, right*). In the parenchymal tissue (**b**), a major vessel connected with a surface vessel was located in a direction perpendicular to the image plane, which makes an ellipse shape on the image. Therefore, the diameter was measured as a length of the minor axis (*a black line*) for an approximated ellipse (*gray area, right*)

(i) An image does not show more than ± 3 % variations for a location of the gravity point of the binarized vessels over ranges of ± 5 % of the applied threshold intensity changes.
(ii) An image does not show more than ± 0.2 μm variations for the measured dilation over rages of ± 5 % of the applied threshold intensity changes.

Data were represented as mean ± standard deviation (a number of animals measured), and a statistical significance was evaluated with parametric Tukey or Bonferroni tests among the data obtained over depths of 0–400 μm.

56.3 Results and Discussion

Typical vessel images observed with either confocal or two-photon microscopy are represented for the cortical surface (Fig. 56.2a) and the parenchymal tissue (Fig. 56.2c), respectively. Mean time courses of the diameter changes within selected ROIs (rectangles in Fig. 56.2a and b) are also represented in Fig. 56.2b and d. Following the onset of whisker stimulation, the vessel significantly dilated for both

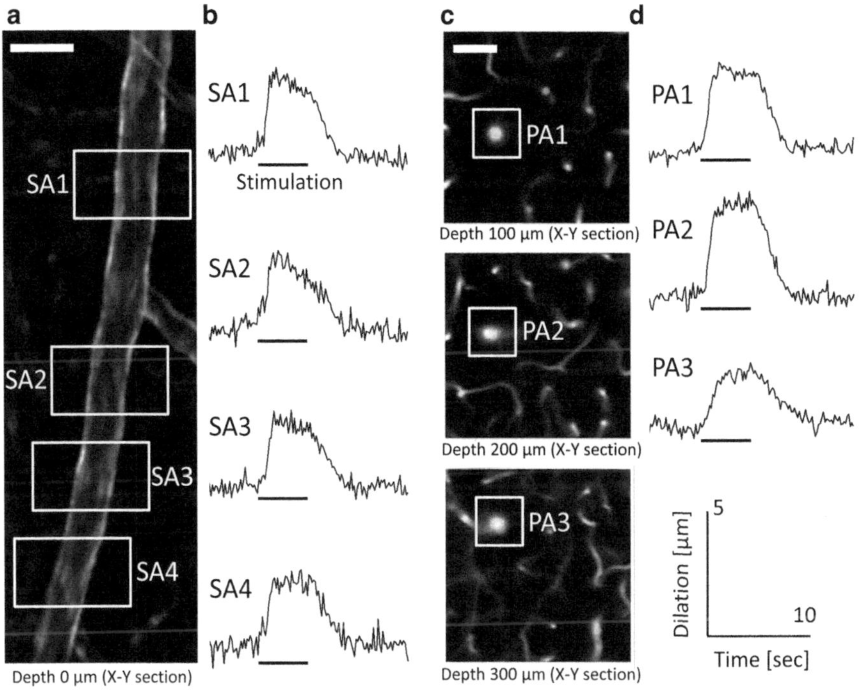

Fig. 56.2 A measurement for temporal dynamics of the vessel dilation induced by whisker stimulation. (**a**) A representative image showing a raw image of the GFP-expressed vascular endothelium measured at a depth of 0 μm (i.e., a cortical surface). The vascular image delineates the edge of the target surface artery (*SA*). (**b**) Stimulation-induced dilation of the surface artery measured for each rectangular ROI (**a**). (**c**) Images show parenchymal microvessels (i.e., sulforhodamine 101-labeled blood plasma) captured at cortical depths of 100, 200, and 300 μm (from *top to bottom*). A target penetrating artery (*PA*) is indicated as a rectangular ROI. (**d**) Stimulation-induced dilation of the penetrating arteries is consistently seen across the cortical depths examined. Scale bar: 50 μm

the surface and parenchymal arteries (Fig. 56.2b and d), and after cessation of the stimulation, the vessel returned to the pre-stimulus baseline level. We observed that a mean arterial diameter during a pre-stimulus baseline state was 39 ± 7 μm (N = 4), 19 ± 1 μm (N = 4), 16 ± 4 μm (N = 7), 17 ± 4 μm (N = 9), and 14 ± 3 μm (N = 9) at depths of 0, 100, 200, 300, and 400 μm, respectively (Fig. 56.3a). And the stimulation-induced dilation was 3.4 ± 0.8 μm, 1.8 ± 0.8 μm, 1.8 ± 0.9 μm, 1.6 ± 0.9 μm, and 1.5 ± 0.6 μm at each depth, respectively (Fig. 56.3b). These changes account for 8.8 %, 9.9 %, 10.9 %, 9.2 %, and 10.3 % relative to their baseline diameters, respectively (Fig. 56.3c). Consequently, no significant differences (P > 0.05) were observed for the vessel dilation rate over depths of 0–400 μm.

Although we used different methods for determining the diameters of the cortical surface vessels (i.e., GFP-expressed endothelium with confocal microscopy) and parenchymal ones (i.e., sulforhodamine 101-labeled blood plasma with two-photon

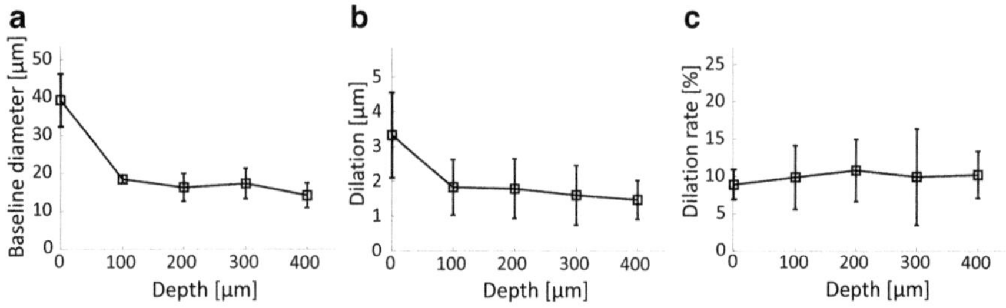

Fig. 56.3 Baseline diameter and stimulation-induced changes compared across different cortical depths (0–400 µm). (**a**) Surface arteries (measured at depth 0 µm) have twofold larger baseline diameters ($P<0.05$ vs. other depths) relative to the parenchymal arteries (depths 100–400 µm), whereas a consistent baseline diameter was observed for parenchymal arteries ($P>0.05$). (**b**) The stimulation-induced arterial dilation was also twofold larger ($P<0.05$) for the surface arteries than those of the parenchymal arteries. (**c**) No significant differences ($P>0.05$) in the dilation rate were observed over depths of 0–400 µm

microscopy), the effects of these methodological differences on the measurements could be negligible. It has been shown that a thickness of endothelial cells was less than 1 µm [10], which corresponds to approximately 2 pixels in our experimental conditions.

A previous report showed that a half of maximum pixel intensity within the vessel cross-sectional image was threshold pixel intensity for determining an edge of the vessel lumen [11]. However, this criterion is subject to a variation of the intensity differences between the maximum and the background non-vessel pixels, such as for the images captured across different cortical depths. Alternatively, we focused on steady points of two parameters (i.e., a location of the gravity point for extracted vessel cross-sectional images and a magnitude of vessel dilation) against changes of the threshold intensity. Because a baseline diameter decreased monotonically with an increase of the threshold intensity, an average error of the diameter measurements was ±0.91 µm within the ±5 % of the selected threshold.

We observed 10 % dilation of the both surface and parenchymal arteries during whisker stimulation. According to the Hagen-Poiseuille law, it can be expected that 10 % increase in the vessel diameter accounts for 46 % increase of the CBF (i.e., flow varies with the fourth power of a vessel diameter). In the present study, however, only 15±6 % ($N=4$) increase of the CBF induced by whisker stimulation was observed in the LDF experiments. These results further suggest that additional mechanisms participate in regulating parenchymal blood flow. Future study is needed to focus on the spatiotemporal propagations of the vasodilatory signals along the cortical surface and parenchymal arteries as well as additional mechanisms that may have a responsible role for the changes of the parenchymal blood flow.

Acknowledgments A part of this work was supported by Special Coordination Funds for Promoting Science and Technology, Japan (K.M.) and JSPS KAKENHI Grant Number 24659578 (I.K.).

References

1. Roggendorf W, Cervos-Navarro J, Matakas F (1976) The ultrastructural criteria of intracerebral arterioles. In: Cervos-Navarro J (ed) The cerebral vessel wall. Raven, New York, pp 23–31
2. Jones EG (1970) On the mode of entry of blood vessels into the cerebral cortex. J Anat 106(Pt 3):507–520
3. Faraci FM, Heistad DD (1990) Regulation of large cerebral arteries and cerebral microvascular pressure. Circ Res 66(1):8–17
4. Nishimura N, Schaffer CB, Friedman B, Lyden PD, Kleinfeld D (2007) Penetrating arterioles are a bottleneck in the perfusion of neocortex. Proc Natl Acad Sci USA 104(1):365–370
5. Kawaguchi H, Masamoto K, Ito H, Kanno I (2012) Image-based vessel-by-vessel analysis for red blood cell and plasma dynamics with automatic segmentation. Microvasc Res 84(2):178–187
6. Takuwa H, Autio J, Nakayama H et al (2011) Reproducibility and variance of a stimulation-induced hemodynamic response in barrel cortex of awake behaving mice. Brain Res 1369: 103–111
7. Motoike T, Loughna S, Perens E et al (2000) Universal GFP reporter for the study of vascular development. Genesis 28(2):75–81
8. Tomita Y, Kubis N, Calando Y et al (2005) Long-term in vivo investigation of mouse cerebral microcirculation by fluorescence confocal microscopy in the area of focal ischemia. J Cereb Blood Flow Metab 25(7):858–867
9. Masamoto K, Tomita Y, Toriumi H et al (2012) Repeated longitudinal in vivo imaging of neuro-glio-vascular unit at the peripheral boundary of ischemia in mouse cerebral cortex. Neuroscience 212:190–200
10. Rhodin JA (1967) The ultrastructure of mammalian arterioles and precapillary sphincters. J Ultrastruct Res 18(1):181–223
11. Drew PJ, Shih AY, Kleinfeld D (2011) Fluctuating and sensory-induced vasodynamics in rodent cortex extend arteriole capacity. Proc Natl Acad Sci USA 108(20):8473–8478

Chapter 57
Simultaneous Imaging of Cortical Blood Flow and Haemoglobin Concentration with LASCA and RGB Reflectometry

André Steimers, M. Gramer, M. Takagaki, R. Graf, U. Lindauer, and Matthias Kohl-Bareis

Abstract We demonstrate a system for the simultaneous imaging of cortical blood flow and haemoglobin oxygenation by laser speckle contrast analysis (LASCA) and RGB reflectometry. The sensitivity of the system was tested by observing changes of haemoglobin oxygenation and blood flow in rats in response to ischaemic stroke, hypercapnia, hyperoxia, hypoxia, cortical spreading depression and cortical activation following forepaw stimulation.

57.1 Introduction

Optical imaging is widely used as a tool for the assessment of brain function and pathological tissue. Two dominant aims in neurological research are the quantification of cortical haemoglobin oxygenation and blood flow changes as the key parameters for an understanding of neurometabolic-vascular coupling [1]. Illuminating the exposed cortex with a continuous wave light source and observing the backscattered light with a CCD camera is a suitable approach for mapping of haemoglobin changes [2]. Another more reliable spectroscopic solution is accomplished by combining a RGB (red, green, blue)-LED and a RGB-(colour) camera. It has already been shown that RGB reflectometry is able to record small changes in haemoglobin oxygenation associated with neurometabolic and neurovascular coupling [3].

A. Steimers (✉) • M. Kohl-Bareis
RheinAhr Campus Remagen, University of Applied Sciences Koblenz, Remagen, Germany
e-mail: steimers@rheinahrcampus.de

M. Gramer • M. Takagaki • R. Graf
Max Planck Institute for Neurological Research, Cologne, Germany

U. Lindauer
Department of Neurosurgery, Klinikum rechts der Isar, Technical University Munich, Munich, Germany

S. Van Huffel et al. (eds.), *Oxygen Transport to Tissue XXXV*, Advances in Experimental Medicine and Biology 789, DOI 10.1007/978-1-4614-7411-1_57,

We now extend this by integrating laser speckle contrast analysis (LASCA) to enable a simultaneous measurement of both parameters, haemoglobin and blood flow changes. For this aim we exploit the fact that these two methods use different spectral ranges. A commercial 2-CCD camera separates the reflected NIR (near infrared) laser light (for LASCA) and the reflected visible light (for haemoglobin) by means of a dichroic prism and images; these different wavelength ranges on two separate CCDs sensors. By the use of GPU (graphical processor unit), computing a temporal resolution of up to 15 Hz is achieved.

The sensitivity of the combined system was tested by imaging haemoglobin oxygenation and blood flow changes in cortical tissue of rats.

57.2 Methods and Setup

57.2.1 RGB Reflectometry

The standard approach for the analysis of reflectance spectra is based on the Lambert-Beer equation. When both the (RGB-) light source and the (CCD-) detector have broad, overlapping spectra, the extinction coefficients matrix of haemoglobin can be expressed by integrating the product of the known parameters of the tissue and spectroscopy system over the wavelength range used:

$$E_{ji} = \int \varepsilon_j(\lambda) \cdot D_i(\lambda) \cdot S(\lambda) \cdot L(\lambda) \cdot d\lambda \tag{57.1}$$

with the index i for the colour sensors of the camera, j for the single chromophores and ε the extinction coefficients of oxygenated and deoxygenated haemoglobin (oxyHb and deoxyHb). $D_i(\lambda)$ is the sensitivity spectra of the camera, $S(\lambda)$ the normalised intensity spectrum of the light source and $L(\lambda)$ the mean optical path length. The unknown concentration changes Δc can then be obtained by matrix inversion from the experimental attenuation changes ΔA: $\Delta c_j = E_{ji}^{-1} \cdot \Delta A_i$.

57.2.2 Laser Speckle Contrast Analysis

When illuminating a surface with coherent laser light, it will appear granular to the observer. This effect is commonly known as speckle effect. A scattering of the laser light by moving particles like blood cells will cause fluctuations of the speckle pattern. If these fluctuations are observed by an integrating system like a camera with fixed exposure time, the pattern will appear blurred. This blurring is generally quantified as speckle contrast SC and defined as the quotient of the standard deviation of the averaged intensity of a small pixel window inside the pattern. Goodman gave a statistical description of speckle patterns and defined the correlation between the speed of the moving particle and the speckle contrast [4]. More information about the methods used can be found in [5].

57.2.3 Setup of the Imaging System

A 2-CCD camera (AD-080GE, Jai A/S, Denmark; 1,024×768 pixels of 4.65^2 μm^2 size, 12 bit dynamic, 30 frames/s) in conjunction with a standard zoom-camera lens with f=5.6–32 mm and f/#=1:5.6 (Computar Corp., Japan) allowed simultaneous imaging of the RGB and NIR wavelength bands.

A ring of six RGB-LEDs (LZ4-00MC10, LedEngin Inc., USA) served as illumination for RGB reflectometry and an AlGaAs laser diode (ADL78901TL, Arima Lasers Corp., Taiwan) with a wavelength of 785 nm and 90 mW for LASCA. The speckle contrast was calculated as the ratio of standard deviation divided by the averaged intensity within sub-windows of 7×7 pixels. Image acquisition and control of all camera functions was programmed in LabVIEW 2010 (National Instruments Inc., USA). The software running on a workstation (i7 950, Intel; GTX 480, NVIDIA) allowed an on-line calculation and display of the haemoglobin and flow images at up to 15 Hz.

57.3 Evaluation Measurements

For evaluation of the combined imaging system, changes in cortical blood flow and haemoglobin concentration were mapped with the combined imaging system. Information about the animal preparations can be found in [5, 6].

57.3.1 Ischaemic Stroke Caused by Injection of Macro-spheres

An ischaemic stroke was induced by injecting two macro-spheres of 300–360 μm in diameter into the carotid artery [6]. Figure 57.1 displays false colour images of changes in deoxyHb, oxyHb and flow for four times (frames) during the first minute after occlusion onset. Haemoglobin parameters and flow are displayed as changes with respect to the baseline before occlusion onset. There is a clear increase in deoxyHb concentration and a decrease both in oxyHb and blood flow for two different areas. A dominating PCAO (posterior cerebral artery occlusion) and a smaller ACAO (anterior cerebral artery occlusion) located nearby could be identified due to the spatially resolved images. The observed pattern of cerebral haemoglobin and blood flow changes is consistent with the physiological concept of an ischaemic reaction following an artery occlusion by the injection of the macro-spheres.

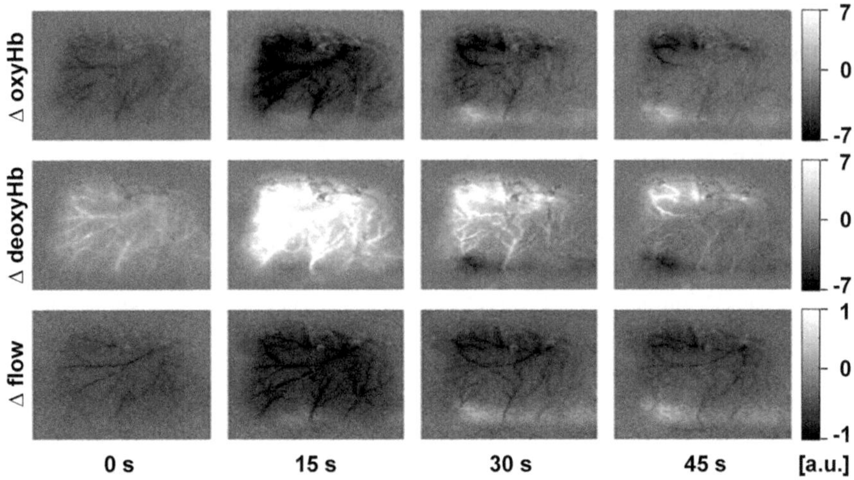

Fig. 57.1 False colour images of deoxyHb, oxyHb and flow changes following anterior cerebral artery occlusion and posterior cerebral artery occlusion caused by two macro-spheres

57.3.2 Hypercapnia, Hyperoxia and Cortical Spreading Depression

The imaging system was further evaluated under different oxygen supply conditions. The experiment was separated into four parts for hypercapnia, hyperoxia, cortical spreading depression (CSD) and hypoxia (Fig. 57.2). The haemoglobin parameters were averaged for three regions of interest (ROIs), with the first two on the left hemisphere close to the excitation point of the CSD and the third on the right hemisphere. During hypercapnia, CO_2 concentration in the supplied air mix was increased to 5 % with a similar systemic effect on the blood parameters in all ROIs. Due to its triggering effect on cortical blood flow, CO_2 and flow are correlated and concurrent with an increase in oxyHb and a decrease in deoxyHb. During the ventilation conditions of hyperoxia and hypoxia (100 % and 0 % O_2 in the supplied air, respectively), there are large changes in the haemoglobin concentrations while the effect on blood flow is small. This experiment demonstrates that the three measurement parameters are independent with no simple correlation between them. While hypercapnia, hyperoxia and hypoxia induce systemic changes, a CSD was triggered by a needle prick to generate a self-propagating wave of depolarization of neurons and its glial cells. This wave spreads over the left hemisphere, which gives rise to a time shift for the signal of the first two ROIs. For the third ROI on the right hemisphere, only very small changes were observed. During CSD the increased blood flow is correlated with an increase in oxyHb and a decrease in deoxyHb.

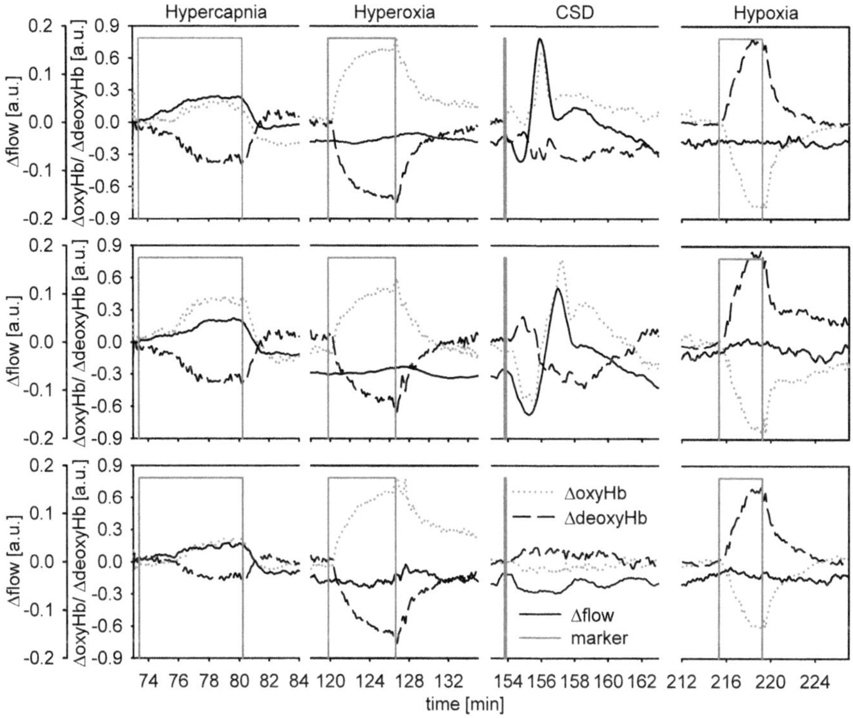

Fig. 57.2 Time course of changes in blood flow as well as oxyHb and deoxyHb averaged for ROIs on the left (*upper* and *middle traces*) and right (*lower part*) hemisphere of a rat during hypercapnia, hyperoxia, cortical spreading depression (*CSD*) and hypoxia

57.3.3 Cortical Activation

To activate a somatosensory stimulation, two needle electrodes were inserted into the skin of the left forepaw between digits 2 and 4 and rectangle pulses (0.3 ms duration at amplitude of 1.6–1.8 mA at 3 Hz) delivered in stimulation trains (length 16–20 s). The interstimulus interval was 45 s to 2 min.

In Fig. 57.3 the haemoglobin and blood flow response following cortical activation are shown for ten stimuli. The p-values (superimposed on the anatomical images) indicate a very high sensitivity combined with a high spatial resolution in the images. When averaging over the ROI, each stimulus is linked to prominent and strong changes in oxyHb, deoxyHb and flow. The observed pattern of a decreased deoxyHb concentration and an increase in both oxyHb and flow is consistent with the concept of neurometabolic-vascular coupling which links blood flow as well as blood volume and oxygenation to the higher oxygen extraction after cortical activation.

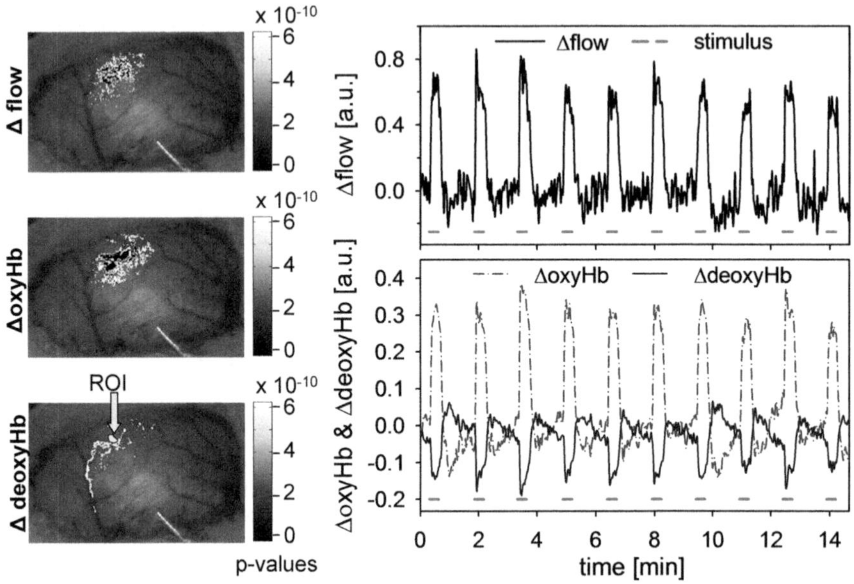

Fig. 57.3 Haemoglobin and blood flow response following cortical activation as observed with the system. *Left*: anatomical images with a selected region of interest (*ROI*) and *p*-values. *Right*: time courses of changes in oxyHb, deoxyHb and flow as averaged values from the ROI for ten single stimuli

57.4 Conclusion

In conclusion, a system for simultaneous spatially and temporally resolved imaging of haemoglobin oxygenation and blood flow with RGB reflectometry and LASCA was presented. It consists of a RGB-LED and a low power diode laser as light sources. The system is based on a two-CCD camera to separate the VIS and NIR spectral bands, allowing a continuous illumination and observation of signals from both light sources. We demonstrate that the system allows haemoglobin and blood flow changes to be imaged in cortical tissue for different conditions like ischaemic stroke, hypercapnia, hyperoxia, hypoxia, cortical spreading depression and cortical activation. The new system is demonstrated to have a high sensitivity combined with good temporal and spatial resolution. Its design offers significant advantages in terms of technical simplicity, reliability and robustness compared to other approaches.

References

1. Hillman EMC (2007) Optical brain imaging in vivo: techniques and applications from animal to man. J Biomed Opt 12(5):051402
2. Malonek D, Grinvald A (1966) Interactions between electrical activity and cortical microcirculation revealed by imaging spectroscopy: implications for functional brain mapping. Science 272(5261):551–554

3. Steimers A, Gramer M, Ebert B et al (2009) Imaging of cortical haemoglobin concentration with RGB reflectometry. Proc SPIE 7368–30
4. Goodman JW (1984) Laser speckle and related phenomena (Topics in applied physics), vol 9. Springer, Berlin, pp 9–75
5. Steimers A, Gramer M, Takagaki M, Graf R, Kohl-Bareis M (2011) Simultaneous imaging of haemoglobin oxygenation and blood flow with RGB reflectometry and LASCA during stroke in rats. Proc SPIE 8088–08, Diffuse Optical Imaging III, 808808
6. Gerriets T, Li F, Silva MD et al (2003) The macrosphere model evaluation of a new stroke model for permanent middle cerebral artery occlusion in rats. J Neurosci Methods 122(2): 201–211

Chapter 58
Quality Evaluation Method for Rat Brain Cryofixation on the Basis of NADH Fluorescence

Nannan Sun, Weihua Luo, Anle Wang, and Qingming Luo

Abstract The goal of biological samples' cryofixation is to trap a metabolic state as it exists in vivo by rapidly stopping internal reactions. However, obtaining perfect quality of cryofixation for large and high hypermetabolism organ/tissue (such as brain, heart) remains a challenge. The aim of this study was to develop and display a comprehensive and direct method to evaluate cryofixation's process and quality. Here, we adopt a delicate combination of homemade cryo-imaging system with a rat cardiac arrest model that can control cryofixation time optionally. We successfully evaluate the cryofixation time-related nicotinamide adenine dinucleotide (NADH) fluorescence pattern of several coronal sections in rat's brain that suffered from directional funnel cryofixation procedure. Through quantitative analysis of the distribution map of NADH fluorescence, we could obtain a relationship between cryofixation time and well cryofixation volume and then could deduce the cryofixation rates and quality at different time points. Our results also demonstrated that dissection of the temporal muscle of rat could significantly optimize the classical direct funnel cryofixation protocol.

Nannan Sun and Weihua Luo have contributed equally.

N. Sun • W. Luo • A. Wang • Q. Luo (✉)
Britton Chance Center for Biomedical Photonics, Wuhan National Laboratory for Optoelectronics-Huazhong University of Science and Technology, Wuhan 430074, P. R. China

Key Laboratory of Biomedical Photonics of Ministry of Education, Department of Biomedical Engineering, Huazhong University of Science and Technology, Wuhan 430074, P. R. China
e-mail: qluo@mail.hust.edu.cn

S. Van Huffel et al. (eds.), *Oxygen Transport to Tissue XXXV*, Advances in Experimental Medicine and Biology 789, DOI 10.1007/978-1-4614-7411-1_58,

58.1 Introduction

It is very important to monitor the metabolic states in many pathophysiological processes accurately, such as tumor or neurodegenerative diseases. The major tools for obtaining metabolic information from whole organs or individuals are in vivo imaging techniques, including optical microscopy, positron emission tomography (PET), or magnetic resonance spectroscopy (MRS). However, those methods have limitations in resolution, sensitivity, or imaging depth [1, 2] and then need to be validated or complemented by ex vivo postmortem studies. As many intracellular substances relating to the respiratory chain, such as ATP, glucose, lactate, flavin adenine dinucleotide (FAD), and nicotinamide adenine dinucleotide (NADH), undergo rapid postmortem alterations, they are prone to present some twisted properties from the natural state except for rapidly and effective tissue fixation. Cryofixation or snap freezing is the most widely used method for rapidly reducing or even completely ceasing postmortem metabolism in optical/electrical microscopy studies [2–6]. However, scientists are cautious about using this method in large hypermetabolism organ or tissue, such as brain and heart, because the cryofixation rate and quality are difficult to predict and control.

We aimed to find a way to precisely evaluate the metabolic state fixation homogeneous distribution and the proximity to the natural state. We study the fixation process parameters, such as fixation time and fixation depth, in order to develop optimized methods for fixation. Here, we introduce an evaluation system on the basis of intrinsic NADH, a cellular endogenous molecule that participates in the respiratory chain of mitochondria. Its unique fluorescence characteristics emerge as a sensitive indicator to evaluate change in tissue metabolism using the fluorescence signal [7, 8]. Using a homemade NADH fluorescence cryo-imaging system we can noninvasively acquire the intensity distribution of endogenous NADH fluorescence in large size sample, we detected the metabolism-related pattern in the rat brain section and then evaluated the cryofixation quality at different cryofixation time points which was controlled by periodic cardiac arrest through intravenous injection of potassium solution. With a rat cardiac arrest model, which can rapidly change the metabolic status from normal state to high reduced state in unfixed brain tissue, we could indicate the metabolic state fixation progress and quality during funnel freezing of whole brain in rats. Moreover, through this evaluation system, we demonstrated that dissection of the temporal muscle of rat brain could profoundly accelerate the fixation rate and improve fixation quality during rat brain funnel freezing.

58.2 Methods

58.2.1 *Animal Preparation*

Male Wistar rats (240 ± 40 g) were anesthetized by 20 % ethyl urethane (0.7 ml/100 g) before operation and then were randomly divided into three experimental groups.

Group I ($n=7$): We monitored the changes of ECG, BP, EEG, cerebral cortex NADH fluorescence, and CBF [8], in vivo during cardiac arrest of rat.

Group II: Twenty-one Wistar rats were divided into seven subgroups exposed to directional funnel freezing [3, 9]. A 1 ml solution of KCl (1 mol/L) was infused starting at 20 s, 40 s, 60 s, 80 s, 100 s, 120 s, and 140 s after the onset of freezing, respectively. After a total freezing time of 3 min, the animals were plunged into liquid nitrogen and were sectioned for NADH imaging by the cryo-imaging system.

Group III: Two Wistar rats were treated in accordance with Group II, with the exception of dissection of the right temporal muscle of rat brain before installing the funnel. 1 ml KCl was infused starting at 20 s after the onset of freezing. After 3 min, the animals were plunged into liquid nitrogen and were sectioned for NADH imaging by the cryo-imaging system.

58.2.2 Imaging System

We built a cryo-imaging system, which has 2 cm × 2 cm field of view and 20 μm plane resolution. Details will be present in another paper. All brain samples were cryofixated previously and sliced coronally in low-temperature environment. The system employs 365 nm LED to capture NADH fluorescence image from the block face. The images can indicate distribution of NADH levels on the block faces.

58.3 Results and Discussion

In the first stage, we detected the systemic circulation stop time after cardiac arrest. We monitor multiparameter responses of the rat during cardiac arrest in vivo. Table 58.1 shows systemic circulation stopped and microcirculation suppressed in less than 5 s after cardiac arrest. The NADH signal presents an early dip and a consistent increase.

Second, using the cryo-imaging system, we evaluated the differentiating cryofixation quality of rat brain coronal sections under different cardiac arrest treated time. Figure 58.1 suggests that NADH fluorescence before cardiac arrest is lower than that after cardiac arrest. The fluorescence intensity is higher in deeper tissue. The metabolic state of the rat brain parietal cortex was fixed in 20 s. The metabolic state of most part of the rat brain will be fixed in 60 s.

The metabolic state cryofixation process is too sensitive to resisting external intervention, such as implantation of thermocouple or injection of vascular contrast medium [10–12]. Using intrinsic NADH fluorescence signal to demonstrate the cryofixation process is a better way to avoid intervention to obtain a better fixation quality.

Table 58.1 Time points of various parameters in rat cardiac arrest model

	Onset time mean ± SD (s)	Peak time mean ± SD (s)
ECG↓	2.0 ± 0.4	60.5 ± 11.8
BP↓	2.4 ± 0.2	62.4 ± 9.6
EEG↓	–	15.9 ± 2.8
NADH-1↓	2.3 ± 0.5	4.0 ± 0.9
NADH-2↑	5.3 ± 0.8	84.6 ± 32.6
CBF↓	2.8 ± 0.3	31.5 ± 4.0

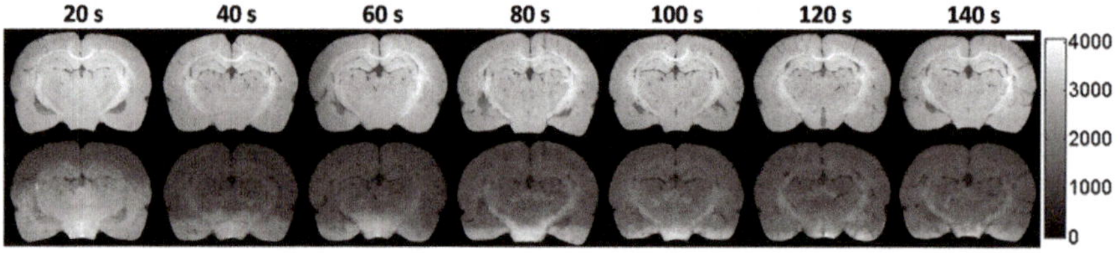

Fig. 58.1 A 1 mol/L solution of KCl was infused starting at 20 s, 40 s, 60 s, 80 s, 100 s, 120 s, or 140 s after the onset of freezing to induce rat cardiac arrest. Scale bar 3 mm

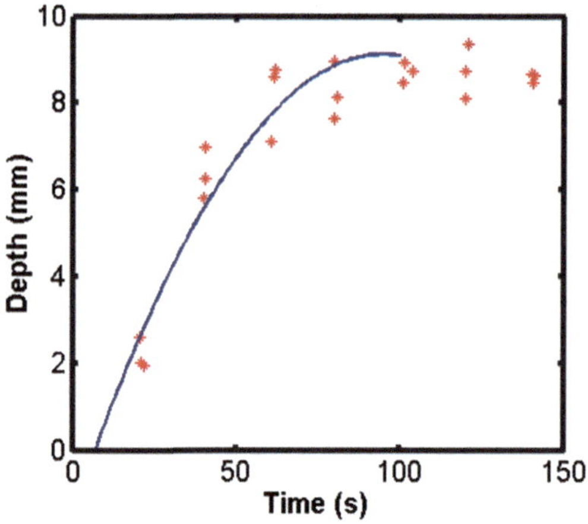

Fig. 58.2 Relationship between cryofixation time and cryofixation depth of metabolic state. *Asterisks* show metabolic state cryofixation depth of the same coronal section with corresponding cryofixation time. The fitting curve was $y = -0.0012X^2 + 0.2232x - 1.4558$

Figure 58.2 shows the relationship between cryofixation time and cryofixation depth of metabolic state. The fitting curve in Fig. 58.2 shows that the steepest slope is between 20 s and 40 s, which correspond to the maximum value in the fixation rate of metabolic state shown in Table 58.2. After 40 s, the fixation rate decreases. The fixation depth of metabolic state reaches the maximum at about 8 mm in 80 s.

In previous studies, the freezing depth changes in time were presented [3, 5]. However, the fixation depth of the metabolic state changing with time was not studied, and hence it is impossible to differentiate the location of the metabolic state

Table 58.2 Metabolic state cryofixation depth and rate at different cryofixation time points

Time (s)	Mean depth ± SD (mm)	Rate (mm/s)
20	2.21 ± 0.36	0.11
40	6.36 ± 0.60	0.21
60	8.17 ± 0.92	0.15
80	8.25 ± 0.67	0.10
100	8.71 ± 0.22	0.08
120	8.73 ± 0.62	0.07
140	8.60 ± 0.10	0.05

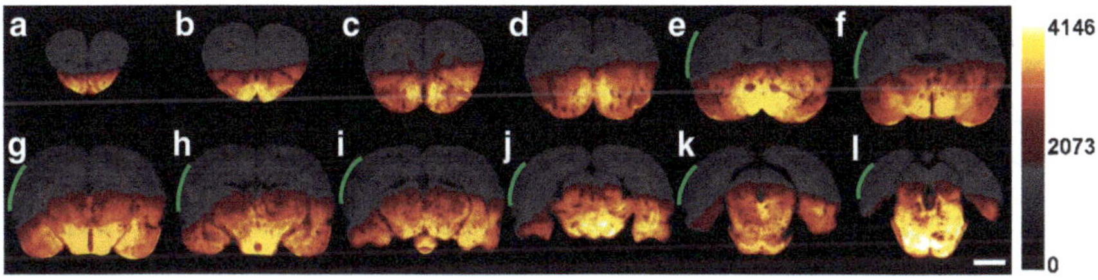

Fig. 58.3 Dissection of the right temporal muscle of rat brain. 1 ml KCl was infused starting at 20 s after the onset of freezing. After 3 min, the animals were plunged into liquid nitrogen. The interval between each two images is 1.5 mm. Scale bar 3 mm

Table 58.3 Comparison of the mean cryofixation depths in left and right side of rat brain

	Left (mm)	Right (mm)
G	5.05	4.52
H	5.27	4.55
I	5.09	4.55

between shallow tissue and deep tissue precisely. Understanding the metabolic state fixation process and characterization will help us to control or optimize the procedures for sample preparation.

We verified the validity of the cryofixation evaluation method on the basis of NADH signal (Fig. 58.3). We dissected the right temporal muscle of rat brain; the right-side metabolic state fixation depth is about 0.5–0.7 mm deeper than the left side (shown in Table 58.3). The above results confirmed that dissection of the temporal muscle can improve the metabolic state fixation rate. Therefore we can use NADH fluorescence to evaluate the cryofixation quality.

58.4 Conclusions

In summary, we described the relationship between cryofixation time and cryofixation depth of metabolic state. We developed a cryofixation evaluation method for rat brain on the basis of NADH endogenous signal. This method can accurately reveal the unchanged metabolism profile of well cryofixation samples and therefore guiding further improvement of the cryofixation method.

Acknowledgments This work was supported by the National Major Scientific Research Program of China (Grant No. 2011CB910401) and the Science Fund for Creative Research Group of China (Grant No.61121004) and the Director Fund of Wuhan National Laboratory for Optoelectronics and the Specific International Scientific Cooperation (Grant No. 2010DFR30820).

References

1. Ntziachristos V (2010) Going deeper than microscopy: the optical imaging frontier in biology. Nat Methods 7(8):603–614
2. de Graaf RA, Chowdhury GM, Brown PB, Rothman DL, Behar KL (2009) In situ 3D magnetic resonance metabolic imaging of microwave-irradiated rodent brain: a new tool for metabolomics research. J Neurochem 109(2):494–501
3. Pontén U, Ratcheson RA, Salford LG, Siesjö BK (1973) Optimal freezing conditions for cerebral metabolites in rats. J Neurochem 21(5):1127–1138
4. Swaab DF (1971) Pitfalls in the use of rapid freezing for stopping brain and spinal cord metabolism in rat and mouse. J Neurochem 18(11):2085–2092
5. Roos MW, Johansson A, Sperber GO (1997) In situ freezing of the rabbit and rat brain. Cryobiology 35(2):187–191
6. Roos MW (2002) Theoretical local freezing times of small rodent brains submerged in situ in liquid nitrogen. Ups J Med Sci 107(2):71–75
7. Chance B, Cohen P, Jobsis F, Schoener B (1962) Intracellular oxidation-reduction states in vivo. Science 137(3529):499–508
8. Mayevsky A, Chance B (1982) Intracellular oxidation-reduction state measured in situ by a multichannel fiber-optic surface fluorometer. Science 217(4559):537–540
9. Shiino A, Haida M, Beauvoit B, Chance B (1999) Three-dimensional redox image of the normal gerbil brain. Neuroscience 91(4):1581–1585
10. Lowry OH, Passonneau JV, Hasselberger FX, Schulz DW (1994) Effect of ischemia on known substrates and cofactors of the glycolytic pathway in brain. J Biol Chem 239:18–30
11. Ferrendelli JA, Gay MH, Sedgwick WG, Chang MM (1972) Quick freezing of the murine CNS: comparison of regional cooling rates and metabolite levels when using liquid nitrogen of Freon-12. J Neurochem 19(4):979–987
12. Hatakeyama T, Sakaki S, Nakamura K, Furuta S, Matsuoka K (1992) Improvement in local cerebral blood flow measurement in gerbil brains by prevention of postmortem diffusion of [14C]iodoantipyrine. J Cereb Blood Flow Metab 12(2):296–300

Chapter 59
Cerebral Cortex Activation Mapping upon Electrical Muscle Stimulation by 32-Channel Time-Domain Functional Near-Infrared Spectroscopy

Rebecca Re, Makii Muthalib, Davide Contini, Lucia Zucchelli, Alessandro Torricelli, Lorenzo Spinelli, Matteo Caffini, Marco Ferrari, Valentina Quaresima, Stephane Perrey, and Graham Kerr

Abstract The application of different EMS current thresholds on muscle activates not only the muscle but also peripheral sensory axons that send proprioceptive and pain signals to the cerebral cortex. A 32-channel time-domain fNIRS instrument was employed to map regional cortical activities under varied EMS current intensities applied on the right wrist extensor muscle. Eight healthy volunteers underwent four EMS at different current thresholds based on their individual maximal tolerated intensity (MTI), i.e., 10 %<50 %<100 %<over 100 % MTI. Time courses of the

R. Re (✉) • D. Contini • L. Zucchelli • A. Torricelli
Dipartimento di Fisica, Politecnico di Milano, Piazza Leonardo da Vinci, 32, Milan I-20133, Italy
e-mail: rebecca.re@mail.polimi.it; rebecca.re84@gmail.com

M. Muthalib (✉)
Movement To Health (M2H) Laboratory, Euromov, Montpellier-1 University, Montpellier, France

Movement Neuroscience Program, Queensland University of Technology, Brisbane, Australia
e-mail: makii.muthalib@univ-montp1.fr; makii.muthalib@gmail.com

L. Spinelli
Istituto di Fotonica e Nanotecnologie, CNR, Milan, Italy

M. Caffini
Dipartimento di Fisica, Politecnico di Milano, Piazza Leonardo da Vinci, 32, Milan, I-20133, Italy

Department of Health Sciences, University of L'Aquila, L'Aquila, Italy

M. Ferrari • V. Quaresima
Department of Health Sciences, University of L'Aquila, L'Aquila, Italy

S. Perrey
Movement To Health (M2H) Laboratory, Euromov, Montpellier-1 University, Montpellier, France

G. Kerr
Movement Neuroscience Program, Queensland University of Technology, Brisbane, Australia

S. Van Huffel et al. (eds.), *Oxygen Transport to Tissue XXXV*, Advances in Experimental Medicine and Biology 789, DOI 10.1007/978-1-4614-7411-1_59,

absolute oxygenated and deoxygenated hemoglobin concentrations primarily over the bilateral sensorimotor cortical (SMC) regions were extrapolated, and cortical activation maps were determined by general linear model using the NIRS-SPM software. The stimulation-induced wrist extension paradigm significantly increased activation of the contralateral SMC region according to the EMS intensities, while the ipsilateral SMC region showed no significant changes. This could be due in part to a nociceptive response to the higher EMS current intensities and result also from increased sensorimotor integration in these cortical regions.

59.1 Introduction

Electrical muscle stimulation (EMS) is widely used during the treatment of muscle atrophy and neurorehabilitation. Although the effects of the application of different EMS current intensities on muscle function are well known, the functional cortical correlates of EMS-evoked movements have not been yet clarified. Blickenstorfer et al. [1] used fMRI to measure cortical and subcortical activation responses during EMS of the wrist extensors/flexors at 150 % of the individual motor threshold intensity and showed a cortical activation pattern comprising the contralateral sensorimotor cortex (SMC) and premotor cortex, the bilateral secondary somatosensory cortex, the supplementary motor area, and the anterior cingulate cortex. Smith et al. [2] previously showed a relationship between changes in SMC activation using fMRI and different levels of increasing current intensity up to motor threshold during EMS of the knee extensors. However, since neurorehabilitation commonly utilizes EMS current intensities at the maximum tolerated current intensity (MTI), it is important to know the SMC correlates of increasing EMS current intensity from low (10 % MTI) to over motor threshold (50 % MTI), at MTI (100 % MTI), and further increases in current intensity (over 100 % MTI). Our previous study [3], performed with one-channel continuous-wave functional near-infrared spectroscopy (fNIRS) over the contralateral prefrontal cortex (PFC), demonstrated EMS current intensity-related increases during EMS of the elbow flexors. The aim of the present study was to employ multichannel time-domain fNIRS (TD-fNIRS) and general linear model methods, using NIRS-SPM software, to map regional absolute SMC and PFC oxygenation with increasing EMS current intensities.

59.2 Methods

59.2.1 Subjects

Eight healthy volunteers (40.3 ± 13.5 year) participated in this study. All the subjects had no known health problems or any upper extremity muscle or joint injuries. The study conformed to the recommendations of the local Human Research Ethics Committee in accordance with the Declaration of Helsinki.

59.2.2 Equipment

EMS was carried out with the portable system CEFAR Physio 5 (DJO France SAS, Mouguerre, France). The device is a microcontroller-based system with 2+2 current-regulated stimulation channels and is controlled with a chip card. The right wrist extensor muscles were stimulated with a pair of 5×5 cm adhesive electrodes. The negative electrode was located on the motor point of the right wrist extensor muscles, the positive one on the distal end of the muscle near the wrist. Symmetric, biphasic, and charged balanced rectangular pulse shapes were applied. The depolarizing pulses had a width of 200 μs and frequency of 30 Hz.

For fNIRS mapping a multichannel, dual-wavelength (690 nm and 829 nm) time-domain class I medical device, developed at the Department of Physics of Politecnico di Milano, was used [4]. Optodes were positioned over the bilateral SMC, with 2 channels extending onto the PFC regions: on each hemisphere five sources and eight detection points for a total of 32 measurement channels, according to the international 10–20 system for the EEG electrode placement. Sequential illumination of pairs of light sources in the left and right hemispheres every 0.2 s allowed for the acquisition of fNIRS signal from the 32 channels at 1 Hz. In this way a signal, suitable for the extrapolation of the optical parameters, was acquired.

59.2.3 Protocol

At least 10 min prior to the EMS protocol, MTI was determined for each subject by using a series of 6–8 brief (3–5 s) electrically stimulated contractions with increasing intensity. After each increase in intensity that included a visibly stronger contraction, the participants were asked if they could tolerate any further increase in intensity. MTI was then defined as the intensity of stimulation received when the subjects were no more able to tolerate an increase in intensity.

Each subject underwent four EMS current intensity thresholds, at 10 %<50 %<100 % and over 100 % of the MTI. Each experiment consisted of an initial baseline (10 s), 10 stimulation blocks (10 s baseline, 20 s EMS, 10 s of recovery), and a final recovery (10 s), for a total experiment length of 420 s. Each 20 s EMS condition was composed of 1 s stimulation and 1 s rest alternatively.

During each experiment, pain rating was also monitored using a 10 cm visual analogue scale (VAS).

59.2.4 fNIRS Signal Analysis

For each wavelength λ, a reference time-domain fNIRS curve $R_0(t;\lambda)$ is derived by averaging the tracks recorded during the initial baseline period. Fitting of $R_0(t;\lambda)$

yields the reference absorption value $\mu_{a0}(\lambda)$. Then, at each recording time T during the experiment, changes in the absorption coefficient are derived as [5]

$$\Delta\mu_a(\lambda;T) = -\frac{1}{vt}\ln\left(\frac{R(t;\lambda;T)}{R_0(t;\lambda)}\right)$$

where v is the speed of light in the medium, t is the arrival time of photons, and $R(t,\lambda,T)$ is the time-domain fNIRS curve at the recording time T. To enhance the contribution from deep layers and to remove possible disturbances caused by superficial ones, a correction method based on the use of late time windows (t=1,750–2,500 ps) was also applied [5]. It is well known that depth information in TD-fNIRS is encoded in photons time of flight: early photons probe the superficial layers of the head, while late photons have a higher probability to visit deeper layers. Finally, the absorption coefficient is derived from corrected late gate intensities as

$$\mu_a(\lambda;T) = \mu_{a0}(\lambda) + \Delta\mu_a(\lambda;T)$$

On the assumptions that, in the wavelengths applied here, oxygenated (O_2Hb), deoxygenated (HHb) hemoglobin, and water (H_2O) are the main chromophores contributing to absorption and also that H_2O concentration is unlikely subjected to relevant modifications and thus can be considered constant, O_2Hb and HHb concentration changes are then derived by Lambert-Beer law. Then, changes in total hemoglobin content (tHb = HHb + O_2Hb) and oxygen saturation (SO_2 = O_2Hb/tHb) are calculated.

Cortical activation maps for each subject and for the whole population in each experiment were obtained by applying general linear model (GLM) methods, by means of the NIRS-SPM software (http://bisp.kaist.ac.kr/NIRS-SPM.html, [6]) For any channel a design matrix was created that modeled the O_2Hb and HHb time series as a linear combination of regressors plus an error term. Hypotheses on predictor variables could be tested constructing the t-statistic. Within the analysis protocol, a good trade-off between statistical reliability and temporal resolution was found by analyzing data with constant regressors. The task was modeled as a series of consecutive boxcars that represented an equal number of activation periods, chosen accordingly to the different activities performed during the experiment.

59.3 Results

During the EMS conditions, the VAS pain values were 0.2 ± 0.3 (10 % MTI), 1.6 ± 1 (50 % MTI), 4.7 ± 0.5 (100 % MTI), and 7.1 ± 1 (over 100 % MTI). Figure 59.1 shows the typical cortical oxygenation time course changes (average of 10 blocks) during the 100 % MTI EMS condition for Subject 1. Activation of the contralateral SMC channels and of bilateral PFC channels is observed.

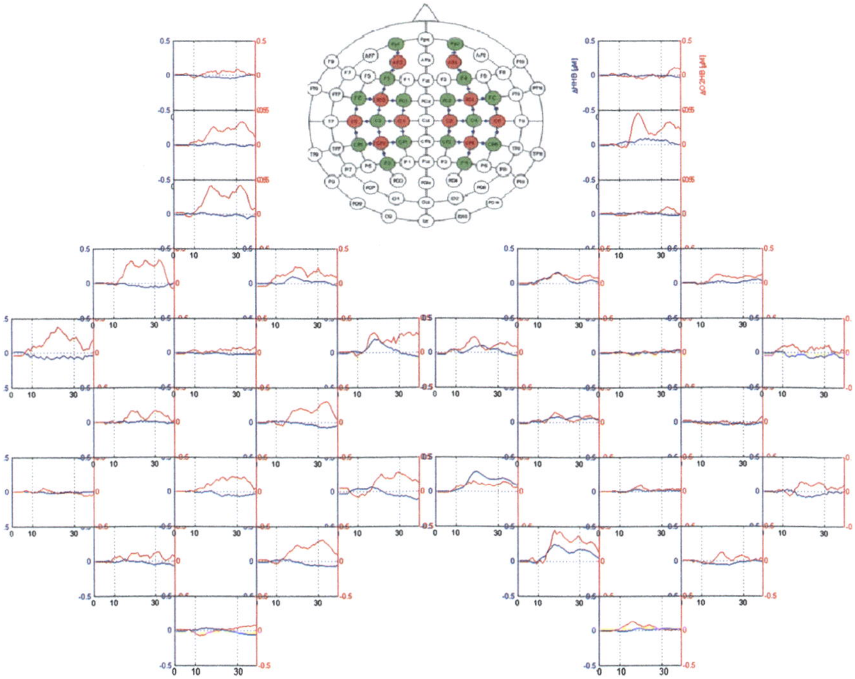

Fig. 59.1 Block-averaged cortical oxy- and deoxyhemoglobin time course changes on Subject 1 during the 100 % MTI EMS condition. *Left-side panels* represent the contralateral hemisphere to the EMS of the right wrist extensors. The *inset* represents the optodes location on the 10–20 international system

Figure 59.2 shows the cortical activation maps for three EMS conditions (group level t-statistic, $P<0.001$ uncorrected). The GLM analysis indicated that 50 % MTI, 100 % MTI, and over 100 % MTI conditions significantly increased the contralateral SMC activation (increase in O_2Hb and decrease in HHb) compared to rest; no significant activation was found at 10 % MTI. Moreover, in Fig. 59.2, a larger area of activation was observed (particularly for the HHb maps) with the increase of EMS current intensity. The ipsilateral SMC region showed no significant changes at the group level, but some subjects had ipsilateral SMC activation in some channels. The same was observed for the changes in the two PFC channels.

59.4 Discussion

The present study, utilizing 32-channel TD-fNIRS, confirms the fMRI findings of Blickenstorfer et al. [1] by showing contralateral SMC activation during EMS-evoked wrist extensor muscle contractions at different current thresholds. Furthermore, the results of this study show distinct patterns of cerebral cortex

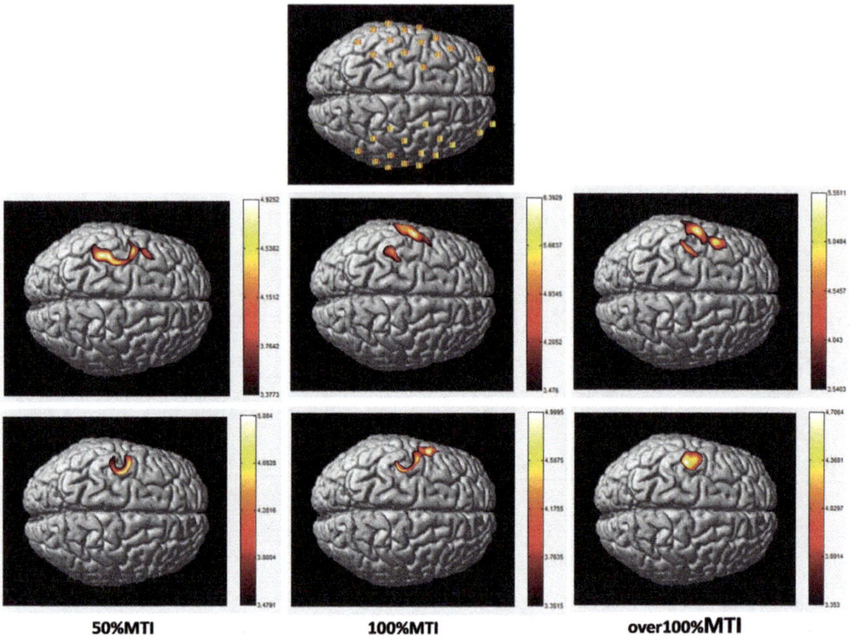

Fig. 59.2 Group t-statistic ($P<0.001$ uncorrected) cortical activation maps during the EMS conditions (no activation for 10 % MTI). Each map depicts the significant difference between task and baseline. *Top panel*: projection of the 32 measurement points over the cortical area. *Middle panels*, O_2Hb maps, *lower panels*, HHb maps

activation when the EMS intensity is increased from not painful (10 % MTI), mild pain (50 % MTI), to extremely painful (100 % MTI and over 100 % MTI), which extends the findings of Smith et al. [2]. It is possible that the higher contralateral SMC activation maps during painful EMS stimulation are due in part to a nociceptive response to the higher EMS current intensities and result also from increased sensorimotor integration in these cortical regions. Due to the fNIRS probe arrangement on the scalp (see inset in Fig. 59.1), the bilateral secondary somatosensory cortical activation could not be confirmed [1]. It is worth noting that when the individual subject t-statistical activation maps were analyzed at a less constrictive threshold ($P<0.05$ EC correction), as utilized by Ye et al. [6]), a larger and bilateral area of the SMC and PFC showed activation in many subjects, suggesting a much greater activation of sensorimotor and pain networks [1].

59.5 Conclusion

In conclusion, the results of the present study showed that the EMS-evoked wrist extension stimulation paradigm increased activation of contralateral SMC region according to the EMS intensities (over 100 %>100 %>50 %>10 % MTI), while the ipsilateral SMC region showed no significant changes. Although several studies demonstrated that EMS can be safely performed in the MRI scanner [1, 2, 7], fNIRS can be more conveniently utilized to assess the time course of the neuroplastic changes associated with EMS rehabilitation treatment in patients with stroke directly in the rehabilitation clinic.

References

1. Blickenstorfer A, Kleiser R, Keller T et al (2009) Cortical and subcortical correlates of functional electrical stimulation of wrist extensor and flexor muscles revealed by fMRI. Hum Brain Mapp 30(3):963–975
2. Smith GV, Alon G, Roys GS, Gullapalli RP (2003) Functional MRI determination of a dose-response relationship to lower extremity neuromuscular electrical stimulation in healthy subjects. Exp Brain Res 150(1):33–39
3. Muthalib M, Ferrari M, Quaresima V, Nosaka K (2012) Frontal cortex activation during electrical muscle stimulation as revealed by functional near-infrared spectroscopy. Adv Exp Med Biol 737:45–49
4. Contini D, Torricelli A, Pifferi A, Spinelli L, Paglia F, Cubeddu R (2006) Multi-channel time-resolved system for functional near infrared spectroscopy. Opt Express 14(12):5418–5432
5. Contini D, Spinelli L, Torricelli A et al (2007) Novel method for depth-resolved brain functional imaging by time-domain NIRS. Proc. SPIE, 6629: 662908
6. Ye JC, Tak S, Jang KE, Jung J, Jang J (2009) NIRS-SPM: statistical parametric mapping for near-infrared spectroscopy. Neuroimage 44(2):428–447
7. Han BS, Jang SH, Chang Y, Byun WM, Lim SK, Kang DS (2003) Functional magnetic resonance image finding of cortical activation by neuromuscular electrical stimulation on wrist extensor muscles. Am J Phys Med Rehabil 82(1):17–20

Chapter 60
NIRS-Based Neurofeedback Learning Systems for Controlling Activity of the Prefrontal Cortex

Kaoru Sakatani, N. Takemoto, T. Tsujii, K. Yanagisawa, and H. Tsunashima

Abstract The aim of this study was to develop a NIRS-based neurofeedback system to modulate activity in the prefrontal cortex (PFC). We evaluated the effectiveness of the system in terms of separability of changes in oxy-Hb and its derivative. Training with neurofeedback resulted in higher separability than training without neurofeedback or no training, suggesting that the neurofeedback system could enhance self-control of PFC activity. Interestingly, the dorsolateral PFC exhibited enhanced activity and high separability after neurofeedback training. These observations suggest that the neurofeedback system might be useful for training subjects to regulate emotions by self-control of dorsolateral PFC activity.

60.1 Introduction

Neurofeedback is a specific form of biofeedback, which feeds back information about brain activity to allow for training of subjects to achieve voluntary regulation of brain activity. Data for neurofeedback has been obtained by using electroencephalography [1], magnetoencephalography [2], real-time fMRI [3, 4], and NIRS [5]. The EEG feedback system has been successfully used clinically, for example, in epilepsy or as a brain-computer interface (BCI); however, EEG provides only unreliable localization of active brain areas. MEG and real-time fMRI have high spatial resolution, but the systems are bulky and expensive. In contrast, NIRS is

K. Sakatani, M.D., Ph.D. (✉) • N. Takemoto • T. Tsujii
Department of Neurological Surgery, Division of Optical Brain Engineering, Nihon University School of Medicine, Tokyo, Japan
e-mail: sakatani@med.nihon-u.ac.jp

K. Yanagisawa • H. Tsunashima
Department of Mechanical Engineering, College of Industrial Technology, Nihon University, Tokyo, Japan

S. Van Huffel et al. (eds.), *Oxygen Transport to Tissue XXXV*, Advances in Experimental Medicine and Biology 789, DOI 10.1007/978-1-4614-7411-1_60,

compact and less expensive than MEG or fMRI and should be more suitable for practical use.

In the present study, we developed a NIRS-based neurofeedback system to modulate activity in the prefrontal cortex (PFC). We evaluated the effect of the neurofeedback system in terms of separability of changes in oxy-Hb and its derivative. Separability is an index that shows how easily groups can be distinguished [6, 7]. It can be calculated from the variance of each class (in this study, activation or deactivation trajectory of oxy-Hb and its derivative) and the distance between the classes as described in Sect. 2.2.

60.2 Materials and Methods

60.2.1 Experimental Settings

We monitored concentration changes of oxyhemoglobin (oxy-Hb) in the bilateral PFC employing a multichannel NIRS (OMM 3000, Shimadzu, Japan) connected to a computer. This system consists of 16 light-source fibers and 16 detectors resulting in 48 source-detector pairs; each light source has three laser diodes with wavelengths of 780, 805, and 830 nm. The optodes for the NIRS were placed on the skull to cover the bilateral frontal lobes, including the dorsolateral PFC, employing a holder cap to avoid motion-related artifacts; the source-detector distance was 30 mm.

The subjects were instructed to concentrate on the display so as to make the color on the screen become red. The display showed a red color when oxy-Hb increased above the baseline, whereas a blue color was displayed when oxy-Hb decreased below the baseline.

We studied 26 young adult subjects, who were classified into groups A, B, and C.

In group A ($n=14$), subjects were trained to change the color on the display to red for 7 days using the neurofeedback system. In group B ($n=5$), subjects were trained to do so for 7 days without the system. In group C ($n=7$), subjects did not receive training. The subjects were trained for about 1 h a day for 1 week.

60.2.2 Data Analysis

NIRS signals are relative values; thus, it is difficult to compare changes of NIRS signals among subjects. We therefore calculated the Z-score of the NIRS signals and averaged the Z-scores (Fig. 60.1a):

$$Z = \frac{X - \mu}{\sigma} \tag{60.1}$$

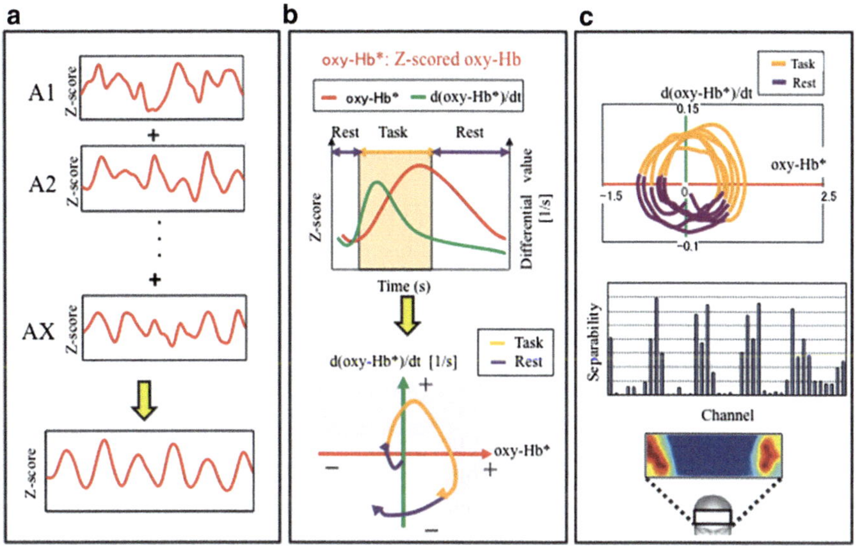

Fig. 60.1 Analysis of NIRS signals. Averaging using Z-score (**a**). Trajectory of Z-scored oxy-Hb and differential value (**b**). Evaluation of separability (**c**)

where X, μ, σ denote measurement value, mean value, and standard deviation of the NIRS signals, respectively.

In order to define the activated cortical area, we calculate differential values of oxy-Hb (i.e., d(oxy-Hb)/dt) during rest and during the task. The differential value was computed by finite difference with a sampling time 0.205 s. Then, we plotted a phase plane of oxy-Hb and its derivative (Fig. 60.1b). We defined the activated cortical area as the area where both oxy-Hb concentration and differential value increased.

The feature quantity is identified using a vector of oxy-Hb (p_{oxy}) and its derivative ($\dot{p}_{oxy}$) as $\mathbf{p} = (p_{oxy}, \dot{p}_{oxy})$. Within-class variance, σ_W^2, and between-class variance, σ_B^2, are calculated by

$$\sigma_W^2 = \left(\sum_{\mathbf{p} \in X_{task}} (\mathbf{p} - \mathbf{m}_{task})^T (\mathbf{p} - \mathbf{m}_{task}) + \sum_{\mathbf{p} \in X_{rest}} (\mathbf{p} - \mathbf{m}_{rest})^T (\mathbf{p} - \mathbf{m}_{rest}) \right) / n, \quad (60.2)$$

$$\sigma_B^2 = \left(n_{task} (\mathbf{m}_{task} - \mathbf{m})^T (\mathbf{m}_{task} - \mathbf{m}) + n_{rest} (\mathbf{m}_{rest} - \mathbf{m})^T (\mathbf{m}_{rest} - \mathbf{m}) \right) / n \quad (60.3)$$

where n, n_{task}, n_{rest} are a total number of signals of all, the task, and the rest, respectively. m, m_{task}, m_{rest} are an averaged vector of all signals, the task signal, and the rest signal, respectively.

The separability can be defined as

$$J_\sigma = \frac{\sigma_B^2}{\sigma_W^2} \tag{60.4}$$

Within-class variance represents an average spread in the class, and between-class variance represents a spread between classes [6, 7]. When the value of J_σ is large, groups can be easily distinguished.

In order to assess the reproducibility of PFC activation, we evaluated the separability of NIRS signals during the task and rest periods (Fig. 60.1c).

60.3 Results

Figure 60.2 shows pseudo-color 2D images of the averaged Z-score of oxy-Hb in groups A, B, and C. In group A, the bilateral lateral PFC exhibited higher activation than other PFC regions after neurofeedback training. In group B, the training without neurofeedback increased activity only in the left lateral PFC. No effect on PFC activity was observed in group C.

Figure 60.3a shows pseudo-color 2D images of separability of oxy-Hb changes in group A. Separability increased from 1.037 to 3.449, indicating that the subjects were well able to control PFC activity after training. The training effect was observed mainly in the lateral PFC. Figure 60.3b shows the results in group B. Separability increased from 0.729 to 2.080. Thus, training without neurofeedback was less effective than training with neurofeedback. In addition, the training effect was observed only in the left lateral PFC. In group C, no change was observed (Fig. 60.3c).

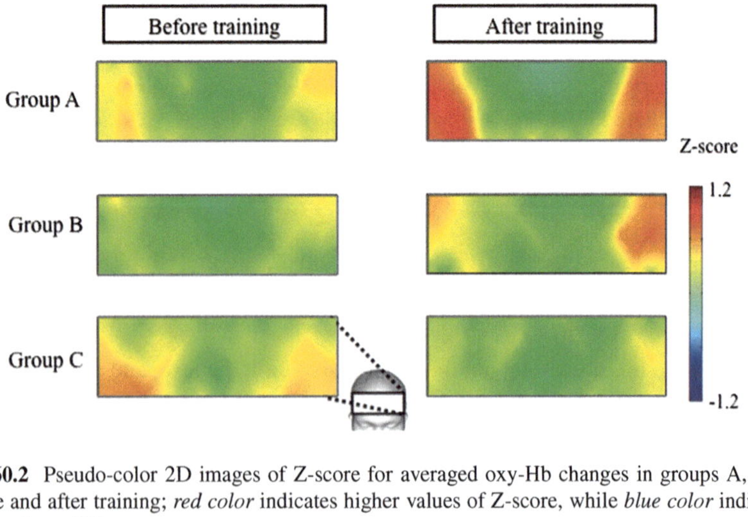

Fig. 60.2 Pseudo-color 2D images of Z-score for averaged oxy-Hb changes in groups A, B, C before and after training; *red color* indicates higher values of Z-score, while *blue color* indicates lower values of Z-score

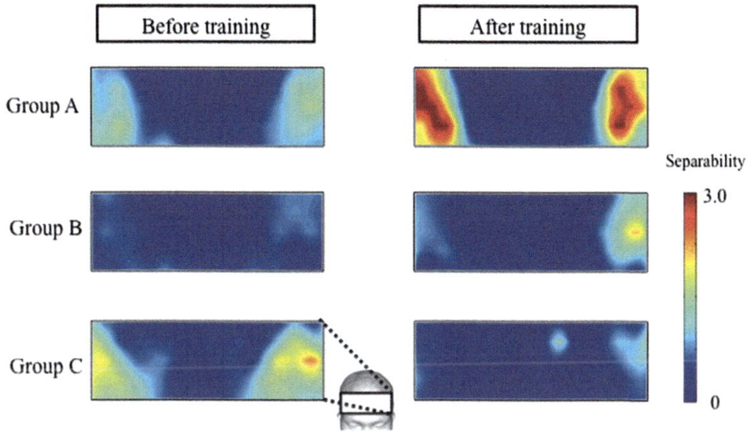

Fig. 60.3 Pseudo-color 2D images of separability of oxy-Hb changes in groups A, B, C before and after training; *red color* indicates higher values of separability, while *blue color* indicates lower values of separability

60.4 Discussion

We evaluated the effect of neurofeedback training on self-control of PFC activation by examining the separability of NIRS signals, which allows assessment of the reproducibility of PFC activation. We found that subjects trained with neurofeedback exhibited higher separability than those trained without neurofeedback or those who received no training. These results suggest that our neurofeedback system could enhance self-control of PFC activity compared with training without neurofeedback and no training. Interestingly, the lateral PFC exhibited enhanced activity and high separability after neurofeedback training.

The PFC plays a key role in both negative and positive emotional regulation via connections with subcortical nuclei, including the amygdala [8–10]. Specifically, the dorsolateral PFC is one of the brain regions implicated in emotional processing, particularly during downregulation of negative emotional conditions [11]. It has been reported that increased activity in the dorsolateral PFC was associated with suppression of fearful stimuli [12] and processing of positive emotional stimuli [13]. In addition, increased activity in the dorsolateral PFC was observed during modification of the intensity of emotional stimuli using cognitive strategies [14, 15]. These observations suggest that neurofeedback training using the present system might be useful for regulatory control of impulses and emotions by self-control of dorsolateral PFC activity.

Finally, it should be emphasized that the present neurofeedback system with NIRS is compact and practical for use in both normal subjects and patients with mental disorders. However, further studies are necessary to evaluate the psychological effects of the system and to establish its usefulness in self-control of emotions.

References

1. Friel PN (2007) EEG biofeedback in the treatment of attention deficit hyperactivity disorder. Altern Med Rev 12(2):146–151
2. Buch E, Weber C, Cohen LG et al (2008) Think to move: a neuromagnetic brain-computer interface (BCI) system for chronic stroke. Stroke 39(3):910–917
3. deCharms RC (2007) Reading and controlling human brain activation using real-time functional magnetic resonance imaging. Trends Cogn Sci 11(11):473–481
4. Johnston SJ, Boehm SG, Healy D, Goebel R, Linden DE (2010) Neurofeedback: a promising tool for the self-regulation of emotion networks. Neuroimage 49(1):1066–1072
5. Nagaoka T, Sakatani K, Awano T et al (2010) Development of a new rehabilitation system based on a brain-computer interface using near-infrared spectroscopy. Adv Exp Med Biol 662:497–503
6. Cios KJ (2007) Data mining: a knowledge discovery approach. Springer, New York
7. Ozawa S, Sakaguchi Y, Kotani M (2001) A study of feature extraction using supervised independent component analysis. In: Proceedings of the international joint conference on neural networks, vol 4, pp 2958–2963
8. Davidson RJ, Fox A, Kalin NH (2007) Neural bases of emotion regulation in nonhuman primates and humans. In: Gross JJ (ed) Handbook of emotion regulation. Guilford, New York, pp 47–68
9. Wager TD, Davidson ML, Hughes BL, Lindquist MA, Ochsner KN (2008) Prefrontal-subcortical pathways mediating successful emotion regulation. Neuron 59(6):1037–1050
10. Sotres-Bayon F, Quirk GJ (2010) Prefrontal control of fear: more than just extinction. Curr Opin Neurobiol 20(2):231–235
11. Davidson RJ, Putman KM, Larson CL (2000) Dysfunction in neural circuitry of emotion regulation-a possible prelude to violence. Science 289(5479):591–594
12. Amting JM, Greening SG, Mitchell DG (2010) Multiple mechanisms of consciousness: the neural correlates of emotional awareness. J Neurosci 30(30):10039–10047
13. Dolcos F, LaBar KS, Cabeza R (2004) Dissociable effects of arousal and valence on prefrontal activity indexing emotional evaluation and subsequent memory: an event-related fMRI study. Neuroimage 23(1):64–74
14. Ochsner KN, Bunger S, Gros JJ, Gabriel JD (2002) Rethinking feelings: an fMRI study of the cognitive regulation of emotion. J Cogn Neurosci 14(8):1215–1229
15. Kim SH, Hamann S (2002) Neural correlates of positive and negative emotion regulation. J Cogn Neurosci 19(5):776–798

Chapter 61
Cortical Mapping of 3D Optical Topography in Infants

Maria D. Papademetriou, John Richards, Teresa Correia, Anna Blasi, Declan G. Murphy, Sarah Lloyd-Fox, Mark H. Johnson, and Clare E. Elwell

Abstract Precise localisation of cortical activation in the early development of the infant brain remains unclear. It is challenging to co-register haemodynamic responses during functional activation in infants with the underlying anatomy of the brain. We used a multispectral imaging algorithm to reconstruct 3D optical topographic images of haemodynamic responses in an infant during voice processing. In this chapter, we present a method for co-registering 3D optical topography images reconstructed from functional activation data in infants onto anatomical brain images obtained from MRI structurals of the individual infants.

61.1 Introduction

Optical topography (OT) is a noninvasive technique which uses near-infrared light to study functional brain imaging in infants. Over the last 10 years, progress and development of this technique has led to an increase in interest in the application for

M.D. Papademetriou (✉) • T. Correia • C.E. Elwell
Biomedical Optics Research Laboratory, Medical Physics and Bioengineering, University College London, London, UK
e-mail: m.papademetriou@ucl.ac.uk

J. Richards
Department of Psychology, University of South Carolina, Columbia, SC, USA

A. Blasi
Centre for Brain and Cognitive Development, Birkbeck College, London, UK

Department of Neurodevelopmental Science, Institute of Psychiatry, Kings College, London, UK

D.G. Murphy
Department of Neurodevelopmental Science, Institute of Psychiatry, Kings College, London, UK

S. Lloyd-Fox • M.H. Johnson
Centre for Brain and Cognitive Development, Birkbeck College, London, UK

S. Van Huffel et al. (eds.), *Oxygen Transport to Tissue XXXV*, Advances in Experimental Medicine and Biology 789, DOI 10.1007/978-1-4614-7411-1_61,

investigating cortical activity in the developing brain [1]. Even though in the early years OT was used to detect the response to basic stimuli activating the primary cortical areas (auditory and visual cortex), recent advances in technology shifted the focus to study awake infants and address topics such as biological motion processing, face processing, and voice processing [2–4]. In these studies, OT was used to map haemodynamic responses during activation with specific regions of the cortex. However, co-registration between the haemodynamic response measured at the surface of the head and the underlying cortical anatomy remains a challenge. This is primarily due to the lack in capacity for measuring brain structure for anatomical reference and the lack of common control points between the optical array, the head, and the anatomical image.

Furthermore, OT studies treat the data as single source–detector (channel) haemodynamic responses even if the data are measured from multiple channels at different separations. Numerous research groups have produced optical images of brain activity. The first method, pioneered by researchers at Hitachi Medical Co., assumes that a change in intensity measured by a given source–detector pair has its origin midway between the source and detector and there is no attempt to resolve in the depth direction [5]. Conventional image reconstruction involves finding the optical properties of the medium from a set of measurements, which are then combined to calculate the chromophore concentrations. We use a multispectral imaging algorithm to reconstruct haemodynamic responses and obtain 3D optical topography images. In order to make a precise statement about cerebral specialisation during cortical activation, it is necessary to co-register functional activation onto an anatomical atlas. In this chapter we describe a method for co-registering 3D optical topography images on 3D brain volumes rendered from MRI scans.

61.2 Methods

61.2.1 Protocol

Data from one healthy, 6-month-old infant were taken from the previously published data where the experimental paradigm and psychological rationale for the work are explained in detail [4]. The infant sat on the parent's lap while the stimuli were displayed on a screen. The session began with a rest period (30 s) to get familiar with the experimental setup. Following this, the trials alternated one after the other, beginning with a 10-s baseline trial followed by a 10-s experimental trial. Three types of auditory experimental trials (*voice, non-voice, and silence*) were presented pseudorandomly to prevent anticipatory effects and to ensure the infant was presented with an equal number of trials per condition after every 12 trials. The *voice* condition included nonspeech adult vocalisations (coughing, yawning, throat clearing, laughing, and crying). The *non-voice* trials included naturalistic environmental sounds (water running and toys such as rattles, squeaky toy, spinning balls).

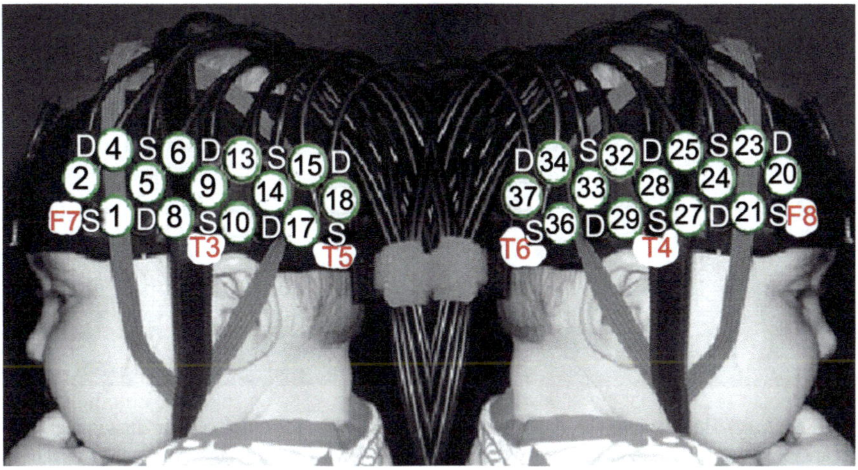

Fig. 61.1 The custom-made headgear consisting of two arrays. Each array has five detectors (*D*) and five sources (*S*). This configuration allows a total of 19 channels for each array, 13 at S-D separation of 2 cm (*numbered in circles*) and 6 at S-D separation of 4.5 cm (Provided with permission from Sarah Lloyd-Fox, Birkbeck, University of London)

Each stimulus sequence lasted 8 s and consisted of four different sounds (of voice or non-voice stimuli) presented for 0.37–2.92 s each, interleaved by short silence periods (of 0.16–0.24 s).

61.2.2 Registering the Optode Positions onto a Brain Volume

The UCL OT system which emits at wavelengths of 770 and 850 nm was used for the data acquisition [6]. The sources are frequency modulated and illuminated simultaneously. The infant wore a custom-built OT headgear consisting of two optical imaging arrays placed on each temporal lobe. Each array consists of five sources and five detectors, providing a total of 19 source–detector pairs per array. Each array had 13 source–detector separations at 2 cm and 6 at 4.5 cm (Fig. 61.1).

Before the infants began the study, measurements of their head circumference, the distance between glabella and ears and glabella and inion, were taken, and the location of the channels and arrays relative to these anatomical landmarks were recorded. The distance from the midpoint of the headband over the forehead (the glabella) to the midpoint of the temporal arrays (channel 9, left hemisphere; 28, right hemisphere) is fixed at 11 cm and is aligned approximately with T3 and T4 of the 10–20 system on an average 5-month-old infant head. A fiducial was placed over the glabella and photographs of the infant facing sideways and towards the camera were taken while wearing the headgear.

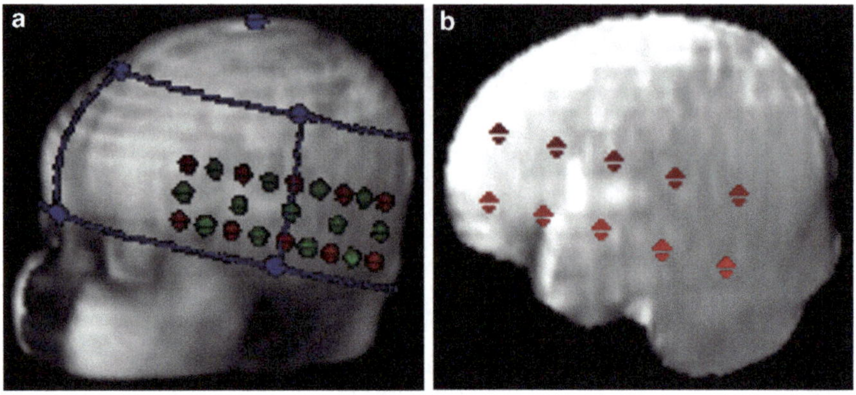

Fig. 61.2 (**a**) Head volume co-registered with the optical headgear. The main anatomical landmarks (left pre-auricular, nasion, vertex – *blue dots*), optodes (*red dots*), and channels (*green dots*) can be identified. (**b**) Brain volume extracted from the head volume co-registered with the optode positions (*red dots*)

T2-weighted MRI structural of the head was obtained with a fiducial placed over the glabella. A 3D reconstruction of the head was obtained from the T2-weighted MRI structural scan. The coordinates of the fiducials and anatomical marks according to the 10/20 coordinate system were obtained in relation to anterior commissure (AC) and were marked on the 3D head images. These marks along with the photographs were used to reconstruct the placement of the headgear on the 3D head and subsequently mark the optode and channel positions (Fig. 61.2a). The brain was then extracted from the individual MRI scans and channels, and optode positions were also projected on the 3D brain reconstruction (Fig. 61.2b).

61.2.3 Optical Image Reconstruction

Data were acquired and low-pass filtered (cut-off frequency of 1.8 Hz) for all the trials. Then each trial was separated into blocks consisting of 4-s baseline, followed by a 10-s stimulus and a 10-s poststimulus baseline period. Each block was detrended using a linear fit between the last 4 s of the pre-stimulus baseline and the last 4 s of the poststimulus baseline period to remove any effects of baseline drift throughout the experiment [3]. Trials for each experimental condition were averaged for each subject. This resulted in a time course for each of the three stimuli per infant.

A multispectral method involving reconstructing chromophore concentrations using all measurements at both wavelengths simultaneously was used to obtain 3D OT images during activation. We assumed that the true chromophore concentrations were close to an initial estimate and any changes were small, and used a linear approximation to reconstruct the difference in chromophore concentrations[7]. For two measurement wavelengths ($\lambda 1$, $\lambda 2$) and two chromophores (c_1, c_2), the linear multispectral method solves the matrix equation $\Delta A\lambda = J.\Delta C$, where $\Delta A_{\lambda} = [\Delta A_{\lambda 1} \; \Delta A_{\lambda 2}]^T$ represents the changes in the logarithm of the measured data and $\Delta c = [\Delta c_1 \; \Delta c_2]^T$ the chromophore concentration changes in μMolar (μM). The matrix $J = [J_{\lambda 1 \varepsilon 1, \lambda 1} \; J_{\lambda 1 \varepsilon 2, \lambda 1}; \; J_{\lambda 2 \varepsilon 1, \lambda 2} \; J_{\lambda 2 \varepsilon 2, \lambda 2}]$ is the Jacobian or sensitivity matrix (in units of mm) and ε is the extinction coefficient of the corresponding chromophore (in mm^{-1} μM^{-1}). It has dimensions $2M \times 2N$, where M represents the number of measurements and N is the number of image pixels. The Jacobian was calculated for each wavelength using the software package TOAST (temporal optical absorption and scattering tomography) [8]. It uses the finite element method (FEM) to model the propagation of light through highly scattering tissue using the diffusion equation, given an estimate of the medium optical properties and optode positions. The medium is considered to be a homogeneous slab with refractive index $n = 1.4$, absorption coefficient $\mu a = 0.038$ mm^{-1}, and reduced scattering coefficient $\mu = 0.75$ mm^{-1} at wavelength 770 nm and $\mu a = 0.042$ mm^{-1} and $\mu s = 0.71$ mm^{-1} at wavelength 850 nm. A FEM mesh with 32,615 nodes, 21,845 elements, and dimensions $120 \times 60 \times 40$ mm was generated with the meshing software NETGEN [9].

61.3 Results

Figure 61.3 shows the 3D optical topography image reconstructions obtained during one of the experimental conditions (voice). Images show 3D reconstructions of HbO_2 changes as a function of time obtained at a depth of approximately 1.5 cm. Changes in HbO_2 are obtained for both the right and left hemispheres. In the right hemisphere, the images show an evident increase in HbO_2 upon the onset of the stimulus gradually decreasing when the stimulus ceases. In this case the images are in good agreement with the 2D time course activation data. The time course data for

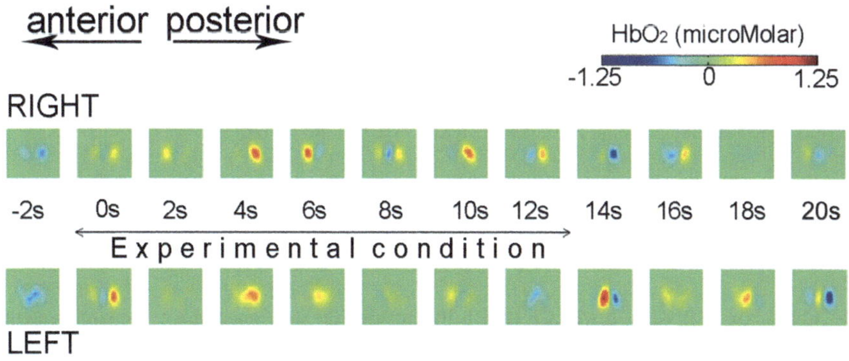

Fig. 61.3 Optical topography image reconstructions showing changes in HbO_2 as a function of time during one of the experimental conditions (voice) for the right hemisphere and left hemisphere

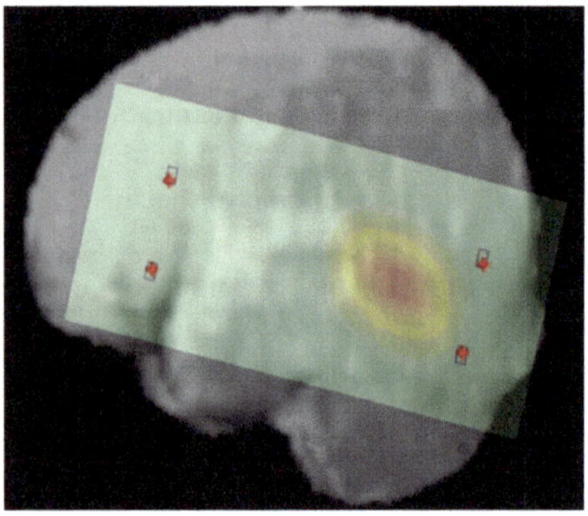

Fig. 61.4 Optical topography image reconstruction at the peak of activation (4 s after the initiation of the experimental condition) co-registered onto the infant's own brain (right hemisphere)

the left hemisphere appear to be noisy, and this is reflected in the 3D images showing an increase in HbO_2 before and after the onset of the stimulus.

Figure 61.4 shows one of the 3D optical topography reconstruction images co-registered onto the brain of the individual infant. The 3D brain extracted from the MRI structural was converted into a 2D image. The optical topography reconstruction was scaled, translated, and rotated accordingly and was co-registered on the brain image.

61.4 Conclusions

We have used optical topography and reconstructed changes in HbO_2 as a function of time during functional activation in a 4-month-old infant. We have developed a method for co-registering optical topography reconstruction images onto anatomical brain extracted from the individuals MRI structural. Co-registration of functional activation could lead to better and more accurate interpretation of the information provided by brain imaging techniques. This could be used to delineate cognitive development of the infant brain.

Acknowledgments This work was supported by a grant from the Simons Foundation (SFARI 201287 M.J) and the National Institutes of Health, NICHD, R37 HD19842.

References

1. Lloyd-Fox S, Blasi A, Elwell CE (2010) Illuminating the developing brain: the past, present and future of functional near infrared spectroscopy. Neurosci Biobehav Rev 34(3):269–284
2. Lloyd-Fox S, Blasi A, Volein A, Everdell N, Elwell CE, Johnson MH (2009) Social perception in infancy: a near infrared spectroscopy study. Child Dev 80(4):986–999
3. Blasi A, Lloyd-Fox S, Everdell N et al (2007) Investigation of depth dependent changes in cerebral haemodynamics during face perception in infants. Phys Med Biol 52(23):6849–6864

4. Lloyd-Fox S, Blasi A, Mercure E, Elwell CE, Johnson MH (2011) The emergence of cerebral specialisation for the human voice over the first months of life. Soc Neurosci 7:317–330
5. Taga G, Asakawa A, Maki A, Konishi Y, Koizumi H (2003) Brain imaging in the awake infants by near infra-red optical topography. Proc Natl Acad Sci USA 100(19):10722–10727
6. Everdell NL, Gibson AP, Tullis IDC, Vaithianathan T, Hebden JC, Delpy DT (2005) A frequency multiplexed near-infrared topography system for imaging functional activation in the brain. Rev Sci Instrum 76:093705
7. Correia T, Lloyd-Fox S, Everdell N et al (2012) Three-dimensional optical topography of brain activity in infants watching videos of human movement. Phys Med Biol 57(5):1135–1146
8. Arridge SR, Hebden JC, Schweiger M et al (2000) A method for three dimensional time-resolved optical tomography. Int J Imaging Syst Technol 11(1):2–11
9. Schöberl J (1997) NETGEN an advancing front 2D/3D-mesh generator based on abstract rules. Comput Vis Sci 1(1):41–52

Chapter 62
Monitoring of Hemodynamic Change in Patients with Carotid Artery Stenosis During the Tilt Test Using Wearable Near-Infrared Spectroscopy

Takahiro Igarashi, Kaoru Sakatani, Norio Fujiwara, Yoshihiro Murata, Takeshi Suma, Tadashi Shibuya, Teruyasu Hirayama, and Yoichi Katayama

Abstract Transient ischemic attack (TIA) is a major complication in patients with carotid artery stenosis. Patients with severe stenosis sometimes complain of orthostatic dizziness, such as syncope. The purpose of this study was to examine the usefulness of near-infrared spectroscopy (NIRS) for evaluating cerebral circulation in patients with carotid artery stenosis during head-up tilt test (HUTT). Fourteen patients with carotid artery stenosis and nine normal control subjects participated. In addition to blood pressure monitoring, hemoglobin (Hb) values (oxy-Hb, deoxy-Hb, and total Hb) were recorded by a wearable NIRS instrument with a high time resolution during HUTT. Oxy-Hb, which decreased initially when the test table was elevated, subsequently increased in normal volunteers and patients with carotid artery stenosis and did not differ significantly between the two groups. However, the oxy-Hb reduction in the carotid artery stenosis group (-0.02 ± 0.03 a.u.) at 30 s after elevation of the table was significantly larger than in the normal group (0.02 ± 0.02 a.u., $P < 0.01$). Our results indicate that oxy-Hb reduction in patients with carotid artery stenosis may be related to orthostatic dizziness. We concluded that NIRS monitoring is useful for evaluating cerebral autoregulation in patients with severe carotid artery stenosis.

T. Igarashi, M.D., Ph.D. (✉) • N. Fujiwara • Y. Murata • T. Suma
T. Hirayama • Y. Katayama
Department of Neurological Surgery, Division of Neurosurgery, Nihon University School of Medicine, 30-1 Oyaguchi-kamimachi, Itabashi-ku, Tokyo 173-8610, Japan
e-mail: igarashi.takahiro@nihon-u.ac.jp

K. Sakatani
Department of Neurological Surgery, Division of Neurosurgery, Nihon University School of Medicine, 30-1 Oyaguchi-kamimachi, Itabashi-ku, Tokyo 173-8610, Japan

Department of Neurological Surgery, Division of Optical Brain Engineering, Nihon University School of Medicine, Tokyo, Japan

T. Shibuya
Department of Neuroendovascular Therapy, Sagamihara Kyodo Hospital, Kanagawa, Japan

S. Van Huffel et al. (eds.), *Oxygen Transport to Tissue XXXV*, Advances in Experimental Medicine and Biology 789, DOI 10.1007/978-1-4614-7411-1_62,

62.1 Introduction

Transient ischemic attack (TIA) is a major complication in patients with carotid artery stenosis, affecting approximately 7 %/year of patients with severe carotid artery stenosis. Patients with severe stenosis also sometimes complain of orthostatic dizziness, such as syncope, though the reason for this is unclear. The baroreflex is the most important regulatory mechanism in the short-term control of circulation, and some reports indicate that baroreflex sensitivity (BRs) of patients with carotid artery stenosis is low, compared with normal subjects [1–3]. The head-up tilt test (HUTT) has been used to examine physiological events during graded orthostatic challenge in individuals with significant handicap owing to neurocardiogenic syncope (NCS).

Near-infrared spectroscopy (NIRS), a noninvasive optical method, utilizes the characteristic absorption spectra of hemoglobin (Hb) in the near-infrared range to provide information on hemodynamic changes and cerebral blood oxygenation changes. However, the influence of postural stresses on cerebral circulation in patients with carotid artery stenosis is unclear.

We hypothesized that impaired cerebral oxygenation may be related to postural symptoms, and in this study we investigated whether NIRS monitoring would be a suitable approach to examine whether patients with carotid artery stenosis and healthy volunteers show differential hemodynamic changes during HUTT.

62.2 Methods

We investigated 14 patients with carotid artery stenosis (10 men and 4 women; mean age, 69.4 ± 7.2 years) and nine normal control subjects (5 men and 4 women; mean age, 26.0 ± 1.1 years). A HUTT procedure was started by having the patient lie supine on a tilt table while connected to a precordial lead electrocardiogram and an automated intermittent oscillometric blood pressure monitor. Each subject lay in the supine position on the tilt table for at least 15 min before the procedure. In the first step, the table was elevated to 30° during 10 min. Then, it was elevated to 70° during 10 min. We measured the concentration changes of each Hb in the bilateral frontal lobe using a newly developed NIRS device (Pocket NIRS, Hamamatsu Photonics K.K., Japan) (Fig. 62.1). This device employs a wireless communication system (Bluetooth®) instead of optical fibers, so that the subject's movement is not constrained. In the present study, we used continuous-wave NIRS. Measurement of continuous-wave NIRS is dependent on the modified Beer-Lambert law, which includes optical pathlength (PL) as an essential parameter. Therefore, we believe that it is better to used arbitrary units rather than micro molar. This study was approved by the Committee for Clinical Trials and Research at Nihon University School of Medicine and Sagamihara Kyodo Hospital. All patients or their relatives gave written informed consent.

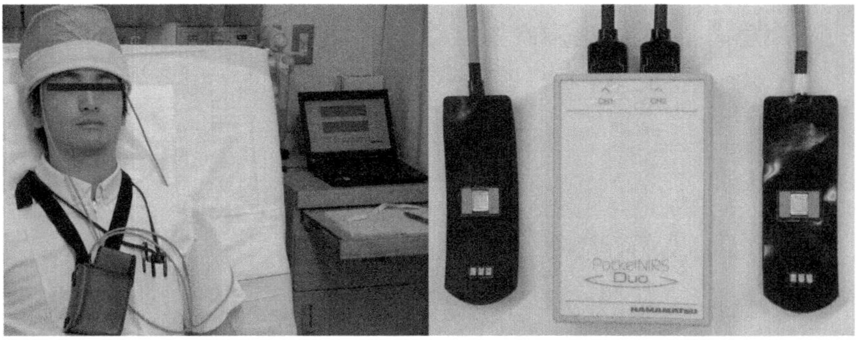

Fig. 62.1 The NIRS system uses LEDs of three different wavelengths (735, 810, and 850 nm) as light sources and one photodiode as a detector; it has two channels. The total weight of the controller and probes is only 160 g. The sampling rate was 61.3 Hz (i.e., the sampling time was about 16.3 ms)

62.3 Results

There was no significant difference in systemic circulation change during HUTT between the two groups. In normal volunteers, oxy-Hb decreased initially after the table was elevated. However, it rapidly recovered, then remained constant (Fig. 62.2a). In contrast, oxy-Hb of patients with carotid artery stenosis decreased and did not subsequently recover to the initial level (Fig. 62.2b). The oxy-Hb reduction in the carotid artery stenosis group at 30 s after elevation of the table was significantly larger than that in the normal group at 30° ($P<0.05$) and at 70° ($P<0.01$) (Fig. 62.3a). Deoxy-Hb slightly increased during HUTT in the carotid artery stenosis group, but was not significantly different from that of normal volunteers (Fig. 62.3b). When the tilt table was brought back to the supine position, each Hb concentration returned to baseline.

62.4 Discussion

Static cerebral circulation in patients with carotid artery stenosis has been evaluated using single-photon emission computed tomography (SPECT) [4] and positron emission tomography (PET) [5]. However, these techniques cannot measure dynamic changes in cerebral blood flow (CBF). On the other hand, NIRS can measure dynamic changes in both CBF and cerebral metabolism. Therefore, in the present study, we examined whether NIRS monitoring is suitable for studying hemodynamic changes in patients with carotid artery stenosis during HUTT.

We observed a sustained reduction of oxy-Hb during HUTT in patients with severe carotid artery stenosis, and this might be associated with orthostatic

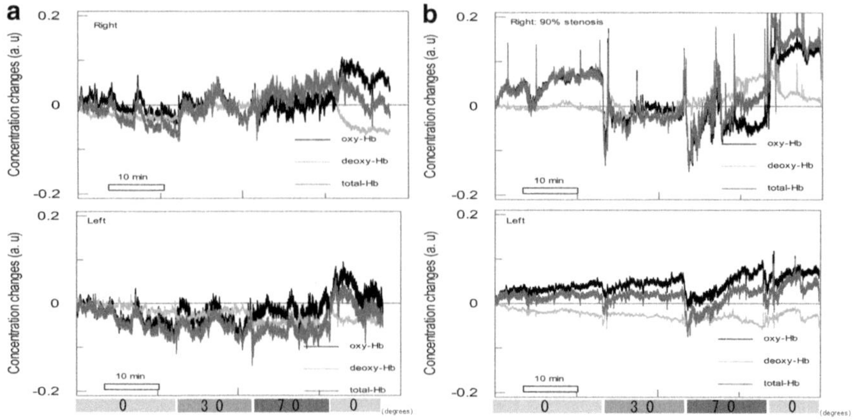

Fig. 62.2 Hb concentration changes during HUTT in normal volunteers (**a**, *left*) and patients with carotid artery stenosis (**b**, *right*). The ordinates indicate concentration changes of oxy-Hb (*black*), deoxy-Hb (*light gray*), and total Hb (*dark gray*) in arbitrary units (a.u.). The abscissa indicates time (min); *horizontal bars* indicate 10 min. The *horizontal bars* indicate the time schedule of HUTT

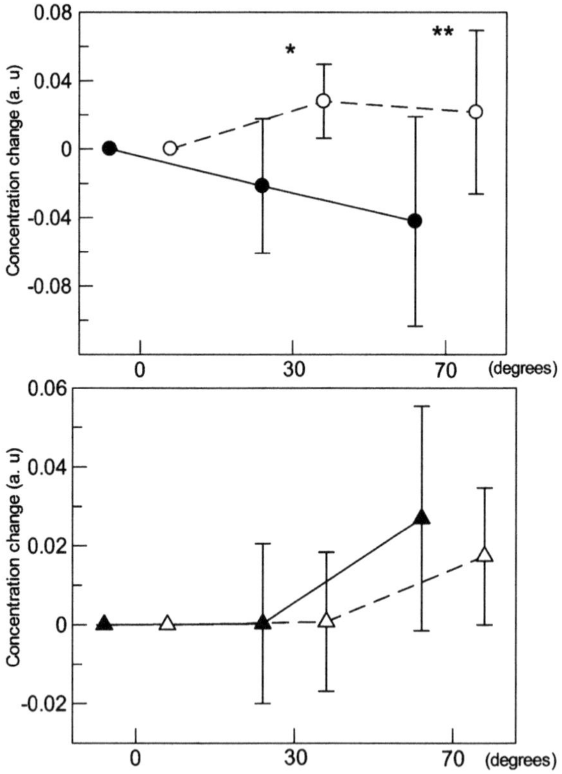

Fig. 62.3 (**a**) Change of oxy-Hb during HUTT in normal volunteers and patients with carotid artery stenosis. *Open circles* ($n=9$) indicate the mean value of normal volunteers; *solid circles* ($n=14$) indicate the mean value of patients with carotid artery stenosis. (**b**) Change of deoxy-Hb during HUTT in normal volunteers and patients with carotid artery stenosis. *Open triangles* ($n=9$) indicate the mean value of normal volunteers; *solid triangles* ($n=14$) indicate the mean value of patients with carotid artery stenosis. The ordinates indicate concentration changes of Hb in arbitrary units (a.u.)

dizziness. On the other hand, NIRS monitoring showed only a transient reduction of oxy-Hb in patients with mild or moderate carotid artery stenosis and in healthy volunteers during postural stresses, followed by a rapid recovery. None of these patients or volunteers had experienced orthostatic dizziness.

It is important to note that we examined only a relatively small number of patients and normal volunteers. Also, it is not yet clear whether NIRS parameters measured in the frontal lobe fully reflect hemodynamic changes in the cerebral hemisphere. Nevertheless, our results indicate that the present NIRS system, which is compact enough to attach to patient's clothes and is equipped with telemetric data transfer, is suitable for investigating and evaluating cerebral autoregulation in patients with severe carotid artery stenosis during HUTT.

62.5 Conclusions

We concluded that NIRS monitoring is useful for evaluating cerebral circulation and autoregulation in patients with severe carotid artery stenosis.

Acknowledgments This work was supported by Grants-in-Aid from the Ministry of Education, Culture, Sports, Science and Technology of Japan (Strategic Research Program for Brain Sciences).

References

1. Yakhou L, Constant I, Merle JC, Laude D, Becquemin JP, Duvaldestin P (2006) Noninvasive investigation of autonomic activity after carotid stenting or carotid endarterectomy. J Vasc Surg 44(3):472–479
2. Parlow J, Viale JP, Annat G, Hughson R, Quintin L (1995) Spontaneous cardiac baroreflex in humans. Comparison with drug-induced responses. Hypertension 25(5):1058–1068
3. Constant I, Abbas M, Boucheseiche S, Laude D, Murat I (2002) Non-invasive assessment of cardiovascular autonomic activity induced by brief exposure to 50 % nitrous oxide in children. Br J Anaesth 88(5):637–643
4. Kuroda S, Kamiyama H, Abe H, Houkin K, Isobe M, Mitsumori K (1993) Acetazolamide test in detecting reduced cerebral perfusion reserve and predicting long-term prognosis in patients with internal carotid artery occlusion. Neurosurgery 32(6):912–918, discussion 918-9
5. Powers WJ, Press GA, Grubb RL Jr, Gado M, Raichle ME (1987) The effect of hemodynamically significant carotid artery disease on the hemodynamic status of the cerebral circulation. Ann Intern Med 106(1):27–34

Name Index

S. Van Huffel et al. (eds.), *Oxygen Transport to Tissue XXXV*, Advances in Experimental Medicine and Biology 789, DOI 10.1007/978-1-4614-7411-1,

C

T

Subject Index

S. Van Huffel et al. (eds.), *Oxygen Transport to Tissue XXXV*, Advances in Experimental Medicine and Biology 789, DOI 10.1007/978-1-4614-7411-1,

N

V

GPSR Compliance

The European Union's (EU) General Product Safety Regulation (GPSR) is a set of rules that requires consumer products to be safe and our obligations to ensure this.

If you have any concerns about our products, you can contact us on ProductSafety@springernature.com

In case Publisher is established outside the EU, the EU authorized representative is:

Springer Nature Customer Service Center GmbH
Europaplatz 3
69115 Heidelberg, Germany

Batch number: 10755324

Printed by Printforce, the Netherlands